Generalized Linear Models and Extensions

Fourth Edition

T0132599

Generalized Linear Models and Extensions

Fourth Edition

James W. Hardin
Department of Epidemiology and Biostatistics
University of South Carolina

Joseph M. Hilbe
Statistics, School of Social and Family Dynamics
Arizona State University

A Stata Press Publication
StataCorp LLC
College Station, Texas

Published by Stata Press, 4905 Lakeway Drive, College Station, Texas 77845
Typeset in LaTeX 2_ε
Printed in the United States of America

10 9 8 7 6 5 4 3 2 1

Print ISBN-10: 1-59718-225-7
Print ISBN-13: 978-1-59718-225-6
ePub ISBN-10: 1-59718-226-5
ePub ISBN-13: 978-1-59718-226-3
Mobi ISBN-10: 1-59718-227-3
Mobi ISBN-13: 978-1-59718-227-0

Library of Congress Control Number: 2018937959

In all editions of this text, this dedication page was written as the final deliverable to the editor—after all the work and all the requested changes and additions had been addressed. In every previous edition, Joe and I co-dedicated this book to our wives and children. This time, I write the dedication alone.

In memory of

Joseph M. Hilbe

who passed away after responding to the editor's final change requests, but before the specification of our dedication. Joe was a dear friend and colleague. We worked closely on this new edition, and he was as cheerful and tireless as always. We worked a long time to put this latest edition together, and he would want the reader to know that he is very proud of our collaboration, but even more proud of his family: Cheryl, Heather, Michael, and Mitchell.

Contents

VI   Extensions to the GLM   385

17   Extending the likelihood   387

18   Clustered data   401

Figures

Tables

Listings

Preface

We have added several new models to the discussion of extended generalized linear models (GLMs). We have included new software and discussion of extensions to negative binomial regression because of Waring and Famoye. We have also added discussion of heaped data and bias-corrected GLMs because of Firth. There are two new chapters on multivariate outcomes and Bayes GLMs. In addition, we have expanded the clustered data discussion to cover more of the commands available in Stata.

We now include even more examples using synthetically created models to illustrate estimation results, and we illustrate to readers how to construct synthetic Monte Carlo models for binomial and major count models. Code for creating synthetic Poisson, negative binomial, zero-inflated, hurdle, and finite mixture models is provided and further explained. We have enhanced discussion of marginal effects and discrete change for GLMs.

This fourth edition of *Generalized Linear Models and Extensions* is written for the active researcher as well as for the theoretical statistician. Our goal has been to clarify the nature and scope of GLMs and to demonstrate how all the families, links, and variations of GLMs fit together in an understandable whole.

In a step-by-step manner, we detail the foundations and provide working algorithms that readers can use to construct and better understand models that they wish to develop. In a sense, we offer readers a workbook or handbook of how to deal with data using GLM and GLM extensions.

This text is intended as a textbook on GLMs and as a handbook of advice for researchers. We continue to use this book as the required text for a web-based short course through *Statistics.com* (also known as the *Institute for Statistical Education*); see http://www.statistics.com. The students of this six-week course include university professors and active researchers from hospitals, government agencies, research institutes, educational concerns, and other institutions across the world. This latest edition reflects the experiences we have had in communicating to our readers and students the relevant materials over the past decade.

Many people have contributed to the ideas presented in the new edition of this book. John Nelder has been the foremost influence. Other important and influential people include Peter Bruce, David Collett, David Hosmer, Stanley Lemeshow, James Lindsey, J. Scott Long, Roger Newson, Scott Zeger, Kung-Yee Liang, Raymond J. Carroll, H. Joseph Newton, Henrik Schmiediche, Norman Breslow, Berwin Turlach, Gordon Johnston, Thomas Lumley, Bill Sribney, Vince Wiggins, Mario Cleves, William

Greene, Andrew Robinson, Heather Presnal, and others. Specifically, for this edition, we thank Tammy Cummings, Chelsea Deroche, Xinling Xu, Roy Bower, Julie Royer, James Hussey, Alex McLain, Rebecca Wardrop, Gelareh Rahimi, Michael G. Smith, Marco Geraci, Bo Cai, and Feifei Xiao.

As always, we thank William Gould, president of StataCorp, for his encouragement in this project. His statistical computing expertise and his contributions to statistical modeling have had a deep impact on this book.

We are grateful to StataCorp's editorial staff for their equanimity in reading and editing our manuscript, especially to Patricia Branton and Lisa Gilmore for their insightful and patient contributions in this area. Finally, we thank Kristin MacDonald and Isabel Canette-Fernandez, Stata statisticians at StataCorp, for their expert assistance on various programming issues, and Nikolay Balov, Senior Statistician and Software Developer at StataCorp, for his helpful assistance with chapter 20 on Bayesian GLMs. We would also like to thank Rose Medeiros, Senior Statistician at StataCorp, for her assistance in the final passes of this edition.

Stata Press allowed us to dictate some of the style of this text. In writing this material in other forms for short courses, we have always included equation numbers for all equations rather than only for those equations mentioned in text. Although this is not the standard editorial style for textbooks, we enjoy the benefits of students being able to communicate questions and comments more easily (and efficiently). We hope that readers find this practice as beneficial as our short-course participants have found it.

Errata, datasets, and supporting Stata programs (do-files and ado-files) may be found at the publisher's site http://www.stata-press.com/books/generalized-linear-models-and-extensions/. We also maintain these materials on the author sites at http://www.thirdwaystat.com/jameshardin/ and at https://works.bepress.com/joseph_hilbe/. We are very pleased to be able to produce this newest edition. Working on this text from the first edition in 2001 over the past 17 years has been a tremendously satisfying experience.

James W. Hardin
Joseph M. Hilbe

March 2018

1 Introduction

Contents

In updating this text, our primary goal is to convey the practice of analyzing data via generalized linear models to researchers across a broad spectrum of scientific fields. We lay out the framework used for describing various aspects of data and for communicating tools for data analysis. This initial part of the text contains no examples. Rather, we focus on the lexicon of generalized linear models used in later chapters. These later chapters include examples from fields such as biostatistics, economics, and survival analysis.

In developing analysis tools, we illustrate techniques via their genesis in estimation algorithms. We believe that motivating the discussion through the estimation algorithms clarifies the origin and usefulness of all generalized linear models. Instead of detailed theoretical exposition, we refer to texts and papers that present such material so that we may focus our detailed presentations on the algorithms and their justification. Our detailed presentations are mostly algebraic; we have minimized matrix notation whenever possible.

We often present illustrations of models using data that we synthesize. Although it is preferable to use real data to illustrate interpretation in context, there is a distinct advantage to examples using simulated data. The advantage of worked examples relying on data synthesis is that the data-generating process offers yet another glimpse into the associations between variables and outcomes that are to be captured in the procedure. The associations are thus seen from both the results of the model and the origins of data generation.

1.1   Origins and motivation

We wrote this text for researchers who want to understand the scope and application of generalized linear models while being introduced to the underlying theory. For brevity's sake, we use the acronym GLM to refer to the generalized linear model, but we acknowledge that GLM has been used elsewhere as an acronym for the general linear model. The latter usage, of course, refers to the area of statistical modeling based solely on the normal or Gaussian probability distribution.

We take GLM to be the generalization of the general, because that is precisely what GLMs are. They are the result of extending ordinary least-squares (OLS) regression, or the normal model, to a model that is appropriate for a variety of response distributions, specifically to those distributions that compose the single parameter exponential family of distributions. We examine exactly how this extension is accomplished. We also aim to provide the reader with a firm understanding of how GLMs are evaluated and when their use is appropriate. We even advance a bit beyond the traditional GLM and give the reader a look at how GLMs can be extended to model certain types of data that do not fit exactly within the GLM framework.

Nearly every text that addresses a statistical topic uses one or more statistical computing packages to calculate and display results. We use Stata exclusively, though we do refer occasionally to other software packages—especially when it is important to highlight differences.

Some specific statistical models that make up GLMs are often found as standalone software modules, typically fit using maximum likelihood methods based on quantities from model-specific derivations. Stata has several such commands for specific GLMs including `poisson`, `logistic`, and `regress`. Some of these procedures were included in the Stata package from its first version. More models have been addressed through commands written by users of Stata's programming language leading to the creation of highly complex statistical models. Some of these community-contributed commands have since been incorporated into the official Stata package. We highlight these commands and illustrate how to fit models in the absence of a packaged command; see especially chapter 14.

Stata's `glm` command was originally created as a community-contributed command (Hilbe 1993b) and then officially adopted into Stata two years later as part of Stata 4.0. Examples of the `glm` command in this edition reflect StataCorp's continued updates to the command.

Readers of technical books often need to know about prerequisites, especially how much math and statistics background is required. To gain full advantage from this text and follow its every statement and algorithm, you should have an understanding equal to a two-semester calculus-based course on statistical theory. Without a background in statistical theory, the reader can accept the presentation of the theoretical underpinnings and follow the (mostly) algebraic derivations that do not require more than a mastery of simple derivatives. We assume prior knowledge of multiple regression but no other specialized knowledge is required.

We believe that GLMs are best understood if their computational basis is clear. Hence, we begin our exposition with an explanation of the foundations and computation of GLMs; there are two major methodologies for developing algorithms. We then show how simple changes to the base algorithms lead to different GLM families, links, and even further extensions. In short, we attempt to lay the GLM open to inspection and to make every part of it as clear as possible. In this fashion, the reader can understand exactly how and why GLM algorithms can be used, as well as altered, to better model a desired dataset.

Perhaps more than any other text in this area, we alternatively examine two major computational GLM algorithms and their modifications:

1. Iteratively reweighted least squares

2. Newton–Raphson

Interestingly, some of the models we present are calculated only by using one of the above methods. Iteratively reweighted least squares is the more specialized technique and is applied less often. Yet it is typically the algorithm of choice for quasilikelihood models such as generalized estimating equations (GEEs). On the other hand, truncated models that do not fit neatly into the exponential family of distributions are modeled using Newton–Raphson methods—and for this, too, we show why. Again, focusing on the details of calculation should help the reader understand both the scope and the limits of a particular model.

Whenever possible, we present the log likelihood for the model under discussion. In writing the log likelihood, we include offsets so that interested programmers can see how those elements enter estimation. In fact, we attempt to offer programmers the ability to understand and write their own working GLMs, plus many useful extensions. As programmers ourselves, we believe that there is value in such a presentation; we would have much enjoyed having it at our fingertips when we first entered this statistical domain.

1.2   Notational conventions

We use L to denote the likelihood and the script $\mathcal{L}$ to denote the log likelihood. We use X to denote the design matrix of independent (explanatory) variables. When appropriate, we use boldface type $\mathbf{X}$ to emphasize that we are referring to a matrix; a lowercase letter with a subscript $\mathbf{x}_i$ will refer to the ith row from the matrix $\mathbf{X}$.

We use Y to denote the dependent (response) variable and refer to the vector $\boldsymbol{\beta}$ as the coefficients of the design matrix. We use $\widehat{\boldsymbol{\beta}}$ when we wish to discuss or emphasize the fitted coefficients. Throughout the text, we discuss the role of the (vector) linear predictor $\eta = \mathbf{X}\boldsymbol{\beta}$. In generalizing this concept, we also refer to the augmented (by an offset) version of the linear predictor $\xi = \eta + \text{offset}$.

Finally, we use the $E(\cdot)$ notation to refer to the expectation of a random variable and the $V(\cdot)$ notation to refer to the variance of a random variable. We describe other notational conventions at the time of their first use.

1.3   Applied or theoretical?

A common question regarding texts concerns their focus. Is the text applied or theoretical? Our text is both. However, we would argue that it is basically applied. We show enough technical details for the theoretician to understand the underlying basis of GLMs. However, we believe that understanding the use and limitations of a GLM includes understanding its estimation algorithm. For some, dealing with formulas and algorithms appears thoroughly theoretical. We believe that it aids in understanding the scope and limits of proper application. Perhaps we can call the text a bit of both and not worry about classification. In any case, for those who fear formulas, each formula and algorithm is thoroughly explained. We hope that by book's end the formulas and algorithms will seem simple and meaningful. For completeness, we give the reader references to texts that discuss more advanced topics and theory.

1.4   Road map

Part I of the text deals with the basic foundations of GLM. We detail the various components of GLM, including various family, link, variance, deviance, and log-likelihood functions. We also provide a thorough background and detailed particulars of both the Newton–Raphson and iteratively reweighted least-squares algorithms. The chapters that follow highlight this discussion, which describes the framework through which the models of interest arise.

We also give the reader an overview of GLM residuals, introducing some that are not widely known, but that nevertheless can be extremely useful for analyzing a given model's worth. We discuss the general notion of goodness of fit and provide a framework through which you can derive more extensions to GLM. We conclude this part with discussion and illustration of simulation and data synthesis.

We often advise participants only interested in the application and interpretation of models to skip this first part of the book. Even those interested in the theoretical underpinnings will find that this first part of the book can serve more as an appendix. That is, the information in this part often turns out to be most useful in subsequent readings of the material.

Part II addresses the continuous family of distributions, including the Gaussian, gamma, inverse Gaussian, and power families. We derive the related formulas and relevant algorithms for each family and then discuss the ancillary or scale parameters appropriate to each model. We also examine noncanonical links and generalizations to the basic model. Finally, we give examples, showing how a given dataset may be analyzed using each family and link. We give examples dealing with model application, including discussion of the appropriate criteria for the analysis of fit. We have expanded the number of examples in this new edition to highlight both model fitting and assessment.

Part III addresses binomial response models. It includes exposition of the general binomial model and of the various links. Major links described include the canonical logit, as well as the noncanonical links probit, log-log, and complementary log-log. We also cover other links. We present examples and criteria for analysis of fit throughout the part. This new edition includes extensions to generalized binomial regression resulting from a special case of building a regression model from the generalized negative binomial probability function.

We also give considerable space to overdispersion. We discuss the problem's nature, how it is identified, and how it can be dealt with in the context of discovery and analysis. We explain how to adjust the binomial model to accommodate overdispersion. You can accomplish this task by internal adjustment to the base model, or you may need to reformulate the base model itself. We also introduce methods of adjusting the variance–covariance matrix of the model to produce robust standard errors. The problem of dealing with overdispersion continues in the chapters on count data.

Part IV addresses count response data. We include examinations of the Poisson, the geometric, and the negative binomial models. With respect to the negative binomial, we show how the standard models can be further extended to derive a class called heterogeneous negative binomial models. There are several "brands" of negative binomial, and it is wise for the researcher to know how each is best used. The distinction of these models is typically denoted NB-1 and NB-2 and relates to the variance-to-mean ratio of the resulting derivation of the model. We have updated this discussion to include the generalized Poisson regression model, which is similar to NB-1. In this edition, we present several other variations of negative binomial regression.

Part V addresses categorical response regression models. Typically considered extensions to the basic GLM, categorical response models are divided into two general varieties: unordered response models, also known as multinomial models, and ordered response models. We begin by considering ordered response models. In such models, the discrete number of outcomes are ordered, but the integer labels applied to the ordered levels of outcome are not necessarily equally spaced. A simple example is the set of outcomes "bad", "average", and "good". We also cover unordered multinomial responses, whose outcomes are given no order. For an example of an unordered outcome, consider choosing the type of entertainment that is available for an evening. The following choices may be given as "movie", "restaurant", "dancing", or "reading". Ordered response models are themselves divisible into two varieties: 1) ordered binomial including ordered logit, ordered probit, ordered complementary log-log, or ordered log-log and 2) the generalized ordered binomial model with the same links as the nongeneralized parameterization. We have expanded our discussion to include more ordered outcome models, generalized ordered outcome models, and continuation-ratio models.

Finally, part VI is about extensions to GLMs. In particular, we examine the following models:

1. Fixed-effects models
2. Random-effects models
3. Quasilikelihood models
4. GEEs
5. Generalized additive models

We give the reader a thorough outline or overview of GLMs. We cover nearly every major development in the area. We also show the direction in which statistical modeling is moving, hence laying a foundation for future research and for ever-more-appropriate GLMs. Moreover, we have expanded each section of the original version of this text to bring new and expanded regression models into focus. Our attempt, as always, is to illustrate these new models within the context of the GLM.

1.5   Installing the support materials

All the data used in this book are freely available for you to download from the Stata Press website, http://www.stata-press.com. In fact, when we introduce new datasets, we merely load them into Stata the same way that you would. For example,

```
. use http://www.stata-press.com/data/hh4/medpar
```

To download the datasets and do-files for this book, type

```
. net from http://www.stata-press.com/data/hh4/
. net get glme4
```

The datasets and do-files will be downloaded to your current working directory. We suggest that you create a new directory into which the materials will be downloaded.

Part I

Foundations of Generalized Linear Models

2 GLMs

Contents

Nelder and Wedderburn (1972) introduced the theory of GLMs. The authors derived an underlying unity for an entire class of regression models. This class consisted of models whose single response variable, the variable that the model is to explain, is hypothesized to have the variance that is reflected by a member of the single-parameter exponential family of probability distributions. This family of distributions includes the Gaussian or normal, binomial, Poisson, gamma, inverse Gaussian, geometric, and negative binomial.

To establish a basis, we begin discussion of GLMs by initially recalling important results on linear models, specifically those results for linear regression. The standard linear regression model relies on several assumptions, among which are the following:

1. Each observation of the response variable is characterized by the normal or Gaussian distribution; $y_i \sim N(\mu_i, \sigma_i^2)$.

2. The distributions for all observations have a common variance; $\sigma_i^2 = \sigma^2$ for all i.

3. There is a direct or "identical" relationship between the linear predictor (linear combination of covariate values and associated parameters) and the expected values of the model; $E(\mathbf{x}_i\boldsymbol{\beta}) = \mu_i$.

The purpose of GLMs, and the linear models that they generalize, is to specify the relationship between the observed response variable and some number of covariates. The outcome variable is viewed as a realization from a random variable.

Nelder and Wedderburn showed that general models could be developed by relaxing the assumptions of the linear model. By restructuring the relationship between the linear predictor and the fit, we can "linearize" relationships that initially seem to be nonlinear. Nelder and Wedderburn accordingly dubbed these models "generalized linear models".

Most models that were placed under the original GLM framework were well established and popular—some more than others. However, these models had historically been fit using maximum likelihood (ML) algorithms specific to each model. ML algorithms, as we will call them, can be hard to implement. Starting or initial estimates for parameters must be provided, and considerable work is required to derive model-specific quantities to ultimately obtain parameter estimates and their standard errors. In the next chapter, we show much effort is involved.

Ordinary least squares (OLS) extends ML linear regression such that the properties of OLS estimates depend only on the assumptions of constant variance and independence. ML linear regression carries the more restrictive distributional assumption of normality. Similarly, although we may derive likelihoods from specific distributions in the exponential family, the second-order properties of our estimates are shown to depend only on the assumed mean–variance relationship and on the independence of the observations rather than on a more restrictive assumption that observations follow a particular distribution.

The classical linear model assumes that the observations that our dependent variable y represents are independent normal variates with constant variance σ^2. Also covariates are related to the expected value of the independent variable such that

$$E(y) \;\; = \;\; \mu \tag{2.1}$$
$$\mu \;\; = \;\; \mathbf{X}\boldsymbol{\beta} \tag{2.2}$$

This last equation shows the "identical" or identity relationship between the linear predictor $\mathbf{X}\boldsymbol{\beta}$ and the mean μ.

Whereas the linear model conceptualizes the outcome y as the sum of its mean μ and a random variable ϵ, Nelder and Wedderburn linearized each GLM family member by means of a link function. They then altered a previously used algorithm called *iterative weighted least squares*, which was used in fitting weighted least-squares regression models. Aside from introducing the link function relating the linear predictor to the fitted values, they also introduced the variance function as an element in the weighting of the regression. The iterations of the algorithm updates parameter estimates to produce appropriate linear predictors, fitted values, and standard errors. We will clarify exactly how all this falls together in the section on the iteratively reweighted least-squares (IRLS) algorithm.

The estimation algorithm allowed researchers to easily fit many models previously considered to be nonlinear by restructuring them into GLMs. Later, it was discovered that an even more general class of linear models results from more relaxations of assumptions for GLMs.

However, even though the historical roots of GLMs are based on IRLS methodology, many generalizations to the linear model still require Newton–Raphson techniques common to ML methods. We take the position here that GLMs should not be constrained to those models first discussed by Nelder and Wedderburn but rather that they encompass all such linear generalizations to the standard model.

Many other books and journal articles followed the cornerstone article by Nelder and Wedderburn (1972) as well as the text by McCullagh and Nelder (1989) (the original text was published in 1983). Lindsey (1997) illustrates the application of GLMs to biostatistics, most notably focusing on survival models. Hilbe (1994) gives an overview of the GLM and its support from various software packages. Software was developed early on. In fact, Nelder was instrumental in developing the first statistical program based entirely on GLM principles—generalized linear interactive modeling (GLIM). Published by the Numerical Algorithms Group (NAG), the software package has been widely used since the mid-1970s. Other vendors began offering GLM capabilities in the 1980s, including GENSTAT and S-Plus. Stata and SAS included it in their software offerings in 1993 and 1994, respectively.

This text covers much of the same foundation material as other books. What distinguishes our presentation of the material is twofold. First, we focus on the estimation of various models via the estimation technique. Second, we present our derivation of the methods of estimation in a more accessible manner than which is presented in other sources. In fact, where possible, we present complete algebraic derivations that include nearly every step in the illustrations. Pedagogically, we have found that this manner of exposition imparts a more solid understanding and "feel" of the area than do other approaches. The idea is this: if you can write your own GLM, then you are probably more able to know how it works, when and why it does not work, and how it is to be evaluated. Of course, we also discuss methods of fit assessment and testing. To model data without subjecting them to evaluation is like taking a test without checking the answers. Hence, we will spend considerable time dealing with model evaluation as well as algorithm construction.

2.1   Components

Cited in various places such as Hilbe (1993b) and Francis, Green, and Payne (1993), GLMs are characterized by an expanded itemized list given by the following:

1. A random component for the response, $\mathbf{y}$, which has the characteristic variance of a distribution that belongs to the exponential family.

2. A linear systematic component relating the linear predictor, $\eta = \mathbf{X}\boldsymbol{\beta}$, to the product of the design matrix $\mathbf{X}$ and the parameters $\boldsymbol{\beta}$.

3. A known monotonic, one-to-one, differentiable link function $g(\cdot)$ relating the linear predictor to the fitted values. Because the function is one-to-one, there is an inverse function relating the mean expected response, $E(y) = \mu$, to the linear predictor such that $\mu = g^{-1}(\eta) = E(y)$.

4. The variance may change with the covariates only as a function of the mean.

5. There is one IRLS algorithm that suffices to fit all members of the class.

Item 5 is of special interest. The traditional formulation of the theory certainly supposed that there was one algorithm that could fit all GLMs. We will see later how this was implemented. However, there have been extensions to this traditional viewpoint. Adjustments to the weight function have been added to match the usual Newton–Raphson algorithms more closely and so that more appropriate standard errors may be calculated for noncanonical link models. Such features as scaling and robust variance estimators have also been added to the basic algorithm. More importantly, sometimes a traditional GLM must be restructured and fit using a model-specific Newton–Raphson algorithm. Of course, one may simply define a GLM as a model requiring only the standard approach but doing so would severely limit the range of possible models. We prefer to think of a GLM as a model that is ultimately based on the probability function belonging to the exponential family of distributions, but with the proviso that this criterion may be relaxed to include quasilikelihood models as well as certain types of multinomial, truncated, censored, and inflated models. Most of the latter type require a Newton–Raphson approach rather than the traditional IRLS algorithm.

Early GLM software development constrained GLMs to those models that could be fit using the originally described estimation algorithm. As we will illustrate, the traditional algorithm is relatively simple to implement and requires little computing power. In the days when RAM was scarce and expensive, this was an optimal production strategy for software development. Because this is no longer the case, a wider range of GLMs can more easily be fit using a variety of algorithms. We will discuss these implementation details at length.

In the classical linear model, the observations of the dependent variable $\mathbf{y}$ are independent normal variates with constant variance σ^2. We assume that the mean value of $\mathbf{y}$ may depend on other quantities (predictors) denoted by the column vectors $\mathbf{X}_1, \mathbf{X}_2, \ldots, \mathbf{X}_{p-1}$. In the simplest situation, we assume that this dependency is linear and write

$$E(\mathbf{y}) = \beta_0 + \beta_1 \mathbf{X}_1 + \cdots + \beta_{p-1} \mathbf{X}_{p-1} \tag{2.3}$$

and attempt to estimate the vector $\boldsymbol{\beta}$.

GLMs specify a relationship between the mean of the random variable $\mathbf{y}$ and a function of the linear combination of the predictors. This generalization admits a model specification allowing for continuous or discrete outcomes and allows a description of the variance as a function of the mean.

2.2 Assumptions

The link function relates the mean $\mu = E(y)$ to the linear predictor $\mathbf{X}\boldsymbol{\beta}$, and the variance function relates the variance as a function of the mean $V(y) = a(\phi)v(\mu)$, where $a(\phi)$ is the scale factor. For the Poisson, binomial, and negative binomial variance models, $a(\phi) = 1$.

Breslow (1996) points out that the critical assumptions in the GLM framework may be stated as follows:

1. Statistical independence of the n observations.

2. The variance function $v(\mu)$ is correctly specified.

3. $a(\phi)$ is correctly specified (1 for Poisson, binomial, and negative binomial).

4. The link function is correctly specified.

5. Explanatory variables are of the correct form.

6. There is no undue influence of the individual observations on the fit.

As a simple illustration, in table 2.1 we demonstrate the effect of the assumed variance function on the model and fitted values of a simple GLM.

Table 2.1. Predicted values for various choices of variance function

Observed (y)	1.00	2.00	9.00	
Predicted [Normal: $v(\mu) = \phi$]	0.00	4.00	8.00	$\widehat{y} = -4.00 + 4.00x$
Predicted [Poisson: $v(\mu) = \mu$]	0.80	4.00	7.20	$\widehat{y} = -2.40 + 3.20x$
Predicted [Gamma: $v(\mu) = \phi\mu^2$]	0.94	3.69	6.43	$\widehat{y} = -1.80 + 2.74x$
Predicted [Inverse Gaussian: $v(\mu) = \phi\mu^3$]	0.98	3.33	5.69	$\widehat{y} = -1.37 + 2.35x$

Note: The models are all fit using the identity link, and the data consist of three observations $(y, x) = \{(1,1), (2,2), (9,3)\}$. The fitted models are included in the last column.

2.3 Exponential family

GLMs are traditionally formulated within the framework of the exponential family of distributions. In the associated representation, we can derive a general model that may be fit using the scoring process (IRLS) detailed in section 3.3. Many people confuse the estimation method with the class of GLMs. This is a mistake because there are many estimation methods. Some software implementations allow specification of more diverse models than others. We will point this out throughout the text.

The exponential family is usually (there are other algebraically equivalent forms in the literature) written as

$$f_y(y; \theta, \phi) = \exp\left\{\frac{y\theta - b(\theta)}{a(\phi)} + c(y, \phi)\right\} \tag{2.4}$$

where θ is the canonical (natural) parameter of location and ϕ is the parameter of scale. The location parameter (also known as the canonical link function) relates to the means, and the scalar parameter relates to the variances for members of the exponential family of distributions including Gaussian, gamma, inverse Gaussian, and others. Using the notation of the exponential family provides a means to specify models for continuous, discrete, proportional, count, and binary outcomes.

In the exponential family presentation, we construe each of the y_i observations as being defined in terms of the parameters θ. Because the observations are independent, the joint density of the sample of observations y_i, given parameters θ and ϕ, is defined by the product of the density over the individual observations (review section 2.2). Interested readers can review Barndorff-Nielsen (1976) for the theoretical justification that allows this factorization:

$$f_{y_1, y_2, \ldots, y_n}(y_1, y_2, \ldots, y_n; \theta, \phi) = \prod_{i=1}^{n} \exp\left\{\frac{y_i\theta_i - b(\theta_i)}{a(\phi)} + c(y_i, \phi)\right\} \qquad (2.5)$$

Conveniently, the joint probability density function may be expressed as a function of θ and ϕ given the observations y_i. This function is called the likelihood, L, and is written as

$$L(\theta, \phi; y_1, y_2, \ldots, y_n) = \prod_{i=1}^{n} \exp\left\{\frac{y_i\theta_i - b(\theta_i)}{a(\phi)} + c(y_i, \phi)\right\} \qquad (2.6)$$

We wish to obtain estimates of (θ, ϕ) that maximize the likelihood function. Given the product in the likelihood, it is more convenient to work with the log likelihood,

$$\mathcal{L}(\theta, \phi; y_1, y_2, \ldots, y_n) = \sum_{i=1}^{n} \left\{\frac{y_i\theta_i - b(\theta_i)}{a(\phi)} + c(y_i, \phi)\right\} \qquad (2.7)$$

because the values that maximize the likelihood are the same values that maximize the log likelihood.

Throughout the text, we will derive each distributional family member from the exponential family notation so that the components are clearly illustrated. The log likelihood for the exponential family is in a relatively basic form, admitting simple calculations of first and second derivatives for ML estimation. The IRLS algorithm takes advantage of this form of the log likelihood.

First, we generalize the log likelihood to include an offset to the linear predictor. This generalization will allow us to investigate simple equality constraints on the parameters.

The idea of an offset is simple. To fit models with covariates, we specify that θ is a function of specified covariates, $\mathbf{X}$, and their associated coefficients, $\boldsymbol{\beta}$. Within the linear combination of the covariates and their coefficients $\mathbf{X}\boldsymbol{\beta}$, we may further wish to constrain a particular subset of the coefficients β_i to particular values. For example, we may know or wish to test that $\beta_3 = 2$ in a model with a constant, X_0, and three covariates X_1, X_2, and X_3. If we wish to enforce the $\beta_3 = 2$ restriction on the estimation, then we will want the optimization process to calculate the linear predictor as

$$\eta = \widehat{\beta}_0 + \widehat{\beta}_1 X_1 + \widehat{\beta}_2 X_2 + 2X_3 \qquad (2.8)$$

at each step. We know (or wish to enforce) that the linear predictor is composed of a linear combination of the unrestricted parameters plus two times the X_3 covariate. If we consider that the linear predictor is generally written as

$$\eta = \mathbf{X}\boldsymbol{\beta} + \text{offset} \tag{2.9}$$

then we can appreciate the implementation of a program that allows an offset. We could generate a new variable equal to two times the variable containing the X_3 observations and specify that generated variable as the offset. By considering this issue from the outset, we can include an offset in our derivations, which will allow us to write programs that include this functionality.

The offset is a given (nonstochastic) component in the estimation problem. By including the offset, we gain the ability to fit (equality) restricted models without adding unnecessary complexity to the model; the offset plays no role in derivative calculations. If we do not include an offset in our derivations and subsequent programs, we can still fit restricted models, but the justification is less clear; see the arguments of Nyquist (1991) for obtaining restricted estimates in a GLM.

2.4 Example: Using an offset in a GLM

In subsequent chapters (especially chapter 3), we illustrate the two main components of the specification of a GLM. The first component of a GLM specification is a function of the linear predictor, which substitutes for the location (mean) parameter of the exponential family. This function is called the link function because it links the expected value of the outcome to the linear predictor comprising the regression coefficients; we specify this function with the `link()` option. The second component of a GLM specification is the variance as a scaled function of the mean. In Stata, this function is specified using the name of a particular member distribution of the exponential family; we specify this function with the `family()` option. The example below highlights a log-link Poisson GLM.

For this example, it is important to note the treatment of the offset in the linear predictor. The particular choices for the link and variance functions are not relevant to the utility of the offset.

Below, we illustrate the use of an offset with Stata's `glm` command. From an analysis presented in chapter 12, consider the output of the following model:

```
. use http://www.stata-press.com/data/hh4/medpar
. glm los hmo white type2 type3, family(poisson) link(log) nolog
```

Generalized linear models		No. of obs	=	1,495
Optimization	: ML	Residual df	=	1,490
		Scale parameter =		1
Deviance	= 8142.666001	(1/df) Deviance =		5.464877
Pearson	= 9327.983215	(1/df) Pearson =		6.260391
Variance function: V(u) = u		[Poisson]		
Link function	: g(u) = ln(u)	[Log]		
		AIC	=	9.276131
Log likelihood	= -6928.907786	BIC	=	-2749.057

los	Coef.	OIM Std. Err.	z	P>|z|	[95% Conf. Interval]
hmo	-.0715493	.023944	-2.99	0.003	-.1184786 -.02462
white	-.153871	.0274128	-5.61	0.000	-.2075991 -.100143
type2	.2216518	.0210519	10.53	0.000	.1803908 .2629127
type3	.7094767	.026136	27.15	0.000	.6582512 .7607022
_cons	2.332933	.0272082	85.74	0.000	2.279606 2.38626

We would like to test whether the coefficient on `white` is equal to -0.20. We could use Stata's `test` command to obtain a Wald test

```
. test white=-.20
 ( 1)  [los]white = -.2
          chi2( 1) =     2.83
        Prob > chi2 =    0.0924
```

which indicates that -0.15 (coefficient on `white`) is not significantly different at a 5% level from -0.20. However, we want to use a likelihood-ratio test, which is usually a more reliable test of parameter estimate significance. Stata provides a command that stores the likelihood from the unrestricted model (above) and then compares it with a restricted model. Having fit the unrestricted model, our attention now turns to fitting a model satisfying our specific set of constraints. Our constraint is that the coefficient on `white` be restricted to the constant value -0.20.

First, we store the log-likelihood value from the unrestricted model, and then we generate a variable indicative of our constraint. This new variable contains the restrictions that we will then supply to the software for fitting the restricted model. In short, the software will add our restriction any time that it calculates the linear predictor $\mathbf{x}_i\beta$. Because we envision a model for which the coefficient of `white` is equal to -0.20, we need to generate a variable that is equal to -0.20 times the variable `white`.

```
. estimates store Unconstrained
. generate offvar = -.20*white
. glm los hmo type2 type3, family(poisson) link(log) offset(offvar) nolog
```

Generalized linear models			No. of obs	=	1,495
Optimization : ML			Residual df	=	1,491
			Scale parameter =		1
Deviance = 8145.531652			(1/df) Deviance =		5.463133
Pearson = 9334.640731			(1/df) Pearson =		6.260658
Variance function: V(u) = u			[Poisson]		
Link function : g(u) = ln(u)			[Log]		
			AIC	=	9.27671
Log likelihood = -6930.340612			BIC	=	-2753.502

los	Coef.	OIM Std. Err.	z	P>|z|	[95% Conf. Interval]	
hmo	-.0696133	.0239174	-2.91	0.004	-.1164906	-.022736
type2	.218131	.020951	10.41	0.000	.1770677	.2591942
type3	.7079687	.0261214	27.10	0.000	.6567717	.7591658
_cons	2.374881	.0107841	220.22	0.000	2.353744	2.396017
offvar	1	(offset)				

```
. lrtest Unconstrained
```

Likelihood-ratio test	LR chi2(1) =	2.87
(Assumption: . nested in Unconstrained)	Prob > chi2 =	0.0905

Because we restricted one coefficient from our full model, the likelihood-ratio statistic is distributed as a chi-squared random variable with one degree of freedom. We fail to reject the hypothesis that the coefficient on white is equal to -0.20 at the 5% level.

Restricting coefficients for likelihood-ratio tests is just one use for offsets. Later, we discuss how to use offsets to account for exposure in count-data models.

2.5 Summary

The class of GLMs extends traditional linear models so that a linear predictor is mapped through a link function to model the mean of a response characterized by any member of the exponential family of distributions. Because we are able to develop one algorithm to fit the entire class of models, we can support estimation of such useful statistical models as logit, probit, and Poisson.

The traditional linear model is not appropriate when it is unreasonable to assume that data are normally distributed or if the response variable has a limited outcome set. Furthermore, in many instances in which homoskedasticity is an untenable requirement, the linear model is inappropriate. The GLM allows these extensions to the linear model.

A GLM is constructed by first selecting explanatory variables for the response variable of interest. A probability distribution that is a member of the exponential family is selected, and an appropriate link function is specified, for which the mapped range of values supports the implied variance function of the distribution.

GLM estimation algorithms

Contents

This chapter covers the theory behind GLMs. We present the material in a general fashion, showing all the results in terms of the exponential family of distributions. We illustrate two computational approaches for obtaining the parameter estimates of interest and discuss the assumptions that each method inherits. Parts II through VI of this text will then illustrate the application of this general theory to specific members of the exponential family.

The goal of our presentation is a thorough understanding of the underpinnings of the GLM method. We also wish to highlight the assumptions and limitations that algorithms inherit from their associated framework.

Traditionally, GLMs are fit by applying Fisher scoring within the Newton–Raphson method applied to the entire single-parameter exponential family of distributions. After making simplifications (which we will detail), the estimation algorithm is then referred to as iteratively reweighted least squares (IRLS). Before the publication of this algorithm, models based on a member distribution of the exponential family were fit using distribution-specific Newton–Raphson algorithms.

GLM theory showed how these models could be unified and fit using one IRLS algorithm that does not require starting values for the coefficients $\widehat{\beta}$; rather, it substitutes easy-to-compute fitted values $\widehat{y}_i$. This is the beauty and attraction of GLM. Estimation and theoretical presentation are simplified by addressing the entire family of distributions. Results are valid regardless of the inclusion of Fisher scoring in the Newton–Raphson computations.

In what follows, we highlight the Newton–Raphson method for finding the zeros (roots) of a real-valued function. In the simplest case, we view this problem as changing the point of view from maximizing the log likelihood to that of determining the root of the derivative of the log likelihood. Because the values of interest are obtained by setting the derivative of the log likelihood to zero and solving, that equation is referred to as the *estimating equation*.

We begin our investigation by assuming that we have a measured response variable that follows a member distribution of the exponential family. We also assume that the responses are independent and identically distributed (i.i.d.). Focusing on the exponential family of distributions admits two powerful generalizations. First, we are allowed a rich collection of models that includes members for binary, discrete, and continuous outcomes. Second, by focusing on the form of the exponential family, we derive general algorithms that are suitable for all members of the entire distribution family; historically, each member distribution was addressed in separate distribution-specific model specifications.

A probability function is a means to describe data on the basis of given parameters. Using the same formula for the probability function, we reverse the emphasis to the description of parameters on the basis of given data. Essentially, given a dataset, we determine those parameters that would most likely admit the given data within the scope of the specific probability function. This reversal of emphasis on the probability function of what is given is then called the likelihood function.

Although the joint i.i.d. probability function is given as

$$f(\mathbf{y}; \theta, \phi) = \prod_{i=1}^{n} f(y_i; \theta, \phi) \tag{3.1}$$

the likelihood is understood as

$$L(\theta, \phi; \mathbf{y}) = \prod_{i=1}^{n} f(\theta, \phi; y_i) \tag{3.2}$$

where the difference appears in that the probability function is a function of the unknown data $\mathbf{y}$ for the given parameters θ and ϕ, whereas the likelihood is a function of the unknown parameters θ and ϕ for the given data $\mathbf{y}$.

Moreover, because the joint likelihood of the function is multiplicative, we log-transform the likelihood function, L, to obtain a function on an additive scale. This function is known as the log likelihood, $\mathcal{L}$, and it is the focus of maximum likelihood (ML) theory. Thus, ML is a reasonable starting point in the examination of GLMs because the estimating equation of interest (equivalence of the derivative of the log-likelihood function and zero) provides the same results as those that maximize the log-likelihood function.

We begin with the specification of the log likelihood under standard likelihood theory, cf. Kendall and Stuart (1979), for the exponential family.

$$\mathcal{L} = \sum_{i=1}^{n} \left\{ \frac{y_i \theta_i - b(\theta_i)}{a(\phi)} + c(y_i, \phi) \right\} \tag{3.3}$$

The individual elements of the log likelihood are identified through descriptive terms: θ is the canonical parameter (it is the parameter vector of interest), $b(\theta)$ is the cumulant (it describes moments), ϕ is the dispersion parameter (it is a scale or ancillary parameter), and $c(\cdot)$ is a normalization term. The normalization term is not a function of θ and simply scales the range of the underlying density to integrate (or sum) to one.

To obtain the ML estimate $\hat{\theta}$ of θ, we first rely on the facts

$$E\left(\frac{\partial \mathcal{L}}{\partial \theta}\right) = 0 \tag{3.4}$$

$$E\left\{\frac{\partial^2 \mathcal{L}}{\partial \theta^2} + \left(\frac{\partial \mathcal{L}}{\partial \theta}\right)^2\right\} = 0 \tag{3.5}$$

See Gould, Pitblado, and Poi (2010) for examples wherein the two previous equations are presented as a lemma.

Using (3.4) for our sample of i.i.d. observations, we can write

$$0 = \frac{E(y_i) - b'(\theta_i)}{a(\phi)} \tag{3.6}$$

which implies that

$$b'(\theta_i) = E(y_i) = \mu_i \tag{3.7}$$

where μ_i is the mean value parameter, and from (3.5), we can write

$$0 = -\frac{b''(\theta_i)}{a(\phi)} + \frac{1}{a(\phi)^2} E\left\{y_i - b'(\theta_i)\right\}^2 \tag{3.8}$$

$$= -\frac{b''(\theta_i)}{a(\phi)} + \frac{1}{a(\phi)^2} E\left(y_i - \mu_i\right)^2 \tag{3.9}$$

$$= -\frac{b''(\theta_i)}{a(\phi)} + \frac{1}{a(\phi)^2} V(y_i) \tag{3.10}$$

Because the observations are i.i.d., we can drop the subscript and see that

$$b''(\theta) = \frac{1}{a(\phi)} V(y) \tag{3.11}$$

Solving for $V(y)$, we see that

$$\begin{aligned} V(y) &= b''(\theta)a(\phi) & (3.12) \\ &= v(\mu)a(\phi) & (3.13) \end{aligned}$$

because the variance is a function of the mean (expected value of y) via $b''(\theta)$. As such, we refer to $b''(\theta)$ as the variance function. That is, the variance of the outcome $V(y)$ is the product of the scalar parameter and a function of the mean, $v(\mu)$. From (3.7) and (3.11)–(3.13), we know that

$$v(\mu) = \frac{\partial \mu}{\partial \theta} \tag{3.14}$$

Using only the properties (3.4) and (3.5) of the ML estimator for θ, we then specify models that allow a parameterization of the mean μ in terms of covariates $\mathbf{X}$ with associated coefficients $\boldsymbol{\beta}$. Covariates are introduced via a known (invertible) function that links the mean μ to the linear predictor $\eta = \mathbf{X}\boldsymbol{\beta} + \text{offset}$ (the sum of the products of the covariates and coefficients).

$$\begin{aligned} g(\mu) &= \eta & (3.15) \\ g^{-1}(\eta + \text{offset}) &= \mu & (3.16) \end{aligned}$$

Finally, we also know that because $\eta = \sum_{j=1}^{p} x_j \beta_j + \text{offset}$,

$$\frac{\partial \eta}{\partial \beta_j} = x_j \tag{3.17}$$

The linear predictor is not constrained to lie within a specific range. Rather, we have $\eta \in \Re$. Thus, one purpose of the link function is to map the linear predictor to the range of the response variable. Mapping to the range of the response variable ensures nonnegative variances associated with a particular member distribution of the exponential family. The unconstrained linear model in terms of the form of the coefficients and associated covariates is constrained through the range-restricting properties of the link function.

The natural or canonical link is the result of equating the canonical parameter to the linear predictor, $\eta = \theta$; it is the link function obtained relating η to μ when $\eta = \theta$. In practice, we are not limited to this link function. We can choose any monotonic link function that maps the unrestricted linear predictor to the range implied by the variance function (see table A.1). In fact, in section 10.5 we investigate the use of link functions that could theoretically map the linear predictor to values outside the range implied by the variance function. This concept is somewhat confusing at first glance. After all, why would one want to consider different parameterizations? We demonstrate various interpretations and advantages to various parameterizations throughout the rest of the book.

Summarizing the current points, we can express the scalar derivatives for the vector $\boldsymbol{\beta}$ (and thus find the estimating equation) by using the chain rule for our sample of n observations,

$$\frac{\partial \mathcal{L}}{\partial \beta_j} = \sum_{i=1}^{n} \left(\frac{\partial \mathcal{L}_i}{\partial \theta_i} \right) \left(\frac{\partial \theta_i}{\partial \mu_i} \right) \left(\frac{\partial \mu_i}{\partial \eta_i} \right) \left(\frac{\partial \eta_i}{\partial \beta_j} \right) \tag{3.18}$$

$$= \sum_{i=1}^{n} \left\{ \frac{y_i - b'(\theta_i)}{a(\phi)} \right\} \left\{ \frac{1}{v(\mu_i)} \right\} \left(\frac{\partial \mu}{\partial \eta} \right)_i (x_{ji}) \tag{3.19}$$

$$= \sum_{i=1}^{n} \frac{y_i - \mu_i}{a(\phi)v(\mu_i)} \left(\frac{\partial \mu}{\partial \eta} \right)_i x_{ji} \tag{3.20}$$

where $i = 1, \ldots, n$ indexes the observations and x_{ji} is the ith observation for the jth covariate $\mathbf{X}_j$, $j = 1, \ldots, p$.

To find estimates $\widehat{\boldsymbol{\beta}}$, we can use Newton's method. This method is a linear Taylor series approximation where we expand the derivative of the log likelihood (the gradient) in a Taylor series. In the following, we use $\mathcal{L}' = \partial \mathcal{L}/\partial \boldsymbol{\beta}$ for the gradient and $\mathcal{L}'' = \partial^2 \mathcal{L}/(\partial \boldsymbol{\beta} \partial \boldsymbol{\beta}^T)$. We wish to solve

$$\mathcal{L}'(\boldsymbol{\beta}) = \mathbf{0} \tag{3.21}$$

Solving this equation provides estimates of $\boldsymbol{\beta}$. Thus, this is called the estimating equation. The first two terms of a Taylor series expansion approximates the estimating equation to the linear equation given by

$$\mathbf{0} \approx \mathcal{L}'(\boldsymbol{\beta}^{(0)}) + (\boldsymbol{\beta} - \boldsymbol{\beta}^{(0)})\mathcal{L}''(\boldsymbol{\beta}^{(0)}) \tag{3.22}$$

such that we may write (solving for $\boldsymbol{\beta}$)

$$\boldsymbol{\beta} \approx \boldsymbol{\beta}^{(0)} - \left\{ \mathcal{L}''(\boldsymbol{\beta}^{(0)}) \right\}^{-1} \mathcal{L}'(\boldsymbol{\beta}^{(0)}) \tag{3.23}$$

We may iterate this estimation using

$$\boldsymbol{\beta}^{(r)} = \boldsymbol{\beta}^{(r-1)} - \left\{ \mathcal{L}''(\boldsymbol{\beta}^{(r-1)}) \right\}^{-1} \mathcal{L}'(\boldsymbol{\beta}^{(r-1)}) \tag{3.24}$$

for $r = 1, 2, \ldots$ and a reasonable vector of starting values $\boldsymbol{\beta}^{(0)}$.

This linearized Taylor series approximation is exact if the function is truly quadratic. This is the case for the linear regression model (Gaussian variance with the identity link) as illustrated in an example in section 5.8. As such, only one iteration is needed. Other models require iteration from a reasonable starting point. The benefit of the derivation so far is that we have a general solution applicable to the entire exponential family of distributions. Without the generality of considering the entire family of distributions, separate derivations would be required for linear regression, Poisson regression, logistic regression, etc.

In the next two sections, we present derivations of two estimation approaches. Both are Newton–Raphson algorithms. In fact, there are many so-called Newton–Raphson algorithms. In (3.24), we see that the second derivative matrix of the log likelihood is required to obtain an updated estimate of the parameter vector $\boldsymbol{\beta}$. This matrix is, numerically speaking, difficult. Clearly, one approach is to form the second derivatives from analytic specification or from numeric approximation using the score function. Instead, we can use the outer matrix product of the score function [see (3.5)] as is done in IRLS.

There are many other techniques that Stata has conveniently made available through the `ml` command. Users of this command can specify a score function and investigate the performance of several numeric algorithms for a given command. Larger investigations are also possible through simulation, which is also conveniently packaged for users in the `simulate` command. We recommend that Stata users take time to study these commands, which enable quick development and deployment of software to support estimation for new models. The best source of information for investigating built-in support for optimization within Stata is Gould, Pitblado, and Poi (2010).

The obvious choice for the required matrix term in (3.24) is to use the second-derivative matrix, the so-called observed Hessian matrix. When we refer to the Newton–Raphson algorithm, we mean the algorithm using the observed Hessian matrix. The second obvious choice for the required matrix term is (Fisher scoring) the outer product of the score function, the so-called expected Hessian matrix. When we refer to IRLS, we mean the Newton–Raphson algorithm using the expected Hessian matrix. We discuss these choices in section 3.6.

Although there are conditions under which the two estimation methods (Newton–Raphson and IRLS) are equivalent, finding small numeric differences in software (computer output) is common, even for problems with unique solutions. These small discrepancies are due to differences in starting values and convergence paths. Such differences from the two estimation algorithms should not concern you. When the two algorithms are not numerically equivalent, they are still equal in the limit (they have the same expected value).

If we choose the canonical link in a model specification, then there are certain simplifications. For example, because the canonical link has the property that $\eta = \theta$, the score function may be written as

$$
\frac{\partial \mathcal{L}}{\partial \beta_j} = \sum_{i=1}^{n} \left(\frac{\partial \mathcal{L}_i}{\partial \theta_i} \right) \left(\frac{\partial \theta_i}{\partial \mu_i} \right) \left(\frac{\partial \mu_i}{\partial \eta_i} \right) \left(\frac{\partial \eta_i}{\partial \beta_j} \right) \tag{3.25}
$$

$$
= \sum_{i=1}^{n} \left(\frac{\partial \mathcal{L}_i}{\partial \eta_i} \right) \left(\frac{\partial \eta_i}{\partial \mu_i} \right) \left(\frac{\partial \mu_i}{\partial \eta_i} \right) \left(\frac{\partial \eta_i}{\partial \beta_j} \right) \tag{3.26}
$$

$$
= \sum_{i=1}^{n} \left(\frac{\partial \mathcal{L}_i}{\partial \eta_i} \right) \left(\frac{\partial \eta_i}{\partial \beta_j} \right) \tag{3.27}
$$

$$
= \sum_{i=1}^{n} \frac{y_i - \mu_i}{a(\phi)} x_{ji} \tag{3.28}
$$

Thus, using the canonical link leads to setting (3.28) to zero, such that $\widehat{\mu}_i$ is seen to play the role of $\widehat{y}_i$. Interpretation of predicted values (transformed via the inverse canonical link) are the predicted outcomes. Interpretation of predicted values under other (noncanonical) links is more difficult, as we will see in the discussion of negative binomial regression in chapter 13.

3.1   Newton–Raphson (using the observed Hessian)

ML begins by specifying a likelihood. The likelihood is a restatement of the joint probability density or joint probability mass function in which the data are taken as given and the parameters are estimated.

To find ML estimates, we proceed with the derivation originating with the log likelihood given in (3.3) and use the first derivatives given in (3.20). We now derive the (observed) matrix of the second derivatives, which will then give us the necessary ingredients for the Newton–Raphson algorithm to find the ML estimates.

This algorithm solves or optimizes (3.21), using variations on (3.24). The Newton–Raphson algorithm implements (3.24) without change. Research in optimization has found that this algorithm can be improved. If there are numeric problems in the optimization, these problems are usually manifested as singular (estimated) second-derivative matrices. Various algorithms address this situation. The other modifications to the Newton–Raphson algorithm include the Marquardt (1963) modification used in the heart of Stata's ml program. This software includes many other approaches. For a brief description, see, for example, the documentation on the bfgs and bhhh options in [R] ml (or help ml); for a more in-depth explanation, see Gould, Pitblado, and Poi (2010).

The observed matrix of second derivatives (Hessian matrix) is given by

$$\left(\frac{\partial^2 \mathcal{L}}{\partial \beta_j \partial \beta_k}\right) = \sum_{i=1}^{n} \frac{1}{a(\phi)} \left(\frac{\partial}{\partial \beta_k}\right) \left\{ \frac{y_i - \mu_i}{v(\mu_i)} \left(\frac{\partial \mu}{\partial \eta}\right)_i x_{ji} \right\} \tag{3.29}$$

$$= \sum_{i=1}^{n} \frac{1}{a(\phi)} \left[\left(\frac{\partial \mu}{\partial \eta}\right)_i \left\{ \left(\frac{\partial}{\partial \mu}\right)_i \left(\frac{\partial \mu}{\partial \eta}\right)_i \left(\frac{\partial \eta}{\partial \beta_k}\right)_i \right\} \frac{y_i - \mu_i}{v(\mu_i)} \right.$$

$$\left. + \frac{y_i - \mu_i}{v(\mu_i)} \left\{ \left(\frac{\partial}{\partial \eta}\right)_i \left(\frac{\partial \eta}{\partial \beta_k}\right)_i \right\} \left(\frac{\partial \mu}{\partial \eta}\right)_i \right] x_{ji} \tag{3.30}$$

$$= -\sum_{i=1}^{n} \frac{1}{a(\phi)} \left[\frac{1}{v(\mu_i)} \left(\frac{\partial \mu}{\partial \eta}\right)_i^2 \right.$$

$$\left. - (\mu_i - y_i) \left\{ \frac{1}{v(\mu_i)^2} \left(\frac{\partial \mu}{\partial \eta}\right)_i^2 \frac{\partial v(\mu_i)}{\partial \mu} - \frac{1}{v(\mu_i)} \left(\frac{\partial^2 \mu}{\partial \eta^2}\right)_i \right\} \right] x_{ji} x_{ki} \tag{3.31}$$

Armed with the first two derivatives, one can easily implement a Newton–Raphson algorithm to obtain the ML estimates. Without the derivatives written out analytically, one could still implement a Newton–Raphson algorithm by programming numeric derivatives (calculated using difference equations).

After optimization is achieved, one must estimate a suitable variance matrix for β. An obvious and suitable choice is based on the estimated observed Hessian matrix. This is the most common default choice in software implementations because the observed Hessian matrix is a component of the estimation algorithm. As such, it is already calculated and available. We discuss in section 3.6 that there are other choices for estimated variance matrices.

Thus, the Newton–Raphson algorithm provides

1. an algorithm for estimating the coefficients for all single-parameter exponential family GLM members and

2. estimated standard errors of estimated coefficients: square roots of the diagonal elements of the inverse of the estimated observed negative Hessian matrix.

Our illustration of the Newton–Raphson algorithm could be extended in several ways. The presentation did not show the estimation of the scale parameter, ϕ. Other implementations could include the scale parameter in the derivatives and cross derivatives to obtain ML estimates.

3.2   Starting values for Newton–Raphson

To implement an algorithm for obtaining estimates of β, we must have an initial guess for the parameters. There is no global mechanism for good starting values, but there is a reasonable solution for obtaining starting values when there is a constant in the model.

If the model includes a constant, then a common practice is to find the estimates for a constant-only model. For ML, this is a part of the model of interest, and knowing the likelihood for a constant-only model then allows a likelihood-ratio test for the parameters of the model of interest.

Often, the ML estimate for the constant-only model may be found analytically. For example, in chapter 12 we introduce the Poisson model. That model has a log likelihood given by

$$\mathcal{L} = \sum_{i=1}^{n} \{y_i(\mathbf{x}_i\boldsymbol{\beta}) - \exp(\mathbf{x}_i\boldsymbol{\beta}) - \ln\Gamma(y_i+1)\} \tag{3.32}$$

If we assume that there is only a constant term in the model, then the log likelihood may be written

$$\mathcal{L} = \sum_{i=1}^{n} \{y_i\beta_0 - \exp(\beta_0) - \ln\Gamma(y_i+1)\} \tag{3.33}$$

The ML estimate of β_0 is found by setting the derivative

$$\frac{\partial\mathcal{L}}{\partial\beta_0} = \sum_{i=1}^{n} \{y_i - \exp(\beta_0)\} \tag{3.34}$$

to zero and solving

$$0 = \sum_{i=1}^{n} \left\{y_i - \exp(\widehat{\beta}_0)\right\} \tag{3.35}$$

$$n\exp(\widehat{\beta}_0) = \sum_{i=1}^{n} y_i \tag{3.36}$$

$$\widehat{\beta}_0 = \ln(\bar{y}) \tag{3.37}$$

We can now use $\boldsymbol{\beta} = (\widehat{\beta}_0, 0, \ldots, 0)^T$ as the starting values (initial vector) in the Newton–Raphson algorithm.

Using this approach offers us two advantages. First, we start our iterations from a reasonable subset of the parameter space. Second, for ML, because we know the solution for the constant-only model, we may compare the log likelihood obtained for our model of interest with the log likelihood obtained at the initial step in our algorithm. This comparison is a likelihood-ratio test that each of the covariates (except the constant) are zero.

If there is not a constant in the model of interest or if we cannot analytically solve for the constant-only model, then we must use more sophisticated methods. We can either iterate to a solution for the constant-only model or use a search method to look for reasonable points at which to start our Newton–Raphson algorithm.

3.3 IRLS (using the expected Hessian)

Here we discuss the estimation algorithm known as IRLS. We begin by rewriting the (usual) updating formula from the Taylor series expansion presented in (3.24) as

$$\Delta\beta^{(r-1)} = -\left\{\frac{\partial^2 \mathcal{L}}{\partial(\beta^{(r-1)})^T \partial\beta^{(r-1)}}\right\}^{-1} \frac{\partial \mathcal{L}}{\partial\beta^{(r-1)}} \tag{3.38}$$

and we replace the calculation of the observed Hessian (second derivatives) with its expectation. This substitution is known as the method of Fisher scoring. Because we know that $E\{(y_i - \mu_i)^2\} = v(\mu_i)a(\phi)$, we may write

$$-E\left(\frac{\partial^2 \mathcal{L}}{\partial\beta_j \partial\beta_k}\right) = E\left(\frac{\partial\mathcal{L}}{\partial\beta_j}\frac{\partial\mathcal{L}}{\partial\beta_k}\right) \tag{3.39}$$

$$= \sum_{i=1}^{n}\left(\frac{\partial\mu}{\partial\eta}\right)_i^2 \frac{1}{v(\mu_i)a(\phi)} x_{ji}x_{ki} \tag{3.40}$$

Substituting (3.40) and (3.20) into (3.38) and rearranging, we see that $\delta\beta^{(r-1)}$ is the solution to

$$\left\{\sum_{i=1}^{n}\frac{1}{v(\mu_i)a(\phi)}\left(\frac{\partial\mu}{\partial\eta}\right)_i^2 x_{ji}x_{ki}\right\}\Delta\beta^{(r-1)} = \sum_{i=1}^{n}\frac{y_i - \mu_i}{v(\mu_i)a(\phi)}\left(\frac{\partial\mu}{\partial\eta}\right)_i \mathbf{x}_i^T \tag{3.41}$$

Using the $(r-1)$ superscript to emphasize calculation with $\beta^{(r-1)}$, we may refer to the linear predictor as

$$\eta_i^{(r-1)} - \text{offset}_i = \sum_{k=1}^{p} x_{ki}\beta_k^{(r-1)} \tag{3.42}$$

which may be rewritten as

$$\left\{\sum_{i=1}^{n}\frac{1}{v(\mu_i)a(\phi)}\left(\frac{\partial\mu}{\partial\eta}\right)_i^2 x_{ji}x_{ki}\right\}\beta^{(r-1)} =$$

$$\sum_{i=1}^{n}\frac{1}{v(\mu_i)a(\phi)}\left(\frac{\partial\mu}{\partial\eta}\right)_i^2 \left(\eta_i^{(r-1)} - \text{offset}_i\right)\mathbf{x}_i^T \tag{3.43}$$

Summing (3.41) and (3.43) and then substituting (3.38), we obtain

$$\left\{ \sum_{i=1}^{n} \frac{1}{v(\mu_i)a(\phi)} \left(\frac{\partial \mu}{\partial \eta}\right)_i^2 x_{ji}x_{ki} \right\} \left(\Delta\boldsymbol{\beta}^{(r-1)} + \boldsymbol{\beta}^{(r-1)}\right) =$$

$$\sum_{i=1}^{n} \frac{1}{v(\mu_i)a(\phi)} \left(\frac{\partial \mu}{\partial \eta}\right)_i^2 \left\{ (y_i - \mu_i) \left(\frac{\partial \eta}{\partial \mu}\right)_i + \left(\eta_i^{(r-1)} - \text{offset}_i\right) \right\} \mathbf{x}_i^T \quad (3.44)$$

$$\left\{ \sum_{i=1}^{n} \frac{1}{v(\mu_i)a(\phi)} \left(\frac{\partial \mu}{\partial \eta}\right)_i^2 x_{ji}x_{ki} \right\} \boldsymbol{\beta}^{(r)} =$$

$$\sum_{i=1}^{n} \frac{1}{v(\mu_i)a(\phi)} \left(\frac{\partial \mu}{\partial \eta}\right)_i^2 \left\{ (y_i - \mu_i) \left(\frac{\partial \eta}{\partial \mu}\right)_i + \left(\eta_i^{(r-1)} - \text{offset}_i\right) \right\} \mathbf{x}_i^T \quad (3.45)$$

Now, again using the $(r-1)$ superscript to emphasize calculation with $\boldsymbol{\beta}^{(r-1)}$, we let

$$\mathbf{W}^{(r-1)} = \text{diag} \left\{ \frac{1}{v(\mu)a(\phi)} \left(\frac{\partial \mu}{\partial \eta}\right)^2 \right\}_{(n \times n)} \quad (3.46)$$

$$\mathbf{z}^{(r-1)} = \left\{ (y - \mu) \left(\frac{\partial \eta}{\partial \mu}\right) + \left(\eta^{(r-1)} - \text{offset}\right) \right\}_{(n \times 1)} \quad (3.47)$$

so that we may rewrite (3.45) in matrix notation as

$$\left(\mathbf{X}^T \mathbf{W}^{(r-1)} \mathbf{X}\right) \boldsymbol{\beta}^{(r)} = \mathbf{X}^T \mathbf{W}^{(r-1)} \mathbf{z}^{(r-1)} \quad (3.48)$$

$$\boldsymbol{\beta}^{(r)} = \left(\mathbf{X}^T \mathbf{W}^{(r-1)} \mathbf{X}\right)^{-1} \mathbf{X}^T \mathbf{W}^{(r-1)} \mathbf{z}^{(r-1)} \quad (3.49)$$

This algorithm then shows that a new estimate of the coefficient vector is obtained via weighted least squares (weighted linear regression).

After optimization is achieved, one must estimate a suitable variance matrix for $\boldsymbol{\beta}$. An obvious and suitable choice is the estimated expected Hessian matrix. This is the most common default choice in software implementations because the expected Hessian matrix is a component of the estimation algorithm. As such, it is already calculated and available. We discuss in section 3.6 that there are other choices for estimated variance matrices.

Thus, the IRLS algorithm provides

1. an algorithm for estimating the coefficients for all single-parameter exponential family GLM members,

2. an algorithm that uses weighted least squares and thus may be easily incorporated into almost any statistical analysis software, and

3. estimated standard errors of estimated coefficients: square roots of the diagonal elements of $(\mathbf{X}^T\mathbf{W}\mathbf{X})^{-1}$.

The usual estimates of standard errors are constructed using the expected Hessian. These are equal to the observed Hessian when the canonical link is used. Otherwise, the calculated standard errors will differ from usual Newton–Raphson standard error estimates. Because the expected Hessian is commonly used by default for calculating standard errors, some refer to these as naïve standard errors. They are naïve in that the estimation assumes that the conditional mean is specified correctly.

In the end, our algorithm needs specification of the variance as a function of the mean $v(\mu)$, the derivative of the inverse link function, the dispersion parameter $a(\phi)$, the linear predictor (the systematic component), and the random variable y. These are the components listed in section 2.1.

This algorithm may be generalized in several important ways. First, the link function may be different for each observation (may be subscripted by i). The algorithm may include an updating step for estimating unknown ancillary parameters and include provisions for ensuring that the range of the fitted linear predictor does not fall outside specified bounds. A program that implements these extensions may do so automatically or via user-specified options. As such, you may see in the literature somewhat confusing characterizations of the power of GLM, when in fact the characterizations are of the IRLS algorithm, which supports estimation of GLMs. This substitution of the class of models for the estimation algorithm is common.

John Nelder, as one of the responders in the discussion of Green (1984), included an "etymological quibble" in which he argued that the use of "reweighted" over "weighted" placed unnecessary emphasis on the updated weights. That we update both the weights and the (synthetic) dependent variable at each iteration is the heart of Nelder's quibble in naming the algorithm IRLS. As such, although we refer to this estimation algorithm as IRLS, we will also see reference in various places to iterative weighted least squares. This is the term originally used in Nelder and Wedderburn (1972).

Wedderburn's (1974) extension shows that the specification of variance as a function of the mean implies a quasilikelihood. We will use this result throughout the text in developing various models, but see chapter 17 for motivating details.

The important fact from Wedderburn's paper is that the applicability of the IRLS algorithm is extended to specifying the ingredients of the algorithm and that we do not have to specify a likelihood. Rather, a quasilikelihood is implied by the assumed (specified) moments. Estimates from the IRLS algorithm are ML when the implied quasilikelihood is a true likelihood; otherwise, the estimates are maximum quasilikelihood (MQL). Thus, we can think of IRLS as an algorithm encompassing the assumptions and basis of ML (Newton–Raphson method) or as an algorithm encompassing the assumptions and basis of the methodology of MQL.

In the appendix (tables A.1, A.2, and A.4), we present a collection of tabular material showing typical variance and link functions. Most of these are derived from likelihoods. These derivations will be highlighted in later chapters of the book. The power family of the variance functions listed in the tables is not derived from a likelihood, and its use is a direct result of the extension given by Wedderburn (1974).

3.4   Starting values for IRLS

Unlike Newton–Raphson, the IRLS algorithm does not require an initial guess for the parameter vector of interest β. Instead, we need initial guesses for the fitted values $\hat{\mu}_i$. This is much easier to implement.

A reasonable approach is to initialize the fitted values to the inverse link of the mean of the response variable. We must ensure that the initial starting values are within the implied range of the variance function specified in the GLM that we are fitting. For example, this prevents the situation where the algorithm may try to compute the logarithm (or square root) of a nonpositive value.

Another approach is to set the initial fitted values to $(y_i + \bar{y})/2$ for a nonbinomial model and $k_i(y_i + 0.5)/(k_i + 1)$ for a binomial(k_i, p_i) model. In fact, this is how initial values are assigned in the Stata `glm` command.

3.5   Goodness of fit

In developing a model, we hope to generate fitted values $\hat{\mu}$ that are close to the data y. For a dataset with n observations, we may consider candidate models with one to n parameters. The simplest model would include only one parameter. The best one-parameter model would result in $\hat{\mu}_i = \mu$ (for all i). Although the model is parsimonious, it does not estimate the variability in the data. The saturated model (with n parameters) would include one parameter for each observation and result in $\hat{\mu}_i = y_i$. This model exactly reproduces the data but is uninformative because there is no summarization of the data.

We define a measure of fit for the model as twice the difference between the log likelihoods of the model of interest and the saturated model. Because this difference is a measure of the deviation of the model of interest from a perfectly fitting model, the measure is called the deviance. Our competing goals in modeling are to find the simplest model (fewest parameters) that has the smallest deviance (reproduces the data).

The deviance, D, is given by

$$D = \sum_{i=1}^{n} 2\left[y_i\{\theta(y_i) - \theta(\mu_i)\} - b\{\theta(y_i)\} + b\{\theta(\mu_i)\}\right] \tag{3.50}$$

where the equation is given in terms of the mean parameter μ instead of the canonical parameter θ. In fitting a particular model, we seek the values of the parameters that minimize the deviance. Thus, optimization in the IRLS algorithm is achieved when the difference in deviance calculations between successive iterations is small (less than some chosen tolerance). The values of the parameters that minimize the deviance are the same as the values of the parameters that maximize the likelihood.

The quasideviance is an extension of this concept for models that do not have an underlying likelihood. We discuss this idea more fully in chapter 17.

3.6   Estimated variance matrices

It is natural to ask how the Newton–Raphson (based on the observed Hessian) variance estimates compare with the usual (based on the expected Hessian) variance estimates obtained using the IRLS algorithm outlined in the preceding section. The matrix of second derivatives in the IRLS algorithm is equal to the first term in (3.31). As Newson (1999) points out, the calculation of the expected Hessian is simplified from that of the observed Hessian because we assume that $E(\mu - y) = 0$ or, equivalently, the conditional mean of y given $\mathbf{X}$ is correct. As such, the IRLS algorithm assumes that the conditional mean is specified correctly. Both approaches result in parameter estimates that differ only because of numeric roundoff or because of differences in optimization criteria.

This distinction is especially important in the calculation of sandwich estimates of variance. The Hessian may be calculated as given above in (3.31) or may be calculated using the more restrictive (naïve) assumptions of the IRLS algorithm as

$$
E\left(\frac{\partial^2 \mathcal{L}}{\partial \beta_j \partial \beta_k}\right) = -\sum_{i=1}^{n} \frac{1}{a(\phi)} \frac{1}{v(\mu_i)} \left(\frac{\partial \mu}{\partial \eta}\right)_i^2 x_{ji} x_{ki} \tag{3.51}
$$

Equations (3.31) and (3.51) are equal when the canonical link is used. This occurs because for the canonical link we can make the substitution that $\theta = \eta$ and thus that $v(\mu) = \partial \mu / \partial \theta = \partial \mu / \partial \eta$. The second term in (3.31) then collapses to zero because

$$
(\mu_i - y_i) \left\{ \frac{1}{v(\mu_i)^2} \left(\frac{\partial \mu}{\partial \eta}\right)_i^2 \frac{\partial v(\mu_i)}{\partial \mu} - \frac{1}{v(\mu_i)} \left(\frac{\partial^2 \mu}{\partial \eta^2}\right)_i \right\}
$$

$$
= (\mu_i - y_i) \left\{ \frac{1}{(\partial \mu / \partial \eta)_i^2} \left(\frac{\partial \mu}{\partial \eta}\right)_i^2 \frac{\partial}{\partial \mu_i} \left(\frac{\partial \mu}{\partial \eta}\right)_i - \frac{1}{(\partial \mu / \partial \eta)_i} \left(\frac{\partial^2 \mu}{\partial \eta^2}\right)_i \right\} \tag{3.52}
$$

$$
= (\mu_i - y_i) \left\{ \frac{\partial}{\partial \mu_i} \left(\frac{\partial \mu}{\partial \eta}\right)_i - \left(\frac{\partial \eta}{\partial \mu}\right)_i \left(\frac{\partial^2 \mu}{\partial \eta^2}\right)_i \right\} \tag{3.53}
$$

$$
= (\mu_i - y_i) \left\{ \left(\frac{\partial^2 \mu}{\partial \mu \partial \eta}\right)_i - \left(\frac{\partial^2 \mu}{\partial \mu \partial \eta}\right)_i \right\} \tag{3.54}
$$

$$
= 0 \tag{3.55}
$$

An estimation algorithm is not limited to using one version of the Hessian or the other. Users of PROC GENMOD in SAS, for example, should be familiar with the SCORING option. This option allows the user to specify how many of the initial iterations are performed using the Fisher scoring method, expected Hessian calculated using (3.51), before all subsequent iterations are performed using the observed Hessian calculated with (3.31). The overall optimization is still the Newton–Raphson method (optimizes the log likelihood). The Stata software described in association with this text also includes this capability; see the documentation of the fisher() option in the *Stata Base Reference Manual* for the glm entry.

Although there is an obvious choice for the estimated variance matrix for a given implementation of Newton–Raphson, one is not limited to that specific estimator used in the optimization (estimation algorithm). The IRLS algorithm could calculate the observed Hessian after convergence and report standard errors based on the observed Hessian.

Later, we will want to alter IRLS algorithms to substitute this observed information matrix for the IRLS expected information matrix. To do so, we merely have to alter the $\mathbf{W}$ matrix by the factor listed in (3.52). Here we identify the ingredients of this factor in general form. For specific applications, the individual terms are specified in the tables of appendix A.

$$-(\mu_i - y_i) \left\{ \frac{1}{v(\mu_i)^2} \left(\frac{\partial \mu}{\partial \eta} \right)_i^2 \frac{\partial v(\mu_i)}{\partial \mu} - \frac{1}{v(\mu_i)} \left(\frac{\partial^2 \mu}{\partial \eta^2} \right)_i \right\}$$

$$= (y_i - \mu_i) \left\{ \frac{1}{v(\mu_i)^2} \left(\frac{\partial \mu}{\partial \eta} \right)_i^2 \frac{\partial v(\mu_i)}{\partial \mu} - \frac{1}{v(\mu_i)} \left(\frac{\partial^2 \mu}{\partial \eta^2} \right)_i \right\} \tag{3.56}$$

$$= (y_i - \mu_i) \left(\frac{v'(\mu_i)}{v(\mu_i)^2} \left[\{g^{-1}(\eta_i)\}' \right]^2 - \frac{1}{v(\mu_i)} \{g^{-1}(\eta_i)\}'' \right) \tag{3.57}$$

$$= (y_i - \mu_i) \left[\frac{v'(\mu_i)}{v(\mu_i)^2} \frac{1}{\{g'(\mu_i)\}^2} - \frac{1}{v(\mu_i)} \frac{-g''(\mu_i)}{\{g'(\mu_i)\}^3} \right] \tag{3.58}$$

$$= (y_i - \mu_i) \left[\frac{v'(\mu_i)g'(\mu_i) + v(\mu_i)g''(\mu_i)}{v^2(\mu_i)\{g'(\mu_i)\}^3} \right] \tag{3.59}$$

Tables A.4 and A.6 provide the information for using (3.57). We have also provided the derivatives in tables A.3 and A.5 for using (3.59). Which set of derivatives you choose in a given application is motivated by how easily they might be calculated. For discussion on the derivation of the GLM theory, we focus on the derivatives of the inverse link function, but for developing algorithms, we often use the derivatives of the link function. The derivation from (3.57) to (3.59) is justified (leaving off the subscripts) by

$$g \;=\; g(\mu) = \eta \tag{3.60}$$

$$g' \;=\; g'(\mu) = \frac{\partial \eta}{\partial \mu} \tag{3.61}$$

$$g'' \;=\; g''(\mu) = \frac{\partial}{\partial \mu} g' = \frac{\partial^2 \eta}{\partial \mu^2} \tag{3.62}$$

$$g^{-1} \;=\; g^{-1}(\eta) = \mu \tag{3.63}$$

$$(g^{-1})' \;=\; \{g^{-1}(\eta)\}' = \frac{\partial \mu}{\partial \eta} = \frac{1}{g'} \tag{3.64}$$

$$(g^{-1})'' \;=\; \{g^{-1}(\eta)\}'' = \frac{\partial^2 \mu}{\partial \eta^2} = \frac{\partial}{\partial \eta} \frac{1}{g'} = -\frac{1}{(g')^2} \frac{\partial}{\partial \eta} g'$$

$$\;=\; -\frac{1}{(g')^2} \frac{\partial}{\partial \mu} \frac{\partial \mu}{\partial \eta} g' = -\frac{1}{(g')^2} \frac{\partial \mu}{\partial \eta} \frac{\partial}{\partial \mu} g' = -\frac{1}{(g')^2} \frac{1}{g'} g'' = -\frac{g''}{(g')^3} \tag{3.65}$$

Davidian and Carroll (1987) provide details for estimating variance functions in general, and Oakes (1999) shows how to calculate the information matrix (see section 3.6.1) by using the EM algorithm. In the following subsections, we present details on calculating various variance matrix estimates. Throughout, $\partial \mu / \partial \eta$ is calculated at $\mu = \widehat{\mu}$, and $\widehat{\phi}$ is an estimator of the dispersion parameter $a(\phi)$. We will also use the hat diagonals defined later in (4.5).

3.6.1   Hessian

The usual variance estimate in statistical packages is calculated (numerically or analytically) as the inverse matrix of (negative) second derivatives. For GLMs, we can calculate the Hessian in two ways.

Assuming that the conditional mean is specified correctly, we may calculate the expected Hessian as

$$\widehat{V}_{\text{EH}} = \left\{ E\left(-\frac{\partial^2 \mathcal{L}}{\partial \beta \partial \beta^T} \right) \right\}^{-1} \tag{3.66}$$

where the matrix elements $-\partial^2 \mathcal{L}/(\partial \beta_j \partial \beta_k)$ are calculated using (3.51). Sandwich estimates of variance formed using this calculation of the Hessian are called semirobust or semi-Huber because they are not robust to misspecification of the conditional mean.

More generally, we may calculate the observed Hessian without the additional assumption of correct conditional mean specification using

$$\widehat{V}_{\text{OH}} = \left\{ \left(-\frac{\partial^2 \mathcal{L}}{\partial \beta \partial \beta^T} \right) \right\}^{-1} \tag{3.67}$$

where the matrix elements $-\partial^2 \mathcal{L}/(\partial \beta_j \partial \beta_k)$ are calculated using (3.31). Sandwich estimates of variance formed using this calculation of the Hessian are called robust or full Huber.

Because the two calculations are equal when using the canonical link, the result in those cases is always robust or full Huber. Throughout the remaining subsections, we refer generally to $\widehat{V}_H$ and make clear in context whether the inference is affected by using the observed versus expected Hessian.

3.6.2 Outer product of the gradient

Berndt et al. (1974) describe the conditions under which we may use the outer product of the gradient (OPG) vector in optimization. Known in the economics literature as the "B-H-cubed" method (from the initials of the last names of the authors of the previously cited paper), we can estimate the variance matrix without calculating second derivatives. Instead, we use the OPG. The gradient is the score function or the vector of first derivatives when the derivation is the result of a likelihood approach. This is similar to the expected Hessian matrix except that we are not taking the expected values of the outer product. In the equation, $\mathbf{x}_i$ refers to the ith row of the matrix $\mathbf{X}$.

$$\widehat{V}_{\text{OPG}} = \sum_{i=1}^{n} \mathbf{x}_i^T \left\{ \frac{y_i - \widehat{\mu}_i}{v(\widehat{\mu}_i)} \left(\frac{\partial \mu}{\partial \eta} \right)_i \widehat{\phi} \right\}^2 \mathbf{x}_i \tag{3.68}$$

3.6.3 Sandwich

The validity of the covariance matrix based on the Hessian depends on the correct specification of the variance function (3.31) or the correct specification of both the variance and link functions (3.51). The sandwich estimate of variance provides a consistent estimate of the covariance matrix of parameter estimates even when the specification of the variance function is incorrect.

The sandwich estimate of variance is a popular estimate of variance that is known by many different names; see Hardin (2003). The formula for calculating this variance estimate (either in algebraic or matrix form) has a score factor "sandwiched" between two copies of the Hessian. Huber (1967) is generally credited with first describing this variance estimate. Because his discussion addressed ML estimates under misspecification, the variance estimate is sometimes called the robust variance estimate or Huber's estimate of variance.

White (1980) independently showed that this variance estimate is consistent under a model including heteroskedasticity, so it is sometimes labeled the heteroskedasticity covariance matrix or the heteroskedasticity-consistent covariance matrix. Because of this same reference, the sandwich estimate of variance is also called White's estimator.

From the derivation of the variance estimate and the discussion of its properties in survey statistics literature papers such as Binder (1983), Kish and Frankel (1974), and Gail, Tan, and Piantadosi (1988), the sandwich estimate of variance is sometimes called the survey variance estimate or the design-based variance estimate. The statistics literature has also referred to the sandwich estimate of variance as the empirical estimate of variance. This name derives from the point of view that the variance estimate is not relying on the model being correct. We have heard in private conversation the descriptive "model-agnostic variance estimate", though we have not yet seen this phrase in the literature.

The estimated score is calculated using properties of the variance function (family) and the link function:

$$\text{score}(\beta)_i = \left(\frac{\partial \mathcal{L}}{\partial \eta}\right)_i = \left(\frac{\partial \mathcal{L}}{\partial \mu}\right)_i \left(\frac{\partial \mu}{\partial \eta}\right)_i = \frac{y_i - \widehat{\mu}_i}{v(\widehat{\mu}_i)} \left(\frac{\partial \mu}{\partial \eta}\right)_i \tag{3.69}$$

The contribution from the estimated scores is then calculated, where $\mathbf{x}_i$ refers to the ith row of the matrix $\mathbf{X}$ using

$$\widehat{B}_S = \sum_{i=1}^{n} \mathbf{x}_i^T \left\{ \frac{y_i - \widehat{\mu}_i}{v(\widehat{\mu}_i)} \left(\frac{\partial \mu}{\partial \eta}\right)_i \widehat{\phi} \right\}^2 \mathbf{x}_i \tag{3.70}$$

and the (usual) sandwich estimate of variance is then

$$\widehat{V}_S = \widehat{V}_H^{-1} \widehat{B}_S \widehat{V}_H^{-1} \tag{3.71}$$

The OPG estimate of variance is the "middle" of the sandwich estimate of variance

$$\widehat{V}_{\text{OPG}} = \widehat{B}_S \tag{3.72}$$

Royall (1986) discusses the calculation of robust confidence intervals for ML estimates using the sandwich estimate of variance.

3.6.4 Modified sandwich

If observations are grouped because they are correlated (perhaps because the data are really panel data), then the sandwich estimate is calculated, where n_i refers to the observations for each panel i and $\mathbf{x}_{ij}$ refers to the row of the matrix $\mathbf{X}$ associated with the jth observation for subject i, using

$$\widehat{B}_{\text{MS}} = \sum_{i=1}^{n} \left\{ \sum_{j=1}^{n_i} \mathbf{x}_{ij}^T \frac{y_{ij} - \widehat{\mu}_{ij}}{v(\widehat{\mu}_{ij})} \left(\frac{\partial \mu}{\partial \eta}\right)_{ij} \widehat{\phi} \right\} \left\{ \sum_{j=1}^{n_i} \frac{y_{ij} - \widehat{\mu}_{ij}}{v(\widehat{\mu}_{ij})} \left(\frac{\partial \mu}{\partial \eta}\right)_{ij} \widehat{\phi} \mathbf{x}_{ij} \right\} \tag{3.73}$$

as the modified (summed or partial) scores.

In either case, the calculation of $\widehat{V}_H$ is the same and the modified sandwich estimate of variance is then

$$\widehat{V}_{\text{MS}} = \widehat{V}_H^{-1} \widehat{B}_{\text{MS}} \widehat{V}_H^{-1} \tag{3.74}$$

The problem with referring to the sandwich estimate of variance as "robust" and the modified sandwich estimate of variance as "robust cluster" is that it implies that robust standard errors are bigger than usual (Hessian) standard errors and that robust cluster standard errors are bigger still. This is a false conclusion. See Carroll et al. (1998) for a lucid comparison of usual and robust standard errors. A comparison of the sandwich estimate of variance and the modified sandwich estimate of variance depends on the within-panel correlation of the score terms. If the within-panel correlation is negative, then the panel score sums of residuals will be small, and the panel score sums will have less variability than the variability of the individual scores. This will lead to the modified sandwich standard errors being smaller than the sandwich standard errors.

Strictly speaking, the calculation and application of the modified sandwich estimate of variance is a statement that the fitted model is not represented by a true likelihood, even though the chosen family and link might otherwise imply a true likelihood. The fitted model is optimized assuming that the distribution of the sample may be calculated as the product of the distribution for each observation. However, when we calculate a modified sandwich estimate of variance, we acknowledge that the observations within a panel are not independent but that the sums over individual panels are independent. Our assumed or implied likelihood does not account for this within-panel correlation and therefore is not the true likelihood.

In the end, we use an inefficient (compared with the model-based variance when the assumptions of the model are valid) but unbiased point estimate and a variance matrix from which standard errors ensure inference is robust to any particular within-panel correlation.

3.6.5   Unbiased sandwich

The sandwich estimate of variance has been applied in many cases and is becoming more common in statistical software. One area of current research concerns the small-sample properties of this variance estimate. There are two main modifications to the sandwich estimate of variance in constructing confidence intervals. The first is a degrees-of-freedom correction (scale factor), and the second is the use of a more conservative distribution ("heavier-tailed" than the normal).

Acknowledging that the usual sandwich estimate is biased, we may calculate an unbiased sandwich estimate of variance with improved small-sample performance in coverage probability. This modification is a scale factor multiplier motivated by the knowledge that the variance of the estimated residuals are biased on terms of the ith diagonal element of the hat matrix, h_i, defined in (4.5):

$$V(\widehat{\epsilon}_i) = \sigma^2(1 - h_i) \tag{3.75}$$

We can adjust for the bias of the contribution from the scores, where $\mathbf{x}_i$ refers to the ith row of the matrix $\mathbf{X}$, using

$$\widehat{B}_{\mathrm{US}} = \sum_{i=1}^{n} \mathbf{x}_i^T \left\{ \frac{y_i - \widehat{\mu}_i}{v(\widehat{\mu}_i)} \left(\frac{\partial \mu}{\partial \eta}\right)_i \widehat{\phi} \right\}^2 \frac{\mathbf{x}_i}{1 - \widehat{h}_i} \tag{3.76}$$

where the (unbiased) sandwich estimate of variance is then

$$\widehat{V}_{\mathrm{US}} = \widehat{V}_H^{-1} \widehat{B}_{\mathrm{US}} \widehat{V}_H^{-1} \tag{3.77}$$

The Stata command `regress` calculates this variance matrix in linear regression using the `vce(hc2)` option and cites MacKinnon and White (1985). This reference includes three scale factor adjustments. MacKinnon and White label the first adjustment HC1. The HC1 adjustment applies the scale factor $n/(n-k)$ to the estimated variance matrix. This is a degree-of-freedom correction from Hinkley (1977). The second adjustment is labeled HC2 and is a scale factor adjustment of $1/(1-h_i)$. For linear regression, this scale factor is discussed by Belsley, Kuh, and Welsch (1980) and by Wu (1986). The final adjustment is labeled HC3 and is an application of the scale factor $1/(1-h_i)^2$. This scale factor is a jackknife approximation discussed by Efron in the Wu paper and is available using the `vce(hc3)` option with the `regress` command. The unadjusted sandwich variance matrix is labeled HC0—which makes 11 different names if you are counting.

We use the descriptive name "unbiased sandwich estimate of variance" found in Carroll et al. (1998) rather than the HC2 designation. Another excellent survey of the sandwich estimator may be found in the working paper by Long and Ervin (1998).

3.6.6   Modified unbiased sandwich

If observations may be grouped because they are correlated (perhaps because the data are really panel data), then the unbiased sandwich estimate is calculated, where n_i refers to the observations for each panel i and $\mathbf{x}_{ij}$ refers to the row of the matrix $\mathbf{X}$ associated with the jth observation for subject i, using

$$
\begin{aligned}
\widehat{B}_{\mathrm{MUS}} \;=\; \sum_{i=1}^{n} &\left[\left\{ \sum_{j=1}^{n_i} \mathbf{x}_{ij}^T \frac{y_{ij} - \widehat{\mu}_{ij}}{v(\widehat{\mu}_{ij})} \left(\frac{\partial \mu}{\partial \eta} \right)_{ij} \frac{\widehat{\phi}}{\sqrt{1 - \widehat{h}_{ij}}} \right\} \right. \\
&\left. \times \left\{ \sum_{j=1}^{n_i} \frac{y_{ij} - \widehat{\mu}_{ij}}{v(\widehat{\mu}_{ij})} \left(\frac{\partial \mu}{\partial \eta} \right)_{ij} \frac{\widehat{\phi}}{\sqrt{1 - \widehat{h}_{ij}}} \mathbf{x}_{ij} \right\} \right]
\end{aligned}
\tag{3.78}
$$

as the modified (summed or partial) scores. This is the same adjustment as was made for the modified sandwich estimate of variance. Here h_{ij} is the ijth diagonal element of the hat matrix described in (4.5). The modified unbiased sandwich estimate of variance is then

$$\widehat{V}_{\mathrm{MUS}} = \widehat{V}_H^{-1} \widehat{B}_{\mathrm{MUS}} \widehat{V}_H^{-1} \tag{3.79}$$

3.6.7   Weighted sandwich: Newey–West

These variance estimators are referred to as HAC variance estimators because they are heteroskedasticity- and autocorrelation-consistent estimates of the variances (of parameter estimators). A weighted sandwich estimate of variance calculates a (possibly) different middle of the sandwich. Instead of using only the usual score contributions, a weighted sandwich estimate of variance calculates a weighted mean of score contributions and lagged score contributions.

Newey and West (1987) discuss a general method for combining the contributions for each considered lag. The specific implementation then assigns a weight to each lagged score contribution. In related sections, we present various weight functions for use with this general approach.

In the following, let n be the number of observations, p be the number of predictors, G be the maximum lag, C be an overall scale factor, and q be the prespecified bandwidth (number of lags for which the correlation is nonzero). The overall scale factor is usually defined as one but could be defined as $n/(n-p)$ to serve as a small sample scale factor adjustment.

$$\widehat{V}_{\mathrm{NW}} = \widehat{V}_H^{-1}\widehat{B}_{\mathrm{NW}}\widehat{V}_H^{-1} \tag{3.80}$$

$$\widehat{B}_{\mathrm{NW}} = C\left\{\widehat{\Omega}_0 + \sum_{j=1}^{G}\omega\left(\frac{j}{q+1}\right)\left(\widehat{\Omega}_j + \widehat{\Omega}_j'\right)\right\} \tag{3.81}$$

$$\widehat{\Omega}_j = \sum_{i=j+1}^{n}\mathbf{x}_i\widehat{r}_i^S\,\widehat{r}_{i-j}^S\mathbf{x}_i^T \tag{3.82}$$

$$\omega(z) = \text{sandwich weights} \tag{3.83}$$

$$\widehat{r}_i^S = \text{score residuals (see section 4.4.9)} = \nabla_i(y_i - \mu_i)/v_i \tag{3.84}$$

To calculate the Newey–West variance–covariance matrix for a given model, we need the link function, the derivative of the link function, the mean-variance function, and the sandwich weights

$$\{\mu_i, \nabla_i, v_i, \omega(z)\} = \{\mu_i, (\partial\mu/\partial\eta)_i, v(\mu_i), w_i\} \tag{3.85}$$

The required quantities are defined in tables A.1, A.2, A.4, and A.8.

Specifically for $G = 0$ (no autocorrelations), $\widehat{B}_{\mathrm{NW}} = \widehat{\Omega}_0 = \widehat{B}_S$ from (3.70) so that $\widehat{V}_{\mathrm{NW}}$ is equal to a robust sandwich estimate of variance $\widehat{V}_S$ (see section 3.6.3) for all the above weight functions.

In table A.8, the Newey and West (1987) estimator uses weights derived from the Bartlett kernel for a user-specified bandwidth (number of autocorrelation lags). These are the usual kernel weights in software packages implementing this weighted sandwich estimate of variance. For $G = 0$, the kernel is more appropriately called the truncated kernel, resulting in the sandwich variance estimator.

In table A.8, the Gallant (1987, 533) estimator uses weights derived from the Parzen (1957) kernel for a user-specified bandwidth (number of autocorrelation lags).

In table A.8, the Anderson estimator uses weights derived from the quadratic spectral kernel. These weights are based on a spectral decomposition of the variance matrix.

A general class of kernels is given in Andrews (1991), where he refers to using the Tukey–Hanning kernel

$$\omega(z) = \begin{cases} \{1 + \cos(\pi z)\}/2 & \text{if } |z| \leq 1 \\ 0 & \text{otherwise} \end{cases} \tag{3.86}$$

Andrews mentions an entire class of admissible kernels and his paper investigates bandwidth selection.

Automatic lag selection is investigated in Newey and West (1994). In further research, Lumley and Heagerty (1999) provide details on weighted sandwich estimates of variance where the weights are data driven (rather than kernel driven). These are presented under the name weighted empirical adaptive variance estimates (WEAVEs), given that the weights are computed based on estimated correlations from the data instead of on fixed weights based on user-specified maximum lag value.

3.6.8   Jackknife

The jackknife estimate of variance estimates variability in fitted parameters by comparing results from leaving out one observation at a time in repeated estimations. Jackknifing is based on a data resampling procedure in which the variability of an estimator is investigated by repeating an estimation with a subsample of the data. Subsample estimates are collected and compared with the full sample estimate to assess variability. Introduced by Quenouille (1949), an excellent review of this technique and extensions is available in Miller (1974).

The sandwich estimate of variance is related to the jackknife. Asymptotically, it is equivalent to the one-step and iterated jackknife estimates, and as shown in Efron (1981), the sandwich estimate of variance is equal to the infinitesimal jackknife.

There are two general methods for calculating jackknife estimates of variance. One approach is to calculate the variability of the individual estimates from the full sample estimate. We supply formulas for this approach. A less conservative approach is to calculate the variability of the individual estimates from the average of the individual estimates. You may see references to this approach in other sources. Generally, we prefer the approach outlined here because of its more conservative nature.

3.6.8.1   Usual jackknife

A jackknife estimate of variance may be formed by estimating the coefficient vector for $n - 1$ observations n different times. Each of the n estimates is obtained for a different subsample of the original dataset. The first estimate considers a subsample where only the first observation is deleted from consideration. The second estimate considers a subsample where only the second observation is deleted from consideration. The process continues until n different subsamples have been processed and the n different estimated coefficient vectors are collected. The n estimates are then used as a sample from which the variability of the estimated coefficient vector may be estimated.

The jackknife estimate of variance is then calculated as

$$\widehat{V}_J = \frac{n - p}{n} \sum_{i=1}^{n} \left(\widehat{\beta}_{(i)} - \widehat{\beta}\right) \left(\widehat{\beta}_{(i)} - \widehat{\beta}\right)^T \tag{3.87}$$

where $\widehat{\beta}_{(i)}$ is the estimated coefficient vector leaving out the ith observation and p is the number of predictors (possibly including a constant).

.6.8.2   One-step jackknife

The updating equations for one-step jackknife-type estimates come from

$$\widehat{\beta}_{(i)}^* = \widehat{\beta} - \frac{(\mathbf{X}^T \widehat{V} \mathbf{X})^{-1} \mathbf{x}_i^T \widehat{r}_i^R}{1 - \widehat{h}_i} \tag{3.88}$$

where $\widehat{V} = \text{diag}\{v(\widehat{\mu})\}$ is an $(n \times n)$ matrix, $\mathbf{X}$ is the $(n \times p)$ matrix of covariates, $\widehat{h}_i$ is the ith diagonal element of the hat matrix described in (4.5), $\widehat{\beta}$ is the estimated coefficient vector using all observations, and $\widehat{r}_i^R$ is the estimated response residual (see section 4.4.1). In this approach, instead of fitting (to full convergence) each of the n candidate models, we instead use the full observation estimate as a starting point and take only one Newton–Raphson step (iteration).

The one-step jackknife estimate of variance is then calculated as

$$\widehat{V}_{J1} = \frac{n - p}{n} \sum_{i=1}^{n} \left(\widehat{\beta}_{(i)}^* - \widehat{\beta}\right) \left(\widehat{\beta}_{(i)}^* - \widehat{\beta}\right)^T \tag{3.89}$$

where $\widehat{\beta}_{(i)}^*$ is defined in (3.88) and p is the number of predictors (possibly including a constant).

.6.8.3   Weighted jackknife

Just as the subsample estimates may be combined with various weighting schemes representing different underlying assumptions, we may also generalize the jackknife resampling estimate. Hinkley (1977) is credited with introducing this extension. Though an important extension in general, this particular estimate of variance is not included in the associated software for this text.

3.6.8.4 Variable jackknife

The variable jackknife is an extension that allows different subset sizes. For a clear review of this and other resampling methods, see Wu (1986) or Efron (1981). This flexible choice of subset size allows specifying the independent subsamples of the data.

A variable jackknife estimate of variance may be formed by generalizing the strategy for what is left out at each estimation step. We can leave out a group of observations at a time. The usual jackknife variance estimate is then a special case of the variable jackknife where each group is defined as one observation.

The variable jackknife estimate of variance is then calculated as

$$\widehat{V}_{\mathrm{VJ}} = \frac{n-p}{ng} \sum_{i=1}^{g} \left(\widehat{\boldsymbol{\beta}}_{(i)}^{\mathrm{VJ}} - \widehat{\boldsymbol{\beta}} \right) \left(\widehat{\boldsymbol{\beta}}_{(i)}^{\mathrm{VJ}} - \widehat{\boldsymbol{\beta}} \right)^{T} \tag{3.90}$$

where $\widehat{\boldsymbol{\beta}}_{(i)}^{\mathrm{VJ}}$ is the estimated coefficient vector leaving out the ith group of observations, g is the number of groups, and p is the number of predictors (possibly including a constant).

3.6.9 Bootstrap

The bootstrap estimate of variance is based on a data resampling procedure in which the variability of an estimator is investigated by repeating an estimation with a subsample of the data. Subsamples are drawn with replacement from the original sample and subset estimates are collected and compared with the full (original) sample estimate.

There are two major types of bootstrap variance estimates: parametric and nonparametric. Parametric bootstrap variance estimates are obtained by resampling residuals, and nonparametric variance estimates are obtained by resampling cases (as described above). Because the provided software calculates nonparametric bootstrap variance estimates, we describe that in the following subsections.

Within the area of nonparametric bootstrap variance estimates, one may calculate the variability about the mean of the bootstrap samples or about the original estimated coefficient vector. The former is more common, but we generally use (and have implemented) the latter because it is more conservative.

As a note of caution: applying this variance estimate usually calls for art as well as science because there are usually some bootstrap samples that are "bad". They are bad in the sense that estimation results in infinite coefficients, collinearity, or some other difficulty in obtaining estimates. When those conditions are known, they can be specified in the `reject()` option of the `bootstrap` command, which will cause all results to be set to missing for that bootstrap sample.

6.9.1 Usual bootstrap

A bootstrap estimate of variance may be formed by estimating the coefficient vector k different times. Each of the k estimates is obtained by fitting the model using a random sample of the n observations (sampled with replacement) drawn from the original sample dataset.

A bootstrap estimate of variance is then calculated as

$$\widehat{V}_{\mathrm{BS}} = \frac{n-p}{nk} \sum_{i=1}^{k} \left(\widehat{\beta}_i^{\mathrm{BS}} - \widehat{\beta} \right) \left(\widehat{\beta}_i^{\mathrm{BS}} - \widehat{\beta} \right)^T \tag{3.91}$$

where $\widehat{\beta}_i^{\mathrm{BS}}$ is the estimated coefficient vector for the ith sample (drawn with replacement) and p is the number of predictors (possibly including a constant).

6.9.2 Grouped bootstrap

A grouped (or cluster) bootstrap estimate of variance may be formed by estimating the coefficient vector k different times. Each of the k estimates is obtained by fitting the model using a random sample of the g groups of observations (sampled with replacement) drawn from the original sample dataset of g groups. The sample drawn for the groups in this case is such that all members of a group are included in the sample if the group is selected.

A bootstrap estimate of variance is then calculated as

$$\widehat{V}_{\mathrm{GBS}} = \frac{n-p}{nk} \sum_{i=1}^{k} \left(\widehat{\beta}_i^{\mathrm{GBS}} - \widehat{\beta} \right) \left(\widehat{\beta}_i^{\mathrm{GBS}} - \widehat{\beta} \right)^T \tag{3.92}$$

where $\widehat{\beta}_i^{\mathrm{GBS}}$ is the estimated coefficient vector for the ith sample (drawn with replacement) and p is the number of predictors (possibly including a constant).

3.7 Estimation algorithms

Here we present algorithms for fitting GLMs. The algorithms do not illustrate all the details such as choosing starting values or alternative criteria for convergence. Instead, they focus on the major steps of implementing the algorithms. Actual software implementations will have the above modifications, but for those implementations limited to the list of variance functions and link functions that are outlined in the tables in the appendix, these algorithms are an accurate representation. The algorithm may be generalized by including ϕ in the parameter list to obtain ML estimates of all parameters.

In addition to the issues raised previously, other modifications can be made to the algorithms. In particular, we raised the issue of whether we can use the observed information matrix or expected information matrix in the IRLS algorithm. Also, we mentioned that `PROC GENMOD` uses the expected information matrix in initial iterations of the Newton–Raphson ML routines before switching to the use of the observed information matrix in later iterations. These types of modifications are easily implemented in the following algorithms:

Listing 3.1. IRLS algorithm

```
1   Initialize μ
2   Calculate η = g(μ)                                          Table A.2
3   Set dev_old = 0
4   Set dev_new = 1
5   Set Δdev = 1
6   while(abs(Δdev) > tolerance) {
7       Calculate W = 1/{(g')²v(μ)a(φ)}          (3.46), Table A.1, Table A.4
8       Calculate z = (y − μ)g' + η − offset           (3.47), Table A.4
9       Regress z on X with weights W
10      Calculate η = Xβ+offset
11      Calculate μ = g⁻¹(η)                                   Table A.2
12      Set dev_old = dev_new
13      Calculate dev_new                              (3.50), Table A.11
14      Calculate Δdev = dev_new − dev_old
15  }
16  Calculate variance matrix                              Section 3.6
```

Listing 3.2. Newton–Raphson algorithm

```
1   Initialize β
2   while(||β_new − β_old|| > tol and abs(L_new − L_old) > ltol) {
3       Calculate g = ∂L/∂β                          Table A.1, Table A.2
4       Calculate H = ∂²L/∂β²     Table A.1, Table A.2, Table A.4, Table A.6
5       Set β_old = β_new
6       Calculate β_new = β_old − H⁻¹g
7       Set L_old = L_new
8       Calculate L_new                                        Table A.7
9   }
10  Calculate variance matrix                              Section 3.6
```

3.8 Summary

This chapter covers several topics. The IRLS method is the simplification that results when the Newton–Raphson method uses the expected Hessian matrix. Either way, these approaches are the result of theoretical arguments and assumptions that admit a function (the score function) to be optimized.

The Newton–Raphson method provides an updating equation for optimizing the parameter vector of interest. The method may then be codified in algorithms in one of several ways. For example, algorithm development may implement various techniques to choose step sizes and directions for the updating equation. The IRLS method does not specifically rely on an assumed likelihood; it is further justified based only on moment assumptions.

Algorithms are intended to optimize a function. Algorithms derived from methods in ML and MQL will result in the same estimated coefficient vector except for roundoff and differences in convergence criteria (when there is a true underlying likelihood). This statement is true regardless of whether the model uses the canonical link.

Variance matrices are estimated for the parameter vector of interest. The variance matrix is calculated after the score function has been optimized. Because IRLS usually uses V_{EH} as part of its algorithm, it is not surprising that most software implementations provide estimated standard errors based on this matrix. Because Newton–Raphson algorithms usually use V_{OH}, it is no surprise that most software implementations provide standard errors based on that matrix. In both cases, the usual supplied standard errors, or variance matrix, are selected on the basis of an efficiency (in the programming sense rather than the statistical sense) argument; the matrix is already calculated and available from the optimization algorithm.

For either approach, however, we are free to calculate variance matrices described in section 3.6. We do not equate the estimated variance matrix with the method or the algorithm. If we obtain estimates for a model using the expected Hessian for optimization, for example, that does not preclude us from calculating and using V_{OH} to estimate standard errors.

4 Analysis of fit

Contents

This chapter presents various statistics and techniques useful for assessing the fit of a GLM. Our presentation here is a catalog of formulas and reference material. Because illustrations of the concepts and techniques cataloged here are left to the individual chapters later in the book, first-time readers may find glancing ahead at those examples useful as they read through this material. This organization makes introducing the material slightly more inconvenient but makes introducing it as reference material much more convenient.

In assessing a model, we assume that the link and variance functions are correctly specified. Another assumption of our model is that the linear structure expresses the relationship of the explanatory variables to the response variable. The information in this chapter is useful for assessing the validity of these assumptions. One detail of GLMs not covered in this text is correcting for the bias in coefficient vector estimates. Interested readers can find such information in Cordeiro and McCullagh (1991) or Neuhaus and Jewell (1993).

4.1   Deviance

In estimation based on maximum likelihood (ML), a standard assessment is to compare the fitted model with a fully specified model (a model with as many independent parameters as observations). The scaled deviance S is given in terms of the likelihoods of the fitted model m and the full model f by

$$S = -2\ln(L_m/L_f) \tag{4.1}$$

Thus, the statistic has the familiar form (and limiting χ^2 distribution) of a likelihood-ratio test statistic.

We may generalize this concept to GLMs such that

$$S = D/\phi \tag{4.2}$$

where D is the deviance.

A limitation of this statistic is that its distribution is exactly chi-squared for the Gaussian family with an identity link but is asymptotically chi-squared for other variance functions and links.

The deviance is calculated for each family using the canonical parameter, which is the function $\theta(\mu)$ illustrated in the canonical links throughout the text. Our goal in assessing deviance is to determine the utility of the parameters added to the null model.

The deviance for GLMs is calculated as

$$D(u; \widehat{\mu}) = \sum_i 2\left[y\{\theta(y_i) - \theta(\widehat{\mu}_i)\} - b\{\theta(y_i)\} + b\{\theta(\widehat{\mu}_i)\}\right] \tag{4.3}$$

where $\theta(\cdot)$ is the canonical parameter and $b(\cdot)$ is the cumulant. Brown (1992) offers more advice on graphically assessing the deviance, and the deviance concept is generalized to quasilikelihoods in chapter 17. Throughout the text, we derive properties for a given family and then illustrate deriving the deviances (which have an underlying likelihood) in more succinct terms using

$$D = 2\phi\left\{\mathcal{L}(y; y) - \mathcal{L}(y; \mu)\right\} \tag{4.4}$$

where $\mathcal{L}(y; y) = \ln(L_f)$ and $\mathcal{L}(y; \mu) = \ln(L_m)$.

The notation $\mathcal{L}(y; y)$ for the full model is a reflection that the saturated model perfectly captures the outcome variable. Thus, the model's predicted values $\widehat{\mu} = y$. The difference in deviance statistics between the saturated model and the fitted model captures the distance between the predicted values and the outcomes.

Models may be compared using deviance-based statistics. When competing models are nested, this comparison is a straightforward application of likelihood-ratio tests. Wahrendorf, Becher, and Brown (1987) discuss bootstrap comparison of nonnested GLMs.

.2 Diagnostics

We will use $\widehat{\mathbf{W}}$ from (3.46), and $\widehat{h}_{ii}$ is the ith diagonal element of

$$\widehat{\mathbf{H}} = \widehat{\mathbf{W}}^{1/2}\mathbf{X}(\mathbf{X}^T\widehat{\mathbf{W}}\mathbf{X})^{-1}\mathbf{X}^T\widehat{\mathbf{W}}^{1/2} \tag{4.5}$$

and denotes the hat diagonals (that is, diagonal elements of the hat matrix) as a generalization of the same concept in linear regression. Pregibon (1981) presents this definition in assessing "leverage" for logistic distribution. The hat diagonals in regression have many useful properties. The property of leverage (or potential) in linear regression relates to the fact that as the hat diagonal for the ith observation approaches one, the corresponding residual approaches zero. Williams (1987) discusses diagnostics based on the deviance and single case deletions.

.2.1 Cook's distance

By using the one-step approximations to the change in the estimated coefficients when an observation is omitted from the estimation procedure, we approximate Cook's distance using

$$C_i = \left(\widehat{\boldsymbol{\beta}}^*_{(i)} - \widehat{\boldsymbol{\beta}}\right)^T \mathcal{F} \left(\widehat{\boldsymbol{\beta}}^*_{(i)} - \widehat{\boldsymbol{\beta}}\right) \tag{4.6}$$

where $\mathcal{F}$ is the Fisher information (negative Hessian; see section 3.6.1) and $\widehat{\boldsymbol{\beta}}^*_{(i)}$, defined in (3.88), is the one-step approximation to the jackknife-estimated coefficient vector. These measures can be used to detect observations with undue influence. Standard approaches include declaring those observations with measures greater than $4/(n-p-1)$ to be problematic and those measures greater than $4/n$ to be worth investigating; see Bollen and Jackman (1985).

.2.2 Overdispersion

Overdispersion is a phenomenon that occurs when fitting data to discrete distributions, such as the binomial, Poisson, or negative binomial distribution. If the estimated dispersion after fitting is not near the assumed value, then the data may be overdispersed if the value is greater than expected or underdispersed if the value is less than expected. Overdispersion is far more common. For overdispersion, the simplest remedy is to assume a multiplicative factor in the usual implied variance. As such, the resulting covariance matrix will be inflated through multiplication by the estimated scale parameter. Care should be exercised because an inflated, estimated dispersion parameter may result from model misspecification rather than overdispersion, indicating that the model should be assessed for appropriateness by the researcher. Smith and Heitjan (1993) discuss testing and adjusting for departures from the nominal dispersion, Breslow (1990) discusses Poisson regression as well as other quasilikelihood models, and Cox (1983) gives a general overview of overdispersion. This topic is discussed in chapter 11.

Hilbe (2009) devotes an entire chapter to binomial overdispersion, and Hilbe (2011) devotes a chapter to extradispersed count models. Ganio and Schafer (1992), Lambert and Roeder (1995), and Dean and Lawless (1989) discuss diagnostics for overdispersion in GLMs.

A score test effectively compares the residuals with their expectation under the model. A test for overdispersion of the Poisson model, which compares its variance with the variance of a negative binomial, is given by

$$T_1^2 = \frac{\left[\sum_{i=1}^{n}\left\{(y_i - \widehat{\mu}_i)^2 - (1 - \widehat{h}_i)\widehat{\mu}_i\right\}\right]^2}{2\sum_{i=1}^{n}\widehat{\mu}_i^2} \sim \chi_1^2 \tag{4.7}$$

This formula tests the null hypothesis $H_0\colon \phi = 0$, assuming that the variance is described by the function $v(\mu, \phi) = \mu + \phi\mu^2$ (negative binomial variance).

Another test for the Poisson model is given by

$$T_2^2 = \frac{1}{2n}\left\{\sum_{i=1}^{n}\frac{(y_i - \widehat{\mu}_i)^2 - (1 - \widehat{h}_i)\widehat{\mu}_i}{\widehat{\mu}_i}\right\}^2 \sim \chi_1^2 \tag{4.8}$$

This formula tests the null hypothesis $H_0\colon \phi = 1$, assuming that the variance is described by the function $v(\mu, \phi) = \phi\mu$ (scaled Poisson variance).

4.3 Assessing the link function

We may wish to investigate whether the link function is appropriate. For example, in Poisson regression, we may wish to examine whether the usual log-link (multiplicative effects) is appropriate compared with an identity link (additive effects). For binomial regression, we may wish to compare the logit link (symmetric about one half) with the complementary log-log link (asymmetric about one half).

Pregibon (1980) advocates the comparison of two link functions by embedding them in a parametric family of link functions. The Box–Cox family of power transforms

$$g(\mu; \lambda) = \frac{\mu^\lambda - 1}{\lambda} \tag{4.9}$$

and yields the log-link at $\lim_{\lambda \to 0} g(\mu; \lambda)$ and the identity link at $\lambda = 1$. Likewise, the family

$$g(\mu; \lambda) = \ln \left\{ \frac{(1-\mu)^{-\lambda} - 1}{\lambda} \right\} \tag{4.10}$$

gives the logit link at $\lambda = 1$ and the complementary log-log link at $\lim_{\lambda \to 0} g(\mu; \lambda)$.

We may test the hypothesis $H_0 \colon \lambda = \lambda_0$ by fitting the model with the standard link ($\lambda = \lambda_0$) and by obtaining the fitted values $\widehat{\mu}_i$. We then construct the variable

$$z = - \left. \frac{\partial g(\widehat{\mu}; \lambda)}{\partial \lambda} \right|_{\lambda = \lambda_0} \tag{4.11}$$

and use the standard (chi-squared) score test for added variables.

For instance, to test the log link ($\lambda = 0$) embedded in the Box–Cox power family, we see that

$$-\left. \frac{\partial g(\widehat{\mu}; \lambda)}{\partial \lambda} \right|_{\lambda = 0} = - \left. \left\{ \frac{(\ln \widehat{\mu}) \exp(\lambda \ln \widehat{\mu})}{\lambda} - \frac{\exp(\lambda \ln \widehat{\mu}) - 1}{\lambda^2} \right\} \right|_{\lambda = 0} \tag{4.12}$$

$$= - \lim_{\lambda \to 0} \left\{ \frac{(\ln \widehat{\mu}) \exp(\lambda \ln \widehat{\mu})}{\lambda} - \frac{\exp(\lambda \ln \widehat{\mu}) - 1}{\lambda^2} \right\} \tag{4.13}$$

$$= - \lim_{\lambda \to 0} \left\{ (\ln \widehat{\mu})^2 \exp(\lambda \ln \widehat{\mu}) - \frac{(\ln \widehat{\mu})^2}{2} \exp(\lambda \ln \widehat{\mu}) \right\} \tag{4.14}$$

$$= - \frac{(\ln \widehat{\mu})^2}{2} \tag{4.15}$$

(using two steps of l'Hôpital's Rule). The test is then a standard (chi-squared) score test for an omitted variable. That is, the originally hypothesized link function is used and fitted values $\widehat{\mu}$ are obtained. The example above shows how those fitted values are then used to generate a new predictor variable. Adding this predictor variable (which forms the systematic difference between the original link and the hypothesized alternative) is then assessed through inspecting the change in deviance.

4.4 Residual analysis

In assessing the model, residuals measure the discrepancy between our observed and fitted values for each observation. The degree to which one observation affects the estimated coefficients is a measure of influence.

Pierce and Schafer (1986) and Cox and Snell (1968) provide excellent surveys of various definitions for residuals in GLMs. In the following sections, we present the definitions of several residuals that have been proposed for GLMs.

Discussions on residuals are hampered by a lack of uniform terminology throughout the literature, so we will expand our descriptions to facilitate comparison with other books and papers.

We will use the following conventions when naming residuals:

- A residual is named based on the statistic from which it was derived or the author that proposed its use.

- The adjective "modified" means that the denominator of a residual has been modified to be a reasonable estimate of the variance of y. The base residual has been multiplied by the factor $(k/w_i)^{-1/2}$, where k is a scale parameter (defaulting to one if not set).

- The adjective "standardized" means that the variance of the residual has been standardized to account for the correlation between y and $\widehat{\mu}$. The base residual has been multiplied by the factor $(1 - h)^{-1/2}$.

- The adjective "studentized" means that the residuals have been scaled by an estimate of the unknown scale parameter. The base residual has been multiplied by the factor $\widehat{\phi}^{-1/2}$. In many texts and papers, it is the standardized residuals that are then scaled (studentized). However, the result is often called studentized residuals instead of the more accurate standardized studentized residuals.

- The adjective "adjusted" means that the residual has been adjusted (per variance function family) from the original definition. The adjustment has to do with including an extra term to bring higher moments of the resulting distribution in line with expected values for a normal distribution. We have seen this adjustment only with deviance residuals (Pierce and Schafer 1986), so we include a specific section on it.

Using the above lexicon, we prefer the more descriptive "modified standardized Pearson residuals" over "Cox–Snell residuals", even though the latter is fairly common (because of the description mentioned at the beginning of this section).

Under these conventions, we present software at the end of the book that allows generating residuals using these adjectives as option names. This form enables users of the software to exactly specify desired calculations.

A general aim for defining residuals is that their distribution is approximately normal. This goal is accomplished by considering linear equations, transformed linear equations, and deviance contributions. For transformed linear equations, our choice of transformation can be made to ensure first-order skewness of zero (Anscombe residuals) or to ensure that the transformed variance is constant (variance-stabilizing residuals) or at least comparable (Pearson residuals).

For achieving approximate normality, the Anscombe and adjusted deviance residuals are similar and perform well. If we are considering a model using the binomial family and if the denominator is less than 10, the Anscombe residuals are preferred. Otherwise, the adjusted deviance residuals may be used.

In the following sections, we catalog the formulas for residuals of interest. Throughout these formulas, the quantity $\partial\eta/\partial\mu$ is evaluated at $\widehat{\mu}$.

4.4.1 Response residuals

These residuals are simply the difference between the observed and fitted outcomes.

$$r_i^R = y_i - \widehat{\mu}_i \tag{4.16}$$

4.4.2 Working residuals

These residuals are the difference between the working (synthetic) response and the linear predictor at convergence

$$r_i^W = (y_i - \widehat{\mu}_i)\left(\frac{\partial \eta}{\partial \mu}\right)_i \tag{4.17}$$

and are associated with the working response and the linear predictor in the iteratively reweighted least-squares algorithm given in (3.47).

4.4.3 Pearson residuals

Pearson residuals are a rescaled version of the working residuals. The sum of the squared Pearson residuals is equal to the Pearson chi-squared statistic. The scaling puts the residuals on similar scales of variance so that standard plots of Pearson residuals versus an individual predictor or versus the outcome will reveal dependencies of the variance on these other factors (if such dependencies exist).

$$r_i^P = \frac{y_i - \widehat{\mu}_i}{\sqrt{v(\widehat{\mu}_i)}} \tag{4.18}$$

Large absolute values of the residual indicate the model's failure to fit a particular observation. A common diagnostic for detecting outliers is to plot the standardized Pearson residuals versus the observation number. Formulas for the Pearson residuals are given for common families in table A.9.

4.4.4 Partial residuals

Partial residuals are used to assess the form of a predictor and are thus calculated for each predictor. O'Hara Hines and Carter (1993) discuss the graphical use of these residuals in assessing model fit.

$$r_{ki}^T = (y_i - \widehat{\mu}_i)\left(\frac{\partial \eta}{\partial \mu}\right)_i + (x_{ik}\widehat{\beta}_k) \tag{4.19}$$

Here $k = 1, \ldots, p$, where p is the number of predictors, and $(x_{ki}\widehat{\beta}_k)$ refers to the ith observation of the kth predictor times the kth fitted coefficient.

4.4.5 Anscombe residuals

Frank Anscombe, as one of the responders in the discussion of Hotelling (1953), contributed the following definition of residuals. First, let

$$A(\cdot) = \int \frac{d\mu}{v^{1/3}(\mu)} \tag{4.20}$$

where the residual is

$$r_i^A = \frac{A(y_i) - A(\widehat{\mu}_i)}{A'(\widehat{\mu}_i)\sqrt{v(\widehat{\mu}_i)}} \tag{4.21}$$

The choice of the function $A(\cdot)$ was made so that the resulting residuals would be as normal as possible (have a distribution that was approximately normal).

These residuals may be difficult to compute because they involve the hypergeometric $_2F_1(\cdot)$ function (not common in statistical packages). Among the commonly supported variance functions, dependence on this function is limited to the binomial and negative binomial families. However, this function is available in the Stata package, so residuals are fully supported even for these variance functions. Formulas for the Anscombe residuals are given for common families in table A.10.

4.4.6 Deviance residuals

The deviance plays a key role in our derivation of GLM and in the inference of our results. The deviance residual is the increment to the overall deviance for each observation. These residuals are common and are often standardized, studentized, or both. The residuals are based on the χ^2 distribution.

$$r_i^D = \text{sign}(y_i - \widehat{\mu}_i)\sqrt{\widehat{d_i^2}} \tag{4.22}$$

Computation formulas of d_i^2 for individual families are given in table A.11.

In general, the deviance residual (standardized or not) is preferred to the Pearson residual for model checking because its distributional properties are closer to the residuals arising in linear regression models.

4.4.7 Adjusted deviance residuals

The deviance residual may be adjusted (corrected) to make the convergence to the limiting normal distribution faster. The adjustment removes an $O(n^{-1/2})$ term.

$$r_i^{D_a} = r_i^D + \frac{1}{6}\rho_3(\theta) \tag{4.23}$$

Here $\rho_3(\theta)$ is defined for individual families in table A.11.

4.4.8 Likelihood residuals

Likelihood residuals are a combination of standardized Pearson residuals, $r_i^{P'}$, and standardized deviance residuals, $r_i^{D'}$.

$$r_i^L = \operatorname{sign}(y_i - \widehat{\mu}_i) \left\{ h_i (r_i^{P'})^2 + (1 - h_i)(r_i^{D'})^2 \right\}^{1/2} \tag{4.24}$$

4.4.9 Score residuals

These are the scores used in calculating the sandwich estimate of variance. The scores are related to the score function (estimating equation), which is optimized.

$$r_i^S = \frac{y_i - \widehat{\mu}_i}{v(\widehat{\mu}_i)} \left(\frac{\partial \eta}{\partial \mu} \right)_i^{-1} \tag{4.25}$$

4.5 Checks for systematic departure from the model

To check for systematic departure from the model, the standard diagnostic plots include a plot of the standardized deviance residuals versus the linear predictor or versus the linear predictor transformed to the constant-information scale of the error distribution. A normal quantile plot of the deviance or Anscombe residuals can also be used to check for systematic departure from the model.

To check individual covariates, plot residuals versus one of the predictor variables. There should be no trend because the null pattern is the same as that for residuals versus fitted values.

Added variable plots check whether an omitted covariate, u, should be included in the linear predictor. First, we obtain the unstandardized residuals for u as a response, using the same linear predictor and quadratic weights as for y. The unstandardized residuals for y are then plotted against the residuals for u. There should be no trend if u is correctly omitted.

Plot the absolute residuals versus the fitted values as an informal check of the adequacy of the assumed variance function. The constant information scale for the fitted values is usually helpful in spreading out the points on the horizontal scale.

A standard approach to investigating the link function for a given model is to generate a covariate equal to the square of the linear predictor for that model. Refitting the model with the new predictor, one assesses the goodness of the specification by looking at the p-value associated with the additional covariate. If that covariate is significant, then there is evidence of lack of fit. This process is easily done in Stata with the `linktest` command; see the `linktest` manual entry in the *Stata Base Reference Manual*.

Exploratory data analysis plots of the outcome and various predictors are useful. Such plots allow the investigator to assess the linearity assumption as well as to diagnose outliers.

4.6   Model statistics

Here we catalog several commonly used model fit statistics and criterion measures for comparing models. An excellent source for more information on these topics is Long (1997). Most of the following measures are available to Stata users with the `fitstat` command implemented in Long and Freese (2000). Users may also use the official `estat` command, though the Bayesian information criterion (BIC) and Akaike information criterion (AIC) statistics available through these commands are not scaled by the number of observations.

4.6.1   Criterion measures

In the following definitions,

$$p \;=\; \text{number of predictors} \tag{4.26}$$
$$n \;=\; \text{number of observations} \tag{4.27}$$
$$L(M_k) \;=\; \text{likelihood for model } k \tag{4.28}$$
$$\mathcal{L}(M_k) \;=\; \text{log likelihood for model } k \tag{4.29}$$
$$D(M_k) \;=\; \text{deviance of model } k \tag{4.30}$$
$$G^2(M_k) \;=\; \text{likelihood-ratio test of model } k \tag{4.31}$$

Next, we provide the formulas for two criterion measures useful for model comparison. They include terms based on the log likelihood along with a penalty term based on the number of parameters in the model. In this way, the criterion measures seek to balance our competing desires for finding the best model (in terms of maximizing the likelihood) with model parsimony (including only those terms that significantly contribute to the model).

We introduce the two main model selection criterion measures below; also see Hilbe (1994, 2011) for more considerations.

4.6.1.1   AIC

The Akaike (1973) information criterion may be used to compare competing nested or nonnested models. The information criterion is a measure of the information lost in using the associated model. The goal is to find the model that has the lowest loss of information. In that sense, lower values of the criterion are indicative of a preferable model. Furthermore, a difference of greater than 2 indicates a marked preference for the model with the smaller criterion measure. When the measure is defined without scaling by the sample size, marked preference is called when there is a difference greater than two. The (scaled) formula is given by

$$\text{AIC} = -2\mathcal{L}(M_k) + 2p \tag{4.32}$$

Though models should be compared with the same sample, one can compare models with small differences in sample sizes because of some missing values in covariates. In such cases, one should use the scaled version of the criterion.

Including the term involving the number of parameters p is a penalty against larger covariate lists. The criterion measure is especially amenable to comparing GLMs of the same link and variance function but different covariate lists. The penalty term is somewhat analogous to the choice of level in a formal hypothesis test. Some authors have suggested $3p$ or other values in comparing models, including Atkinson (1980). When models are nested, we view the penalty term as the precision required to eliminate a candidate predictor from the model.

One should take care on how this statistic is calculated. The above definition includes the model log likelihood. In GLMs, one usually replaces this with a deviance-based calculation. However, the deviance does not involve the part of the density specification that is free of the parameter of interest, the normalization term $c(y_i, \phi)$ in (3.3). Thus, such calculations would involve different "penalties" if a comparison were made of two models from different families.

Both versions of the statistic are interpreted in the same manner, with the lower value indicating a comparatively better fit. Care must be taken, though, to recognize which version is being used in a particular application and not to mix versions when comparing models.

There are several other versions of the AIC statistic that are used in research. Other than the above primary versions, two alternatives have found substantial use. The first alternative is the AICc, which is a corrected or finite-sample AIC from Sugiura (1978) and Hurvich and Tsai (1989). The second is the AIChq described by Hannan and Quinn (1979). The unscaled versions of these statistics are presented with the unscaled AIC for comparison:

$$\text{AIC} \quad = \quad -2\mathcal{L}(M_k) + 2p \tag{4.33}$$

$$\text{AICc} \quad = \quad -2\mathcal{L}(M_k) + 2p + 2\frac{p(p+1)}{n-p-1} \tag{4.34}$$

$$\text{AIChq} \quad = \quad -2\mathcal{L}(M_k) + 2p \ln\{\ln(n)\} \tag{4.35}$$

How do we decide whether the difference between two AIC statistics is great enough to warrant us concluding that one model is a better fit than another? In particular, that the model having the lower AIC statistic is preferred over another model? Though we know that the model with the smaller AIC is preferable, there is no specific statistical test from which a p-value may be computed. Hilbe (2009) devised a subjective table based on simulation studies that can be used to make such a decision for comparing unscaled AIC measures.

Difference: Model A and B	Decision (assuming $A < B$)
$0.00 < \text{difference} \leq 2.50$	No difference in models
$2.50 < \text{difference} \leq 6.00$	Prefer A if $n > 256$
$6.00 < \text{difference} \leq 9.00$	Prefer A if $n > 64$
$10.00 < \text{difference}$	Prefer A

4.6.1.2 BIC

The initial and most common implementation of the Bayesian information criterion (BIC) is the statistic developed by Schwarz (1978). Some researchers refer to this version of the BIC as the Schwarz criterion. It is defined in terms of the log likelihood, $\mathcal{L}$, as

$$\text{BIC}_{\mathcal{L}} = -2L + p\ln(n) \tag{4.36}$$

where p is the number of predictors, including the intercept. This version of the BIC is the same as that reported by Stata's `estat ic` command.

As with the AIC, the preferred model is the one having the smaller criterion value. The same cautions and guidelines for use of the AIC hold for use of the BIC.

Raftery (1995) presented an alternative version of the BIC based on the deviance $D(\cdot)$ rather than the log likelihood. Raftery's aim was to use the alternative version of the criterion for selecting among GLMs. This deviance-based version is given by

$$\text{BIC}_D = D(M_k) - (\text{df})\ln(n) \tag{4.37}$$

Contrary to the AIC, the BIC includes a penalty term that becomes more severe as the sample size increases. This characteristic reflects the power available to detect significance. Bai et al. (1992) investigate the utility of this measure for the comparison of models, and Lindsey and Jones (1998) do the same for comparisons of distributions.

When comparing nonnested models, one may assess the degree of model preference on the basis of the absolute value of the difference between the BIC statistics of the two models. The scale given by Raftery (1995) for determining the relative preference is

|Difference|	Degree of preference
0–2	Weak
2–6	Positive
6–10	Strong
>10	Very strong

When we compare two models, A and B, if $\text{BIC}_A - \text{BIC}_B < 0$, then model A is preferred. If $\text{BIC}_A - \text{BIC}_B > 0$, then model B is preferred. The degree of preference is outlined above.

4.6.2   The interpretation of R^2 in linear regression

One of the most prevalent model measures is R^2. This statistic is usually discussed in introductory linear regression along with various ad hoc rules on its interpretation. A fortunate student is taught that there are many ways to interpret this statistic and that these interpretations have been generalized to areas outside linear regression.

For the linear regression model, we can define the R^2 measure in the following ways:

$$n \quad = \quad \text{number of observations} \tag{4.38}$$
$$p \quad = \quad \text{number of predictors} \tag{4.39}$$
$$M_\alpha \quad = \quad \text{model with only an intercept} \tag{4.40}$$
$$M_\beta \quad = \quad \text{model with intercept and predictors} \tag{4.41}$$

4.6.2.1   Percentage variance explained

The most popular interpretation is the percentage variance explained, where it can be shown that the R^2 statistic is equal to the ratio of the variance of the fitted values and to the total variance of the fitted values.

$$\text{RSS} \quad = \quad \text{residual sum of squares} = \sum_{i=1}^{n}(y_i - \widehat{y}_i)^2 \tag{4.42}$$

$$\text{TSS} \quad = \quad \text{total sum of squares} = \sum_{i=1}^{n}(y_i - \bar{y})^2 \tag{4.43}$$

$$R^2 \quad = \quad \frac{\text{TSS} - \text{RSS}}{\text{TSS}} = 1 - \frac{\text{RSS}}{\text{TSS}} = 1 - \frac{\sum_{i=1}^{n}(y_i - \widehat{y}_i)^2}{\sum_{i=1}^{n}(y_i - \bar{y})^2} \tag{4.44}$$

4.6.2.2   The ratio of variances

Another interpretation originates from the equivalence of the statistic to the ratio of the variances.

$$R^2 = \frac{\widehat{V}(\widehat{y})}{\widehat{V}(y)} = \frac{\widehat{V}(\widehat{y})}{\widehat{V}(\widehat{y}) + \widehat{V}(\widehat{\epsilon})} \tag{4.45}$$

4.6.2.3   A transformation of the likelihood ratio

R^2 is equal to a transformation of the likelihood ratio between the full and null models. If the errors are assumed to be normal,

$$R^2 = 1 - \left\{ \frac{L(M_\alpha)}{L(M_\beta)} \right\}^{2/n} \tag{4.46}$$

4.6.2.4   A transformation of the F test

In assessing the fitted model, an F test may be performed to assess the adequacy of the additional parameters. The R^2 statistic can be shown to equal a transformation of this test.

The hypothesis that all the independent variables are zero may be tested with an F test and then R^2 may be written in terms of that statistic as

$$R^2 = \frac{Fp}{Fp + (n - p - 1)} \tag{4.47}$$

4.6.2.5   Squared correlation

The statistic is also equal to the squared correlation between the fitted and observed outcomes. This is the reason for the statistic's name.

$$R^2 = \left\{ \mathrm{Corr}(y, \widehat{y}) \right\}^2 \tag{4.48}$$

4.6.3   Generalizations of linear regression R^2 interpretations

Starting from each of the interpretations in the preceding section, one may generalize the associated formula for use with models other than linear regression.

The most important fact to remember is that a generalized version of the R^2 statistic extends the formula associated with the particular interpretation that served as its genesis. As generalizations, these statistics are sometimes called pseudo-R^2 statistics. However, this name makes it too easy to forget which original formula was used in the derivation. Many people make the mistake that the use of any pseudo-R^2 statistic can be interpreted in the familiar and popular "percentage variance explained" manner. Although the various interpretations in linear regression result in the same calculated value, the pseudo-R^2 scalar criteria generalized from different definitions do not result in the same value.

Just as there are adjusted R^2 measures for linear regression, there is current research in adjusting pseudo-R^2 criterion measures. We list only a few of the proposed adjusted measures.

4.6.3.1   Efron's pseudo-R^2

Efron (1978) defines the following measure as an extension to the regression model's "percentage variance explained" interpretation:

$$R_{\text{Efron}}^2 = 1 - \frac{\sum_{i=1}^{n}(y_i - \widehat{y}_i)^2}{\sum_{i=1}^{n}(y_i - \overline{y})^2} \tag{4.49}$$

Efron's presentation was directed at binary outcome models and listed $\widehat{\pi}_i$ for $\widehat{y}_i$. The measure could be used for continuous models. The equation is the same as given in (4.44), and the measure is sometimes called the sum of squares R^2 or R_{SS}^2.

4.6.3.2   McFadden's likelihood-ratio index

Sometimes called the likelihood-ratio index, McFadden (1974) defines the following measure as another extension to the "percentage variance explained" interpretation:

$$R_{\text{McFadden}}^2 = 1 - \frac{\mathcal{L}(M_\beta)}{\mathcal{L}(M_\alpha)} \tag{4.50}$$

Note that this equation can be used for any model fit by ML.

4.6.3.3   Ben-Akiva and Lerman adjusted likelihood-ratio index

Ben-Akiva and Lerman (1985) extended McFadden's pseudo-R^2 measure to include an adjustment. Their adjustment is in the spirit of the adjusted R^2 measure in linear regression, and the formula is given by

$$R_{\text{Ben-Akiva\&Lerman}}^2 = 1 - \frac{\mathcal{L}(M_\beta) - p}{\mathcal{L}(M_\alpha)} \tag{4.51}$$

where p is the number of parameters in the M_β model. Adjusted R^2 measures have been proposed in several forms. The aim of adjusting the calculation of the criterion is to address the fact that R^2 monotonically increases as terms are added to the model. Adjusted R^2 measures include penalty or shrinkage terms so that noncontributory terms will not significantly increase the criterion measure.

Note that this equation can be used for any model fit by ML.

4.6.3.4 McKelvey and Zavoina ratio of variances

McKelvey and Zavoina (1975) define the following measure as an extension of the "ratio of variances" interpretation:

$$R^2_{\text{McKelvey\&Zavoina}} \quad = \quad \frac{\widehat{V}(\widehat{y}*)}{\widehat{V}(y^*)} \tag{4.52}$$

$$= \quad \frac{\widehat{V}(\widehat{y}^*)}{\widehat{V}(\widehat{y}^*) + V(\epsilon)} \tag{4.53}$$

$$V(\epsilon) \quad = \quad \begin{cases} 1 & \text{probit} \\ \pi^2/3 & \text{logit} \end{cases} \tag{4.54}$$

$$\widehat{V}(\widehat{y}^*) \quad = \quad \widehat{\boldsymbol{\beta}}'\widehat{V}\widehat{\boldsymbol{\beta}} \tag{4.55}$$

$$\widehat{V} \quad = \quad \text{variance–covariance matrix of } \boldsymbol{\beta} \tag{4.56}$$

Note that this equation can be used for ordinal, binary, or censored outcomes.

4.6.3.5 Transformation of likelihood ratio

An ML pseudo-R^2 measure extends the "transformation of the likelihood ratio" interpretation by using the formula

$$R^2_{\text{ML}} \quad = \quad 1 - \left\{ \frac{L(M_\alpha)}{L(M_\beta)} \right\}^{2/n} \tag{4.57}$$

$$= \quad 1 - \exp(-G^2/n) \tag{4.58}$$

$$G^2 \quad = \quad -2\ln\{L(M_\alpha)/L(M_\beta)\} \tag{4.59}$$

where the relationship to G^2 (a likelihood-ratio chi-squared statistic) is shown in Maddala (1983).

Note that this equation can be used for any model fit by ML.

4.6.3.6 Cragg and Uhler normed measure

Cragg and Uhler (1970) introduced a normed version of the transformation of the likelihood-ratio pseudo-R^2 measure from the observation that as the fit of the two comparison models coincides, R^2_{ML} does not approach 1. The normed version is given by

$$R^2_{\text{Cragg\&Uhler}} \quad = \quad \frac{R^2_{\text{ML}}}{\max R^2_{\text{ML}}} \tag{4.60}$$

$$= \quad \frac{1 - \{L(M_\alpha)/L(M_\beta)\}^{2/n}}{1 - L(M_\alpha)^{2/n}} \tag{4.61}$$

Note that this equation can be used for any model fit by ML.

.6.4   More R^2 measures

.6.4.1   The count R^2

A measure based on the correct classification of ordinal outcomes may be defined as in Maddala and Lahiri (2009) using

$$R^2_{\text{count}} \quad = \quad \frac{1}{n} \sum_j n_{jj} \tag{4.62}$$

$$n_{jj} \quad = \quad \text{number of observed } j \text{ correctly classified as } j \tag{4.63}$$

Note that this equation is for models with binary or ordinal outcome.

.6.4.2   The adjusted count R^2

Because a poor model might be expected to correctly classify half the observations, an adjusted measure accounts for the largest row margin.

$$R^2_{\text{count}} \quad = \quad \frac{\sum_j n_{jj} - \max(n_{r+})}{n - \max(n_{r+})} \tag{4.64}$$

$$n_{jj} \quad = \quad \text{number of observed } j \text{ correctly classified as } j \tag{4.65}$$

$$n_{r+} \quad = \quad \text{marginal sum for observed outcome } j \tag{4.66}$$

Note that this equation is for models with binary or ordinal outcome.

.6.4.3   Veall and Zimmermann R^2

Veall and Zimmermann (1992) define a goodness-of-fit criterion measure for ordered outcome models given by

$$R^2_{\text{Veall\&Zimmermann}} \quad = \quad \left(\frac{\delta - 1}{\delta - R^2_{\text{McFadden}}} \right) R^2_{\text{McFadden}} \tag{4.67}$$

$$\delta \quad = \quad \frac{n}{2\mathcal{L}(M_\alpha)} \tag{4.68}$$

Note that this equation is for models with ordinal outcome.

4.6.4.4 Cameron–Windmeijer R^2

Cameron and Windmeijer (1997) define a proportionate reduction in uncertainty measure valid for exponential families. The formula is given by

$$R^2_{\text{Cameron\&Windmeijer}} = 1 - \frac{K(y, \mu)}{K(y, \mu_0)} \tag{4.69}$$

where $K(\cdot)$ is the Kullback–Leibler divergence, μ is the predicted values for the fitted model, and μ_0 is the estimated mean for the constant-only model. This measure may be solved for various families, and the associated formulas are given in tables A.12 and A.13. This measure assumes that the model includes a constant term.

4.7 Marginal effects

Econometricians often prefer to interpret regression model coefficients in terms of either marginal effects or discrete change. Marginal effects are calculated and interpreted when a predictor is continuous; discrete change is used for binary or factor variables and refers to a change of values between discrete units of a categorical variable.

The use of marginal effects and discrete change is not common outside the discipline of econometrics. However, both interpretations are applicable in areas such as health outcomes analysis, medical research, transportation studies, insurance projections, fisheries and wildlife management, and most other areas of research. Here we examine them in greater context.

4.7.1 Marginal effects for GLMs

Marginal effects quantify the relationship of how the probability of an outcome changes given a unit (or units) change in the value of the predictor. Marginal effects are calculated for specific continuous predictors, with other predictors being held at their mean (or other) values; the most common "other" statistic is the median.

Marginal effects can also be thought of as the instantaneous rate of change in the probability of an outcome given a specific change in value for a predictor, again holding other predictor values at their mean (or other) values. The concept of instantaneous rate of change is derived from the nature of a slope and is given in terms of a partial derivative. Recall that regression coefficients are themselves slopes. Marginal effects are identical to the coefficients given in ordinary least squares regression. However, this same identity fails to hold for nonlinear models such as logistic and Poisson regression. We shall, therefore, focus our discussion of marginal effects on non-Gaussian GLMs.

We discuss two interpretations of the marginal effect. The first deals with a marginal effect being calculated at the sample means of the remaining model predictors. This interpretation is perhaps the most popular interpretation of the marginal effect. When a predictor is binary, the average marginal effect is also termed a partial effect, although some statisticians allow the partial effect term to be used for any form of predictor. The second interpretation deals with the marginal effect being computed at the average of the predicted response values in the model. We examine each of these two interpretations, beginning with the marginal effect at mean approach.

The marginal effect for a continuous predictor is expressed as

$$\frac{\partial E(y_i|\mathbf{x}_i)}{\partial x_{ki}} = \frac{\partial \mu_i}{\partial x_{ki}} \tag{4.70}$$

This calculation can be further derived for a given link function. For example, under the log link, the calculation is

$$\frac{\partial \mu_i}{\partial x_{ki}} = \frac{\partial \exp(\mathbf{x}_i^T \boldsymbol{\beta})}{\partial x_{ki}} = \beta_k \exp(\mathbf{x}_i^T \boldsymbol{\beta}) \tag{4.71}$$

The marginal effect involves all the predictors, which is why we must choose specific values at which to calculate the marginal effect.

Here we demonstrate the calculation of the marginal effect at the means (MEMs) (substituting the covariate means in the definition of the marginal effects above) using data from the 1984 German health registry. We first fit a logistic regression model on the binary response variable `outwork`, indicating whether a person is unemployed. Covariates include `age` in years and indicator variable `kids`, denoting whether the person has children. Note that the marginal effect for the logit link is given by

$$\frac{\partial \mu_i}{\partial x_{ki}} = \frac{\partial}{\partial x_{ki}} \frac{\exp(\mathbf{x}_i \boldsymbol{\beta})}{1 + \exp(\mathbf{x}_i \boldsymbol{\beta})} = \frac{\exp(\mathbf{x}_i \boldsymbol{\beta})}{\{1 + \exp(\mathbf{x}_i \boldsymbol{\beta})\}^2} \beta_k \tag{4.72}$$

```
. use http://www.stata-press.com/data/hh4/rwm1984
(German health data for 1984; Hardin & Hilbe, GLM and Extensions, 4th ed)
. glm outwork age i.kids, family(binomial) nolog
Generalized linear models                      No. of obs      =       3,874
Optimization       : ML                        Residual df     =       3,871
                                               Scale parameter =           1
Deviance        =    4856.119835               (1/df) Deviance =    1.254487
Pearson         =    3918.926527               (1/df) Pearson  =    1.012381

Variance function: V(u) = u*(1-u)              [Bernoulli]
Link function    : g(u) = ln(u/(1-u))          [Logit]

                                               AIC             =    1.255064
Log likelihood   = -2428.059918                BIC             =   -27126.25
```

outwork	Coef.	OIM Std. Err.	z	P>|z|	[95% Conf. Interval]	
age	.0506817	.003558	14.24	0.000	.0437082	.0576552
1.kids	.2401179	.0777066	3.09	0.002	.0878158	.3924199
_cons	-2.923307	.1835341	-15.93	0.000	-3.283027	-2.563586

so that the marginal effects at the means can be obtained using Stata's `margins` command.

```
. margins, dydx(*) atmeans
Conditional marginal effects                        Number of obs    =      3,874
Model VCE    : OIM

Expression   : Predicted mean outwork, predict()
dy/dx w.r.t. : age 1.kids
at           : age              =    43.99587 (mean)
               0.kids           =     .5508518 (mean)
               1.kids           =     .4491482 (mean)
```

	dy/dx	Delta-method Std. Err.	z	P>|z|	[95% Conf.	Interval]
age	.0116431	.000808	14.41	0.000	.0100594	.0132268
1.kids	.0553017	.0179019	3.09	0.002	.0202146	.0903888

```
Note: dy/dx for factor levels is the discrete change from the base level.
```

The marginal effect of **age** indicates that, at the average age, an extra year of age is associated with a 0.0116 or 1.16% increase in the probability of being out of work when **kids** is set at its mean value.

Effects for a Poisson regression model can be similarly calculated. Here we model the number of doctor visits as a function of being unemployed (binary) and age (continuous).

```
. glm docvis i.outwork age, family(poisson) noheader nolog
```

docvis	Coef.	OIM Std. Err.	z	P>|z|	[95% Conf.	Interval]
1.outwork	.4079314	.0188447	21.65	0.000	.3709965	.4448663
age	.0220842	.0008377	26.36	0.000	.0204423	.0237261
_cons	-.033517	.0391815	-0.86	0.392	-.1103113	.0432773

Rather than calculating these effects in this manner, we can again use the Stata **margins** command to get the identical values.

```
. margins, dydx(*) atmeans
Conditional marginal effects                     Number of obs    =      3,874
Model VCE    : OIM

Expression   : Predicted mean docvis, predict()
dy/dx w.r.t. : 1.outwork age
at           : 0.outwork      =      .6334538 (mean)
               1.outwork      =      .3665462 (mean)
               age            =     43.99587 (mean)
```

	dy/dx	Delta-method Std. Err.	z	P>|z|	[95% Conf. Interval]	
1.outwork	1.287021	.0625526	20.58	0.000	1.16442	1.409622
age	.0655285	.0024232	27.04	0.000	.060779	.0702779

Note: dy/dx for factor levels is the discrete change from the base level.

The marginal effect of **age** indicates that, at the average age, an extra year of age is associated with a 0.066 increase in doctor visits per year when **outwork** is set at its mean value.

Marginal effects can also be calculated as the average marginal effects (AMEs). For example, the AME of age on the number of doctor visits may be calculated using Stata's margins command:

```
. margins, dydx(*)
Average marginal effects                         Number of obs    =      3,874
Model VCE    : OIM

Expression   : Predicted mean docvis, predict()
dy/dx w.r.t. : 1.outwork age
```

	dy/dx	Delta-method Std. Err.	z	P>|z|	[95% Conf. Interval]	
1.outwork	1.327171	.0635453	20.89	0.000	1.202624	1.451718
age	.0698497	.0027237	25.64	0.000	.0645113	.0751881

Note: dy/dx for factor levels is the discrete change from the base level.

The AME of **age** indicates that, for each additional year in age, there are 0.0698 additional doctor visits on average.

AMEs in a logistic regression model are calculated similarly. For example,

```
. glm outwork age i.kids, family(binomial) noheader nolog
```

outwork	Coef.	OIM Std. Err.	z	P>|z|	[95% Conf. Interval]	
age	.0506817	.003558	14.24	0.000	.0437082	.0576552
1.kids	.2401179	.0777066	3.09	0.002	.0878158	.3924199
_cons	-2.923307	.1835341	-15.93	0.000	-3.283027	-2.563586

```
. margins, dydx(*)
```

Average marginal effects   Number of obs   =   3,874
Model VCE   : OIM

Expression   : Predicted mean outwork, predict()
dy/dx w.r.t. : age 1.kids

		Delta-method				
	dy/dx	Std. Err.	z	P>|z|	[95% Conf. Interval]	
age	.0110712	.0007012	15.79	0.000	.0096969	.0124456
1.kids	.0520515	.0166564	3.13	0.002	.0194056	.0846974

Note: dy/dx for factor levels is the discrete change from the base level.

Marginal effects may be extended in a number of ways, including as elasticities and semielasticities that can be calculated for both MEMs and AMEs. See Hilbe (2011) for more details.

4.7.2   Discrete change for GLMs

Discrete change represents the change in the outcome for a change in the predictor (usually $x = 1$ versus $x = 0$). Formally, we write this as

$$\frac{\Delta Y|x}{\Delta x} \tag{4.73}$$

Similar to MEMs and AMEs, discrete change can be calculated at the means or at other values of the predictors.

```
. glm docvis i.outwork i.kids, family(poisson) noheader nolog
```

		OIM				
docvis	Coef.	Std. Err.	z	P>|z|	[95% Conf. Interval]	
1.outwork	.5217068	.0180965	28.83	0.000	.4862383	.5571752
1.kids	-.3395266	.0188449	-18.02	0.000	-.3764619	-.3025913
_cons	1.06355	.0146962	72.37	0.000	1.034746	1.092354

```
. margins, dydx(kids) atmeans
Conditional marginal effects                           Number of obs      =       3,874
Model VCE    : OIM

Expression   : Predicted mean docvis, predict()
dy/dx w.r.t. : 1.kids
at           : 0.outwork       =     .6334538 (mean)
               1.outwork       =     .3665462 (mean)
               0.kids          =     .5508518 (mean)
               1.kids          =     .4491482 (mean)
```

	dy/dx	Delta-method Std. Err.	z	P>|z|	[95% Conf. Interval]
1.kids	-1.009658	.0548741	-18.40	0.000	-1.117209	-.902107

```
Note: dy/dx for factor levels is the discrete change from the base level.
```

The average discrete change (ADC) is sometimes referred to as the partial effect. This measures the average effect of a change in the predictor on the response. Here we illustrate the calculation of the effect along with the use of the **margins** command.

```
. margins, dydx(kids)
Average marginal effects                               Number of obs      =       3,874
Model VCE    : OIM

Expression   : Predicted mean docvis, predict()
dy/dx w.r.t. : 1.kids
```

	dy/dx	Delta-method Std. Err.	z	P>|z|	[95% Conf. Interval]
1.kids	-1.043274	.056514	-18.46	0.000	-1.154039	-.9325084

```
Note: dy/dx for factor levels is the discrete change from the base level.
```

Again, these effects relate to the change in the predicted count for the change in the binary predictor. Thus, having children decreases the average number of visits to the doctor by about one.

Here we illustrate discrete change in logistic regression. First, we fit a logistic regression model, and then we calculate the discrete change in the probability of being out of work.

```
. glm outwork age i.kids, family(binomial) noheader nolog
```

outwork	Coef.	OIM Std. Err.	z	P>|z|	[95% Conf. Interval]
age	.0506817	.003558	14.24	0.000	.0437082	.0576552
1.kids	.2401179	.0777066	3.09	0.002	.0878158	.3924199
_cons	-2.923307	.1835341	-15.93	0.000	-3.283027	-2.563586

```
. margins, dydx(kids)

Average marginal effects                          Number of obs      =      3,874
Model VCE    : OIM

Expression    : Predicted mean outwork, predict()
dy/dx w.r.t. : 1.kids
```

		Delta-method				
	dy/dx	Std. Err.	z	P>|z|	[95% Conf. Interval]	
1.kids	.0520515	.0166564	3.13	0.002	.0194056	.0846974

```
Note: dy/dx for factor levels is the discrete change from the base level.
```

The `margins` command will also produce the discrete change at the mean (DCM) for comparison.

```
. margins, dydx(kids) atmeans

Conditional marginal effects                      Number of obs      =      3,874
Model VCE    : OIM

Expression    : Predicted mean outwork, predict()
dy/dx w.r.t. : 1.kids
at            : age             =      43.99587 (mean)
                0.kids          =      .5508518 (mean)
                1.kids          =      .4491482 (mean)
```

		Delta-method				
	dy/dx	Std. Err.	z	P>|z|	[95% Conf. Interval]	
1.kids	.0553017	.0179019	3.09	0.002	.0202146	.0903888

```
Note: dy/dx for factor levels is the discrete change from the base level.
```

The two calculations are given by

$$\text{ADC} \;=\; \frac{1}{n}\sum_{i=1}^{n}\left\{\text{invlogit}(\text{age}_i\widehat{\beta}_{\text{age}} + \widehat{\beta}_{\text{kids}} + \widehat{\beta}_0) - \text{invlogit}(\text{age}_i\widehat{\beta}_{\text{age}} + \widehat{\beta}_0)\right\} \quad (4.74)$$

$$\text{DCM} \;=\; \text{invlogit}(\overline{\text{age}}\widehat{\beta}_{\text{age}} + \widehat{\beta}_{\text{kids}} + \widehat{\beta}_0) - \text{invlogit}(\overline{\text{age}}\widehat{\beta}_{\text{age}} + \widehat{\beta}_0) \quad (4.75)$$

Several additional statistics have been developed to investigate the relationships between the outcome variables and predictors—for example, elasticities for the percentage change in the outcome for a standard deviation change in the predictor. Texts such as Long and Freese (2014), Cameron and Trivedi (2013), Cameron and Trivedi (2010), and Hilbe (2011) offer additional details.

Part II

Continuous Response Models

The Gaussian family

Contents

In the 1800s, Johann Carl Friedrich Gauss, the eponymous prince of mathematics, described the least-squares fitting method and the distribution that bears his name. Having a symmetric bell shape, the Gaussian density function is often referred to as the normal density. The normal cumulative distribution function is a member of the exponential family of distributions and so may be used as a basis for a GLM. Moreover, because the theory of GLMs was first conceived to be an extension to the normal ordinary least-squares (OLS) model, we will begin with an exposition of how OLS fits into the GLM framework. Understanding how other GLM models generalize this basic form should then be easier.

Regression models based on the Gaussian or normal distribution are commonly referred to as OLS models. This standard regression model is typically the first model taught in beginning statistics courses.

The Gaussian distribution is a continuous distribution of real numbers with support over the real line $\Re = (-\infty, +\infty)$. A model based on this distribution assumes that the response variable, called the *dependent variable* in the social sciences, takes the shape of the Gaussian distribution. Equivalently, we consider that the error term in the equation

$$y = \mathbf{x}\boldsymbol{\beta} + \epsilon \tag{5.1}$$

is normally or Gaussian distributed. Generalized linear interactive modeling (GLIM), the original GLM software program, uses the term **ERROR** to designate the family or distribution of a model. Regardless, we may think of the underlying distribution as that of either the response variable or the error term.

5.1   Derivation of the GLM Gaussian family

The various components of a particular GLM can be obtained from the base probability function. The manner in which the probability function is parameterized relates directly to the algorithm used for estimation, that is, iteratively reweighted least squares (IRLS) or Newton–Raphson (N–R). If we intend to estimate parameters by using the standard or traditional IRLS algorithm, then the probability function is parameterized in terms of the mean μ (estimated fitted value). On the other hand, if the estimation is to be performed in terms of N–R, parameterization is in terms of $\mathbf{x}\boldsymbol{\beta}$ (the linear predictor). To clarify, the IRLS algorithm considers maximizing a function in terms of the mean where the introduction of covariates is delayed until specification of the link function. On the other hand, N–R typically writes the function to be maximized in terms of the specific link function so that derivatives may be clearly seen.

For the Gaussian distribution, μ is the same as $\mathbf{x}\boldsymbol{\beta}$ in its canonical form. We say that the canonical link is the identity link; that is, there is a straightforward identity between the fitted value and the linear predictor. The canonical-link Gaussian regression model is the paradigm instance of a model with an identity link.

The Gaussian probability density function, parameterized in terms of μ, is expressed as

$$f(y; \mu, \sigma^2) = \frac{1}{\sqrt{2\pi\sigma^2}} \exp\left\{-\frac{(y-\mu)^2}{2\sigma^2}\right\} \tag{5.2}$$

where $f(\cdot)$ is the generic form of the density function of y, given parameters μ and σ^2; y is the response variable; μ is the mean parameter; and σ^2 is the scale parameter.

In GLM theory, a density function is expressed in exponential family form. That is, all density functions are members of the exponential family of distributions and can be expressed in the form

$$f(y; \theta, \phi) = \exp\left\{\frac{y\theta - b(\theta)}{a(\phi)} + c(y; \phi)\right\} \tag{5.3}$$

where θ is the canonical parameter, $b(\theta)$ is the cumulant, $a(\phi)$ is the scale parameter, and $c(y, \phi)$ is the normalization function. The normalization function ensures that the probability function sums or integrates to 1. Because GLMs optimize the deviance (difference of two nested ML models), the normalization terms cancel. Thus, the normalization function is ignored in the optimization of GLMs.

5.2   Derivation in terms of the mean

GLM theory traditionally derives the elements of the GLM algorithm in terms of μ. This derivation leads to the standard IRLS algorithm. We first detail this method.

Expressing the Gaussian density function in exponential-family form results in the specification

$$f(y; \mu, \sigma^2) = \exp\left\{-\frac{(y-\mu)^2}{2\sigma^2} - \frac{1}{2}\ln\left(2\pi\sigma^2\right)\right\} \qquad (5.4)$$

$$= \exp\left\{\frac{y\mu - \mu^2/2}{\sigma^2} - \frac{y^2}{2\sigma^2} - \frac{1}{2}\ln\left(2\pi\sigma^2\right)\right\} \qquad (5.5)$$

where we can easily determine that $\theta = \mu$ and $b(\theta) = \mu^2/2$.

The mean (expected value of the outcome y) is determined by taking the derivative of the cumulant, $b(\theta)$, with respect to θ. The variance is the second derivative. Hence,

$$b'(\theta) = \frac{\partial b}{\partial \theta} \qquad (5.6)$$

$$= \frac{\partial b}{\partial \mu}\frac{\partial \mu}{\partial \theta} \qquad (5.7)$$

$$= (\mu)(1) = \mu \qquad (5.8)$$

$$b''(\theta) = \frac{\partial^2 b}{\partial \theta^2} \qquad (5.9)$$

$$= \frac{\partial}{\partial \theta}\left(\frac{\partial b}{\partial \mu}\frac{\partial \mu}{\partial \theta}\right) \qquad (5.10)$$

$$= \frac{\partial}{\partial \theta}(\mu) \qquad (5.11)$$

$$= \frac{\partial}{\partial \mu}\mu\frac{\partial \mu}{\partial \theta} \qquad (5.12)$$

$$= (1)(1) = 1 \qquad (5.13)$$

because $\partial\mu/\partial\theta = 1$ for this case and b is understood to be the function $b(\theta)$.

The mean is simply μ and the variance is 1. The scale parameter, ϕ, is σ^2. The other important statistic required for the GLM–Gaussian algorithm is the derivative of the link, μ.

$$\frac{\partial \theta}{\partial \mu} = \frac{\partial \mu}{\partial \mu} = 1 \qquad (5.14)$$

Finally, the deviance function is calculated as

$$D = 2\phi\left\{\mathcal{L}(y, \sigma^2; y) - \mathcal{L}(\mu, \sigma^2; y)\right\} \qquad (5.15)$$

$\mathcal{L}(y, \sigma^2; y)$ represents the saturated log-likelihood function and is calculated by substituting the value of y_i for each instance of μ_i; the saturated model is one that perfectly reproduces the data, which is why $y_i = \mu_i$. When subtracting the observed $\mathcal{L}$ from the saturated function, we find that the normalization terms cancel from the final deviance equation. The observed log-likelihood function is given by

$$\mathcal{L}(\mu, \sigma^2; y) = \sum_{i=1}^{n}\left\{\frac{y_i\mu_i - \mu_i^2/2}{\sigma^2} - \frac{y_i^2}{2\sigma^2} - \frac{1}{2}\ln\left(2\pi\sigma^2\right)\right\} \qquad (5.16)$$

The saturated log likelihood is

$$\mathcal{L}(y, \sigma^2; y) = \sum_{i=1}^{n} \left\{ \frac{y_i^2 - y_i^2/2}{\sigma^2} - \frac{y_i^2}{2\sigma^2} - \frac{1}{2} \ln\left(2\pi\sigma^2\right) \right\} \tag{5.17}$$

$$= \sum_{i=1}^{n} \left\{ -\frac{1}{2} \ln\left(2\pi\sigma^2\right) \right\} \tag{5.18}$$

The Gaussian deviance is thereby calculated as

$$D = 2\sigma^2 \sum_{i=1}^{n} \left\{ -\frac{1}{2} \ln\left(2\pi\sigma^2\right) - \frac{y_i\mu_i - \mu_i^2/2}{\sigma^2} + \frac{y_i^2}{2\sigma^2} + \frac{1}{2} \ln\left(2\pi\sigma^2\right) \right\} \tag{5.19}$$

$$= 2\sigma^2 \sum_{i=1}^{n} \left(\frac{y_i^2 - 2y_i\mu_i + \mu_i^2}{2\sigma^2} \right) \tag{5.20}$$

$$= \sum_{i=1}^{n} (y_i - \mu_i)^2 \tag{5.21}$$

The Gaussian deviance is identical to the sum of squared residuals. This fact is unique to the Gaussian family. However, because the Gaussian variance is 1, the Pearson χ^2 statistic, defined as

$$\chi^2 = \sum_{i=1}^{n} \frac{(y_i - \mu_i)^2}{v(\mu_i)} \tag{5.22}$$

is equal to the deviance. This too is a unique feature of the Gaussian family.

The IRLS algorithm for the canonical-link Gaussian family is shown in the generic form below. Of course, there are other algorithms, some of which use a DO–WHILE loop (for example, S-Plus and XploRe) rather than a WHILE conditional loop (for example, Stata). Nevertheless, all IRLS algorithms follow the basic logic illustrated here.

5.3   IRLS GLM algorithm (nonbinomial)

With summing understood, the generic IRLS form for nonbinomial models is

Listing 5.1. IRLS algorithm for nonbinomial models

```
1       μ = {y + mean(y)}/2
2       η = g
3       WHILE (abs(ΔDev) > tolerance)  {
4             w = 1/(vg'²)
5             z = η + (y - μ)g'
6             β = (XᵀWX)⁻¹XᵀWz
7             η = Xβ
8             μ = g⁻¹
9             OldDev = Dev
10            Dev = deviance function
11            ΔDev = Dev - OldDev
12      }
13      χ² = ∑(y - μ)²/v
```

where v is the variance function, g is the link function, g' is the derivative of the link, and g^{-1} is the inverse link function. The tolerance specifies the sensitivity of when iterations are to cease. Typically, the tolerance is set to 1.0e–6 (for the difference of the current deviance from previous iteration). Other methods of defining tolerance exist, including the measure of the difference in estimated coefficients between iterations k and $k + 1$, that is, $||\widehat{\boldsymbol{\beta}}_{k+1} - \widehat{\boldsymbol{\beta}}_k||^2$.

Here we use a measure based on the change in deviance as a basis for the canonical Gaussian algorithm:

Listing 5.2. IRLS algorithm for Gaussian models

```
1       Dev = 1
2       μ = {y - mean(y)}/2
3       η = μ                                      /* identity link */
4       WHILE   (abs(ΔDev) > tolerance) {
5             β = (XᵀX)⁻¹Xᵀy
6             η = xβ
7             μ = η
8             OldDev = Dev
9             Dev = ∑(y - μ)²
10            ΔDev = Dev - OldDev
11      }
12      χ² = ∑(y - μ)²
```

Again in the canonical form, we do not typically use the GLM routine for fitting model parameters. The reason is clear from looking at the above algorithm. The GLM

estimation of β is simplified in the canonical Gaussian instance because the weight w
equals 1 and the working response z equals the actual response y. Thus

$$(\mathbf{X}^T\mathbf{W}\mathbf{X})^{-1}\mathbf{X}^T\mathbf{W}\mathbf{z} = (\mathbf{X}^T\mathbf{X})^{-1}\mathbf{X}^T\mathbf{y} \qquad (5.23)$$

In fact, the above algorithm can be further reduced because of the identity between
μ and η.

Listing 5.3. IRLS algorithm (reduced) for Gaussian models

```
1     Dev = 1
2     μ = {y − mean(y)}/2
3     WHILE (abs(ΔDev) > tolerance)  {
4         β = (XᵀX)⁻¹Xᵀy
5         μ = Xβ
6         OldDev = Dev
7         Dev = Σ(y − μ)²
8         ΔDev = Dev − OldDev
9     }
10    χ² = Σ(y − μ)²
```

Moreover, iterations to find estimates are not necessary because estimates can be determined using only one step. Hence, in actuality, only the line estimating β is necessary.
This too is unique among GLMs.

When modeling the canonical Gaussian, it is perhaps best to use the standard linear regression procedure common to most statistical packages. However, using a GLM
routine typically allows access to a variety of diagnostic residuals and to goodness-of-
fit (GOF) statistics. Using GLM when analyzing the Gaussian model also allows easy
comparison with other Gaussian links and to other GLM families.

The χ^2 statistic is traditionally called the Pearson χ^2 GOF statistic. It is the ratio
of the squared residuals to the variance. Because the Gaussian family variance is 1, the
deviance and Pearson χ^2 statistics are identical. This is not the case for other families.
However, on this account, the deviance-based dispersion is the same as the χ^2-based
dispersion.

Dispersion statistics are obtained by dividing the residual degrees of freedom from
the deviance or Pearson χ^2 statistic, respectively. The residual degrees of freedom is
calculated as $n - p$, where n is the number of cases in the model and p is the number
of model predictors (possibly including a constant).

The statistics described above are important to the assessment of model fit. For the
canonical-link Gaussian case, the values are identical to the residual statistics displayed
in an analysis of variance (ANOVA) table, which usually accompanies a linear regression
printout.

$$\text{ERROR SS} \quad = \quad \text{deviance} = \text{Pearson } \chi^2 \qquad (5.24)$$
$$\text{MSE} \quad = \quad \text{dispersion (both deviance and } \chi^2) \qquad (5.25)$$

The scale is defined as 1 when using the IRLS algorithm. The reason for this is simple. A traditional GLM, using the IRLS algorithm, provides only mean parameter estimates; it does not directly estimate the scale or ancillary parameter. If we desired a separate estimate of the scale, we could use an ML algorithm. However, the deviance dispersion can be used to estimate the scale value but without estimating an associated standard error (which would be obtained in an ML approach).

5.4   ML estimation

ML estimation of the canonical-link Gaussian model requires specifying the Gaussian log likelihood in terms of $\mathbf{x}\boldsymbol{\beta}$. Parameterized in terms of $\mathbf{x}\boldsymbol{\beta}$, the Gaussian density function is given by

$$f(y; \boldsymbol{\beta}, \sigma^2) = \frac{1}{\sqrt{2\pi\sigma^2}} \exp\left\{ -\frac{1}{2\sigma^2}(y - \mathbf{x}\boldsymbol{\beta})^2 \right\} \tag{5.26}$$

After we express the joint density in exponential family form, the log likelihood becomes

$$\mathcal{L}(\mu, \sigma^2; y) = \sum_{i=1}^{n} \left\{ \frac{y_i \mathbf{x}_i \boldsymbol{\beta} - (\mathbf{x}_i \boldsymbol{\beta})^2/2}{\sigma^2} - \frac{y_i^2}{2\sigma^2} - \frac{1}{2}\ln\left(2\pi\sigma^2\right) \right\} \tag{5.27}$$

This result implies that

$$\begin{aligned}
\theta &= \mathbf{x}\boldsymbol{\beta} & (5.28) \\
b(\theta) &= (\mathbf{x}\boldsymbol{\beta})^2/2 = \theta^2/2 & (5.29) \\
b'(\theta) &= \mathbf{x}\boldsymbol{\beta} & (5.30) \\
b''(\theta) &= 1 & (5.31) \\
a(\phi) &= \sigma^2 & (5.32)
\end{aligned}$$

The ML algorithm estimates σ^2 in addition to standard parameters and linear predictor estimates. The estimated value of σ^2 should be identical to the dispersion values, but ML provides a standard error and confidence interval. The estimates differ from the usual OLS estimates, which specify estimators for the regression model without assuming normality. The OLS estimate of σ^2 uses $n - p$ in the denominator, whereas the ML and GLM estimates use n; the numerator is the sum of the squared residuals. This slight difference (which makes no difference asymptotically and makes no difference in the estimated regression coefficients) is most noticeable in the estimated standard errors for models based on small samples.

The (partial) derivatives of the canonical-link Gaussian log likelihood are given by

$$\frac{\partial \mathcal{L}}{\partial \beta_j} = \sum_{i=1}^{n} \frac{1}{\sigma^2}(y_i - \mathbf{x}_i \boldsymbol{\beta}) x_{ji} \tag{5.33}$$

$$\frac{\partial \mathcal{L}}{\partial \sigma} = \sum_{i=1}^{n} \frac{1}{\sigma} \left\{ \left(\frac{y_i - \mathbf{x}_i \boldsymbol{\beta}}{\sigma} \right)^2 - 1 \right\} \tag{5.34}$$

and the second (partial) derivatives are given by

$$\frac{\partial^2 \mathcal{L}}{\partial \beta_j \partial \beta_k} = -\sum_{i=1}^{n} \frac{1}{\sigma^2} x_{ji} x_{ki} \tag{5.35}$$

$$\frac{\partial^2 \mathcal{L}}{\partial \beta_j \partial \sigma} = -\sum_{i=1}^{n} \frac{2}{\sigma^3} (y_i - \mathbf{x}_i \boldsymbol{\beta}) x_{ji} \tag{5.36}$$

$$\frac{\partial^2 \mathcal{L}}{\partial \sigma \partial \sigma} = -\sum_{i=1}^{n} \frac{1}{\sigma^2} \left\{ 3 \left(\frac{y_i - \mathbf{x}_i \boldsymbol{\beta}}{\sigma} \right)^2 - 1 \right\} \tag{5.37}$$

5.5   GLM log-Gaussian models

An important and perhaps the foremost reason for using GLM as a framework for model construction is the ability to easily adjust models to fit particular response data situations. The canonical-link Gaussian model assumes a normally distributed response. Although the normal model is robust to moderate deviations from this assumption, it is nevertheless the case that many data situations are not amenable to or appropriate for normal models.

Unfortunately, many researchers have used the canonical-link Gaussian for data situations that do not meet the assumptions on which the Gaussian model is based. Until recently, few software packages allowed users to model data by means other than the normal model; granted, many researchers had little training in nonnormal modeling. Most popular software packages now have GLM capabilities or at least implement many of the most widely used GLM procedures, including logistic, probit, and Poisson regression.

The log-Gaussian model is based on the Gaussian distribution. It uses the log rather than the (canonical) identity link. The log link is generally used for response data that can take only positive values on the continuous scale, or values greater than 0. The data must be such that nonpositive values are not only absent but also theoretically precluded.

Before GLM, researchers usually modeled positive-only data with the normal model. However, they first took the natural log of the response prior to modeling. In so doing, they explicitly acknowledged the need to normalize the response relative to the predictors, thus accommodating one of the assumptions of the Gaussian model. The problem with this method is one of interpretation. Fitted values, as well as parameter estimates, are in terms of the log response. This obstacle often proves to be inconvenient.

A better approach is to internalize within the model itself the log transformation of the response. The log link exponentiates the linear predictor, or $\mathbf{x}\boldsymbol{\beta}$, rather than log-transforming the response to linearize the relationship between the response and predictors. This procedure, implicit within the GLM algorithm, allows easy interpretation of estimates and of fitted values.

The implementation of a log link within the ML algorithm is straightforward—simply substitute $\exp(\mathbf{x}\boldsymbol{\beta})$ for each instance of $\mathbf{x}\boldsymbol{\beta}$ in the log-likelihood function. The canonical-link Gaussian log-likelihood function is thus transformed to a lognormal model. The log-likelihood function is then written in exponential-family form:

$$\mathcal{L}(\mu; y, \sigma^2) = \sum_{i=1}^{n} \left[\frac{y_i \exp(\mathbf{x}_i\boldsymbol{\beta}) - \{\exp(\mathbf{x}_i\boldsymbol{\beta})\}^2/2}{\sigma^2} - \frac{y_i^2}{2\sigma^2} - \frac{1}{2}\ln\left(2\pi\sigma^2\right) \right] \qquad (5.38)$$

Creating a lognormal, or log-linked Gaussian, model using the standard IRLS algorithm is only a bit more complicated. Recalling that parameterization is in terms of μ, we must change the link function from $\eta = \mu$ to $\eta = \ln(\mu)$ and the inverse link function from $\mu = \eta$ to $\mu = \exp(\eta)$.

Finally, we have a value for the derivative of the link. In the canonical form, the derivative is 1, hence excluding it from the algorithm. The derivative of the log link is $1/\mu$ and is placed in the algorithm at line 4 in listing 5.1. All other aspects of the Gaussian-family algorithm remain the same, except that we must extend the algorithm to its standard form.

5.6   Expected versus observed information matrix

To reiterate our discussion from chapter 3, OIM standard errors are more conservative than those based on EIM, and the difference becomes increasingly noticeable in datasets with fewer observations. Again, referring to the discussions of chapter 3, it was demonstrated that the observed information matrix reduces to the naïve expected information matrix for the canonical link. The log link is not the canonical link for the Gaussian. Hence, it may be better to model noncanonical linked data by using a method that provides standard errors based on the observed information matrix (especially in smaller datasets).

One can adjust the IRLS estimating algorithm such that the observed information matrix is used rather than the expected matrix. One need only create a function, $\mathbf{W}_o$, that calculates the observed information matrix given in (3.59).

$$\mathbf{W}_o = \mathbf{W} + (y - \mu) \left\{ \frac{vg'' + v'g'}{v^2(g')^3} \right\} \qquad (5.39)$$

Here $g' = g'(\mu)$, the derivative of the link; $g'' = g''(\mu)$, the second derivative of the link; $v = v(\mu)$, the variance function; $v^2 =$ variance squared; and $\mathbf{W}$ is the standard IRLS weight function given in (3.46).

The generic form of the GLM IRLS algorithm, using the observed information matrix, is then given by

Listing 5.4. IRLS algorithm for GLM using OIM

```
1     μ = {y + mean(y)}/2
2     η = g
3     WHILE (abs(ΔDev) > tolerance) {
4         W = 1/(vg'²)
5         z = η + (y − μ)g' − offset
6         W₀ = W + (y − μ)(vg'' + v'g')/(v²(g')³)
7         β = (XᵀW₀X)⁻¹XᵀW₀z
8         η = Xβ + offset
9         μ = g⁻¹
10        OldDev = Dev
11        Dev = Σ (deviance function)
12        ΔDev = Dev - OldDev
13    }
14    χ² = Σ(y − μ)²/V
```

where V' is the derivative with respect to μ of the variance function.

Using this algorithm, which we will call the IRLS–OIM algorithm, estimates and standard errors will be identical to those produced using ML. The scale, estimated by the ML routine, is estimated as the deviance-based dispersion. However, unlike ML output, the IRLS–OIM algorithm fails to provide standard errors for the scale. Of course, the benefit to using IRLS–OIM is that the typical ML problems with starting values are nearly always avoided. Also, ML routines usually take a bit longer to converge than does a straightforward IRLS–OIM algorithm.

Finally, we need a value for the derivative of the link. In canonical form, the derivative is 1, being in effect excluded from the algorithm. The derivative of the log link, $\ln(\mu)$, is $1/\mu$ and is appropriately placed into the algorithm below at line 4.

Listing 5.5. IRLS algorithm for log-Gaussian models using OIM

```
1    μ = {y + mean(y)}/2
2    η = ln(μ)
3    WHILE (abs(ΔDev) > tolerance {
4        W = 1/{1(1/μ)²} = μ²
5        z = η + (y − μ)/μ − offset
6        Wₒ = W + (y − μ)μ²
7        β = (Xᵀ Wₒ X)⁻¹ Xᵀ Wₒ z
8        η = Xβ + offset
9        μ = exp(η)
10       OldDev = Dev
11       Dev = ∑(y − μ)²
12       ΔDev = Dev − OldDev
13   }
14   χ² = ∑(y − μ)²
```

5.7   Other Gaussian links

The standard or normal linear model, OLS regression, uses the identity link, which is the canonical form of the distribution. It is robust to most minor violations of Gaussian distributional assumptions but is used in more data modeling situations than is appropriate. In fact, it was used to model binary, proportional, and discrete count data. Appropriate software was unavailable, and there was not widespread understanding of the statistical problems inherent in using OLS regression for data modeling.

There is no theoretical reason why researchers using the Gaussian family for a model should be limited to using only the identity or log links. The reciprocal link, $1/\mu$, has been used to model data with a rate response. It also may be desirable to use the power links with the Gaussian model. In doing so, we approximate the Box–Cox ML algorithm, which is used as a normality transform. Some software packages do not have Box–Cox capabilities; therefore, using power links with the Gaussian family may assist in achieving the same purpose.

5.8   Example: Relation to OLS

We provide a simple example showing the relationship of OLS regression output to that of GLM. The example we use comes from a widely used study on prediction of low-birth-weight babies (Hosmer, Lemeshow, and Sturdivant 2013). The response variable, `bwt`, is the birth weight of the baby. The predictors are `lwt`, the weight of the mother; `race`, where a value of 1 indicates white, 2 indicates black, and 3 indicates other; a binary variable, `smoke`, which indicates whether the mother has a history of smoking; another binary variable `ht`, which indicates whether the mother has a history of hypertension; and a final binary variable `ui`, which indicates whether the mother has a history of

uterine irritability. We generated indicator variables for use in model specification `race2`
indicating black and `race3` indicating other. The standard Stata regression output is
given by

```
. use http://www.stata-press.com/data/hh4/lbw
(Hosmer & Lemeshow data)

. regress bwt lwt race2 race3 smoke ht ui

      Source |       SS           df       MS      Number of obs   =       189
-------------+----------------------------------   F(6, 182)       =      9.59
       Model |  23992342.5          6   3998723.74  Prob > F        =    0.0000
    Residual |  75922956.1        182   417159.099  R-squared       =    0.2401
-------------+----------------------------------   Adj R-squared   =    0.2151
       Total |  99915298.6        188   531464.354  Root MSE        =    645.88

------------------------------------------------------------------------------
         bwt |      Coef.   Std. Err.      t    P>|t|     [95% Conf. Interval]
-------------+----------------------------------------------------------------
         lwt |   4.238846   1.675414     2.53   0.012     .9331126    7.544579
       race2 |    -475.22    145.589    -3.26   0.001    -762.4792   -187.9607
       race3 |   -349.669   112.3456    -3.11   0.002    -571.3363   -128.0018
       smoke |  -355.7326   103.4345    -3.44   0.001    -559.8176   -151.6475
          ht |  -584.6279   199.6238    -2.93   0.004    -978.5024   -190.7533
          ui |  -523.8869    134.664    -3.89   0.000    -789.5903   -258.1835
       _cons |   2837.343   243.6806    11.64   0.000      2356.54    3318.145
------------------------------------------------------------------------------
```

The GLM output is given by

```
. glm bwt lwt race2 race3 smoke ht ui

Iteration 0:   log likelihood = -1487.5585

Generalized linear models                          No. of obs      =       189
Optimization     : ML                              Residual df     =       182
                                                   Scale parameter =    417159.1
Deviance         =   75922956.11                   (1/df) Deviance =    417159.1
Pearson          =   75922956.11                   (1/df) Pearson  =    417159.1

Variance function: V(u) = 1                        [Gaussian]
Link function    : g(u) = u                        [Identity]

                                                   AIC             =   15.81543
Log likelihood   = -1487.558492                    BIC             =   7.59e+07

------------------------------------------------------------------------------
             |                 OIM
         bwt |      Coef.   Std. Err.      z    P>|z|     [95% Conf. Interval]
-------------+----------------------------------------------------------------
         lwt |   4.238846   1.675414     2.53   0.011     .9550942    7.522598
       race2 |    -475.22    145.589    -3.26   0.001    -760.5691   -189.8708
       race3 |   -349.669   112.3456    -3.11   0.002    -569.8623   -129.4758
       smoke |  -355.7326   103.4345    -3.44   0.001    -558.4606   -153.0046
          ht |  -584.6279   199.6238    -2.93   0.003    -975.8834   -193.3723
          ui |  -523.8869    134.664    -3.89   0.000    -787.8235   -259.9503
       _cons |   2837.343   243.6806    11.64   0.000     2359.738    3314.948
------------------------------------------------------------------------------
```

Note the similarity of estimate and standard-error values. Parameter p-values differ
because the standard OLS method displays p-values based on the Student's t distribution,
whereas GLM p-values are in terms of the z or normal distribution. Note also the
comparative values of the deviance and Pearson statistics with the residual sum of

squares. Also compare the scaled deviance and Pearson statistics and the ANOVA mean squared error, or residual mean square (MS).

Modeling with GLM provides an estimate of 417,159.1 for the scale value, σ^2. If we used the irls option with the glm command, having estimation depend solely on Fisher scoring, then the scale is not estimated directly and is displayed as 1 on the output heading. Of course, the deviance scale will provide the same information under both methods of estimation, but this will not always be the case with other GLM families.

```
. glm bwt lwt race2 race3 smoke ht ui, irls

Iteration 1:   deviance =  7.59e+07

Generalized linear models                      No. of obs      =         189
Optimization      : MQL Fisher scoring         Residual df     =         182
                    (IRLS EIM)                 Scale parameter =           1
Deviance          =  75922956.11               (1/df) Deviance =    417159.1
Pearson           =  75922956.11               (1/df) Pearson  =    417159.1

Variance function: V(u) = 1                    [Gaussian]
Link function     : g(u) = u                   [Identity]

                                               BIC             =     7.59e+07
```

bwt	Coef.	EIM Std. Err.	z	P>|z|	[95% Conf. Interval]	
lwt	4.238846	1.675414	2.53	0.011	.9550942	7.522598
race2	-475.22	145.589	-3.26	0.001	-760.5691	-189.8708
race3	-349.669	112.3456	-3.11	0.002	-569.8623	-129.4758
smoke	-355.7326	103.4345	-3.44	0.001	-558.4606	-153.0046
ht	-584.6279	199.6238	-2.93	0.003	-975.8834	-193.3723
ui	-523.8869	134.664	-3.89	0.000	-787.8235	-259.9503
_cons	2837.343	243.6806	11.64	0.000	2359.738	3314.948

Finally, looking at the output log (whether we use the IRLS algorithm) demonstrates what was said before. The canonical-link Gaussian algorithm iterates just once, which is unique among GLM specifications.

5.9   Example: Beta-carotene

The URL http://www.stat.cmu.edu/~brian/707/data/prdata.txt (as of writing this text) includes data from an unpublished study.[1]

We repeat analyses on 315 observations of dietary intake. The goal of the study is to investigate whether low dietary intake or low plasma concentrations of retinol, beta-carotene, or other carotenoids might be associated with certain cancers. The investigators collected information over a 3-year period from patients who elected to have a surgical biopsy or removal of a lesion that was found to be noncancerous. Such procedures retrieved tissue from the lung, colon, breast, skin, ovary, or uterus.

---

1. Data and descriptions of the data are used with permission.

The outcome variable `betap` is the plasma concentration reading of beta-carotene. The independent variables of interest include the age of the patient in years, `age`; an indicator of female sex, `female`; an indicator of whether the patient takes vitamins, `pills`; the average daily dietary intake in kilocalories, `kcal`; the average number of fat grams consumed per day, `fat`; the average daily dietary intake of fiber, `fiber`; the average daily dietary intake of alcohol, `alcohol`; a reading of the patient's cholesterol, `chol`; and the average daily intake of dietary beta-carotene, `betad`. We fit a linear model, obtain the Pearson residuals from the fitted model, and draw a histogram in figure 5.1, which clearly shows the skewness of the distribution.

```
. import delimited using data/beta.txt, clear
(14 vars, 315 obs)
. generate female = sex==2
. generate pills = vitamin==1 | vitamin==2
. glm betap age female pills kcal fat fiber alcohol chol betad
Iteration 0:   log likelihood =  -2062.328
Generalized linear models                          No. of obs       =        315
Optimization     : ML                              Residual df      =        305
                                                   Scale parameter =   29393.52
Deviance         =  8965024.147                     (1/df) Deviance =   29393.52
Pearson          =  8965024.147                     (1/df) Pearson  =   29393.52
Variance function: V(u) = 1                        [Gaussian]
Link function    : g(u) = u                        [Identity]
                                                   AIC              =   13.15764
Log likelihood   = -2062.327961                    BIC              =    8963270
```

betap	Coef.	OIM Std. Err.	z	P>|z|	[95% Conf. Interval]	
age	.9515906	.7453438	1.28	0.202	-.5092565	2.412438
female	31.27419	31.99056	0.98	0.328	-31.42615	93.97454
pills	71.89446	21.07482	3.41	0.001	30.58858	113.2003
kcal	-.0358134	.0517117	-0.69	0.489	-.1371666	.0655397
fat	.1546087	.8174872	0.19	0.850	-1.447637	1.756854
fiber	7.771203	2.8335	2.74	0.006	2.217644	13.32476
alcohol	1.228665	1.249511	0.98	0.325	-1.220331	3.677661
chol	-.1388087	.1079521	-1.29	0.199	-.3503909	.0727736
betad	.0162778	.0075557	2.15	0.031	.0014689	.0310867
_cons	15.61038	67.45794	0.23	0.817	-116.6048	147.8255

```
. predict double res if e(sample), pearson
```

```
. histogram res
(bin=17, start=-286.79243, width=82.533625)
```

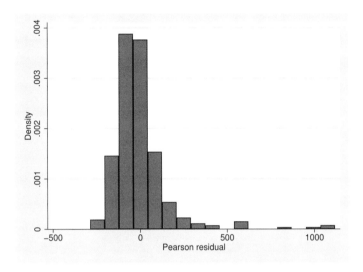

Figure 5.1. Pearson residuals obtained from linear model

In addition to the histogram, we can plot the ordered Pearson residuals versus normal probability points to see how well the two agree. If the Pearson residuals follow a normal distribution, the plot should look like a line along the diagonal from the bottom left of the plot to the top right. This plot, figure 5.2, emphasizes the conclusions already reached looking at figure 5.1. We illustrate the Pearson residuals versus one of the covariates, the daily caloric intake, as an example of how one investigates whether the variance is related to the value of the independent variables.

```
. sort res
. generate nscore = invnormal(_n/316)
. twoway scatter nscore res
```

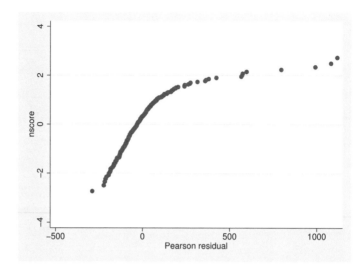

Figure 5.2. Normal scores versus sorted Pearson residuals obtained from linear model

Figure 5.3 demonstrates that the variance does not depend on the dietary caloric intake. However, the positive residuals seem to be much larger (in absolute value) than the negative values. There also appear to be two observations with unusually large daily caloric intake. In a full analysis, one would calculate Cook's distance measures to investigate leverage and would more fully investigate any suspected outliers.

. twoway scatter res kcal, yline(0)

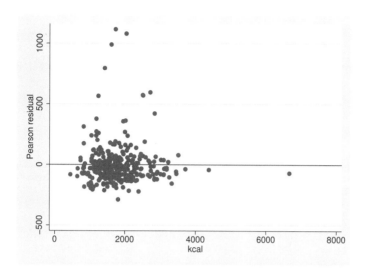

Figure 5.3. Pearson residuals versus kilocalories; Pearson residuals obtained from linear model

For our quick illustration, we will drop from further models the two observations with unusually large average daily intake of kilocalories. Here we switch from a linear regression model to a lognormal model, one that uses the log link to transform the linear predictor to the scale of the observed values.

```
. drop if kcal > 4000
(2 observations deleted)

. glm betap age female pills kcal fat fiber alcohol chol betad, link(log) nolog
Generalized linear models                         No. of obs        =        313
Optimization      : ML                            Residual df       =        303
                                                  Scale parameter =        28332.2
Deviance          =   8584655.935                 (1/df) Deviance =        28332.2
Pearson           =   8584655.935                 (1/df) Pearson  =        28332.2

Variance function: V(u) = 1                       [Gaussian]
Link function     : g(u) = ln(u)                  [Log]

                                                  AIC               =       13.12106
Log likelihood    = -2043.445673                  BIC               =       8582915
```

betap	Coef.	OIM Std. Err.	z	P>|z|	[95% Conf. Interval]	
age	.0056749	.0035317	1.61	0.108	-.0012471	.0125969
female	.3368722	.1966366	1.71	0.087	-.0485284	.7222728
pills	.4761023	.1340172	3.55	0.000	.2134334	.7387712
kcal	-.0000842	.0002087	-0.40	0.687	-.0004933	.0003249
fat	-.000546	.0037247	-0.15	0.883	-.0078462	.0067543
fiber	.042346	.0098439	4.30	0.000	.0230524	.0616397
alcohol	.0222378	.0102478	2.17	0.030	.0021525	.0423232
chol	-.0007851	.0006329	-1.24	0.215	-.0020256	.0004553
betad	.000062	.0000269	2.30	0.021	9.23e-06	.0001147
_cons	3.911164	.3715337	10.53	0.000	3.182971	4.639357

```
. predict double betaphat if e(sample), mu

. predict double resloglink if e(sample), pearson

. histogram resloglink
(bin=17, start=-274.24709, width=80.920857)

. twoway scatter resloglink betaphat, yline(0)
```

Investigating residual plots, including figures 5.4 and 5.5, demonstrates that the residuals are still not close to the normal distribution.

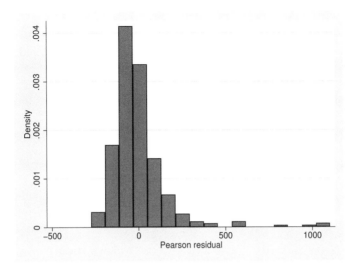

Figure 5.4. Pearson residuals obtained from log-Gaussian model (two outliers removed)

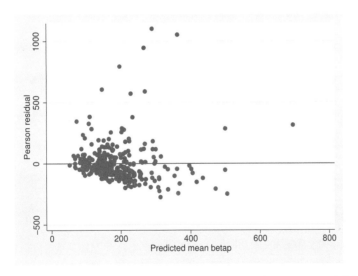

Figure 5.5. Pearson residuals versus fitted values from log-Gaussian model (two outliers removed)

We now switch our focus from transforming the linear predictor to transforming the outcome variable to see if a better model may be found. In this model, we transform the observed variable and leave the linear predictor on the identity link. We call this the lognormal model. However, we admit that this can easily be confused with the log-Gaussian model. The lognormal model is derived from the lognormal distribution, whereas the log-Gaussian model is a GLM specifying the log link and Gaussian variance.

```
. generate double logbetap = log(betap)
(1 missing value generated)

. glm logbetap age female pills kcal fat fiber alcohol chol betad

Iteration 0:   log likelihood = -321.38508

Generalized linear models                         No. of obs      =        312
Optimization       : ML                           Residual df     =        302
                                                  Scale parameter =   .4746678
Deviance         =  143.3496683                   (1/df) Deviance =   .4746678
Pearson          =  143.3496683                   (1/df) Pearson  =   .4746678

Variance function: V(u) = 1                       [Gaussian]
Link function    : g(u) = u                       [Identity]

                                                  AIC             =   2.124263
Log likelihood   = -321.3850781                   BIC             =  -1591.037
```

logbetap	Coef.	OIM Std. Err.	z	P>|z|	[95% Conf. Interval]	
age	.0062599	.0030183	2.07	0.038	.0003442	.0121756
female	.2596354	.1316948	1.97	0.049	.0015184	.5177524
pills	.3338829	.0849924	3.93	0.000	.1673008	.5004649
kcal	-.0002552	.0002092	-1.22	0.223	-.0006652	.0001548
fat	.0016595	.0033158	0.50	0.617	-.0048393	.0081583
fiber	.0397854	.0114408	3.48	0.001	.0173619	.0622089
alcohol	.0102508	.0088814	1.15	0.248	-.0071565	.027658
chol	-.0004819	.0004566	-1.06	0.291	-.0013769	.000413
betad	.000042	.0000305	1.38	0.169	-.0000179	.0001019
_cons	4.019137	.2835851	14.17	0.000	3.46332	4.574953

```
. predict double logbetaphat if e(sample), mu
(1 missing value generated)

. predict double reslognormal if e(sample), pearson
(1 missing value generated)

. histogram reslognormal
(bin=17, start=-2.4730175, width=.26211838)
```

This model shows much better results for the residuals in terms of achieving normality with constant variance. Figure 5.6 shows a histogram of the Pearson residuals from the model. The histogram is much more symmetric than that obtained from earlier models.

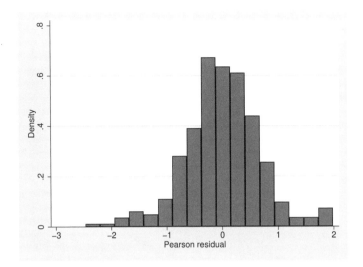

Figure 5.6. Pearson residuals from lognormal model (log-transformed outcome, two outliers removed, and zero outcome removed)

The standard residual plot of the Pearson residuals versus the fitted values illustrated in figure 5.7 shows the classic band of random values with no discernible features.

```
. twoway scatter reslognormal logbetaphat, yline(0)
```

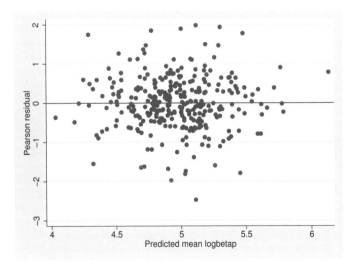

Figure 5.7. Pearson residuals versus fitted values from lognormal model (log-transformed outcome, two outliers removed, and zero outcome removed)

Figure 5.8 shows a better agreement of the percentiles (though the percentiles are not randomly around the reference line) from the normal distribution versus ordered Pearson residuals, whereas figure 5.9 shows no discernible relationship of the Pearson residuals with the average daily kilocaloric intake.

```
. sort reslognormal
. generate nscorelb = invnormal(_n/315)
. twoway scatter nscorelb reslognormal || line reslognormal reslognormal
. twoway scatter reslognormal kcal, yline(0)
```

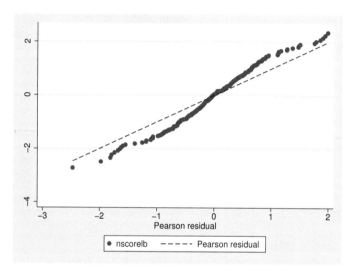

Figure 5.8. Normal scores versus sorted Pearson residuals obtained from lognormal model (log-transformed outcome, two outliers removed, and zero outcome removed)

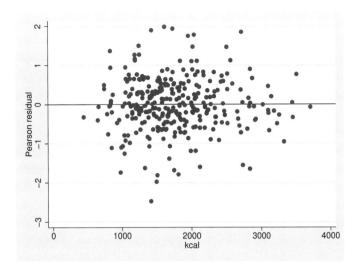

Figure 5.9. Pearson residuals versus kilocalories; Pearson residuals obtained from log-normal model (log-transformed outcome, two outliers removed, and zero outcome removed)

6 The gamma family

Contents

The gamma model is used for modeling outcomes for which the response can take only values greater than or equal to 0. Used primarily with continuous-response data, the GLM gamma family can be used with count data where there are many different count results that in total take the shape of a gamma distribution.

Ideally, the gamma model is best used with positive responses having a constant coefficient of variation. However, the model is robust to wide deviations from the latter criterion. Because the shape of the two-parameter gamma distribution is flexible and can be parameterized to fit many response shapes, it may be preferred over the Gaussian model for many strictly positive response data situations.

The traditional GLM framework limits models to one parameter: that of the mean (μ). Scale parameters are constrained to the value of 1 when they are part of either the variance or link function, or the scale parameter may be assigned a value by the user. However, in the latter case, the model is no longer a straightforward GLM but rather a quasilikelihood model. A prime example of the latter case is the traditional GLM version of the negative binomial model. Examples of the former case include the Gaussian, gamma, and inverse Gaussian models.

Although the traditional iteratively reweighted least-squares (IRLS) GLM algorithm does not formally estimate the scale or ancillary parameter as would an FIML algorithm, the scale may be estimated from the IRLS Pearson-based dispersion or scale statistic. That is, the value for the scale parameter in the gamma regression model using FIML is nearly the same as that for the IRLS Pearson-based estimate.

We demonstrate various GLM-type parameterizations for the gamma model in this chapter. We will not, however, discuss the three-parameter model or give substantial discussion to gamma models that incorporate censoring. The GLM gamma model can be used to directly model exponential regression, but we relegate those types of discussions to the final section in this chapter.

6.1   Derivation of the gamma model

The base density function for the gamma distribution is

$$f(y; \mu, \phi) = \frac{1}{y\Gamma(1/\phi)} \left(\frac{y}{\mu\phi}\right)^{1/\phi} \exp\left(-\frac{y}{\mu\phi}\right) \tag{6.1}$$

In exponential-family form, the above probability density appears as

$$f(y; \mu, \phi) = \exp\left\{\frac{y/\mu - (-\ln \mu)}{-\phi} + \frac{1-\phi}{\phi}\ln y - \frac{\ln \phi}{\phi} - \ln \Gamma\left(\frac{1}{\phi}\right)\right\} \tag{6.2}$$

This equation provides us with the link and the cumulant given by

$$\theta = 1/\mu \tag{6.3}$$

$$b(\theta) = -\ln(\mu) \tag{6.4}$$

From this, we may derive the mean and the variance

$$b'(\theta) = \frac{\partial b}{\partial \mu}\frac{\partial \mu}{\partial \theta} \tag{6.5}$$

$$= \left(-\frac{1}{\mu}\right)(-\mu^2) \tag{6.6}$$

$$= \mu \tag{6.7}$$

$$b''(\theta) = \frac{\partial^2 b}{\partial \mu^2}\left(\frac{\partial \mu}{\partial \theta}\right) + \frac{\partial b}{\partial \mu}\frac{\partial^2 \mu}{\partial \theta^2} \tag{6.8}$$

$$= (1)(-\mu^2) \tag{6.9}$$

$$= -\mu^2 \tag{6.10}$$

The variance here is an ingredient of the variance of y, which is found using $b''(\theta)a(\phi) = -\mu^2(-\phi) = \phi\mu^2$.

From the density function in exponential-family form, we derive the log-likelihood ($\mathcal{L}$) function by dropping the exponential and its associated braces from (6.2). The deviance function may be calculated from $\mathcal{L}$:

$$D = 2\phi\left\{\mathcal{L}(y; y) - \mathcal{L}(\mu; y)\right\} \tag{6.11}$$

$$= 2\phi\sum_{i=1}^{n}\frac{1 + \ln(y_i) - y_i/\mu_i - \ln(\mu_i)}{-\phi} \tag{6.12}$$

$$= 2\sum_{i=1}^{n}\left\{\frac{y_i - \mu_i}{\mu_i} - \ln\left(\frac{y_i}{\mu_i}\right)\right\} \tag{6.13}$$

The other important statistic for the estimating algorithm is the derivative of the gamma link function or g':

$$\frac{\partial g(\mu)}{\partial \mu} = \frac{\partial \theta}{\partial \mu} = -\frac{1}{\mu^2} \tag{6.14}$$

θ is usually parameterized as $-1/\mu$ or as $-y/\mu$. We have placed the negation outside the link to be associated with the scale so that there is a direct relationship between θ and the canonical reciprocal link.

The important statistics that distinguish one GLM from another, in canonical form, are listed below.

$$\text{link} : \qquad \frac{1}{\mu} \tag{6.15}$$

$$\text{inverse link} : \qquad \frac{1}{\eta} \tag{6.16}$$

$$\text{variance} : \qquad \mu^2 \tag{6.17}$$

$$g' : \qquad -\frac{1}{\mu^2} \tag{6.18}$$

$$\text{deviance} : \quad 2 \sum_{i=1}^{n} \left\{ \frac{y_i - \mu_i}{\mu_i} - \ln\left(\frac{y_i}{\mu_i}\right) \right\} \tag{6.19}$$

Inserting these functions into the canonical GLM algorithm yields

Listing 6.1. IRLS algorithm for gamma models

```
1    μ = {y + mean(y)}/2
2    η = 1/μ
3    WHILE (abs(ΔDev) > tolerance) {
4        W = μ²
5        z = η + (y − μ)/μ²
6        β = (XᵀWX)⁻¹XᵀWz
7        η = Xβ
8        μ = 1/η
9        OldDev = Dev
10       Dev = 2Σ{(y − μ)/μ − ln(y/μ)}
11       ΔDev = Dev − OldDev
12   }
13   χ² = Σ(y − μ)²/μ²
```

An OIM-based ML algorithm will give identical answers to those produced by the standard IRLS algorithm because the OIM of the former reduces to the expected information matrix (EIM) of the latter. We have previously mentioned that only for the noncanonical link does this make a difference. We have also shown how one can amend the IRLS algorithm to allow for use of the OIM. Stata's `glm` command uses a default Newton–Raphson algorithm where standard errors are based on the OIM. This is a modification of the standard EIM-based IRLS algorithm and is similar to the default hybrid algorithm used by SAS software. Refer to chapter 3 for details.

6.2   Example: Reciprocal link

We now present a rather famous example of a reciprocal-linked gamma dataset. The example first gained notoriety in McCullagh and Nelder (1989) and later was given a full examination in Hilbe and Turlach (1995). The example deals with car insurance claims (`claims.dta`) and models average claims for damage to an owner's car on the basis of the policy holder's age group (PA 1–8), the vehicle age group (VA 1–4), and the car group (CG 1–4). A frequency weight is given, called `number`, which represents the number of identical covariate patterns related to a particular outcome.

The criterion of coefficient of variation constancy across cell groups is assumed and has been validated in previous studies. Again, the gamma model is robust to deviation from this criterion, but not so much that it should not be assessed. This is particularly the case with the canonical link model.

A schema of the model given the reciprocal link is

$$\eta = (\beta_0 + \beta_1 \text{PA} + \beta_2 \text{CG} + \beta_3 \text{VA})^{-1} \tag{6.20}$$

The main-effects model is displayed below. The levels for the age group, vehicle age group, and car group are included in the model through automated production of indicator variables by using Stata's `glm` command with factor variables.

```
. use http://www.stata-press.com/data/hh4/claims

. glm y i.pa i.cg i.va [fw=number], family(gamma) irls

Iteration 1:    deviance =   135.5555
Iteration 2:    deviance =   124.8845
Iteration 3:    deviance =   124.7828
Iteration 4:    deviance =   124.7828
Iteration 5:    deviance =   124.7828
```

Generalized linear models		No. of obs	=	8,942
Optimization	: MQL Fisher scoring	Residual df	=	8,928
	(IRLS EIM)	Scale parameter	=	1
Deviance	= 124.7827519	(1/df) Deviance	=	.0139766
Pearson	= 131.7861885	(1/df) Pearson	=	.014761
Variance function: V(u) = u^2		[Gamma]		
Link function : g(u) = 1/u		[Reciprocal]		
		BIC	=	-81106.76

y	Coef.	EIM Std. Err.	z	P>|z|	[95% Conf. Interval]
pa					
2	.0001014	.0000482	2.10	0.035	6.91e-06 .0001959
3	.00035	.0000456	7.68	0.000	.0002607 .0004393
4	.0004623	.0000454	10.19	0.000	.0003734 .0005512
5	.00137	.0000463	29.58	0.000	.0012792 .0014608
6	.0009695	.0000447	21.68	0.000	.0008819 .0010571
7	.0009164	.0000451	20.33	0.000	.0008281 .0010048
8	.0009201	.0000459	20.03	0.000	.00083 .0010101
cg					
2	.0000377	.0000186	2.02	0.043	1.12e-06 .0000742
3	-.0006139	.0000188	-32.68	0.000	-.0006507 -.0005771
4	-.0014206	.00002	-71.20	0.000	-.0014597 -.0013815
va					
2	.0003663	.0000111	32.87	0.000	.0003445 .0003881
3	.0016512	.0000251	65.90	0.000	.0016021 .0017003
4	.0041537	.0000489	84.98	0.000	.0040579 .0042495
_cons	.0034105	.0000462	73.85	0.000	.00332 .003501

Using `predict`, we can generate both the Anscombe residuals and the expected value of the model. Using the predicted value, we can generate the log of the variance and illustrate a useful residual plot in figure 6.1.

```
. predict double anscombe, anscombe
(5 missing values generated)

. predict double logvar, mu

. replace logvar = log(logvar*logvar)
(128 real changes made)

. label var logvar "Log(variance)"
```

```
. twoway scatter anscombe logvar, yline(0)
```

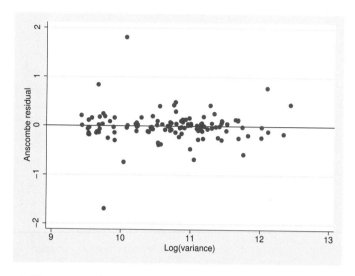

Figure 6.1. Anscombe residuals versus log (variance)

A check (figure 6.1) of the Anscombe residuals versus the log of the variance for the gamma family indicates no problems.

The associated results for the main-effects model may be interpreted as the rate at which each pound sterling (the outcome) is paid for an average claim over a unit of time, which is unspecified in the data; see McCullagh and Nelder (1989, 297). Because the algorithm uses the IRLS method of estimation, the scale is shown as 1. However, as we have seen before, the value of the scale is point-estimated as the Pearson dispersion, or, in this instance, 0.015.

The interactions discussed in the above referenced sources may also be evaluated by using the Stata command

```
. glm y i.pa i.cg i.va i.pa#i.cg i.pa#i.va i.cg#i.va [fw=number],
> family(gamma) irls
```

Both of our reference sources found that the extra interactions in the model were nonproductive, adding no more predictive value to the model. The manner in which the authors showed this to be the case underlies appropriate modeling techniques and warrants discussion here.

Typically, we assess the worth of a predictor, or group of predictors, on the basis of the p-value from a Wald or likelihood-ratio test statistic. However, particularly for large-data situations, many or most of the predictors appear statistically significant. The goal, of course, is to discover the most parsimonious model possible that still possesses maximal explanatory value. One way to organize model construction or selection is through the creation of a deviance table, described by the references cited above, which aids in achieving parsimony while differentiating explanatory power for candidate models.

The method is as follows. The deviances from successive additions of the main-effect term or interaction term are displayed in a column, with values of the difference between deviances, degrees of freedom, and mean differences placed in corresponding columns to the right. Of course, the order the predictors enter the model does matter and manipulation may be required.

The following table for `claims.dta` is reproduced here in part from McCullagh and Nelder (1989) and Hilbe and Turlach (1995):

Model	Deviance	Difference	DF	Mean Difference
Constant	649.9			
Main effects				
PA	567.7	82.2	7	11.7
PA CG	339.4	228.3	3	76.1
PA CG VA	124.8	214.7	3	71.6
Interaction terms				
+PA*CG	90.7	34.0	21	1.62
+PA*VA	71.0	19.7	21	0.94
+CG*VA	65.6	5.4	9	1.13

We have shown the output for the main-effects model above. The header portion for the results from the full interaction model appears as

```
Generalized linear models                      No. of obs      =      8,942
Optimization      : MQL Fisher scoring         Residual df     =      8,877
                    (IRLS EIM)                  Scale param     =          1
Deviance          =  65.58462489               (1/df) Deviance =    .0073882
Pearson           =  64.19830856               (1/df) Pearson  =     .007232
```

The header information is consistent with the deviance table.

Each additional main-effect term results in a substantial difference in deviance values. This is especially the case when adjusted for degrees of freedom. The addition of interaction terms results in a markedly reduced value of the mean differences between hierarchical levels. For this reason, we can accept the main-effects model as effectively representing a parsimonious and sufficiently explanatory model. Interactions are not needed.

6.3 ML estimation

For ML estimation of a model, the log-likelihood function (rather than the deviance) plays a paramount role in estimation. It is the log-likelihood function that is maximized, hence the term "maximum likelihood".

We can illustrate the log-likelihood function for a particular link function (parameterization) by substituting the inverse link function $g^{-1}(\eta)$ for μ. For example, the gamma canonical link (reciprocal function) is

$$\mu = \frac{1}{\mathbf{x}\boldsymbol{\beta}} \tag{6.21}$$

For each instance of a μ in the gamma $\mathcal{L}$ function, we replace μ with the inverse link of the linear predictor $1/\mathbf{x}\boldsymbol{\beta}$.

$$\mathcal{L} = \sum_{i=1}^{n} \left\{ \frac{y_i/\mu_i - (-\ln \mu_i)}{-\phi} + \frac{1-\phi}{\phi} \ln y_i - \frac{\ln \phi}{\phi} - \ln \Gamma \left(\frac{1}{\phi} \right) \right\} \tag{6.22}$$

$$= \sum_{i=1}^{n} \left\{ \frac{y_i \mathbf{x}_i \boldsymbol{\beta} - \ln(\mathbf{x}_i \boldsymbol{\beta})}{-\phi} + \frac{1-\phi}{\phi} \ln y_i - \frac{\ln \phi}{\phi} - \ln \Gamma \left(\frac{1}{\phi} \right) \right\} \tag{6.23}$$

In the previous section, we presented an example related to car insurance claims. We fit the model using the traditional IRLS algorithm. The ML approach results in the same estimates and standard errors when the canonical link is involved. This was the case in the discussion of the Gaussian family of models. We also saw how the IRLS approach accurately estimated the scale parameter, even though the IRLS algorithm assumed a value of 1. In the main-effects model for `claims.dta`, the scale was estimated as 0.014761. To show the scale when modeled using the default modified Newton–Raphson algorithm for LIML, we have the following (header) results:

```
. glm y i.pa i.cg i.va [fw=number], family(gamma) notable nolog
Generalized linear models                    No. of obs      =       8,942
Optimization        : ML                     Residual df     =       8,928
                                             Scale parameter =      .014761
Deviance         =   124.7827519             (1/df) Deviance =     .0139766
Pearson          =   131.7861915             (1/df) Pearson  =      .014761

Variance function: V(u) = u^2                [Gamma]
Link function    : g(u) = 1/u                [Reciprocal]

                                             AIC             =     12.91782
Log likelihood   = -57741.59316              BIC             =    -81106.76
```

6.4   Log-gamma models

We mentioned before that the reciprocal link estimates the rate per unit of the model response, given a specific set of explanatory variables or predictors. The log-linked gamma represents the log-rate of the response. This model specification is identical to exponential regression. Such a specification, of course, estimates data with a negative exponential decline. However, unlike the exponential models found in survival analysis, we cannot use the log-gamma model with censored data. We see, though, that uncensored exponential models can be fit with GLM specifications. We leave that to the end of this chapter.

The log-gamma model, like its reciprocal counterpart, is used with data in which the response is greater than 0. Examples can be found in nearly every discipline. For instance, in health analysis, length of stay (LOS) can generally be estimated using log-gamma regression because stays are always constrained to be positive. LOS data are generally estimated using Poisson or negative binomial regression because the elements of LOS are discrete. However, when there are many LOS elements—that is, many different LOS values—many researchers find the gamma or inverse Gaussian models to be acceptable and preferable.

Before GLM, data that are now estimated using log-gamma techniques were generally estimated using Gaussian regression with a log-transformed response. Although the results are usually similar between the two methods, the log-gamma technique, which requires no external transformation, is easier to interpret and comes with a set of residuals with which to evaluate the worth of the model. Hence, the log-gamma technique is finding increased use among researchers who once used Gaussian techniques.

The IRLS algorithm for the log-gamma model is the same as that for the canonical-link model except that the link and inverse link become $\ln(\mu)$ and $\exp(\eta)$, respectively, and the derivative of g is now $1/\mu$. The ease with which we can change between models is one of the marked beauties of GLM. However, because the log link is not canonical, the IRLS and modified ML algorithms will give different standard errors. But, except in extreme cases, differences in standard errors are usually minimal. Except perhaps when working with small datasets, the method of estimation used generally makes little inferential difference.

Let's use an example to explain. We use the cancer dataset that comes packaged with Stata, `cancer.dta`. The variable `age` is a continuous predictor and `drug` has three levels. The modern approach in Stata to generate an indicator variable is to use the `i.` prefix on factor variables. The indicator variables created in this method are referred to as virtual variables, because they are not added to the dataset. The IRLS model is fit as

```
. sysuse cancer, clear
(Patient Survival in Drug Trial)

. glm studytime age i.drug, family(gamma) link(log) irls nolog
Generalized linear models                          No. of obs      =        48
Optimization      : MQL Fisher scoring             Residual df     =        44
                    (IRLS EIM)                     Scale parameter =         1
Deviance          =  16.17463555                   (1/df) Deviance =  .3676054
Pearson           =  13.99281815                   (1/df) Pearson  =  .3180186

Variance function: V(u) = u^2                      [Gamma]
Link function    : g(u) = ln(u)                    [Log]

                                                   BIC             = -154.1582
```

studytime	Coef.	EIM Std. Err.	z	P>|z|	[95% Conf. Interval]	
age	-.0447763	.014732	-3.04	0.002	-.0736505	-.0159021
drug						
2	.5743482	.1969367	2.92	0.004	.1883594	.9603371
3	1.052072	.1977145	5.32	0.000	.6645583	1.439585
_cons	4.645976	.8353019	5.56	0.000	3.008814	6.283138

Estimation using the default modified ML algorithm displays the following table:

```
. glm studytime age i.drug, family(gamma) link(log) nolog
Generalized linear models                          No. of obs      =        48
Optimization      : ML                             Residual df     =        44
                                                   Scale parameter =  .3180529
Deviance          =  16.17463553                   (1/df) Deviance =  .3676054
Pearson           =  13.99432897                   (1/df) Pearson  =  .3180529

Variance function: V(u) = u^2                      [Gamma]
Link function    : g(u) = ln(u)                    [Log]

                                                   AIC             =  7.403608
Log likelihood    = -173.6866032                   BIC             = -154.1582
```

studytime	Coef.	OIM Std. Err.	z	P>|z|	[95% Conf. Interval]	
age	-.0447789	.015112	-2.96	0.003	-.0743979	-.01516
drug						
2	.5743689	.1986342	2.89	0.004	.185053	.9636847
3	1.0521	.1965822	5.35	0.000	.6668056	1.437394
_cons	4.646108	.8440093	5.50	0.000	2.99188	6.300336

The standard errors are slightly different, as was expected. A scale parameter is estimated and is provided in the output header. That the parameter estimates differ as well may seem strange; they differ only because of different optimization criteria. If the tolerance is altered in either model, the parameter estimates take slightly different values. If the starting values are different, as they are here, then estimates may vary slightly. Because the relative differences in parameter estimates are typically very small, the statistical information remains unchanged.

For a matter of comparison, we can model the same data by using a Stata two-parameter log-gamma program called `lgamma` from Hilbe (2000). This command is available by typing `net install sg126, from(http://www.stata.com/stb/stb53)`. `lgamma` uses a full ML routine for estimation, with no ties to GLM and no IRLS method. `lgamma` treats the additional parameter in an equivalent manner as the regression coefficient parameters. Thus, this command is an FIML command for fitting the log-gamma model.

Previously, we generated "virtual" indicator variables. To use the `lgamma` command, we must generate indicator variables in the dataset. We use the `tabulate` command to generate these indicator variables.

```
. tabulate drug, generate(drug)
```

Drug type (1=placebo)	Freq.	Percent	Cum.
1	20	41.67	41.67
2	14	29.17	70.83
3	14	29.17	100.00
Total	48	100.00	

```
. lgamma studytime age drug2 drug3, nolog
```

Log-gamma model

Number of obs = 48
LR chi2(3) = 28.00

Log likelihood = -160.91053

Prob > chi2 = 0.0000

studytime	Coef.	Std. Err.	z	P>|z|	[95% Conf. Interval]	
age	-.0447789	.0151597	-2.95	0.003	-.0744914	-.0150665
drug2	.5743689	.1992613	2.88	0.004	.183824	.9649138
drug3	1.0521	.1972027	5.34	0.000	.6655893	1.43861
_cons	4.646108	.8466737	5.49	0.000	2.986658	6.305558
/ln_phi	-1.139231	.1942152	-5.87	0.000	-1.519886	-.7585759
phi	.3200651	.0621615			.2187369	.4683329

Here we see that the parameter estimates are identical to those produced using GLM with the modified ML algorithm. Standard errors are also similar to those of the GLM model. The scale estimate is the same to within 0.002 of the previous estimate. Again, different starting values or different tolerances can account for slight differences in output. These differences typically have no bearing on statistical accuracy or on statistical interpretation.

The log-likelihood function for the log-gamma model, parameterized in terms of $\mathbf{x}\boldsymbol{\beta}$, is

$$\mathcal{L} = \sum_{i=1}^{n} \left\{ \frac{y_i / \exp(\mathbf{x}_i\boldsymbol{\beta}) + \mathbf{x}_i\boldsymbol{\beta}}{-\phi} + \frac{\phi + 1}{\phi} \ln y_i - \frac{\ln \phi}{\phi} - \ln \Gamma \left(\frac{1}{\phi} \right) \right\} \tag{6.24}$$

Above, we compared the estimated parameters and standard errors for the various estimation methods. A rather striking difference is that the GLM model fit with the modified ML algorithm produced a log-likelihood value of -173.69, whereas the FIML `lgamma` command produced a value of -160.91. Why are the values so different? The answer lies in how the GLM model alters the deviance-based optimization into one for the log likelihood. The deviance functions used in the GLM algorithms do not include normalization factors or scaling factors, so these terms are lost to calculations for the (limited information) log-likelihood values.

Here the scaled log likelihood is $-173.6866032/0.3180529 = -546.09344$. The normalization term $\{(\phi+1)/\phi\} \ln y_i - \ln \phi/\phi - \ln \Gamma(1/\phi)$ would be 385.18238 such that the comparable log likelihood is $-546.09344 + 385.18238 = -160.91106$.

6.5   Identity-gamma models

The identity link, with respect to the gamma family, models duration data. It assumes that there is a one-to-one relationship between μ and η. In this relationship, it is similar to the canonical-link Gaussian model. However, here it is a noncanonical link.

Choosing between different links of the same model family may sometimes be hard. McCullagh and Nelder (1989) support the method of minimal deviances. They also check the residuals to observe closeness of fit and the normality and independence of the standardized residuals themselves. In this example, we have two noncanonical links that we are interested in comparing: log and identity links. All other factors the same, choosing the model that has the least value for the deviance is preferable. Here it is the identity-linked model. We show both the IRLS and ML outputs below. The log-gamma model has a deviance of 16.17, whereas the identity-gamma model has a deviance of 15.09—a reduction of 1.08 or 6.68%.

We can use other statistical tests to assess the worth of differential models among links. These tests include the BIC and AIC. Models having lower values of these criteria are preferable. With respect to BIC, if the absolute difference between the BIC for two models is less than 2, then there is only weak evidence that the model with the smaller BIC is preferable. Absolute differences between 2 and 6 give positive support for the model with the smaller BIC, whereas absolute differences between 6 and 10 offer strong support. Absolute differences more than 10 are strong support for the model with the smaller BIC.

```
. glm studytime age i.drug, family(gamma) link(identity) irls nolog  /* IRLS */
Generalized linear models                        No. of obs       =        48
Optimization      : MQL Fisher scoring           Residual df      =        44
                   (IRLS EIM)                     Scale parameter  =         1
Deviance          =   15.09376233                (1/df) Deviance  =  .3430401
Pearson           =   12.98530838                (1/df) Pearson   =  .2951206

Variance function: V(u) = u^2                    [Gamma]
Link function     : g(u) = u                     [Identity]

                                                 BIC              = -155.2391
```

| | | EIM | | | | |
studytime	Coef.	Std. Err.	z	P>|z|	[95% Conf.	Interval]
age	-.5369706	.1334085	-4.03	0.000	-.7984464	-.2754948
drug						
2	6.472958	2.134955	3.03	0.002	2.288522	10.65739
3	16.20353	3.874013	4.18	0.000	8.610607	23.79646
_cons	38.96542	8.19225	4.76	0.000	22.9089	55.02193

```
. glm studytime age i.drug, family(gamma) link(identity) nolog         /* N-R */
Generalized linear models                        No. of obs       =        48
Optimization      : ML                           Residual df      =        44
                                                 Scale parameter  =  .2951209
Deviance          =   15.09376233                (1/df) Deviance  =  .3430401
Pearson           =   12.98531951                (1/df) Pearson   =  .2951209

Variance function: V(u) = u^2                    [Gamma]
Link function     : g(u) = u                     [Identity]

                                                 AIC              =   7.38109
Log likelihood    = -173.1461666                 BIC              = -155.2391
```

| | | OIM | | | | |
studytime	Coef.	Std. Err.	z	P>|z|	[95% Conf.	Interval]
age	-.536976	.1256999	-4.27	0.000	-.7833433	-.2906087
drug						
2	6.47296	2.131972	3.04	0.002	2.294372	10.65155
3	16.20346	3.804618	4.26	0.000	8.746548	23.66038
_cons	38.96574	7.747343	5.03	0.000	23.78123	54.15025

6.6 Using the gamma model for survival analysis

We mentioned earlier that exponential regression may be modeled using a log-linked gamma regression. Using the default ML version of glm provides the necessary means by which the observed information matrix (OIM) is used to calculate standard errors. This is the same method used by typical ML implementations of exponential regression. See the documentation for Stata's streg command in *Stata Survival Analysis Reference Manual* for a summary of the exponential regression model.

Why are the exponential regression results the same as the log-gamma model results? The similarity can be seen in the likelihood functions. The exponential probability distribution has the form

$$f\left(\frac{y}{\mu}\right) = \frac{1}{\mu}\exp\left(-\frac{y}{\mu}\right) \tag{6.25}$$

where μ is parameterized as

$$\mu = \exp(\mathbf{x}\boldsymbol{\beta}) \tag{6.26}$$

such that the function appears as

$$f\left(\frac{y}{\mathbf{x}\boldsymbol{\beta}}\right) = \frac{1}{\exp(\mathbf{x}\boldsymbol{\beta})}\exp\left\{-\frac{y}{\exp(\mathbf{x}\boldsymbol{\beta})}\right\} \tag{6.27}$$

The exponential log likelihood is thus

$$\mathcal{L}(\boldsymbol{\beta}; y) = \sum_{i=1}^{n}\left\{-\mathbf{x}_i\boldsymbol{\beta} - \frac{y_i}{\exp(\mathbf{x}_i\boldsymbol{\beta})}\right\} \tag{6.28}$$

Compare this with the log-gamma log likelihood given in (6.24). Drop references to the scale parameter because the exponential distribution assumes a scale of 1. We see that the log-gamma log likelihood in that form is identical to that of the exponential log likelihood, leading to the similarity in output. However, the log link is noncanonical. To obtain the same standard errors between the two models, we must fit the log gamma by using the modified Newton–Raphson method and explicitly specify a scale of one. This is the default parameterization when using Stata, but it may not be the default in other software packages.

Below we display output for an exponential model first using GLM and then using a standard exponential regression model.

```
. glm studytime age i.drug, family(gamma) link(log) scale(1) nolog
Generalized linear models                        No. of obs       =         48
Optimization      : ML                           Residual df      =         44
                                                 Scale parameter =          1
Deviance        =   16.17463553                  (1/df) Deviance =    .3676054
Pearson         =   13.99432897                  (1/df) Pearson  =    .3180529

Variance function: V(u) = u^2                    [Gamma]
Link function     : g(u) = ln(u)                 [Log]

                                                 AIC              =   7.403608
Log likelihood    = -173.6866032                 BIC              =  -154.1582
```

studytime	Coef.	OIM Std. Err.	z	P>|z|	[95% Conf.	Interval]
age	-.0447789	.0267961	-1.67	0.095	-.0972983	.0077404
drug						
2	.5743689	.3522116	1.63	0.103	-.1159532	1.264691
3	1.0521	.348573	3.02	0.003	.368909	1.73529
_cons	4.646108	1.49657	3.10	0.002	1.712886	7.57933

```
(Standard errors scaled using dispersion equal to square root of 1.)
. stset studytime

    failure event:  (assumed to fail at time=studytime)
obs. time interval:  (0, studytime]
exit on or before:  failure
```

```
     48  total observations
      0  exclusions
```

```
     48  observations remaining, representing
     48  failures in single-record/single-failure data
    744  total analysis time at risk and under observation
                                        at risk from t =           0
                               earliest observed entry t =           0
                                   last observed exit t =          39
```

```
. streg age i.drug, nohr distribution(exponential) nolog
        failure _d:  1 (meaning all fail)
   analysis time _t:  studytime

Exponential PH regression
No. of subjects =            48              Number of obs   =          48
No. of failures =            48
Time at risk    =           744
                                             LR chi2(3)      =       11.75
Log likelihood  =   -56.087318              Prob > chi2     =      0.0083
```

_t	Coef.	Std. Err.	z	P>|z|	[95% Conf. Interval]	
age	.0447789	.0267961	1.67	0.095	-.0077404	.0972983
drug						
2	-.5743689	.3522116	-1.63	0.103	-1.264691	.1159532
3	-1.0521	.348573	-3.02	0.003	-1.73529	-.368909
_cons	-4.646108	1.49657	-3.10	0.002	-7.57933	-1.712886

Likewise, the regression parameters are the same because the GLM algorithms are limited information and have the same partial derivative as exponential regression—the two models have the same estimating equation for the regression parameters; they differ only in sign. The log-likelihood values differ because the models contain different normalization and scaling terms.

The gamma model (when fit with GLM specification) cannot incorporate censoring into its routine. However, a right-censored exponential regression model may be duplicated with GLM, as we show in section 12.5.

Gamma regression, in all its parameterizations and forms, has enjoyed little use when compared with the standard OLS or Gaussian model. Researchers often use OLS regression when they should use one of the gamma models. Certainly, one of the historic reasons for this was the lack of supporting software. There has also been a corresponding lack of understanding of the gamma model and all its variations.

The examples illustrated using the cancer data ignored the censoring information that was included in the original study. For comparison purposes in this chapter, we treated the data as if none of the observations were censored.

7 The inverse Gaussian family

Contents

The inverse Gaussian is the most rarely used model of all traditional GLMs. Primary sources nearly always list the inverse Gaussian in tables of GLM families, but it is rarely discussed. Even McCullagh and Nelder (1989) give only passing comment of its existence.

The inverse Gaussian distribution does have a history of use in reliability studies. Typically, however, these are limited to models having only one predictor, much like the traditional nonlinear regression model. We endeavor to add more information to highlight the practical applications of the inverse Gaussian model.

7.1   Derivation of the inverse Gaussian model

The inverse Gaussian probability distribution is a continuous distribution having two parameters given by

$$f(y; \mu, \sigma^2) = \frac{1}{\sqrt{2\pi y^3 \sigma^2}} \exp\left\{-\frac{(y-\mu)^2}{2(\mu\sigma)^2 y}\right\} \tag{7.1}$$

In exponential form, the inverse Gaussian distribution is given by

$$f(y; \mu, \sigma^2) = \exp\left\{-\frac{(y-\mu)^2}{2y(\mu\sigma)^2} - \frac{1}{2}\ln\left(2\pi y^3 \sigma^2\right)\right\} \tag{7.2}$$

$$= \exp\left\{\frac{y/\mu^2 - 2/\mu}{-2\sigma^2} + \frac{1/y}{-2\sigma^2} + \frac{\sigma^2}{-2\sigma^2}\ln\left(2\pi y^3 \sigma^2\right)\right\} \tag{7.3}$$

The log-likelihood function may be written in exponential-family form by dropping the exponential and its associated braces.

$$\mathcal{L} = \sum_{i=1}^{n} \left\{ \frac{y_i/(2\mu_i^2) - 1/\mu_i}{-\sigma^2} + \frac{1}{-2y_i\sigma^2} - \frac{1}{2}\ln\left(2\pi y_i^3\sigma^2\right) \right\} \tag{7.4}$$

GLM theory provides that, in canonical form, the link and cumulant functions are

$$\theta = \frac{1}{2\mu^2} = \frac{1}{2}\mu^{-2} \tag{7.5}$$

$$b(\theta) = \frac{1}{\mu} \tag{7.6}$$

$$a(\phi) = -\sigma^2 \tag{7.7}$$

The sign and coefficient value are typically dropped from the inverse Gaussian link function when inserted into the GLM algorithm. It is normally given the value of $1/\mu^2$, and the inverse link function is normally given the value of $1/\sqrt{\eta}$.

The inverse Gaussian mean and variance functions are derived from the cumulant

$$b'(\theta) = \frac{\partial b}{\partial \mu} \frac{\partial \mu}{\partial \theta} = \left(\frac{-1}{\mu^2}\right)(-\mu^3) = \mu \tag{7.8}$$

$$b''(\theta) = \frac{\partial^2 b}{\partial \mu^2}\left(\frac{\partial \mu}{\partial \theta}\right)^2 + \frac{\partial b}{\partial \mu}\frac{\partial^2 \mu}{\partial \theta^2} \tag{7.9}$$

$$= \left(\frac{2}{\mu^3}\right)(\mu^6) + \left(\frac{-1}{\mu^2}\right)(3\mu^5) = 2\mu^3 - 3\mu^3 = -\mu^3 \tag{7.10}$$

The derivative of the link, $g(\mu)$, is written $g'(\mu)$ or simply g'. Its derivation is given by

$$g' = \frac{\partial \theta}{\partial \mu} = \frac{\partial}{\partial \mu}\frac{1}{2\mu^2} = -\mu^{-3} \tag{7.11}$$

Finally, the deviance function is calculated from the saturated and model log-likelihood formulas

$$D = 2\sigma^2\left\{\mathcal{L}(y,\sigma^2;y) - \mathcal{L}(\mu,\sigma^2;y)\right\} \tag{7.12}$$

$$= 2\sigma^2\sum_{i=1}^{n}\left\{\frac{y_i/(2y_i^2) - 1/y_i}{-\sigma^2} - \frac{y_i/(2\mu_i^2) - 1/\mu_i}{-\sigma^2}\right\} \tag{7.13}$$

$$= \sum_{i=1}^{n}\left(\frac{y_i}{\mu_i^2} - \frac{2}{\mu_i} + \frac{1}{y_i}\right) \tag{7.14}$$

$$= \sum_{i=1}^{n}\frac{(y_i - \mu_i)^2}{y_i\mu_i^2} \tag{7.15}$$

The iteratively reweighted least-squares (IRLS) GLM algorithm calculates the scale value, σ^2, as the Pearson-based dispersion statistic.

.2   Shape of the distribution

The inverse Gaussian probability density function is typically defined as

$$f(y; \mu, \phi) = \left(\frac{\phi}{2\pi y^3} \right)^{1/2} \exp \left\{ \frac{-\phi(y - \mu)^2}{2\mu^2 y} \right\} \tag{7.16}$$

Although analysts rarely use this density function when modeling data, there are times when it fits the data better than other continuous models. It is especially well suited to fit positive continuous data that have a concentration of low-valued data with a long right skew. This feature also becomes useful when mixed with the Poisson distribution to create the Poisson inverse Gaussian mixture model discussed later; see section 14.11.

To illustrate the shape of unadjusted inverse Gaussian density functions, we create a simple set of Stata commands to generate values of the probability density function for specified mean and scale parameters. Graphs of probability density functions for various parameter values demonstrate the flexibility.

We begin with a simple function for which $\mu = 5$ and $\phi = 2$ shown in figure 7.1.

```
. twoway function y = exp((-2.0*(x-5.0)^2)/(2*5.0^2*x)) * sqrt(2.0/(2*_pi*x^3)),
> range(1 10) title("Inverse Gaussian: mu=5.0; phi=2.0")
```

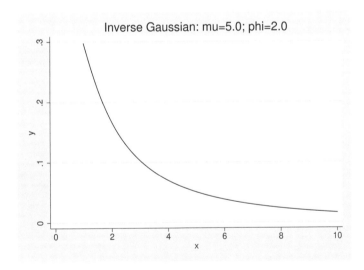

Figure 7.1. Inverse Gaussian ($\mu = 5$, $\phi = 2$)

Figure 7.2 doubles the value of the mean, which makes the initial drop in values that were steeper, but is characterized by a longer right skew. These features are associated with change in values on the vertical scale; note the relative values of the density in the plots.

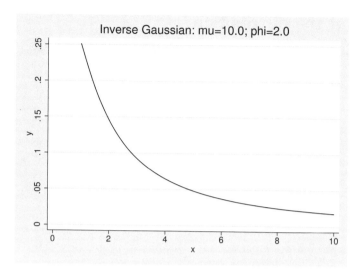

Figure 7.2. Inverse Gaussian ($\mu = 10$, $\phi = 2$)

In figure 7.3, we reduce the value of the scale parameter while keeping the same mean.

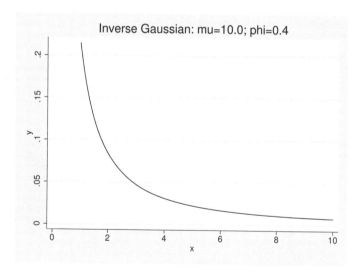

Figure 7.3. Inverse Gaussian ($\mu = 10$, $\phi = 0.4$)

We can see the trend. Next, we reduce the value of the mean to 3 and the scale parameter to 0.1. Remember that the inverse Gaussian values must be greater than 0.

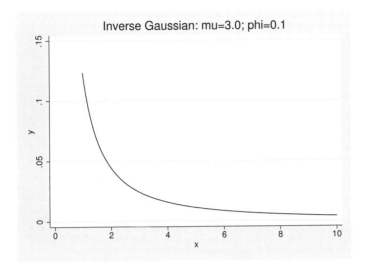

Figure 7.4. Inverse Gaussian ($\mu = 3$, $\phi = 0.1$)

In figure 7.5, we specify a mean of 2 and scale parameter of 8.

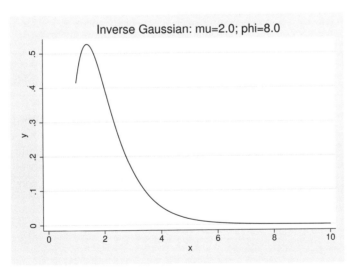

Figure 7.5. Inverse Gaussian ($\mu = 2$, $\phi = 8$)

Figure 7.6 demonstrates what happens when the scale is held constant and the mean is increased (from 2 to 10).

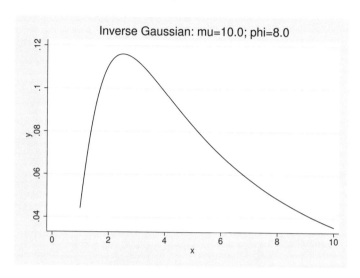

Figure 7.6. Inverse Gaussian ($\mu = 2$, $\phi = 10$)

Shifting the mean to the right (from 2 to 10) considerably changes the shape. Recall in figure 7.3 that with a mean of 10, smaller values of the scale yield a negative exponential shape. The hump shape occurs only when both mean and scale parameters are high.

If we have data to be modeled that have any of the above shapes after predictor adjustment, then canonical inverse Gaussian regression might be a viable model for the data.

Some analysts who use gamma and inverse Gaussian models on continuous data use the log link when the data are only positive. But we can observe that the canonical form is appropriate for positive-only data as well. The caveat is that the underlying data are assumed to take an inverse quadratic form, whereas the log-linked form parameterizes the data to be like count data without the restriction that the data consist of integers.

.3   The inverse Gaussian algorithm

Inserting the appropriate inverse Gaussian formulas into the standard GLM algorithm and excluding prior weights and subscripts, we have

Listing 7.1. IRLS algorithm for inverse Gaussian models

```
1      μ = {y − mean(y)}/2
2      η = 1/μ²
3      WHILE (abs(ΔDev) > tolerance) {
4          W = μ³
5          z = η + (y − μ)/μ³
6          β = (XᵀWX)⁻¹XᵀWz
7          η = Xβ
8          μ = 1/√η
9          OldDev = Dev
10         Dev = Σ(y − μ)²/(yμ²)
11         ΔDev = Dev - OldDev
12     }
13     χ² = σ Σ(y − μ)²/μ³
```

We have mentioned before that the IRLS algorithm may be amended such that the log-likelihood function is used as the basis of iteration rather than the deviance. However, when this is done, the log-likelihood function itself is usually amended. The normalization term is dropped such that terms with no instance of the mean parameter, μ, are excluded. Of course, prior weights are kept. For the inverse Gaussian, the amended log-likelihood function is usually given as

$$\mathcal{L}(\mu, \sigma^2; y) = \sum_{i=1}^{n} \frac{1}{\sigma^2} \left(-\frac{y_i}{2\mu_i^2} + \frac{1}{\mu_i} \right) \tag{7.17}$$

.4   Maximum likelihood algorithm

Maximum likelihood (ML) usually uses some modification of the Newton–Raphson algorithm. We have seen before that the traditional GLM IRLS algorithm may be modified in such a manner that results in identical values to those produced using standard ML methods. Essentially, the IRLS algorithm is modified to produce standard errors from the observed, rather than the expected, information matrix (see the discussion on this topic in chapter 3). Parameter estimates remain unchanged. Moreover, for the canonical link, the two methods mathematically coincide.

ML relies on the log-likelihood function. Using full ML methods instead of those that are amendments to the IRLS algorithm, parameter estimates and standard errors that are based on the observed information matrix are obtained as a direct result of maximization. The method requires use of the full log-likelihood function, which includes the scale parameter. And, in the process, the scale is estimated by ML techniques in the same manner as are other parameter estimates.

We show how the inverse Gaussian algorithm can be amended to simulate ML estimation when we address noncanonical links. We also provide more details about the full ML algorithm.

7.5   Example: The canonical inverse Gaussian

As mentioned earlier, the inverse Gaussian has found limited use in certain reliability studies. In particular, it is most appropriate when modeling a nonnegative response having a high initial peak, rapid drop, and long right tail. If a discrete response has many different values together with the same properties, then the inverse Gaussian may be appropriate for this type of data as well. A variety of other shapes can also be modeled using inverse Gaussian regression.

We can create a synthetic inverse Gaussian dataset using tools available in Stata. This will allow us to observe some of the properties and relationships we have discussed thus far.

Shown below, we create a dataset of 25,000 cases with two random half-normal predictors having coefficients of 0.5 and 0.25, respectively, and a constant of 1. We used Stata's `rigaussian()` function to generate outcomes with inverse Gaussian distributed conditional variances and specified mean. We also specified a shape parameter of the inverse Gaussian distribution of $1/0.25$. We set the random-number seed to allow recreation of these results.

```
. clear
. set seed 12345
. set obs 25000
number of observations (_N) was 0, now 25,000
. generate x1 = abs(rnormal())
. generate x2 = abs(rnormal())
. generate eta = 1 + .5*x1 + .25*x2
. generate mu = 1/sqrt(eta)
. generate xig = rigaussian(mu, 1/0.25)
```

```
. glm xig x1 x2, family(igaussian) nolog
Generalized linear models                     No. of obs        =      25,000
Optimization      : ML                        Residual df       =      24,997
                                              Scale parameter   =     .248692
Deviance        =   6137.219327               (1/df) Deviance   =    .2455182
Pearson         =   6216.553225               (1/df) Pearson    =     .248692

Variance function: V(u) = u^3                 [Inverse Gaussian]
Link function    : g(u) = 1/(u^2)             [Power(-2)]

                                              AIC               =    1.130265
Log likelihood  = -14125.31121                BIC               =   -246998.2
```

xig	Coef.	OIM Std. Err.	z	P>|z|	[95% Conf. Interval]	
x1	.4829143	.0161919	29.82	0.000	.4511787	.5146499
x2	.2398502	.0154139	15.56	0.000	.2096394	.2700609
_cons	1.028708	.017605	58.43	0.000	.9942032	1.063214

Note that the inverse of the shape parameter is approximately equal to the Pearson statistic.

.6 Noncanonical links

The log and identity links are the primary noncanonical links associated with the inverse Gaussian distribution. This is similar to the gamma model.

We usually do not refer to a link function with full ML. The log-likelihood function is simply adjusted such that the inverse link, parameterized in terms of $\mathbf{x}\boldsymbol{\beta}$, is substituted for each instance of μ in the log-likelihood function. Hence, for the log-inverse Gaussian, we have

$$\mathcal{L} = \sum_{i=1}^{n} \left[\frac{y_i/\{2\exp(\mathbf{x}_i\boldsymbol{\beta})^2\} - 1/\exp(\mathbf{x}_i\boldsymbol{\beta})}{-\sigma^2} + \frac{1}{-2y_i\sigma^2} - \frac{1}{2}\ln\left(2\pi y_i^3\sigma^2\right) \right] \quad (7.18)$$

Using the IRLS approach, we can shape the canonical form of the inverse Gaussian to the log link by making the following substitutions:

	Canonical	Substitution
link	$-1/(2\mu^2)$	$\ln(\mu)$
inverse link	$(-2\eta)^{-1/2}$	$\exp(\eta)$
$g'(\mu)$	$1/\mu^3$	$1/\mu$

All other aspects of the algorithm remain the same.

If we wish to amend the basic IRLS algorithm so that the observed information matrix is used—hence, making it similar to ML output—then another adjustment must be made. This has to do with a modification of the weight function, w. The log-inverse Gaussian algorithm, implementing a modified Newton–Raphson approach, appears as

Listing 7.2. IRLS algorithm for log-inverse Gaussian models using OIM

```
1       μ = {y − mean(y)}/2
2       η = ln(μ)
3       WHILE (abs(ΔDev) > tolerance) {
4           W = 1/{μ³(1/μ)²} = 1/μ
5           z = η + (y − μ)/μ
6           W_o = W + 2(y − μ)/μ²
7           β = (XᵀW_oX)⁻¹XᵀW_oz
8           η = Xβ
9           μ = 1/exp(η)
10          OldDev = Dev
11          Dev = Σ(y − μ)²/(yσ²μ²)
12          ΔDev = Dev - OldDev
13      }
14      χ² = Σ(y − μ)²/μ³
```

For an example, we use a dataset from the U.S. Medicare database called MedPar. We model length of stay on the basis of type of admission, `admitype`, which is 0 if the admission is emergency or urgent and 1 if it is elective, and `codes`, which records the number of ICD-9 codes recorded (up to 9). The data are from one hospital and are for one diagnostic code (DRG). The goal is to see whether the type of admission has any bearing on the length of stay, adjusted for the number of codes.

Because inverse Gaussian models fit positively valued outcome measures, the log link is a useful alternative to the canonical link.

```
. use http://www.stata-press.com/data/hh4/medpar1, clear
. glm los admitype codes, family(igaussian) link(log) nolog
Generalized linear models                         No. of obs       =        3,676
Optimization      : ML                            Residual df      =        3,673
                                                  Scale parameter  =      .1024074
Deviance          =    304.271883                 (1/df) Deviance  =      .0828402
Pearson           =    376.1423892                (1/df) Pearson   =      .1024074
Variance function: V(u) = u^3                     [Inverse Gaussian]
Link function    : g(u) = ln(u)                   [Log]
                                                  AIC              =      6.619001
Log likelihood   = -12162.72308                   BIC              =     -29849.52
```

los	Coef.	OIM Std. Err.	z	P>|z|	[95% Conf. Interval]	
admitype	-.1916237	.0259321	-7.39	0.000	-.2424496	-.1407977
codes	.0992712	.0055994	17.73	0.000	.0882964	.1102459
_cons	1.057156	.0452939	23.34	0.000	.9683816	1.14593

Next, we treat codes as a classification variable and investigate the log link, identity link, and the canonical link.

```
. glm los admitype i.codes, family(igaussian) link(log) nolog
Generalized linear models                         No. of obs       =        3,676
Optimization      : ML                            Residual df      =        3,666
                                                  Scale parameter  =      .0984565
Deviance          =    298.4689189                (1/df) Deviance  =      .0814154
Pearson           =    360.9415865                (1/df) Pearson   =      .0984565
Variance function: V(u) = u^3                     [Inverse Gaussian]
Link function    : g(u) = ln(u)                   [Log]
                                                  AIC              =       6.62123
Log likelihood   = -12159.8216                    BIC              =     -29797.85
```

los	Coef.	OIM Std. Err.	z	P>|z|	[95% Conf. Interval]	
admitype	-.2027366	.0253892	-7.99	0.000	-.2524985	-.1529747
codes						
2	.3201292	.2065013	1.55	0.121	-.0846058	.7248643
3	.3337039	.2013326	1.66	0.097	-.0609008	.7283086
4	.3521312	.1977577	1.78	0.075	-.0354667	.7397291
5	.4764281	.1968441	2.42	0.016	.0906208	.8622353
6	.4754921	.1964791	2.42	0.016	.0904001	.8605841
7	.4551619	.1956977	2.33	0.020	.0716014	.8387224
8	.5848797	.1948331	3.00	0.003	.203014	.9667455
9	.9008458	.1924717	4.68	0.000	.5236082	1.278083
_cons	1.103884	.192179	5.74	0.000	.7272204	1.480548

```
. glm los admitype i.codes, family(igaussian) link(identity) nolog
Generalized linear models                          No. of obs      =        3,676
Optimization      : ML                             Residual df     =        3,666
                                                   Scale parameter =      .0971354
Deviance          =   296.9162798                  (1/df) Deviance =      .0809919
Pearson           =   356.0982589                  (1/df) Pearson  =      .0971354

Variance function: V(u) = u^3                      [Inverse Gaussian]
Link function     : g(u) = u                       [Identity]

                                                   AIC             =      6.620808
Log likelihood    = -12159.04528                   BIC             =     -29799.41
```

los	Coef.	OIM Std. Err.	z	P>|z|	[95% Conf.	Interval]
admitype	-1.145623	.1267899	-9.04	0.000	-1.394127	-.8971193
codes						
2	1.013613	.5289403	1.92	0.055	-.0230911	2.050317
3	.9906395	.5086064	1.95	0.051	-.0062107	1.98749
4	1.058073	.4912463	2.15	0.031	.0952479	2.020898
5	1.479535	.4960473	2.98	0.003	.5073004	2.45177
6	1.600658	.4943341	3.24	0.001	.6317811	2.569535
7	1.442416	.4878528	2.96	0.003	.4862419	2.39859
8	2.065912	.4903178	4.21	0.000	1.104907	3.026917
9	3.943232	.4748458	8.30	0.000	3.012551	4.873913
_cons	3.414347	.4690347	7.28	0.000	2.495056	4.333638

```
. glm los admitype i.codes, family(igaussian) nolog
Generalized linear models                          No. of obs      =        3,676
Optimization      : ML                             Residual df     =        3,666
                                                   Scale parameter =      .1011533
Deviance          =   301.6265109                  (1/df) Deviance =      .0822767
Pearson           =   370.8279408                  (1/df) Pearson  =      .1011533

Variance function: V(u) = u^3                      [Inverse Gaussian]
Link function     : g(u) = 1/(u^2)                 [Power(-2)]

                                                   AIC             =      6.622089
Log likelihood    =  -12161.4004                   BIC             =      -29794.7
```

los	Coef.	OIM Std. Err.	z	P>|z|	[95% Conf.	Interval]
admitype	.0084326	.0016685	5.05	0.000	.0051624	.0117028
codes						
2	-.0616845	.0549764	-1.12	0.262	-.1694363	.0460673
3	-.067647	.0543874	-1.24	0.214	-.1742444	.0389504
4	-.0699813	.0541194	-1.29	0.196	-.1760534	.0360909
5	-.0858411	.0539192	-1.59	0.111	-.1915208	.0198385
6	-.0831355	.0539198	-1.54	0.123	-.1888164	.0225454
7	-.0829273	.0538908	-1.54	0.124	-.1885513	.0226967
8	-.0939159	.0538107	-1.75	0.081	-.199383	.0115512
9	-.1111008	.0537499	-2.07	0.039	-.2164487	-.0057529
_cons	.1297928	.0537508	2.41	0.016	.0244432	.2351424

For the inverse Gaussian, the identity link is another common choice that is appropriate when modeling duration-type data. One may fit an identity-link and a log-link model. Graphs of the competing fitted values versus the outcomes can be subjectively compared. Otherwise, we can determine which model is preferable, between the log and identity links, from the value of the deviance function. Of course, we assume that all other aspects of the data and model are the same. If there is a significant difference between the deviance values between two models, then the model with the lower deviance is preferred. If there is little difference between the two, then either may be used. The same logic may be applied with respect to BIC or AIC statistics, as mentioned in chapter 6. Residual analysis may help in determining model preference as well. For the hospital data modeled above, the identity link is nearly the same, in terms of BIC and AIC statistics, as the results of the log model. Neither is preferred over the other.

8 The power family and link

Contents

8.1 Power links

The links that are associated with members of the continuous family of GLM distributions may be thought of in terms of powers. In fact, except for the standard binomial links and the canonical form of the negative binomial, all links are powers of μ. The relationships are listed in the following table:

Link	Function	Power function
Identity	μ	μ^1
Log	$\ln(\mu)$	μ^0
Reciprocal	$1/\mu$	μ^{-1}
Inverse quadratic	$1/\mu^2$	μ^{-2}

A generic power link function can thus be established as

$$\text{Power}(a) = \begin{cases} \mu^a & \text{if } a \neq 0 \\ \ln(\mu) & \text{if } a = 0 \end{cases} \tag{8.1}$$

The corresponding generic inverse link function is

$$\text{Power}(a) = \begin{cases} \eta^{-a} & \text{if } a \neq 0 \\ \exp(\eta) & \text{if } a = 0 \end{cases} \tag{8.2}$$

Variance functions for the continuous distributions as well as for the Poisson distribution can also be thought of in terms of powers. The following table displays the relationships:

Family	Link	Power function
Gaussian	Identity	$\mu^0 = 1$
Poisson	Log	$\mu^1 = \mu$
Gamma	Square	μ^2
Inverse Gaussian	Cube	μ^3

The power link can play an important role in assessing the fit of models based on continuous distributions. For instance, checking the differences in log-likelihood or deviance function for various links within a particular GLM family can help the researcher construct the optimal model. With the gamma models, we can use the canonical or inverse link (power $= -1$), the log link (power $= 0$), or the identity link (power $= 1$). The model having the lower deviance value, or greater log-likelihood value, is the preferred model (all other aspects of the model being equal). However, the optimal model may fall between that of the log and identity links, for example, the square root link, $\mu^{0.5}$. If the fit of residuals is better using this link, and the deviance, BIC, and AIC statistics are less than those of other links, then the choice of the gamma model with power link of $\mu^{0.5}$ is preferred over other models. Of course, the square root link is not one of the standard links, but it is nevertheless valid.

When one specifies a power link within a given family, the algorithm makes the appropriate adjustments to the link, inverse link, and derivative of the link, g'. No adjustments need be made for the deviance or log-likelihood functions or for the variance function. The derivative of the link is given in table A.2. Some of the important power relationships are outlined in table A.14.

8.2   Example: Power link

We used a power analysis on `claims.dta` discussed in chapter 6 on the gamma family. The canonical reciprocal link was used to model the data. In power link terms, we used power $= -1$. Using power links, we can determine whether the canonical inverse reciprocal link is optimal. We can also ascertain whether another distribution may be preferable for the modeling of the data at hand.

We modeled the main effects for `claims.dta` as

```
. glm y i.pa i.cg i.va [fw=number], family(gamma)
```

The algorithm used the default canonical inverse reciprocal link. Using the power link option, we may obtain the same result by specifying

```
. glm y i.pa i.cg i.va [fw=number], family(gamma) link(power -1)
```

The following table shows us that the most preferable link for the gamma family is the canonical reciprocal link. If the data are modeled with the inverse Gaussian family, then the most preferable link is the inverse square root. We do not compare deviance values across families. To make comparisons of models across families, we use the BIC or AIC statistics. One must also evaluate the significance of the predictors. Here the significance of predictors is nearly identical.

Link	Gamma deviance	Inverse Gaussian deviance
−2.0	130.578	0.656
−1.5	126.826	0.638
−1.0	*124.783	0.628
−0.5	124.801	*0.626
0	127.198	0.634
0.5	132.228	0.665
1.0	139.761	0.687

The AIC and BIC statistics for the canonical gamma model are 12.9 and −81106.76, respectively. For the inverse square root–linked inverse Gaussian model, the values are 18.2 and −81230.91. The AIC and BIC statistics may be used to decide whether one family of models is preferable to another for a given set of data. Here the gamma model appears more appropriate than the inverse Gaussian for the main-effects claims data.

8.3 The power family

The power function may be extended so that it is regarded as a family of distributions. That is, we may model data as a power model, like a gamma or logit model. The difference is that the power distribution is not directly based on a probability distribution, that is, a member of the exponential family of distributions.

Although the log-likelihood and deviance functions were derived directly from the exponential family form of a probability distribution, one can derive a working quasideviance from the variance function. See chapter 17 on quasilikelihood for more details. The quasideviance for the power family is

$$\frac{2y}{(1-a)\left(y^{1-a} - \mu^{1-a}\right)} - \frac{2}{(2-a)\left(y^{2-a} - \mu^{2-a}\right)} \tag{8.3}$$

One can then create a power algorithm that uses the above deviance function with the power link, the power inverse link, the derivative of the power link, and the power variance functions. This is an example of a quasideviance model; one that is an extended member of the GLM family.

One use of power links is to assess optimal models within a GLM distribution family. One can use the link function to check various statistics resulting from changes to a in the power link. This gives a wide variety of possible links for each continuous family.

However, we may need to use a general power link model to assess models between families. Of course, we may simply create a series of models using a continuum of links within each of the Gaussian $(a = 1)$, Poisson $(a = 0)$, gamma $(a = -1)$, and inverse Gaussian $(a = -2)$ families. If all other aspects of the models are the same, then the model having the lower deviance value and best-fitting residuals will be the model of choice. However, we are thereby giving wide leniency to the notion of "aspects".

The power family can model data with links that are different from those used in the standard GLM repertoire. Models with positive power values can be created, including powers of 0.5 (square root), 2 (square), 2.5, or 3 (cube). Power links greater than 1 may be appropriate for data having a response with a sharp increase of values.

We know no notable use of the power family algorithm. Typically, power analysis rests with comparative links, as we demonstrated in section 8.2. However, one can use a generalized power family algorithm to model data, but the literature has no evidence that one can derive substantial benefit therefrom.

Part III

Binomial Response Models

9 The binomial–logit family

Contents

Except for the canonical Gaussian model, binomial models are used more often than any other member of the GLM family.

Binomial regression models are used in analyses having discrete (corresponding to the number of successes for k trials) or grouped responses. As shown in table 9.1, models of binary responses are based on the Bernoulli ($n = 1$) or binomial distribution. The Bernoulli distribution is a degenerate case of the binomial, where the number of trials is equal to 1.

Table 9.1. Binomial regression models

Response	Model
Binary: $\{0, 1\}$	Bernoulli = Binomial(1)
Proportional: $\{0, 1, \dots, k\}$	Binomial(k)

Proportional-response data involve two variables corresponding to the number of successes, y_i, of a population of k_i trials, where both variables are indexed by i because grouped data do not require that the number of trials be the same for each observation. Hence, a binomial model can be considered as a weighted binary response model or simply as a two-part response relating the proportion of successes to trials. We discuss the relationship between these two methods of dealing with discrete data. Binary data can be modeled as grouped, but the reverse is not always possible.

Binary responses take the form of 0 or 1, with 1 typically representing a success or some other positive result, whereas 0 is taken as a failure or some negative result. Responses may also take the form of $1/2$, a/b, or a similar pair of values. However, values other than 0 and 1 are translated to a $0/1$ format by the binary regression algorithm before estimation. The reason for this will be apparent when we describe the probability function. Also in Stata, a 0 indicates failure and any other nonzero nonmissing value represents a success in Bernoulli models.

The binomial family is associated with several links where the logit link is the canonical link. The links are associated with either the Bernoulli or the full binomial distribution. The common binomial links (there are others) are given in table 9.2.

Table 9.2. Common binomial link functions

logit	probit	log-log	complementary log-log
identity	log	inverse	log-complement

See table A.2 for the associated formulas.

We begin our exposition of the binomial model by describing it in its canonical form. Because the binary is included within the grouped model, we begin by examining the latter. Conveniently, the canonical link is also the most well known of GLM models (the logistic or logit regression model).

9.1 Derivation of the binomial model

The binomial probability function is expressed as

$$f(y; k, p) = \binom{k}{y} p^y (1-p)^{k-y} \tag{9.1}$$

where p is the probability of success, y is the numerator, and k is the denominator. p is sometimes referred to as π in the literature when the focus is solely on the logit model; however, we use p throughout this text. It is the sole parameter for all binomial models (including the Bernoulli).

In exponential family form, the binomial distribution is written as

$$
\begin{aligned}
f(y; k, p) &= \exp\left\{ y \ln(p) + k \ln(1-p) - y \ln(1-p) + \ln\binom{k}{y} \right\} \tag{9.2}\\
&= \exp\left\{ y \ln\left(\frac{p}{1-p}\right) + k \ln(1-p) + \ln\binom{k}{y} \right\} \tag{9.3}
\end{aligned}
$$

Identifying the link and cumulant is easy when a member of the exponential family of probability distributions is expressed in exponential form as above. The link, θ, and the cumulant, $b(\theta)$, functions are then given by

$$\theta = \ln\left(\frac{p}{1-p}\right) \tag{9.4}$$

$$b(\theta) = -k\ln(1-p) \tag{9.5}$$

and $\phi = 1$ (the variance has no scale component).

The mean and variance functions are calculated as the first and second derivatives, respectively, of the cumulant function.

$$b'(\theta) = \frac{\partial b}{\partial p}\frac{\partial p}{\partial \theta} \tag{9.6}$$

$$= \frac{k}{1-p}p(1-p) \tag{9.7}$$

$$= kp \tag{9.8}$$

$$b''(\theta) = \frac{\partial^2 b}{\partial p^2}\left(\frac{\partial p}{\partial \theta}\right)^2 + \frac{\partial b}{\partial p}\frac{\partial^2 p}{\partial \theta^2} \tag{9.9}$$

$$= \frac{k}{(1-p)^2}(1-p)^2 p^2 + \frac{k}{1-p}(1-p)p(1-2p) \tag{9.10}$$

$$= kp^2 + kp(1-2p) \tag{9.11}$$

$$= kp(1-p) \tag{9.12}$$

The relationship of p and μ is thus given by

$$\mu = kp \tag{9.13}$$

and the variance is given by

$$V(\mu) = \mu\left(1 - \frac{\mu}{k}\right) \tag{9.14}$$

The linear predictor η is expressed in terms of μ. θ is the same as η for the canonical link. Here the linear predictor is also the canonical link function. Hence,

$$\theta = \eta = g(\mu) = g = \ln\left(\frac{\mu}{k-\mu}\right) \tag{9.15}$$

The inverse link can easily be derived from the above as

$$\mu = g^{-1}(\eta) = \frac{k}{1 + \exp(-\eta)} = \frac{k\exp(\eta)}{1 + \exp(\eta)} \tag{9.16}$$

Both forms of the inverse link are acceptable. We primarily use the first form, but there is no inherent reason to prefer one over the other. You will see both forms in the literature.

The derivative of the link is also necessary for the binomial algorithm

$$g'(\mu) = g' = \frac{\partial}{\partial \mu} \ln\left(\frac{\mu}{k - \mu}\right) = \frac{k}{\mu(k - \mu)} \tag{9.17}$$

Finally, to have all the functions required for the binomial regression algorithm, we need to determine the deviance, which is calculated as

$$D = 2\{\mathcal{L}(y; y) - \mathcal{L}(\mu; y)\} \tag{9.18}$$

when $\phi = 1$.

The log-likelihood function is thus required for calculating the deviance. The log likelihood is also important for calculating various fit statistics, and it takes the place of the deviance function when using maximum likelihood methods to estimate model parameters.

We can obtain the log-likelihood function directly from the exponential-family form of the distribution. Removing the exponential and its associated braces, we have

$$\mathcal{L} = \sum_{i=1}^{n} \left\{ y_i \ln\left(\frac{\mu_i}{1 - \mu_i}\right) + k_i \ln(1 - \mu_i) + \ln\binom{k_i}{y_i} \right\} \tag{9.19}$$

Given the log likelihood, the deviance is calculated as

$$D = 2 \sum_{i=1}^{n} \left\{ y_i \ln\left(\frac{y_i}{1 - y_i}\right) + k_i \ln(1 - y_i) - y_i \ln\left(\frac{\mu_i}{1 - \mu_i}\right) - k_i \ln(1 - \mu_i) \right\} \tag{9.20}$$

$$= 2 \sum_{i=1}^{n} \left\{ y \ln\left(\frac{y_i}{\mu_i}\right) + (k_i - y_i) \ln\left(\frac{k_i - y_i}{k_i - \mu_i}\right) \right\} \tag{9.21}$$

The normalization term is dropped from the deviance formula. This dropout occurs for all members of the GLM family and is a prime reason why researchers code the iteratively reweighted least-squares (IRLS) algorithm by using the (simpler) deviance rather than the maximum likelihood algorithm with the log likelihood.

The binomial deviance function is typically given a multiple form depending on the values of y and k. We list the appropriate calculations of the deviance for specific observations; see also table A.11.

Bernoulli

$$D_i(k_i = 1; y_i = 0) \quad = \quad -2\ln(1 - \mu_i) \tag{9.22}$$

$$D_i(k_i = 1; y_i = 1) \quad = \quad -2\ln(\mu_i) \tag{9.23}$$

Binomial(k)

$$D_i(k_i > 1; y_i = 0) \quad = \quad 2k_i \ln\left(\frac{k_i}{k_i - \mu_i}\right) \tag{9.24}$$

$$D_i(k_i > 1; y_i = k_i) \quad = \quad 2k_i \ln\left(\frac{k_i}{\mu_i}\right) \tag{9.25}$$

$$D_i(k_i > 1; 0 < y_i < k_i) \quad = \quad 2y_i \ln\left(\frac{y_i}{\mu_i}\right) + 2(k_i - y_i)\ln\left(\frac{k_i - y_i}{k_i - \mu_i}\right) \tag{9.26}$$

Most commercial software algorithms use the above reductions, but to our knowledge Stata is the only one that documents the full extent of the limiting values.

9.2   Derivation of the Bernoulli model

The binary or Bernoulli response probability distribution is simplified from the binomial distribution. The binomial denominator k is equal to 1, and the outcomes y are constrained to the values $\{0, 1\}$. Again, y may initially be represented in your dataset as 1 or 2, or some other alternative set. If so, you must change your data to the above description.

However, some programs use the opposite default behavior: they use 0 to denote a success and 1 to denote a failure. Transferring data unchanged between Stata and a package that uses this other codification will result in fitted models with reversed signs on the estimated coefficients.

The Bernoulli response probability function is

$$f(y; p) = p^y(1 - p)^{1-y} \tag{9.27}$$

The binomial normalization (combination) term has disappeared, which makes the function comparatively simple.

In canonical (exponential) form, the Bernoulli distribution is written

$$f(y; p) = \exp\left\{ y\ln\left(\frac{p}{1 - p}\right) + \ln(1 - p) \right\} \tag{9.28}$$

Below you will find the various functions and relationships that are required to complete the Bernoulli algorithm in canonical form. Because the canonical form is commonly referred to as the logit model, these functions can be thought of as those of the binary logit or logistic regression algorithm.

$$\theta \;=\; \ln\left(\frac{p}{1-p}\right) = \eta = g(\mu) = \ln\left(\frac{\mu}{1-\mu}\right) \tag{9.29}$$

$$g^{-1}(\theta) \;=\; \frac{1}{1+\exp(-\eta)} = \frac{\exp(\eta)}{1+\exp(\eta)} \tag{9.30}$$

$$b(\theta) \;=\; -\ln(1-p) = -\ln(1-\mu) \tag{9.31}$$

$$b'(\theta) \;=\; p = \mu \tag{9.32}$$

$$b''(\theta) \;=\; p(1-p) = \mu(1-\mu) \tag{9.33}$$

$$g'(\mu) \;=\; \frac{1}{\mu(1-\mu)} \tag{9.34}$$

The Bernoulli log-likelihood and deviance functions are

$$\mathcal{L}(\mu; y) \;=\; \sum_{i=1}^{n}\left\{ y_i \ln\left(\frac{\mu_i}{1-\mu_i}\right) + \ln(1-\mu_i) \right\} \tag{9.35}$$

$$D \;=\; 2\sum_{i=1}^{n}\left\{ y_i \ln\left(\frac{y_i}{\mu_i}\right) + (1-y_i)\ln\left(\frac{1-y_i}{1-\mu_i}\right) \right\} \tag{9.36}$$

As we observed for the binomial probability function, separate values are often given for the log likelihood and deviance when $y = 0$. When this is done, a separate function is generally provided in case $y = 1$. We list these below for individual observations.

$$\mathcal{L}_i(\mu_i; y_i = 0) \;=\; \ln(1-\mu_i) \tag{9.37}$$

$$\mathcal{L}_i(\mu_i; y_i = 1) \;=\; \ln(\mu_i) \tag{9.38}$$

$$D_i(y_i = 0) \;=\; -2\ln(1-\mu_i) \tag{9.39}$$

$$D_i(y_i = 1) \;=\; -2\ln(\mu_i) \tag{9.40}$$

9.3   The binomial regression algorithm

The canonical binomial algorithm is commonly referred to as logistic or logit regression. Traditionally, binomial models have three commonly used links: logit, probit, and complementary log-log (clog-log). There are other links that we will discuss. However, statisticians typically refer to a GLM-based regression by its link function, hence the still-used reference to probit or clog-log regression. For the same reason, statisticians generally referred to the canonical form as logit regression. This terminology is still used.

Over time, some researchers began referring to logit regression as logistic regression. They made a distinction based on the type of predictors in the model. A logit model comprised factor variables. The logistic model, on the other hand, had at least one continuous variable as a predictor. Although this distinction has now been largely discarded, we still find reference to it in older sources. Logit and logistic refer to the same basic model.

In the previous section, we provided all the functions required to construct the binomial algorithm. Because this is the canonical form, it is also the algorithm for logistic regression. We first give the grouped-response form because it encompasses the simpler model.

Listing 9.1. IRLS algorithm for grouped logistic regression

```
1    Dev = 0
2    μ = (y + 0.5)/(k + 1)              /* initialization of μ */
3    η = ln{μ/(k − μ)}                  /* initialization of η */
4    WHILE (abs(ΔDev) > tolerance )  {
5        W = μ(k − μ)
6        z = η + k(y − μ)/{μ(k − μ)}
7        β = (Xᵀ W X)⁻¹ Xᵀ W z
8        η = Xβ
9        μ = k/{1 + exp(−η)}
10       OldDev = Dev
11       Dev = 2 ∑ y ln(y/μ) − (k − y) ln{(k − y)/(k − μ)}
12       ΔDev = Dev − OldDev
13   }
14   χ² = ∑ (y − μ)²/{μ(1 − μ/k)}
```

The Bernoulli or binary algorithm can be constructed by setting k to the value of 1. Because the binary model often reduces the deviance on the basis of the value of y, we show the entire algorithm.

Listing 9.2. IRLS algorithm for binary logistic regression

```
1    Dev = 0
2    μ = (y + 0.5)/2                    /* initialization of μ */
3    η = ln{μ/(1 − μ)}                  /* initialization of η */
4    WHILE (abs(ΔDev) > tolerance )  {
5        W = μ(1 − μ)
6        z = η + (y − μ)/{μ(1 − μ)}
7        β = (Xᵀ W X)⁻¹ Xᵀ W z
8        η = Xβ
9        μ = 1/{1 + exp(−η)}
10       OldDev = Dev
11       Dev = 2 ∑ ln(1/μ) if (y = 1)
12       Dev = 2 ∑ ln{1/(1 − μ)} if (y = 0)
13       ΔDev = Dev − OldDev
14   }
15   χ² = ∑ (y − μ)²/{μ(1 − μ)}
```

There are several ways to construct the various GLM algorithms. Other algorithms may implement a DO-WHILE loop rather than the type we have shown here. Others may use the log-likelihood function as the basis for iteration; that is, iteration stops when there is a defined minimal difference in log-likelihood values. Other algorithms iterate until there is minimal change in parameter estimates. One can find each of these methods in commercial GLM software, but they all result in statistically similar results. The variety we show here is the basis for the Stata IRLS GLM procedure.

9.4   Example: Logistic regression

Our logistic regression example uses a medical dataset. The data come from a subset of the National Canadian Registry of Cardiovascular Disease (FASTRAK), sponsored by Hoffman-La Roche Canada. The goal is to model death within 48 hours by various explanatory variables. For this study, we are interested in the comparative risk for patients who enter a hospital with an anterior infarct (heart attack) rather than an inferior infarct. The terms anterior and inferior refer to the site on the heart where damage has occurred. The model is adjusted by hcabg, history of having had a cardiac bypass surgery (CABG); age1–age4, levels of age; and kk1–kk4, Killip class, levels of risk where the higher the level, the greater the risk. Levels of each variable in the model are shown in table 9.3 below.

Table 9.3. Variables from heart01.dta

death = 1	death within 48 hours of myocardial infarction onset
death = 0	no death
anterior = 1	anterior infarction
anterior = 0	inferior infarction
hcabg = 1	history of CABG
hcabg = 0	no history of CABG
kk1 = 1	Killip class 1
kk2 = 1	Killip class 2
kk3 = 1	Killip class 3
kk4 = 1	Killip class 4
age1 = 1	age less than 60 years
age2 = 1	age between 60 and 69 years
age3 = 1	age between 70 and 79 years
age4 = 1	age greater than or equal to 80 years

We model the data twice, first displaying the coefficients and then showing coefficients (and standard errors) converted to odds ratios. To demonstrate the similarity of results, we also use the slower Newton–Raphson algorithm for the first model and IRLS for the second. The z statistics are identical for both models, which would not necessarily be the case if we used a noncanonical link.

The z statistics are produced by dividing each coefficient by its standard error. When coefficients are exponentiated and standard errors are converted via the delta method to their odds ratio equivalent, the z statistics and p-values remain the same.

GLM theory assumes that the model estimates have asymptotic normal distributions such that most software will provide test statistics based on the Gaussian distribution. Some commercial software routines will display t statistics instead. Slight variations for the p-values will result. However, the differences are usually not great enough to matter.

4.1 Model producing logistic coefficients: The heart data

```
. use http://www.stata-press.com/data/hh4/heart01
. glm death anterior hcabg kk2 kk3 kk4 age2-age4, family(binomial) nolog
```

Generalized linear models		No. of obs	=	4,483
Optimization : ML		Residual df	=	4,474
		Scale parameter =		1
Deviance = 1273.251066		(1/df) Deviance =		.284589
Pearson = 4220.194937		(1/df) Pearson =		.9432711
Variance function: V(u) = u*(1-u)		[Bernoulli]		
Link function : g(u) = ln(u/(1-u))		[Logit]		
		AIC	=	.2880328
Log likelihood = -636.6255331		BIC	=	-36344.35

death	Coef.	OIM Std. Err.	z	P>|z|	[95% Conf. Interval]
anterior	.6425552	.1675538	3.83	0.000	.3141557 .9709547
hcabg	.7444459	.352956	2.11	0.035	.0526648 1.436227
kk2	.8116998	.180502	4.50	0.000	.4579223 1.165477
kk3	.7756968	.2690602	2.88	0.004	.2483486 1.303045
kk4	2.659656	.3559947	7.47	0.000	1.961919 3.357393
age2	.4930224	.3101877	1.59	0.112	-.1149343 1.100979
age3	1.511168	.2662236	5.68	0.000	.9893793 2.032957
age4	2.185288	.2718383	8.04	0.000	1.652494 2.718081
_cons	-5.052069	.2586081	-19.54	0.000	-5.558931 -4.545206

9.4.2   Model producing logistic odds ratios

```
. glm death anterior hcabg kk2 kk3 kk4 age2-age4, family(binomial) eform irls
> nolog
```

```
Generalized linear models                    No. of obs        =       4,483
Optimization        : MQL Fisher scoring     Residual df       =       4,474
                      (IRLS EIM)             Scale parameter =           1
Deviance            =  1273.251066           (1/df) Deviance =      .284589
Pearson             =  4220.186292           (1/df) Pearson  =     .9432692

Variance function: V(u) = u*(1-u)            [Bernoulli]
Link function    : g(u) = ln(u/(1-u))        [Logit]

                                             BIC             =   -36344.35
```

| | | EIM | | | | |
death	Odds Ratio	Std. Err.	z	P>|z|	[95% Conf.	Interval]
anterior	1.901333	.3185755	3.83	0.000	1.369103	2.640464
hcabg	2.105275	.7430692	2.11	0.035	1.054076	4.204801
kk2	2.251732	.4064421	4.50	0.000	1.580786	3.207453
kk3	2.172105	.5844269	2.88	0.004	1.281907	3.680486
kk4	14.29137	5.087652	7.47	0.000	7.112966	28.71423
age2	1.63726	.5078572	1.59	0.112	.8914272	3.007112
age3	4.532029	1.20653	5.68	0.000	2.689572	7.636636
age4	8.893222	2.417514	8.04	0.000	5.219999	15.15123
_cons	.0063961	.0016541	-19.54	0.000	.0038529	.0106179

Note: _cons estimates baseline odds.

We find that patients who enter Canadian hospitals with an anterior infarct have nearly twice the odds of death within 48 hours of the onset of their symptoms as patients who have inferior infarcts. This result is in keeping with what experience has taught cardiologists. Moreover, those who have had a previous CABG appear to have twice the risk as those who have not had the procedure. Those who enter the hospital with a higher Killip-level risk score, or those who are older, are also at greater risk. In particular, those who are 80 and over are at greatest risk. The outlook appears bleak for someone who enters the hospital with an anterior infarct, experienced a previous CABG, is at Killip-level 4, and is over age 80. In fact, we can calculate the probability of death for such a patient directly from parameter results.

The probability of death can be calculated for each observation in the dataset by summing the various estimate terms. First, multiply the value of each appropriate predictor by its coefficient and then sum across terms. The sum is called the linear predictor, or η. Thereafter, apply the inverse link transform to η. Doing so produces a μ value for each observation, which is the probability in the case of binary logistic regression. In Stata, we need only type

```
. predict mu, mu
```

or

```
. predict mu
```

to obtain probabilities after modeling.

The same principle applies to covariate patterns that are not actually in the model dataset, like our high-risk patient above. We can obtain the probability of death for such a patient by implementing the same methods. First, calculate the linear predictor, and then use the inverse link transform. For our high-risk patient, we have

$$\eta = \beta_{\text{anterior}}(1) + \beta_{\text{hcabg}}(1) + \beta_{\text{kk4}}(1) + \beta_{\text{age4}}(1) + \beta_{\text{\_cons}}(1) \tag{9.41}$$
$$= 0.6425552(1) + 0.7444459(1) + 2.659656(1) + 2.185288(1) - 5.052069(1) \tag{9.42}$$
$$= 1.1798761 \tag{9.43}$$

which can also be obtained using

```
. margins, at((zero)_all anterior=1 hcabg=1 kk4=1 age4=1) expression(xb())
```

Applying the inverse link gives

$$\mu = \frac{1}{1 + \exp(-1.1798761)} = 0.7649 \tag{9.44}$$

which can also be obtained by typing

```
. margins, at((zero)_all anterior=1 hcabg=1 kk4=1 age4=1)
> expression(1/(1+exp(-xb())))
```

Hence, a patient who has all the highest risk factors in the model has a 76% probability of death within 48 hours of symptom onset, regardless of treatment. Fortunately, no one in the actual data had all these risk factors.

We can also calculate the odds of a person dying within the given time framework by applying the following formula:

$$\text{Odds} = \frac{p}{1 - p} \tag{9.45}$$

We can thereby calculate the odds of death within 48 hours of symptom onset for a patient entering a Canadian hospital with all the highest risk factors as

$$\text{Odds} = \frac{0.7649}{1 - 0.7649} = 3.25 \tag{9.46}$$

Of course, there are perhaps other important confounding explanatory variables that were not a part of the model. If this were the case, there would probably be a change in the probability and odds ratio. But if the model is well fit and correlates well with the population as a whole, then the added predictors will have little effect on the probability.

.5   GOF statistics

Models can easily be constructed without thought to how well the model actually fits the data. All too often this is seen in publications—logistic regression results with parameter estimates, and standard errors with corresponding p-values and without their associated GOF statistics.

Models are simply that: models. They are not aimed at providing a one-to-one fit, called overfitting. The problem with overfitting is that the model is difficult to adapt to related data situations outside the actual data modeled. If `heart01.dta` were overfit, it would have little value in generalizing to other heart patients.

One measure of fit is called the confusion matrix or classification table. The origins of the term may be more confusing than the information provided in the table. The confusion matrix simply classifies the number of instances where

- $y = 1$, predicted $y = 1$
- $y = 1$, predicted $y = 0$
- $y = 0$, predicted $y = 1$
- $y = 0$, predicted $y = 0$

Predicted values can be calculated by obtaining the linear predictor, $\mathbf{x}\boldsymbol{\beta}$, or η, and converting it by means of the logistic transform, which is the inverse link function. Both of these statistics are available directly after modeling. In Stata, we can obtain the values after modeling by typing

```
. predict xb, xb    /* linear predictor */
```

or

```
. predict mu, mu    /* fitted value   */
```

We may also obtain the linear predictor and then compute the fit for them:

```
. generate mu = 1/(1+exp(-xb))
```

To create a confusion matrix, we simply tabulate the actual 0/1 response with values of $\widehat{\mu}$. The matrix may be cut at different values depending on the distribution of the data. One method that has been used is to cut at the mean value of $\widehat{\mu}$.

Stata's `logistic` command for fitting maximum-likelihood logistic regression models includes many useful postestimation commands. We refit the previous model by typing

```
. logistic death anterior hcabg kk2 kk3 kk4 age2-age4, nolog
```

Various statistics for this model are then presented in summary table form by using the `estat classification` command. This command is useful for investigating the areas in which models are strong or weak in fitting data.

```
. estat classification
Logistic model for death
                   ─────── True ───────
Classified │         D              ~D    │        Total

     +      │         4               7    │          11
     -      │        172            4300   │        4472

   Total    │        176            4307   │        4483
Classified + if predicted Pr(D) >= .5
True D defined as death != 0

Sensitivity                          Pr( +| D)     2.27%
Specificity                          Pr( -|~D)    99.84%
Positive predictive value            Pr( D| +)    36.36%
Negative predictive value            Pr(~D| -)    96.15%

False + rate for true ~D             Pr( +|~D)     0.16%
False - rate for true D              Pr( -| D)    97.73%
False + rate for classified +        Pr(~D| +)    63.64%
False - rate for classified -        Pr( D| -)     3.85%

Correctly classified                              96.01%
```

We find a 96% rate of correct classification for this model.

Area under the receiver operating characteristic (ROC) curve is another measure of fit. Higher values indicate a better fit. The Stata documentation for the `logistic` and `lroc` commands provide excellent discussion of this topic; see the respective entries in the *Stata Base Reference Manual*.

```
. lroc, nograph
Logistic model for death
number of observations =     4483
area under ROC curve   =   0.7965
```

A third measure of fit exploits the notion of the actual response versus predicted value that was given in the confusion matrix. The Hosmer–Lemeshow GOF statistic, with a related table, divides the range of probability values, μ, into groups. The authors recommend 10 groups for relatively large datasets, but the number can be amended to fit the data situation. The point is to provide a count, for each respective group, of actually observed ones, fitted or expected ones, observed zeros, and expected zeros. Theoretically, observed and expected counts should be close. Diagnostically, it is easy to identify groups where deviations from the expected occur. When this is the case, we can go back to the data to see the covariate patterns that gave rise to the discrepancy.

A fit statistic is provided that measures the overall correspondence of counts. Based on the χ^2 distribution, the Hosmer–Lemeshow statistic is such that larger values have smaller p-values indicating worse fit. The smaller the Hosmer–Lemeshow statistic, the less variance in fit and the greater the p-value indicating better fit.

```
. estat gof, group(10) table

Logistic model for death, goodness-of-fit test

(Table collapsed on quantiles of estimated probabilities)
```

Group	Prob	Obs_1	Exp_1	Obs_0	Exp_0	Total
1	0.0064	5	4.7	731	731.3	736
2	0.0104	5	5.0	481	481.0	486
3	0.0120	4	6.5	537	534.5	541
4	0.0137	1	0.5	34	34.5	35
5	0.0195	5	8.3	451	447.7	456
6	0.0282	19	21.3	782	779.7	801
7	0.0429	4	6.4	147	144.6	151
8	0.0538	30	26.8	479	482.2	509
9	0.0976	37	26.3	295	305.7	332
10	0.6072	66	70.2	370	365.8	436

```
            number of observations =       4483
                   number of groups =         10
         Hosmer-Lemeshow chi2(8) =           9.49
                      Prob > chi2 =         0.3025
```

There has been discussion concerning the utility of this test in the general case. The Hosmer–Lemeshow statistic creates 10 groups in the dataset, forming them on the deciles of the fitted values. When there are ties (multiple equal outcomes), choosing the decile induces a randomness to the statistic when the independent variables are different over observations producing the ties. We should ensure that this randomness does not affect a particular application of the test. The original authors have also advised that the proposed test should not be treated as a sole terminal test but that the test be part of a more complete investigation. One concern with the test is that the groups may contain subjects with widely different values of the covariates. As such, one can find situations where one set of fixed groups shows that the model fits, whereas another set of fixed groups shows that the model does not fit.

We provide more discussion of grouped binomial models in chapter 11, where we discuss problems of overdispersion. We demonstrate how certain binary response models can be converted to grouped datasets, as well as how to use a grouped binomial model to assess the fit of its binary case counterpart.

Researchers also evaluate the fit of a model by observing a graph of its residuals. Several residuals are available. Most are listed in the early chapters of this text. One of the most well-used methods of graphing GLM residuals is the graph of standardized deviance residuals versus the fitted values. In the graph, points should be randomly displayed.

9.6   Grouped data

Generally, the models presented throughout this chapter address dichotomous outcomes. In addition, binomial data can be modeled where the user communicates two pieces of information for each row of data. Those information pieces are the number of successes (numerator) and the number of trials (denominator).

In some research trials, an outcome might represent a proportion for which the denominator is unknown or missing from the data. There are a few approaches available to analyze associations with such an outcome measure. Some researchers analyze these outcomes using a linear regression model. The downside is that predicted values could be outside the range of $[0, 1]$; certainly extrapolated values would be outside the range.

In GLMs, the approach is to transform the linear predictor (the so-called right-hand side of the equation), but we could use the link function to transform the outcome (the left-hand side of the equation)

$$\text{logit}(y) = \ln\left(\frac{y}{1-y}\right) = \eta \qquad (9.47)$$

and then analyze the transformed value with linear regression. This approach works well except when there are values of zero or one in the outcome.

Strictly speaking, there is nothing wrong with fitting these outcomes in a GLM using Bernoulli variance and the logit link. In previous versions of the Stata software, the `glm` command would not allow noninteger outcomes, but current software will fit this model. As Papke and Wooldridge (1996) suggest, such models should use the sandwich variance estimator for inference on the coefficients.

9.7   Interpretation of parameter estimates

We have provided an overview of various methods used to assess the logistic model fit. Now we turn to the actual interpretation of parameter estimates.

We know from basic statistical knowledge that linear regression coefficients are all interpreted in the same way. The coefficient represents the amount of change to the response for a one-unit change in the variable associated with the coefficient, with other predictors held to their same values. The same is the case for odds ratios.

We continue with `heart01.dta` to show a table listing of odds ratios:

```
. quietly glm death anterior hcabg kk2 kk3 kk4 age2-age4, family(binomial)
> eform irls nolog
. ereturn disp, eform(Odds Ratio)
```

death	Odds Ratio	EIM Std. Err.	z	P>|z|	[95% Conf. Interval]	
anterior	1.901333	.3185755	3.83	0.000	1.369103	2.640464
hcabg	2.105275	.7430692	2.11	0.035	1.054076	4.204801
kk2	2.251732	.4064421	4.50	0.000	1.580786	3.207453
kk3	2.172105	.5844269	2.88	0.004	1.281907	3.680486
kk4	14.29137	5.087652	7.47	0.000	7.112966	28.71423
age2	1.63726	.5078572	1.59	0.112	.8914272	3.007112
age3	4.532029	1.20653	5.68	0.000	2.689572	7.636636
age4	8.893222	2.417514	8.04	0.000	5.219999	15.15123
_cons	.0063961	.0016541	-19.54	0.000	.0038529	.0106179

```
Note: _cons estimates baseline odds.
```

Now consider the odds ratio for `anterior`. This odds ratio is the ratio of the odds of death for a person with an anterior infarction (`anterior=1`) to the odds of death for a person without an anterior infarction (`anterior=0`)

$$\frac{\Pr(\text{death}|\text{anterior} + 1)/\{1 - \Pr(\text{death}|\text{anterior} + 1)\}}{\Pr(\text{death}|\text{anterior})/\{1 - \Pr(\text{death}|\text{anterior})\}} \tag{9.48}$$

One straightforward way to calculate this in Stata is given by

```
. predict double p                           /* p = predicted probability     */
(option mu assumed; predicted mean death)
(905 missing values generated)
. replace anterior=anterior+1                 /* increment anterior            */
(4,696 real changes made)
. predict double p1                           /* p1= pred prob with anterior+1 */
(option mu assumed; predicted mean death)
(905 missing values generated)
. generate double or=(p1/(1-p1))/(p/(1-p)) /* Calculate ratio of odds        */
(905 missing values generated)
. summarize or                               /* display odds ratio (anterior) */
```

Variable	Obs	Mean	Std. Dev.	Min	Max
or	4,483	1.901333	6.09e-16	1.901333	1.901333

Incrementing the predictor `anterior` by 1 and calculating the odds ratio using the displayed formula illustrates the odds ratio observed in the parameter table. This method of calculating odds ratios can be used for each of the predictors.

Sometimes researchers wish to know the comparative worth of a predictor to the model. Just looking at the odds ratio does little to that end. A common method used to assess the impact of a predictor on the model is to look at the standardized value of its coefficient. The community-contributed `lstand` command (available for

download by typing `net install sg63, from(http://www.stata.com/stb/stb35))`
can be used to obtain standardized coefficients with partial correlations.

Standardized coefficients are calculated as

$$\frac{\beta s}{\sqrt{\pi^2/3}} \tag{9.49}$$

β is the parameter estimate or coefficient, and s is the standard deviation of the predictor. The denominator is the standard deviation of the underlying logistic distribution. Large values of the standardized coefficient indicate greater impact or worth to the model. Here the higher age groups are the most important predictors to the model.

Partial correlations, otherwise known as Atkinson's R, are calculated from

$$R = \sqrt{\frac{W - 2}{2|\mathcal{L}_o|}} \tag{9.50}$$

where W is the Wald statistic and $|\mathcal{L}_o|$ is the absolute value of the intercept-only log likelihood. A partial correlation is the correlation of the response with a predictor, adjusted by the other predictors in the model. The partial correlation for `anterior` can be calculated as

```
. quietly logistic death anterior hcabg kk2 kk3 kk4 age2-age4, nolog
. display "Atkinson's R(anterior) = " as res
> (sqrt((_b[anterior]/_se[anterior]-2)/(2*abs(e(ll_0)))))
Atkinson's R(anterior) = .03515609
```

The data below are from a study conducted by Milicer and Szczotka (1966) on young girls in Warsaw, Poland. Information was collected on the age of each girl on reaching menarche (for 3,918 girls). The dataset is compressed such that the age range of the girls was divided into 25 different categories. For each category (midpoint of the age interval) in the `age` variable, the number of girls having already reached menarche is recorded in the variable `menarche` and the total number of girls in the respective age range is recorded in the `total` variable. Thus, the dataset is a good example of grouped data.

```
. use http://www.stata-press.com/data/hh4/warsaw, clear
. list
```

	age	total	menarche
1.	9.21	376	0
2.	10.21	200	0
3.	10.58	93	0
4.	10.83	120	2
5.	11.08	90	2
6.	11.33	88	5
7.	11.58	105	10
8.	11.83	111	17
9.	12.08	100	16
10.	12.33	93	29
11.	12.58	100	39
12.	12.83	108	51
13.	13.08	99	47
14.	13.33	106	67
15.	13.58	105	81
16.	13.83	117	88
17.	14.08	98	79
18.	14.33	97	90
19.	14.58	120	113
20.	14.83	102	95
21.	15.08	122	117
22.	15.33	111	107
23.	15.58	94	92
24.	15.83	114	112
25.	17.58	1049	1049

```
. generate proportion = menarche/total
. glm proportion age
Iteration 0:   log likelihood =  16.191308
```

Generalized linear models		No. of obs	=	25
Optimization : ML		Residual df	=	23
		Scale parameter =		.0174259
Deviance = .4007947152		(1/df) Deviance =		.0174259
Pearson = .4007947152		(1/df) Pearson =		.0174259
Variance function: V(u) = 1		[Gaussian]		
Link function : g(u) = u		[Identity]		
		AIC	=	-1.135305
Log likelihood = 16.19130842		BIC	=	-73.63335

proportion	Coef.	OIM Std. Err.	z	P>|z|	[95% Conf. Interval]
age	.1852712	.0132578	13.97	0.000	.1592864 .211256
_cons	-1.915139	.1756773	-10.90	0.000	-2.25946 -1.570818

```
. predict phat
(option mu assumed; predicted mean proportion)
```

```
. label var phat "Linear association"
. scatter proportion age || line phat age
```

The above two commands produce the illustration in figure 9.1 for the proportion of girls having reached menarche for each of the 25 age intervals. The figure shows a classic S-shaped curve common to many logistic regression problems.

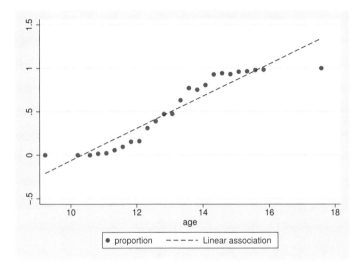

Figure 9.1. Sample proportions of girls reaching menarche for each age category

```
. glm menarche age, family(binomial total)
Iteration 0:   log likelihood = -55.642964
Iteration 1:   log likelihood = -55.377952
Iteration 2:   log likelihood = -55.377628
Iteration 3:   log likelihood = -55.377628
```

Generalized linear models			No. of obs	=	25
Optimization	: ML		Residual df	=	23
			Scale parameter =		1
Deviance	= 26.70345269		(1/df) Deviance =		1.16102
Pearson	= 21.86985435		(1/df) Pearson =		.9508632
Variance function: V(u) = u*(1-u/total)			[Binomial]		
Link function : g(u) = ln(u/(total-u))			[Logit]		
			AIC	=	4.59021
Log likelihood = -55.37762768			BIC	=	-47.33069

menarche	Coef.	OIM Std. Err.	z	P>|z|	[95% Conf. Interval]
age	1.631968	.0589532	27.68	0.000	1.516422 1.747514
_cons	-21.22639	.7706859	-27.54	0.000	-22.73691 -19.71588

Figures 9.2 and 9.3 show two illustrations of the fit of the model to the data. To produce the initial plot, we specify

```
. predict double phat2
(option mu assumed; predicted mean menarche)
. replace phat2 = phat2/total
(25 real changes made)
. label var phat2 "Predicted proportion reaching menarche"
. twoway (scatter phat2 age, msymbol(o)) (scatter proportion age, msymbol(.))
```

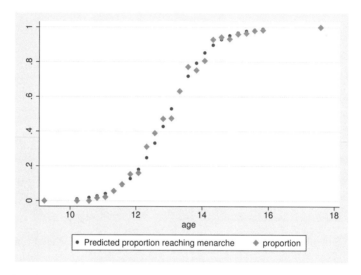

Figure 9.2. Predicted probabilities of girls reaching menarche for each age category

The second illustration first generates a variable that is used to create a diagonal line on the figure. Then, the predicted proportions are plotted versus the sample probabilities. The diagonal line is a reference for equality of the two axis-defining values.

```
. generate equal = (_n-1)/24
. twoway (scatter phat2 proportion) (line equal equal),
> ytitle("Predicted proportion") xtitle("Observed proportion")
```

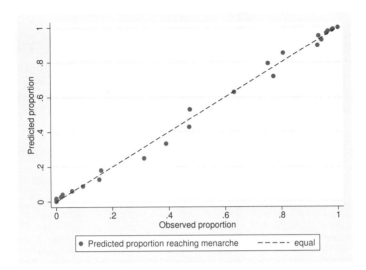

Figure 9.3. Predicted probabilities and sample proportions of girls reaching menarche for each age category

The next example uses data from a survey measuring attitudes about entering the European Economic Community (EEC). These data include whether the respondent views entry into the EEC favorably, yes; whether the respondent is not a union member, unionnone; whether the respondent is working class, classwork; whether the respondent has a minimum of schooling, schoolmin; and whether the respondent's attitude has not been constant, consnot.

```
. use http://www.stata-press.com/data/hh4/eec, clear

. glm yes consnot schoolmin unionnone classwork, family(binomial total)

Iteration 0:   log likelihood = -44.599133
Iteration 1:   log likelihood = -44.577733
Iteration 2:   log likelihood = -44.577733

Generalized linear models                    No. of obs      =         16
Optimization     : ML                        Residual df     =         11
                                             Scale parameter =          1
Deviance         =  15.31299646              (1/df) Deviance =   1.392091
Pearson          =  15.02841402              (1/df) Pearson  =   1.366219

Variance function: V(u) = u*(1-u/total)      [Binomial]
Link function    : g(u) = ln(u/(total-u))    [Logit]

                                             AIC             =   6.197217
Log likelihood   = -44.5777329               BIC             =  -15.18548
```

		OIM				
yes	Coef.	Std. Err.	z	P>|z|	[95% Conf.	Interval]
consnot	-.9705364	.1244174	-7.80	0.000	-1.21439	-.7266827
schoolmin	-.2872516	.1125091	-2.55	0.011	-.5077654	-.0667377
unionnone	.1199609	.1107706	1.08	0.279	-.0971455	.3370673
classwork	-.4568704	.1239988	-3.68	0.000	-.6999036	-.2138372
_cons	1.42455	.1457412	9.77	0.000	1.138903	1.710198

Now we generate the interaction between working class and not constantly working to see whether such interaction might result in a better fitting model.

```
. glm yes consnot schoolmin unionnone classwork consnot#classwork,
> family(binomial total)
note: 1.consnot#0.classwork omitted because of collinearity
note: 1.consnot#1.classwork omitted because of collinearity

Iteration 0:   log likelihood = -44.013849
Iteration 1:   log likelihood = -43.997732
Iteration 2:   log likelihood = -43.997732

Generalized linear models                   No. of obs       =         16
Optimization     : ML                       Residual df      =         10
                                            Scale parameter  =          1
Deviance      =  14.15299404                (1/df) Deviance  =   1.415299
Pearson       =  13.89108844                (1/df) Pearson   =   1.389109

Variance function: V(u) = u*(1-u/total)     [Binomial]
Link function    : g(u) = ln(u/(total-u))   [Logit]
                                            AIC              =   6.249716
Log likelihood   = -43.99773169             BIC              =  -13.57289
```

yes	Coef.	OIM Std. Err.	z	P>|z|	[95% Conf. Interval]	
consnot	-.8076723	.1949602	-4.14	0.000	-1.189787	-.4255574
schoolmin	-.289237	.1126959	-2.57	0.010	-.5101169	-.068357
unionnone	.1210439	.1109298	1.09	0.275	-.0963745	.3384622
classwork	-.5518749	.1527046	-3.61	0.000	-.8511704	-.2525794
consnot# classwork						
0 1	.2685823	.2493185	1.08	0.281	-.220073	.7572376
1 0	0	(omitted)				
1 1	0	(omitted)				
_cons	1.334048	.1660547	8.03	0.000	1.008587	1.659509

The coefficient table lists the associated large-sample Wald-type test for the interaction variable consnot#classwork along with a p-value of 0.281. An analysis of deviance, which is equivalent to a likelihood-ratio test, may also be performed.

The model with the interaction has deviance 14.153, and the model without the interaction has deviance 15.313. The difference in deviance is 1.16, and the difference in the models is 1 degree of freedom (one covariate difference). This difference in deviance is a $\chi^2(1)$ variate with associated p-value given by

```
. display chi2tail(1,1.16)
.28146554
```

The resulting analysis of deviance p-value is close to the Wald test p-value listed in the table of parameter estimates. The interaction term does not significantly contribute to the model and therefore should not be in the final model. Further information on logistic regression models can be found in Hilbe (2009).

10 The general binomial family

Contents

10.1 Noncanonical binomial models

In the previous chapter, we presented the canonical binomial model, including its derivation and separation into grouped and binary parameterizations. Here we discuss many of the noncanonical links associated with the binomial family.

Noncanonical links that have been used with the binomial family are given in table 10.1.

Table 10.1. Common binomial noncanonical link functions

probit	log	log-complement	inverse
clog-log	log-log	identity	

See table A.2 for the associated formulas.

The first two, probit and clog-log, are perhaps the most often used. In fact, the clog-log regression model began in the same year (Fisher 1922) that likelihood theory was formulated (Fisher 1922). Twelve years later, Fisher (1934) discovered the exponential family of distributions and identified its more well-known characteristics.

That same year, Bliss (1934a,b) used the new maximum likelihood (ML) approach to develop a probit model for bioassay analysis. It was not until 1944 that Berkson (1944) designed the grouped logit model to analyze bioassay data. Later, Dyke and Patterson (1952) used the same model to analyze cancer survey data.

Rasch (1960) then developed the Bernoulli or binary parameterization of the logit model. He used it in his seminal work on item analysis. Later in that decade and in the early 1970s, statisticians like Lindsey, Nelder, and Wedderburn began unifying various GLMs. The last two theorists, Nelder and Wedderburn (1972), published a unified theory of GLMs and later led a group that designed the first GLM software program, called GLIM. The software went through four major versions before development was discontinued. Version 4 came out in 1993, about the same period that GENSTAT and S-Plus added support for GLM procedures and when both Stata and SAS were developing their respective GLM programs.

The GLM framework spurred more links that may not have been conceived outside its scope. For example, the exponential link was provided as an option to later versions of GLIM. It is defined as

$$\eta = (r_2 + \mu)^{r_1} \tag{10.1}$$

where r_1 and r_2 are user-specified constants. Because the link has found little use in the literature, we only mention it here. It is a modification to the power link that we discussed in an earlier chapter. Instead, some have used the odds power link. Wacholder (1986) brought log-binomial and identity-binomial links to the forefront where Hardin and Cleves (1999) introduced the techniques into Stata. They have been used for health risk and risk difference analysis. For example, several major infectious disease studies have effectively used the log-binomial model.

Again, traditional GLM uses the iteratively reweighted least-squares (IRLS) algorithm to determine parameter estimates and associated standard errors. We have shown how one can amend the basic algorithm to produce standard errors based on the observed information matrix (OIM), and we have shown how one can fit GLMs by using standard ML techniques. Similarly, we now investigate noncanonical binomial models.

10.2   Noncanonical binomial links (binary form)

We begin by providing table 10.2 of noncanonical links, inverse links, and derivatives. We display them again to allow easy access to them when considering their position within the binomial algorithm.

Table 10.2. Noncanonical binomial link functions ($\eta = \mathbf{x}\boldsymbol{\beta} + \text{offset}$)

Link name	Link function $\eta = g(\mu)$	Inverse link $\mu = g^{-1}(\eta)$	First derivatives $\nabla = \partial\mu/\partial\eta$
Identity	μ	η	1
Log	$\ln(\mu)$	$\exp(\eta)$	μ
Log-complement	$\ln(1-\mu)$	$1-\exp(\eta)$	$\mu - 1$
Log-log	$-\ln\{-\ln(\mu)\}$	$\exp\{-\exp(-\eta)\}$	$-\mu\ln(\mu)$
Clog-log	$\ln\{-\ln(1-\mu)\}$	$1-\exp\{-\exp(\eta)\}$	$(\mu-1)\ln(1-\mu)$
Probit	$\Phi^{-1}(\mu)$	$\Phi(\eta)$	$\phi(\eta)$
Odds power ($\alpha = 0$)	$\ln\left(\dfrac{\mu}{1-\mu}\right)$	$\dfrac{e^{\eta}}{1+e^{\eta}}$	$\mu(1-\mu)$
Odds power ($\alpha \neq 0$)	$\dfrac{\mu/(1-\mu)^{\alpha}-1}{\alpha}$	$\dfrac{(1+\alpha\eta)^{1/\alpha}}{1+(1+\alpha\eta)^{1/\alpha}}$	$\dfrac{\mu(1-\mu)}{1+\alpha\eta}$

You can incorporate table 10.2's formulas into the basic GLM algorithm to provide different linked models. However, noncanonical links offer a problem with standard errors. When modeling datasets that have relatively few observations, we have found that it is better to either adjust the weights in the IRLS algorithm to construct an OIM or use ML methods to model the data (which also use the OIM). If neither adjustment is done, we are left with an expected information matrix (EIM) standard errors based on the EIM V_{EH} given in (3.66).

Using the EIM to produce standard errors will have no significant impact on standard errors when the number of observations in the model is relatively large. Just how large depends on a variety of factors including the number of predictors in the model and how the data are balanced (for example, the comparative number of ones to zeros for binary responses). However, even for smaller unbalanced datasets having many predictors, the resulting standard errors will usually not vary too much between the two estimating processes. Because the enhanced accuracy of standard errors is a paramount concern to researchers, using OIM as the basis of standard errors is preferable whenever possible.

We demonstrate in each of the following sections how one incorporates these formulas into the basic GLM algorithm. We also show how one can construct ML algorithms to model several of the above links. Finally, we will give examples in each section to demonstrate how to use and interpret the models.

10.3   The probit model

The probit model was first used in bioassay or quantal analysis. Typically, the probability of death was measured against the log of some toxin (for example, an insecticide). Whether death occurred depended on the tolerance the subject had to the toxic agent. Subjects with a lower tolerance were more likely to die. Tolerance was assumed to be distributed normally, hence the use of the probit transform, $\mu = \Phi(\mathbf{x}\boldsymbol{\beta})$. Researchers still use probit for bioassay analysis, and they often use it to model other data situations.

The probit link function is the inverse of Φ, the cumulative normal distribution defined as

$$\mu = \int_{-\infty}^{\eta} \phi(u)du \tag{10.2}$$

where $\phi(u)$ is the standard normal density function

$$\frac{1}{\sqrt{2\pi}} \exp\left(-\frac{1}{2}u^2\right) \tag{10.3}$$

Using a probit regression model on binary or grouped binomial data typically results in output similar to that from logistic regression. Logistic regression is usually preferred because of the wide variety of fit statistics associated with the model. Moreover, exponentiated probit coefficients cannot be interpreted as odds ratios, but they can be interpreted as odds ratios in logistic models (see section 10.6.2). However, if normality is involved in the linear relationship, as it often is in bioassay, then probit may be the appropriate model. It may also be used when the researcher is not interested in odds but rather in prediction or classification. Then if the deviance of a probit model is significantly lower than that of the corresponding logit model, the former is preferred. This dictum holds when comparing any of the links within the binomial family.

As shown in figure 10.1, both the logit and probit transforms produce a sigmoidal curve mapping the probability of μ to the linear predictor, η. Of course, the probability interval is $(0, 1)$, and the linear predictor is in $(-\infty, \infty)$. Both functions are symmetric with a mean of 0.5.

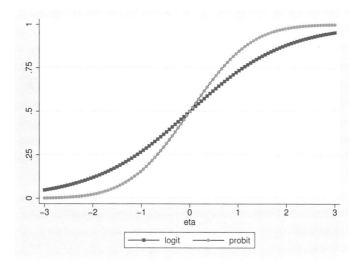

Figure 10.1. Probit and logit functions

One can easily change the traditional Fisher scoring IRLS logit algorithm to a probit algorithm. However, doing so implies that we maintain the same basis for producing standard errors. The substitution of the probit link, inverse link, and derivative of the link into the algorithm is all that is required. Because the link is a bit more complicated than that of the logit link, we add an initial line before calculating w. The probit algorithm, parameterized for binary response models, is

Listing 10.1. IRLS algorithm for binary probit regression

```
1   Dev = 0
2   μ = (y + 0.5)/2              /* initialization of μ */
3   η = Φ⁻¹(μ)                   /* initialization of η */
4   WHILE (abs(ΔDev) > tolerance ) {
5       φ = exp(−η²/2)/√2π   /* standard normal density */
6       W = φ²/{μ(1 − μ)}
7       z = η + (y − μ)/φ
8       β = (XᵀWX)⁻¹XᵀWz
9       η = Xβ
10      μ = Φ(η)
11      OldDev = Dev
12      Dev = 2∑ ln(1/μ) if (y = 1)
13      Dev = 2∑ ln{1/(1 − μ)} if (y = 0)
14      ΔDev = Dev − OldDev
15   }
16   χ² = ∑ (y − μ)²/{μ(1 − μ)}
```

We have stressed throughout this book the importance of standard errors. Certainly, assessing the comparative worth of predictors to a model is crucial to the modeling process. It is one of the prime reasons we engage in data modeling, hence our emphasis on methods that use (3.67) as the basis of OIM standard errors. In the past, software designers and researchers used the EIM standard errors because most GLM software was implemented using Fisher scoring. Because Fisher scoring uses the expected Hessian, that matrix is available for standard-error construction with no extra code or computer cycles.

The IRLS algorithm can be amended to produce an OIM. Standard errors are then identical to those produced using ML algorithms. They are implemented in the algorithm below.

Listing 10.2. IRLS algorithm for binary probit regression using OIM

```
1    Dev = 0
2    μ = (y + 0.5)/2              /* initialization of μ */
3    η = Φ⁻¹(μ)                   /* initialization of η */
4    WHILE (abs(ΔDev) > tolerance )  {
5         φ = exp(-η²/2)/√2π
6         v = μ(1 - μ)
7         v' = 1 - 2μ
8         W = φ²/v
9         g' = 1/φ
10        g'' = ηφ²
11        Wₒ = W + (y - μ)(vg'' + v'g')/(v²(g')³)
12        z = η + (y - μ)g'
13        β = (XᵀWₒX)⁻¹XᵀWₒz
14        η = Xβ
15        μ = Φ(η)
16        OldDev = Dev
17        Dev = 2∑ln(1/μ) if (y = 1)
18        Dev = 2∑ln{1/(1 - μ)} if (y = 0)
19        ΔDev = Dev - OldDev
20    }
21   χ² = ∑(y - μ)²/{μ(1 - μ)}
```

Finally, sometimes ML programs are required for modeling endeavors. We list the link functions for both logit and probit parameterized in the requisite $\mathbf{x}\boldsymbol{\beta}$ format. The binomial log-likelihood function, in terms of μ, can be parameterized to the logit or probit model by substituting the inverse link functions in place of μ. That is, each μ value in the log-likelihood function is replaced by the inverse links below. With appropriate initial or starting values, ML software should be able to use the resulting log-likelihood functions to derive parameter estimates and standard errors based on the observed Hessian.

Binary logit and probit models assume that the response comprises fairly equal numbers of ones compared with zeros. When there is a significant disparity, clog-log or log-log models may provide better models because of their asymmetric nature (as indicated by smaller BIC statistics). The asymmetry in these links will sometimes be a better fit to rare outcomes.

We model heart01.dta using a probit link to see whether there is any significant reduction in deviance from the logit-link model. The binomial or grouped logistic model yielded a deviance of 1,273.251. The probit model below produces a deviance of 1,268.626. Also, the BIC statistic for the probit model is less than that for the logistic model, suggesting a preference for the probit model. If we are not interested in assessing odds ratios but are interested in the predictive and classification value of the model, then the probit model is preferable to that of logit.

```
. use http://www.stata-press.com/data/hh4/heart01, clear
. glm death anterior hcabg kk2 kk3 kk4 age2-age4, family(binomial) link(probit)
> nolog
Generalized linear models                          No. of obs        =      4,483
Optimization     : ML                              Residual df       =      4,474
                                                   Scale parameter =          1
Deviance         =  1268.626075                    (1/df) Deviance =    .2835552
Pearson          =  4469.044561                    (1/df) Pearson  =    .9988924
Variance function: V(u) = u*(1-u)                  [Bernoulli]
Link function    : g(u) = invnorm(u)               [Probit]
                                                   AIC               =    .2870011
Log likelihood   = -634.3130373                    BIC               =  -36348.98
```

death	Coef.	OIM Std. Err.	z	P>|z|	[95% Conf. Interval]	
anterior	.3002858	.0765854	3.92	0.000	.1501812	.4503903
hcabg	.3489019	.1770991	1.97	0.049	.0017941	.6960097
kk2	.3757268	.0854744	4.40	0.000	.2082	.5432535
kk3	.4104931	.130373	3.15	0.002	.1549667	.6660195
kk4	1.428809	.1968568	7.26	0.000	1.042977	1.814642
age2	.1833151	.1266684	1.45	0.148	-.0649504	.4315805
age3	.6487667	.1102819	5.88	0.000	.4326182	.8649152
age4	1.002231	.1164327	8.61	0.000	.7740269	1.230435
_cons	-2.56975	.1046419	-24.56	0.000	-2.774845	-2.364656

The GOF for a probit model may follow the same logic as many fit tests for logistic regression models. A Hosmer–Lemeshow table of observed versus predicted values for both ones and zeros can be constructed for fitted probit statistics. Graphs of standardized deviance or Anscombe residuals versus μ statistics can tell analysts whether the fit of the model is appropriate for the data. Residual analysis is vital to modeling.

In chapter 9, we examined the fit of logistic regression to the grouped Warsaw data. Here we compare that fit with specification of the probit link.

```
. use http://www.stata-press.com/data/hh4/warsaw
. glm menarche age, family(binomial total) link(probit) nolog
Generalized linear models                        No. of obs      =          25
Optimization      : ML                           Residual df     =          23
                                                 Scale parameter =           1
Deviance          =  22.88743324                 (1/df) Deviance =    .9951058
Pearson           =  21.90102423                 (1/df) Pearson  =    .9522184
Variance function: V(u) = u*(1-u/total)          [Binomial]
Link function     : g(u) = invnorm(u/total)      [Probit]
                                                 AIC             =    4.437569
Log likelihood    = -53.46961796                 BIC             =   -51.14671
```

menarche	Coef.	OIM Std. Err.	z	P>|z|	[95% Conf. Interval]	
age	.9078231	.0295303	30.74	0.000	.8499447	.9657015
_cons	-11.81894	.3873598	-30.51	0.000	-12.57815	-11.05973

Although a comparison of the predicted values in figure 10.2 does not indicate strong subjective evidence for distinguishing models on the basis of link, the AIC statistic is smaller for the probit link.

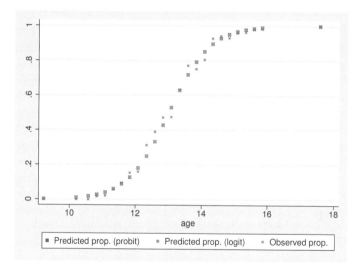

Figure 10.2. Predicted probabilities for probit and logit link function in grouped binary models. The observed (sample) proportions are included as well.

10.4   The clog-log and log-log models

Clog-log and log-log links are asymmetrically sigmoidal. For clog-log models, the upper part of the sigmoid is more elongated or stretched out than the logit or probit. Log-log models are based on the converse. The bottom of the sigmoid is elongated or skewed to the left.

We provide the link, inverse link, and derivative of the link for both clog-log and log-log models in table 10.2. As we showed for the probit model, we can change logit or probit to clog-log or log-log by replacing the respective functions in the GLM–binomial algorithm.

Figure 10.3 shows the asymmetry of the clog-log and log-log links. We investigate the consequences of this asymmetry on the fitted values and discuss the relationship of the coefficient estimates from models using these links.

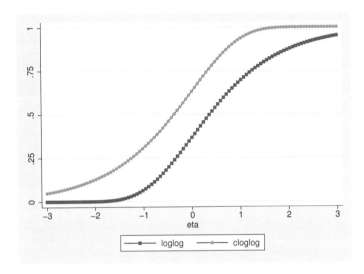

Figure 10.3. Complementary log-log and log-log functions

We may wish to change the focus of modeling from the positive to the negative outcomes. Assume that we have a binary outcome variable y and the reverse outcome variable $z = 1 - y$. We have a set of covariates $\mathbf{X}$ for our model. If we use a symmetric link, then we can fit a binary model to either outcome. The resulting coefficients will differ in sign for the two models and the predicted probabilities (fitted values) will be complementary. They are complementary in that the fitted probabilities for the model on y plus the fitted probabilities for the model on z add to one. See our previous warning note on the sign change of the coefficients in section 9.2.

We say that probit(y) and probit(z) form a complementary pair and that logit(y) and logit(z) form a complementary pair. Although the equivalence of the link in the complementary pair is true with symmetric links, it is not true for the asymmetric links. The log-log and clog-log links form a complementary pair. This means that loglog(y) and cloglog(z) form a complementary pair and that loglog(z) and cloglog(y) form a complementary pair.

Algebraically, we can illustrate the change in coefficient signs for complementary pairs:

$$\text{probit}(z) = \Phi^{-1}(z) = \Phi^{-1}(1-y) = -\Phi^{-1}(y) \tag{10.4}$$
$$= -\text{probit}(y) \tag{10.5}$$
$$\text{logit}(z) = \ln\left(\frac{z}{1-z}\right) = \ln\left\{\frac{1-y}{1-(1-y)}\right\} = \ln\left(\frac{1-y}{y}\right) = -\ln\left(\frac{y}{1-y}\right) \tag{10.6}$$
$$= -\text{logit}(y) \tag{10.7}$$
$$\text{loglog}(z) = -\ln\{-\ln(z)\} = -\ln\{-\ln(1-y)\} = -[\ln\{-\ln(1-y)\}] \tag{10.8}$$
$$= -\text{cloglog}(y) \tag{10.9}$$
$$\text{cloglog}(z) = \ln\{-\ln(1-z)\} = \ln[-\ln\{1-(1-y)\}] = -[-\ln\{-\ln(y)\}] \tag{10.10}$$
$$= -\text{loglog}(y) \tag{10.11}$$

Fitted values are functions of the inverse links given in table A.2. Using this information along with a superscript to identify the dependent variable and a subscript to denote the inverse link, we know that the linear predictors will differ in sign, $\eta^y_{\text{link}} = -\eta^z_{\text{complementary link}}$. We can algebraically illustrate the complementary fitted values for complementary pairs:

$$\mu^z_{\text{probit}} = \Phi(\eta^z_{\text{probit}}) = \Phi(-\eta^y_{\text{probit}}) = 1 - \Phi(\eta^y_{\text{probit}}) \tag{10.12}$$
$$= 1 - \mu^y_{\text{probit}} \tag{10.13}$$
$$\mu^z_{\text{logit}} = \frac{\exp(\eta^z_{\text{probit}})}{1+\exp(\eta^z_{\text{probit}})} = \frac{\exp(-\eta^y_{\text{probit}})}{1+\exp(-\eta^y_{\text{probit}})} = 1 - \frac{\exp(\eta^y_{\text{probit}})}{1+\exp(\eta^y_{\text{probit}})} \tag{10.14}$$
$$= 1 - \mu^y_{\text{logit}} \tag{10.15}$$
$$\mu^z_{\text{loglog}} = \exp\{-\exp(-\eta^z_{\text{loglog}})\} = \exp\{-\exp(\eta^y_{\text{cloglog}})\} \tag{10.16}$$
$$= 1 - \mu^y_{\text{cloglog}} \tag{10.17}$$
$$\mu^z_{\text{cloglog}} = 1 - \exp\{-\exp(\eta^z_{\text{cloglog}})\} = 1 - \exp\{-\exp(-\eta^y_{\text{loglog}})\} \tag{10.18}$$
$$= 1 - \mu^y_{\text{loglog}} \tag{10.19}$$

Because of the asymmetric nature of the links,

$$\mu^y_{\text{loglog}} + \mu^z_{\text{loglog}} \neq 1 \tag{10.20}$$
$$\mu^y_{\text{cloglog}} + \mu^z_{\text{cloglog}} \neq 1 \tag{10.21}$$

The proper relationship illustrates the complementary nature of the clog-log link.

$$\mu^y_{\text{loglog}} + \mu^z_{\text{cloglog}} = 1 \tag{10.22}$$
$$\mu^z_{\text{loglog}} + \mu^y_{\text{cloglog}} = 1 \tag{10.23}$$

The clog-log inverse link function, which defines the fitted value for the model, is also used to predict the probability of death in data that are otherwise modeled using Cox proportional hazards models. In a Cox model, given that death is the failure event, the probability of death given predictors is a function of time. However, if we interpret the data only as those observations having reached the end point (death in this example) by a specific time, then we can estimate the likelihood of death by that specific time given the various predictors in the model as $\Pr(\text{death}) = 1 - \exp\{-\exp(\mathbf{x}\boldsymbol{\beta})\}$. One can thus use the log-log inverse link to predict the probability of survival over the same time period.

The binary clog-log algorithms for EIM and OIM variance calculations are displayed (respectively) below.

Listing 10.3. IRLS algorithm for binary clog-log regression using EIM

```
1    Dev = 0
2    μ = (y + 0.5)/2                    /* initialization of μ */
3    η = ln{- ln(1 - μ)}               /* initialization of η */
4    WHILE (abs(ΔDev) > tolerance )  {
5        W = {(1 - μ) ln(1 - μ)}²/{μ(1 - μ)}
6        z = η - (y - μ)/{(1 - μ) ln(1 - μ)}
7        β = (XᵀWX)⁻¹XᵀWz
8        η = Xβ
9        μ = 1 - exp{- exp(η)}
10       OldDev = Dev
11       Dev = 2 ∑ ln(1/μ) if (y = 1)
12       Dev = 2 ∑ ln{1/(1 - μ)} if (y = 0)
13       ΔDev = Dev - OldDev
14   }
15   χ² = ∑ (y - μ)²/{μ(1 - μ)}
```

Listing 10.4. IRLS algorithm for binary clog-log regression using OIM

```
1    Dev = 0
2    μ = (y + 0.5)/2                    /* initialization of μ */
3    η = ln{- ln(1 - μ)}               /* initialization of η */
4    WHILE (abs(ΔDev) > tolerance )  {
5        v = μ(1 - μ)
6        v' = 1 - 2μ
7        W = {(1 - μ) ln(1 - μ)}²/{μ(1 - μ)}
8        g' = -1/{(1 - μ) ln(1 - μ)}
9        g'' = {-1 - ln(1 - μ)}/{(μ - 1)² ln(1 - μ)²}
10       Wₒ = w + (y - μ)(vg'' + v'g')/(v²(g')³)
11       z = η + (y - μ)g'
12       β = (XᵀWₒX)⁻¹XᵀWₒz
13       η = Xβ
14       μ = 1 - exp{- exp(η)}
15       OldDev = Dev
```

16   `Dev = ` $2\sum \ln(1/\mu)$ ` if ` $(y = 1)$

17   `Dev = ` $2\sum \ln\{1/(1-\mu)\}$ ` if ` $(y = 0)$

18   Δ`Dev = Dev - OldDev`

19   `}`

20     $\chi^2 = \sum (y - \mu)^2 / \{\mu(1 - \mu)\}$

If an ML routine is being used to model clog-log data, then the likelihood function needs to be converted to $\mathbf{x\beta}$ parameterization. The formulas for both the clog-log and log-log log likelihoods are

Inverse link	Formula	$y = 0$	$y = 1$
Clog-log	$1 - \exp\{-\exp(\mathbf{x\beta})\}$	$-\exp(\mathbf{x\beta})$	$\ln[1 - \exp\{-\exp(\mathbf{x\beta})\}]$
Log-log	$\exp\{-\exp(-\mathbf{x\beta})\}$	$\ln[1 - \exp\{-\exp(-\mathbf{x\beta})\}]$	$-\exp(-\mathbf{x\beta})$

Finally, for completeness, we provide an algorithm for the simple log-log model.

Listing 10.5. IRLS algorithm for binary log-log regression

1   `Dev = 0`

2      $\mu = (y + 0.5)/2$ `/* initialization of ` μ ` */`

3      $\eta = -\ln\{-\ln(\mu)\}$ `/* initialization of ` η ` */`

4   `WHILE (abs(`Δ`Dev) > tolerance ) {`

5         $W = \{\mu \ln(\mu)\}^2$

6         $z = \eta + (y - \mu)/\{\mu \ln(\mu)\}$

7         $\beta = (X^T W X)^{-1} X^T W z$

8         $\eta = X\beta$

9         $\mu = \exp\{-\exp(-\eta)\}$

10   `OldDev = Dev`

11   `Dev = ` $2\sum \ln(1/\mu)$ ` if ` $(y = 1)$

12   `Dev = ` $2\sum \ln\{1/(1-\mu)\}$ ` if ` $(y = 0)$

13   Δ`Dev = Dev - OldDev`

14   `}`

15     $\chi^2 = \sum (y - \mu)^2 / \{\mu(1 - \mu)\}$

Continuing with `heart01.dta` as our example, we now use a log-log model to fit the data. There is good reason to suspect that the log-log model is preferable to other binomial links. The following table gives the respective frequency of ones and zeros in the binary model:

```
. use http://www.stata-press.com/data/hh4/heart01, clear
. quietly glm death anterior hcabg kk2 kk3 kk4 age2-age4, family(binomial)
> link(probit) nolog
. tabulate death if e(sample)
```

death	Freq.	Percent	Cum.
0	4,307	96.07	96.07
1	176	3.93	100.00
Total	4,483	100.00	

We have used Stata's e(sample) function to restrict the tabulation sample to be the same sample used in the previous estimation command.

Modeling the grouped data with the log-log link assumes that the data have significantly more zeros than ones in the binary form. Accordingly, the resulting log-log model should have a lower deviance statistic than that of other links.

```
. glm death anterior hcabg kk2 kk3 kk4 age2-age4, family(binomial) link(loglog)
> nolog
```

Generalized linear models		No. of obs	=	4,483
Optimization : ML		Residual df	=	4,474
		Scale parameter =		1
Deviance = 1265.944904		(1/df) Deviance =		.2829559
Pearson = 4719.701375		(1/df) Pearson =		1.054918
Variance function: V(u) = u*(1-u)		[Bernoulli]		
Link function : g(u) = -ln(-ln(u))		[Log-log]		
		AIC	=	.2864031
Log likelihood = -632.972452		BIC	=	-36351.66

death	Coef.	OIM Std. Err.	z	P>|z|	[95% Conf. Interval]	
anterior	.2041431	.0519702	3.93	0.000	.1022834	.3060028
hcabg	.2318145	.1296487	1.79	0.074	-.0222923	.4859213
kk2	.2523179	.0597734	4.22	0.000	.1351642	.3694715
kk3	.3149235	.0952601	3.31	0.001	.1282171	.5016299
kk4	1.18085	.1900758	6.21	0.000	.8083082	1.553392
age2	.104686	.0782599	1.34	0.181	-.0487006	.2580726
age3	.4162827	.0700609	5.94	0.000	.2789658	.5535995
age4	.6921546	.0785702	8.81	0.000	.5381598	.8461494
_cons	-1.699495	.0640278	-26.54	0.000	-1.824987	-1.574003

We find the deviance to be 1,265.945, which is less than that produced by previously investigated links. Ranking of the respective deviance statistics (the clog-log model output does not appear in the text) gives

Model	Deviance	BIC
Log-log	1265.945	−36351.66
Probit	1268.626	−36348.98
Logit	1273.251	−36344.35
Clog-log	1276.565	−36341.04

We do, in fact, determine that the log-log model is preferable to the others on the basis of the minimum-deviance criterion for model selection. (We will not address here whether the difference in log-log and probit deviance values is significant.) However, before putting a final blessing on the model, we would need to evaluate the fit by using residual analysis. Particularly, we advise the use of Anscombe residuals versus μ plots. One can also use many other residuals to assess fit and goodness of link; see chapter 4. We may also construct a table analogous to that of the Hosmer–Lemeshow GOF table. The table, if grouped appropriately, helps to identify areas in the distribution of μ where fit may be lacking.

We can similarly assess the choice of link with the Warsaw data (described in chapter 9). Here we illustrate the header information for each of four candidate link functions.

```
. use http://www.stata-press.com/data/hh4/warsaw
. glm menarche age, family(binomial total) link(logit) notable nolog
Generalized linear models                      No. of obs     =          25
Optimization     : ML                          Residual df    =          23
                                               Scale parameter =          1
Deviance         =  26.70345269                (1/df) Deviance =    1.16102
Pearson          =  21.86985435                (1/df) Pearson  =   .9508632
Variance function: V(u) = u*(1-u/total)        [Binomial]
Link function    : g(u) = ln(u/(total-u))      [Logit]
                                               AIC            =     4.59021
Log likelihood   = -55.37762768                BIC            =   -47.33069

. glm menarche age, family(binomial total) link(probit) notable nolog
Generalized linear models                      No. of obs     =          25
Optimization     : ML                          Residual df    =          23
                                               Scale parameter =          1
Deviance         =  22.88743324                (1/df) Deviance =   .9951058
Pearson          =  21.90102423                (1/df) Pearson  =   .9522184
Variance function: V(u) = u*(1-u/total)        [Binomial]
Link function    : g(u) = invnorm(u/total)     [Probit]
                                               AIC            =    4.437569
Log likelihood   = -53.46961796                BIC            =   -51.14671

. glm menarche age, family(binomial total) link(loglog) notable nolog
Generalized linear models                      No. of obs     =          25
Optimization     : ML                          Residual df    =          23
                                               Scale parameter =          1
Deviance         =  34.6387326                 (1/df) Deviance =    1.506032
Pearson          =  33.90973603                (1/df) Pearson  =    1.474336
Variance function: V(u) = u*(1-u/total)        [Binomial]
Link function    : g(u) = -ln(-ln(u/total))    [Log-log]
                                               AIC            =    4.907621
Log likelihood   = -59.34526764                BIC            =   -39.39541
```

```
. glm menarche age, family(binomial total) link(cloglog) notable nolog
Generalized linear models                          No. of obs      =        25
Optimization      : ML                             Residual df     =        23
                                                   Scale parameter =         1
Deviance          =   118.8207752                  (1/df) Deviance =  5.166121
Pearson           =   228.6307867                  (1/df) Pearson  =  9.940469
Variance function: V(u) = u*(1-u/total)            [Binomial]
Link function     : g(u) = ln(-ln(1-u/total))      [Complementary log-log]
                                                   AIC             =  8.274903
Log likelihood    =  -101.436289                   BIC             =  44.78663
```

Again listing in order based on deviance values, we can assess the models with the following table:

Model	Deviance	BIC
Probit	22.887	−51.147
Logit	26.703	−47.331
Log-log	34.639	−39.395
Clog-log	118.821	44.787

Here we find evidence (minimum deviance and minimum BIC) that the probit model is the most desirable. The clog-log link function provides the least desirable fit to the data.

10.5   Other links

Various GLM implementations, including the **glm** command in Stata, will allow specifying other links with the binomial family. Wacholder (1986) describes using the **log**, **logc**, and **identity** links with binomial data to obtain measures of the risk ratio (**log**), health ratio (**logc**), and risk difference (**identity**).

Although this is occasionally possible, we cannot always use these links with the binomial family. Table A.1 specifies that the range of μ is limited to $(0,1)$ for the Bernoulli case and to $(0,k)$ for grouped data. The **glm** command may take steps to ensure that the values of μ are within the appropriate range (as Wacholder suggests), but μ is calculated from the linear predictor. The linear predictor is simply the product $x\beta$, and β is a vector of free parameters. There is no parameterization to ensure that the estimation of β limits the outcomes of the linear predictor and thus the fitted values μ. Our experience suggests that we should be grateful when a binomial model with these links converges to an answer and that we should not be surprised when the fitting algorithm fails. We can graphically illustrate why this is the case, as shown in figure 10.4. Consider the identity link and compare it with the logit and probit functions depicted before:

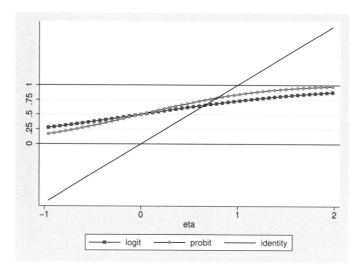

Figure 10.4. Probit, logit, and identity functions

The identity function is outside the required range for most of the domain. The same is true for the log and log complement links.

10.6   Interpretation of coefficients

In the previous chapter, we illustrated the interpretation of coefficients for logistic regression (binomial family with logit link). Here we expand that discussion and investigate the interpretation of coefficients and exponentiated coefficients for other links with the binomial family.

Interpreting coefficients is tenuous because of using link functions. Of course, interpretation is easier when the link is identity. Throughout the following, we assume that we have a binary outcome $\mathbf{y}$ and two covariates $\mathbf{x}_1$ and $\mathbf{x}_2$ (and a constant). The probability of a successful outcome ($\mathbf{y} = 1$) is given by

$$\Pr(y_i = 1) = p_i = g^{-1}(\beta_0 + x_{1i}\beta_1 + x_{2i}\beta_2) \qquad (10.24)$$

where $g^{-1}(\cdot)$ is the inverse link function. We will focus here on interpreting the β_1 coefficient.

In health studies, a successful outcome is generally negative (denotes death or onset of disease). The terminology used to refer to coefficients reflects this property.

10.6.1 Identity link

For the identity link, interpreting coefficients is straightforward. For a one-unit increase in x_1, the β_1 coefficient represents the difference or change in the average of the outcomes (conditional on the other covariates being held constant). Because the outcome represents risk, the coefficients are sometimes called risk differences. We can see that the risk differences are constant and independent of the values of the covariates by looking at the change in the outcome when we increase x_1 by one and hold the other covariates constant. This assumes, of course, that it makes sense to talk about populations that differ only in the value of x_1.

$$
\begin{align}
\Delta y_i &= \{\beta_0 + \beta_1(x_{1i} + 1) + \beta_2 x_{2i}\} - (\beta_0 + \beta_1 x_{1i} + \beta_2 x_{2i}) \tag{10.25} \\
&= \beta_1 \tag{10.26}
\end{align}
$$

Therefore, for the identity link in a binomial family model, the coefficients are interpreted as risk differences.

10.6.2 Logit link

In the previous chapter, we discussed the odds ratio. Here we expand that discussion and illustrate the origin of the odds-ratio term.

Because the inverse link is nonlinear, interpreting the coefficient is difficult.

$$
\Delta y_i = \frac{\exp\{\beta_0 + (x_{1i} + 1)\beta_1 + x_{2i}\beta_2\}}{1 + \exp\{\beta_0 + (x_{1i} + 1)\beta_1 + x_{2i}\beta_2\}} - \frac{\exp(\beta_0 + x_{1i}\beta_1 + x_{2i}\beta_2)}{1 + \exp(\beta_0 + x_{1i}\beta_1 + x_{2i}\beta_2)} \tag{10.27}
$$

The difficulty arises because the difference is not constant and depends on the values of the covariates.

Transforming the coefficients makes interpretation easier. We motivate this transformation by introducing an alternative metric.

The odds of a successful outcome is the ratio of the probability of success to the probability of failure. We write the odds as

$$
\text{Odds} = \frac{p}{1 - p} \tag{10.28}
$$

For the logit inverse link, we have for our specific model

$$
\begin{align}
\text{Odds} = \frac{p}{1 - p} &= \frac{\exp(\mathbf{x}\boldsymbol{\beta})/\{1 + \exp(\mathbf{x}\boldsymbol{\beta})\}}{1/\{1 + \exp(\mathbf{x}\boldsymbol{\beta})\}} \tag{10.29} \\
&= \exp(\mathbf{x}\boldsymbol{\beta}) \tag{10.30} \\
&= \exp(\beta_0 + x_{1i}\beta_1 + x_{2i}\beta_2) \tag{10.31}
\end{align}
$$

The odds may be calculated for any specific values for the covariate vector $\mathbf{x}$.

The odds ratio is the ratio of two odds. We calculate the odds ratio for each of the specific covariates in the model. The numerator odds is calculated such that the specific covariate is incremented by one relative to the value used in the denominator. The odds ratio for $\mathbf{x}_1$ is thus

$$\text{Odds ratio for } \mathbf{x}_1 \quad = \quad \frac{\exp\{\beta_0 + (x_{1i} + 1)\beta_1 + x_{2i}\beta_2\}}{\exp(\beta_0 + x_{1i}\beta_1 + x_{2i}\beta_2)} \tag{10.32}$$

$$= \quad \exp(\beta_1) \tag{10.33}$$

The odds ratio simplifies such that there is no dependence on a particular set of values for the other covariates. The odds ratio is therefore constant—it is independent of the particular values of the covariates. This interpretation does not hold for other inverse links.

Interpretation for binomial family models using the logit link is straightforward for exponentiated coefficients. These exponentiated coefficients represent odds ratios. An odds ratio of 2 indicates that a successful outcome has twice the odds if the associated covariate is increased by one.

10.6.3   Log link

Because the inverse link is nonlinear, interpreting the coefficient is difficult.

$$\Delta y_i \quad = \quad \exp\{\beta_0 + (x_{1i} + 1)\beta_1 + x_{2i}\beta_2\} - \exp(\beta_0 + x_{1i}\beta_1 + x_{2i}\beta_2) \tag{10.34}$$

The difficulty arises because the difference is not constant and depends on the values of the covariates.

Transforming the coefficients makes interpretation easier. We motivate this transformation by introducing an alternative metric.

The risk ratio is a ratio for each covariate of the probabilities of a successful outcome. A successful outcome under the log inverse link is characterized by

$$\Pr(y_i = 1) \quad = \quad g^{-1}(\beta_0 + x_{1i}\beta_1 + x_{2i}\beta_2) \tag{10.35}$$

$$= \quad \exp(\beta_0 + x_{1i}\beta_1 + x_{2i}\beta_2) \tag{10.36}$$

For the risk ratio, the numerator probability is calculated such that the specific covariate is incremented by one relative to the value used in the denominator probability. The risk ratio for $\mathbf{x}_1$ is calculated as

$$\text{Risk ratio for } \mathbf{x}_1 \quad = \quad \frac{\exp\{\beta_0 + (x_{1i} + 1)\beta_1 + x_{2i}\beta_2\}}{\exp(\beta_0 + x_{1i}\beta_1 + x_{2i}\beta_2)} \tag{10.37}$$

$$= \quad \exp(\beta_1) \tag{10.38}$$

There is no dependence on a particular observation. The risk ratio is therefore constant; that is, it is independent of the particular values of the covariates. This interpretation does not hold for other inverse links.

Interpretation for binomial family models using the log link is straightforward for exponentiated coefficients. These exponentiated coefficients represent risk ratios. A risk ratio of 2 indicates that a successful outcome is twice as likely (the risk is doubled) if the associated covariate is increased by one.

10.6.4 Log complement link

Because the inverse link is nonlinear, interpreting the coefficient is difficult.

$$\Delta y_i = 1 - \exp\{\beta_0 + (x_{1i} + 1)\beta_1 + x_{2i}\beta_2\} - \{1 - \exp(\beta_0 + x_{1i}\beta_1 + x_{2i}\beta_2)\} \quad (10.39)$$

The difficulty arises because the difference is not constant and depends on the values of the covariates.

Transforming the coefficients makes interpretation easier. We motivate this transformation by introducing an alternative metric.

Instead of calculating the ratio of probabilities of success to get a measure of risk, we can calculate the ratio of probabilities of failure to get a measure of health. Under the log complement inverse link, an unsuccessful outcome is characterized by

$$
\begin{aligned}
\Pr(y_i = 0) &= 1 - g^{-1}(\beta_0 + x_{1i}\beta_1 + x_{2i}\beta_2) & (10.40) \\
&= 1 - \{1 - \exp(\beta_0 + x_{1i}\beta_1 + x_{2i}\beta_2)\} & (10.41) \\
&= \exp(\beta_0 + x_{1i}\beta_1 + x_{2i}\beta_2) & (10.42)
\end{aligned}
$$

For the health ratio, the numerator probability is calculated such that the specific covariate is incremented by one relative to the value used in the denominator probability. The health ratio for $\mathbf{x}_1$ is calculated as

$$
\begin{aligned}
\text{Health ratio for } \mathbf{x}_1 &= \frac{\exp\{\beta_0 + (x_{1i} + 1)\beta_1 + x_{2i}\beta_2\}}{\exp(\beta_0 + x_{1i}\beta_1 + x_{2i}\beta_2)} & (10.43) \\
&= \exp(\beta_1) & (10.44)
\end{aligned}
$$

There is no dependence on a particular observation. The health ratio is therefore constant—it is independent of the particular values of the covariates. This interpretation does not hold for other inverse links.

Interpretation for binomial family models using the log complement link is straightforward for exponentiated coefficients. These exponentiated coefficients represent health ratios. A health ratio of 2 indicates that a successful outcome is half as likely (the risk is cut in half) if the associated covariate is increased by one.

10.6.5 Log-log link

Because the inverse link is nonlinear, interpreting the coefficient is difficult.

$$
\begin{aligned}
\Delta y_i = {}& \exp(-\exp\{-[\beta_0 + (x_{1i} + 1)\beta_1 + x_{2i}\beta_2]\}) - \\
& \exp(-\exp\{-[\beta_0 + x_{1i}\beta_1 + x_{2i}\beta_2]\}) \quad (10.45)
\end{aligned}
$$

The difficulty arises because the difference is not constant and depends on the values of the covariates.

Transforming the coefficients makes interpretation easier. We motivate this transformation by introducing an alternative metric.

Instead of calculating the ratio of probabilities of success to get a measure of risk, we can calculate the ratio of the logs of probabilities of success to get a measure of log probability of risk. Under the log-log inverse link, a successful outcome is characterized by

$$
\begin{align}
\log[\Pr(y_i = 1)] &= \log[g^{-1}(\beta_0 + x_{1i}\beta_1 + x_{2i}\beta_2)] \tag{10.46}\\
&= \log\exp[-\{\exp(-[\beta_0 + x_{1i}\beta_1 + x_{2i}\beta_2])\}] \tag{10.47}\\
&= -\exp(-[\beta_0 + x_{1i}\beta_1 + x_{2i}\beta_2]) \tag{10.48}
\end{align}
$$

For the inverse ratio of the log of risk, the numerator probability is calculated such that the specific covariate is decremented by one relative to the value used in the denominator probability. The inverse ratio of the log of risk for $\mathbf{x}_1$ is calculated as

$$
\begin{align}
\text{Inverse ratio of log risk } \mathbf{x}_1 &= \frac{-\exp(-[\beta_0 + x_{1i}\beta_1 + x_{2i}\beta_2)]}{-\exp\{-[\beta_0 + (x_{1i}+1)\beta_1 + x_{2i}\beta_2]\}} \tag{10.49}\\
&= \exp(\beta_1) \tag{10.50}
\end{align}
$$

There is no dependence on a particular observation. The inverse ratio of the log risk is therefore constant—it is independent of the particular values of the covariates. This interpretation does not hold for other inverse links.

Interpretation for binomial family models using the log-log link is straightforward for exponentiated coefficients. These exponentiated coefficients represent inverse ratios of the log risk. An inverse ratio of the log risk of 2 indicates that a successful outcome is half as likely on the log scale (the log risk is cut in half) if the associated covariate is increased by one.

10.6.6 Complementary log-log link

Because the inverse link is nonlinear, interpreting the coefficient is difficult.

$$
\begin{align}
\Delta y_i &= 1 - \exp(-\exp\{\beta_0 + (x_{1i}+1)\beta_1 + x_{2i}\beta_2\})\\
&\quad -1 + \exp(-\exp\{\beta_0 + x_{1i}\beta_1 + x_{2i}\beta_2\}) \tag{10.51}
\end{align}
$$

The difficulty arises because the difference is not constant and depends on the values of the covariates.

Transforming the coefficients makes interpretation easier. We motivate this transformation by introducing an alternative metric.

Instead of calculating the ratio of probabilities of success to get a measure of risk, we can calculate the ratio of the logs of probabilities of success to get a measure of log probability of risk. Under the log-log inverse link, a successful outcome is characterized by

$$
\begin{aligned}
\log[\Pr(y_i = 0)] &= \log[g^{-1}(\beta_0 + x_{1i}\beta_1 + x_{2i}\beta_2)] & (10.52)\\
&= \log \exp[-\{\exp(\beta_0 + x_{1i}\beta_1 + x_{2i}\beta_2)\}] & (10.53)\\
&= -\exp(\beta_0 + x_{1i}\beta_1 + x_{2i}\beta_2) & (10.54)
\end{aligned}
$$

For the ratio of the log of health, the numerator probability is calculated such that the specific covariate is incremented by one relative to the value used in the denominator probability. The ratio of the log of health for $\mathbf{x}_1$ is calculated as

$$
\begin{aligned}
\text{Inverse ratio of log risk } \mathbf{x}_1 &= \frac{-\exp\{\beta_0 + (x_{1i}+1)\beta_1 + x_{2i}\beta_2\}}{-\exp(\beta_0 + x_{1i}\beta_1 + x_{2i}\beta_2)} & (10.55)\\
&= \exp(\beta_1) & (10.56)
\end{aligned}
$$

There is no dependence on a particular observation. The ratio of the log health is therefore constant—it is independent of the particular values of the covariates. This interpretation does not hold for other inverse links.

Interpretation for binomial family models using the complementary log-log link is straightforward for exponentiated coefficients. These exponentiated coefficients represent ratios of the log health. A ratio of the log health of 2 indicates that an unsuccessful outcome is twice as likely on the log scale (the log health is doubled) if the associated covariate is increased by one.

10.6.7 Summary

As we have illustrated, exponentiated coefficients for models using the binomial family offer a much clearer interpretation for certain inverse link functions. Notwithstanding our desire for clearer interpretation, the calculation of risk differences, health ratios, and risk ratios are subject to data that support optimization using links that do not necessarily restrict fitted values to the supported range of the binomial family variance function.

10.7   Generalized binomial regression

Jain and Consul (1971) introduced the generalized negative binomial distribution. Then, Famoye (1995) showed that by restricting the ancillary parameter to integer values, the regression model could be used for grouped binomial data, and Famoye and Kaufman (2002) further investigated the regression model.

The generalized negative binomial distribution is given by

$$
\Pr(Y = y) = \frac{m}{m + \phi y}\binom{m + \phi y}{y}\theta^y(1 - \theta)^{m + \phi y - y} \tag{10.57}
$$

for $y = 0, 1, 2, \ldots,\ 0 < \theta < 1,\ 1 \le \phi < \theta^{-1}$, for $m > 0$. The distribution reduces to the binomial distribution when $\phi = 0$ and reduces to the negative binomial distribution when $\phi = 1$.

A generalized binomial distribution results when we take the m parameter and treat it as the denominator (known values) in a grouped-data binomial model. Such a model may be specified as

$$\Pr(Y_i = y_i) = \frac{m_i}{m_i + \phi y_i} \binom{m_i + \phi y_i}{y_i} \left(\frac{\pi_i}{1 + \phi \pi_i} \right)^{y_i} \left(1 - \frac{\pi_i}{1 + \phi \pi_i} \right)^{m_i + \phi y_i - y_i} \tag{10.58}$$

We see that the expected value and variance are related as

$$E(Y_i) = \mu_i = m_i \theta (1 - \theta \phi)^{-1} = m_i \pi_i \quad V(Y_i) = m_i \pi_i (1 + \phi \pi_i)(1 + \phi \pi_i - \pi_i) \tag{10.59}$$

In the usual GLM approach, we can introduce covariates into the model by substituting a link function of the linear predictor for the expected value. The probability mass function has a log likelihood given by

$$\begin{aligned}
\mathcal{L}(\boldsymbol{\beta}, \phi) \;=\; & \sum_{i=1}^{n} \Big\{ \log m_i + y_i \log \pi_i + (m_i + \phi y_i - y_i) \log(1 + \phi \pi_i - \pi_i) \\
& - (m_i + \phi y_i) \log(1 + \phi \pi_i) + \log \Gamma(m_i + \phi y_i + 1) \\
& - \log \Gamma(y_i + 1) - \log \Gamma(m_i + \phi y_i - y_i + 1) \Big\}
\end{aligned} \tag{10.60}$$

Wheatley and Freeman (1982) discuss a model for the plant-to-plant distribution of carrot fly damage on carrots and parsnips in experiments (fields and microplots). They noted that the distribution of larvae was likely to require two descriptive parameters. Our use of these data is meant to illustrate the relationship to GLMs and the utility of Stata's `ml` command capabilities. Because the authors used a clog-log link within a binomial model to describe the data, we will mimic that approach here. The experiment yielding the data includes three replicates (`R1`, `R2`, and `R3` indicator variables) and 11 different treatment combinations of insecticide and depth (`T1`, ..., `T11` indicator variables) given in table 10.3; the two rows for no insecticide constitute the last treatment, `T11`.

Table 10.3. 1964 microplot data of carrot fly damage

Treatment		Replicate 1		Replicate 2		Replicate 3	
		No.	No.	No.	No.	No.	No.
Insecticide	Depth	damaged	examined	damaged	examined	damaged	examined
Diazinon	1.0	120	187	145	181	123	184
	2.5	60	184	85	191	113	171
	5.0	35	179	23	181	33	179
	10.0	5	178	40	184	18	171
	25.0	66	182	104	186	106	181
Disulfoton	1.0	97	187	128	173	112	179
	2.5	85	186	95	167	137	194
	5.0	72	190	89	184	92	192
	10.0	49	188	50	187	65	186
	25.0	132	179	122	182	120	179
Nil	I	138	187	169	184	163	175
	II	156	176	169	190	159	175

Reprinted from Wheatley and Freeman (1982).

Observed proportions of damage for each treatment are illustrated in figure 10.5.

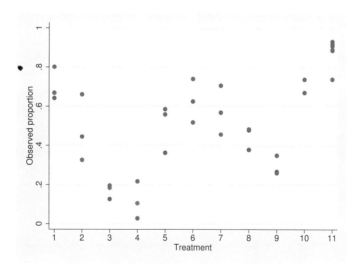

Figure 10.5. Observed proportion of carrot fly damage for each treatment (see table 10.3)

The clog-log binomial regression estimates are easily obtained using `glm`:

```
. glm damage R2 R3 T1 T2 T3 T4 T5 T6 T7 T8 T9 T10, family(binomial examine)
> link(cloglog)
Iteration 0:   log likelihood = -158.43213
Iteration 1:   log likelihood = -156.76371
Iteration 2:   log likelihood = -156.75711
Iteration 3:   log likelihood = -156.75711
```

```
Generalized linear models                         No. of obs      =         36
Optimization      : ML                            Residual df     =         23
                                                  Scale parameter =          1
Deviance          =   123.5631356                 (1/df) Deviance =    5.37231
Pearson           =   121.3150965                 (1/df) Pearson  =   5.274569

Variance function: V(u) = u*(1-u/examine)         [Binomial]
Link function     : g(u) = ln(-ln(1-u/examine))   [Complementary log-log]
                                                  AIC             =    9.43095
Log likelihood    = -156.7571088                  BIC             =    41.1422
```

damage	Coef.	OIM Std. Err.	z	P>|z|	[95% Conf.	Interval]
R2	.3330675	.0453895	7.34	0.000	.2441058	.4220293
R3	.3765869	.0452898	8.32	0.000	.2878204	.4653533
T1	-.5672674	.0669523	-8.47	0.000	-.6984915	-.4360433
T2	-1.20193	.0746685	-16.10	0.000	-1.348277	-1.055582
T3	-2.462982	.1121606	-21.96	0.000	-2.682812	-2.243151
T4	-2.840763	.1321047	-21.50	0.000	-3.099683	-2.581842
T5	-1.117579	.0730595	-15.30	0.000	-1.260773	-.9743852
T6	-.7731088	.0691728	-11.18	0.000	-.908685	-.6375326
T7	-.8988995	.0701799	-12.81	0.000	-1.03645	-.7613494
T8	-1.289745	.0750403	-17.19	0.000	-1.436822	-1.142669
T9	-1.831744	.0878473	-20.85	0.000	-2.003922	-1.659567
T10	-.6123426	.0675081	-9.07	0.000	-.744656	-.4800291
_cons	.5233536	.0466372	11.22	0.000	.4319464	.6147608

After fitting the binomial model, one can store the coefficient estimates for use as starting values in the fitting of the generalized binomial model.

```
. matrix b0 = e(b)
. local ll0 = `e(ll)´
. matrix z0 = (0)
. matrix b0 = b0, z0
```

If our goal were to provide full support for the generalized binomial model, we would program several support routines allowing user specification of useful options. However, our purpose here is to demonstrate how quickly one can obtain answers for specific problems. We will therefore specify the log likelihood only for our particular analysis. Stata's `ml` program allows such specification and will calculate and use numeric derivatives to optimize the objective function. This ability is enormously useful. Thus, we write a rather simple program that accepts arguments with the current parameter values and the name of a scalar in which we will store the log likelihood.

```
. program gbinreg_ll
  1.      version 15
  2.      args todo b lnf
  3.      tempvar  eta pi
  4.      tempname phi
  5.      mleval `eta´ = `b´, eq(1)
  6.      mleval `phi´ = `b´, eq(2) scalar
  7.      local y  "$ML_y1"   /* Dep var name is in global macro by ml   */
  8.      local M  "examine"  /* Specify directly -- problem specific    */
  9.      generate double `pi´    = 1-exp(-exp(`eta´))  /* Cloglog link here */
 10.      mlsum `lnf´ = log(`M´) + `y´*log(`pi´) +                   /*
        */ (`M´+`phi´*`y´-`y´)*log(1+`phi´*`pi´-`pi´) - /*
        */ (`M´+`phi´*`y´)*log(1+`phi´*`pi´) - log(`M´+`phi´*`y´) + /*
        */ lngamma(`M´+`phi´*`y´+1) - lngamma(`y´+1) -  /*
        */ lngamma(`M´+`phi´*`y´-`y´+1)
 11. end
```

After having specified the program that should be called by the `ml` command, we need only start the optimization process by using `ml model`, wherein we specify the functionality of our program (we wrote a `d0` or log-likelihood-only function), the starting values, and the nature of the parameter subvectors (what Stata calls "equations"). After convergence is achieved, we use `ml display` to display our estimates. These results mimic those given in Famoye (1995).

```
. ml model d0 gbinreg_ll (gbr: damage = R2 R3 T1 T2 T3 T4 T5 T6 T7 T8 T9 T10)
> (phi:), search(off) init(b0, copy) max iter(50)
Iteration 0:   log likelihood = -156.75711
Iteration 1:   log likelihood = -142.40792
Iteration 2:   log likelihood = -142.22799
Iteration 3:   log likelihood = -142.22626
Iteration 4:   log likelihood = -142.22626
. ml display, title(Generalized binomial regression)
```

Generalized binomial regression

	Number of obs	=	36
	Wald chi2(12)	=	475.22
Log likelihood = -142.22626	Prob > chi2	=	0.0000

damage	Coef.	Std. Err.	z	P>|z|	[95% Conf. Interval]	
gbr						
R2	.346261	.0780868	4.43	0.000	.1932136	.4993083
R3	.4157763	.0794801	5.23	0.000	.2599982	.5715545
T1	-.54981	.1677757	-3.28	0.001	-.8786443	-.2209757
T2	-1.224663	.1573275	-7.78	0.000	-1.53302	-.9163072
T3	-2.455202	.1735342	-14.15	0.000	-2.795323	-2.115081
T4	-2.852996	.1868674	-15.27	0.000	-3.219249	-2.486743
T5	-1.134379	.1574367	-7.21	0.000	-1.442949	-.8258085
T6	-.7764593	.1618621	-4.80	0.000	-1.093703	-.4592154
T7	-.9148208	.1593118	-5.74	0.000	-1.227066	-.6025755
T8	-1.288835	.1562914	-8.25	0.000	-1.59516	-.9825092
T9	-1.830209	.1599901	-11.44	0.000	-2.143784	-1.516634
T10	-.5553261	.1696358	-3.27	0.001	-.8878061	-.2228461
_cons	.5070029	.1279714	3.96	0.000	.2561835	.7578222
phi						
_cons	.7330808	.2476195	2.96	0.003	.2477555	1.218406

This example demonstrates the tools available to Stata users to extend the binomial model by using the suite of `ml` commands. See the reference manual or the excellent book by Gould, Pitblado, and Poi (2010) for documentation and more examples. One can also use the community-contributed `gbin` command (Harris, Hilbe, and Hardin 2014) for the above model. This command is available by typing `net install st0351_1,` `from(http://www.stata-journal.com/software/sj18-1)`.

```
. gbin damage R2 R3 T1 T2 T3 T4 T5 T6 T7 T8 T9 T10, n(examine) link(cloglog)
> nolog

Generalized binomial regression               Number of obs    =         36
Link          = cloglog                        LR chi2(12)      =     105.75
Dispersion    = generalized binomial           Prob > chi2      =     0.0000
Log likelihood = -142.22626                    Pseudo R2        =     0.2710
```

damage	Coef.	Std. Err.	z	P>|z|	[95% Conf. Interval]	
R2	.3462657	.0780889	4.43	0.000	.1932141	.4993172
R3	.4157871	.0794825	5.23	0.000	.2600042	.5715701
T1	-.5497988	.1677834	-3.28	0.001	-.8786481	-.2209494
T2	-1.22467	.1573334	-7.78	0.000	-1.533038	-.9163022
T3	-2.455205	.173539	-14.15	0.000	-2.795336	-2.115075
T4	-2.852999	.1868719	-15.27	0.000	-3.219261	-2.486737
T5	-1.134384	.1574426	-7.21	0.000	-1.442966	-.8258022
T6	-.7764608	.1618685	-4.80	0.000	-1.093717	-.4592044
T7	-.914827	.1593179	-5.74	0.000	-1.227084	-.6025697
T8	-1.288836	.1562972	-8.25	0.000	-1.595173	-.9824994
T9	-1.830212	.1599955	-11.44	0.000	-2.143797	-1.516626
T10	-.5553061	.1696444	-3.27	0.001	-.887803	-.2228091
_cons	.5070008	.1279769	3.96	0.000	.2561707	.7578308
/lnsigma	-.3104314	.3377714			-.9724511	.3515884
sigma	.7331307	.2476306			.378155	1.421323

```
Likelihood-ratio test of sigma=0:  chibar2(01) =    29.06 Prob>=chibar2 = 0.000
```

The results of the community-contributed command are identical to those from the `ml` command. The likelihood-ratio test included in the output is a comparison of the binomial model to the beta binomial. That is, it assesses the null hypothesis $H_0 : \sigma = 0$ versus $H_a : \sigma \neq 0$. Because this test is assessed on the boundary of the parameter space, the distribution of the test statistic is not $\chi^2(1)$ but rather a mixture distribution called the $\overline{\chi}^2(1)$—pronounced "chi-bar squared".

10.8   Beta binomial regression

As has been discussed, a binomial outcome is characterized by the probability function

$$\Pr(Y = y) = \binom{n}{y} \mu^y (1 - \mu)^{n-y} \tag{10.61}$$

where the expected value of the outcome is $n\mu$ for n independent trials, each with probability of success given by μ. In binomial regression, we model the probability of success μ with a link function $g(\cdot)$ of a linear combination of covariates $\mathbf{x\beta}$. The compound beta-binomial distribution results from assuming the probability of success μ is a random variable that follows a Beta(α, β) distribution. The resulting probability mass function is then written as

$$\Pr(Y = y) = \binom{n}{y} \frac{B(y + \alpha, n - y + \beta)}{B(\alpha, \beta)} \tag{10.62}$$

where

$$B(a, b) = \frac{\Gamma(a)\Gamma(b)}{\Gamma(a + b)} \tag{10.63}$$

such that the mean of Y is given by $n\alpha/(\alpha+\beta)$ and the variance by $n\alpha\beta(\alpha+\beta+n)/\{(\alpha+\beta)^2(\alpha + \beta + 1)\}$. Substituting $p = \alpha/(\alpha + \beta)$ and $\sigma = 1/(\alpha + \beta)$, the probability mass function is given by

$$\Pr(Y = y) = \binom{n}{y} \frac{B\{y + p/\sigma, n - y + (1 - p)/\sigma\}}{B\{p/\sigma, (1 - p)/\sigma\}} \tag{10.64}$$

with mean np and variance $np(1 - p)(1 + n\sigma)/(1 + \sigma)$ so that the variance exceeds that of the binomial model when $(1 + n\sigma)/(1 + \sigma) > 1$. Thus, extra binomial dispersion is accounted for only when $n > 1$; that is, overdispersion can be modeled only when there are more than one trial for each individual.

To highlight the beta-binomial and zero-inflated beta binomial model, we use data from Hilbe (2009) on survivors among passengers on the Titanic. For pedagogical reasons, we have altered the data from the source by changing some survivor numbers to ease examination of the data by the zero-inflated models.

The data are organized in 12 passenger types comprising the outcome for 1,316 passengers. Passenger types are defined by whether the members are adult, whether they are male, and the class of their passage (first, second, or third class).

Table 10.4. Survivors among different categorizations of passengers on the Titanic

Survive	N	Adult	Male	Class	Survive	N	Adult	Male	Class
0	1	0	0	1	140	144	1	0	1
0	13	0	0	2	80	93	1	0	2
14	31	0	0	3	76	165	1	0	3
0	5	0	1	1	57	175	1	1	1
0	11	0	1	2	14	168	1	1	2
0	48	0	1	3	75	462	1	1	3

Because we are interested in the predictors of passenger survival, we use the community-contributed `betabin` command (Hardin and Hilbe 2014a)[1] to fit a beta-binomial
regression model with the outcome variable `survive` as the numerator of the binomial
response and the variable `N` as the denominator of the binomial response. The `N` variable
represents the number of passengers having the same values for `adult`, `male`, and `class`.
There are 12 separate sets of covariate patterns in the data. Explanatory predictors,
`adult` and `male` are binary; `class` is a categorical variable with three values or levels.
We generate three indicator variables for each level of `class`, using `class1` (first-class
passengers) as the reference level.

```
. use http://www.stata-press.com/data/hh4/titanic_altered, clear

. betabin survive adult male class2 class3, n(N) link(cloglog) eform nolog

Beta-binomial regression                    Number of obs    =          12
Link           = cloglog                    LR chi2(4)       =       18.71
Dispersion     = beta-binomial              Prob > chi2      =      0.0022
Log likelihood = -31.990675                 Pseudo R2        =      0.2263
```

survive	exp(b)	Std. Err.	z	P>|z|	[95% Conf. Interval]	
adult	9.762171	8.616112	2.58	0.010	1.730909	55.05778
male	.1380966	.0661646	-4.13	0.000	.0539954	.3531909
class2	.495203	.216559	-1.61	0.108	.2101558	1.166877
class3	.3748033	.195718	-1.88	0.060	.1346839	1.043016
_cons	.2738491	.2611104	-1.36	0.174	.0422576	1.774669
/lnsigma	-2.164766	.7019773	-3.08	0.002	-3.540616	-.7889156
sigma	.1147768	.0805707			.0289955	.4543372

```
Note: Estimates are transformed only in the first equation.
Likelihood-ratio test of sigma=0:  chibar2(01) =   34.08 Prob>=chibar2 = 0.000
```

The corresponding binomial generalized linear model[2] has a dispersion statistic
greater than 11, which indicates overdispersion of the data relative to the conditional
variance allowed for by the binomial model. Thus, the beta-binomial should provide a
better fit. That is, the beta-binomial should have similar conditional means but should
allow for greater conditional variance.

10.9   Zero-inflated models

Lambert (1992) first described this type of mixture model in the context of process
control in manufacturing. It has since been used in many applications and is now
discussed in nearly every book or article dealing with count response models.

For the zero-inflated model, the probability of observing a zero outcome equals the
probability that an individual is in the always-zero group plus the probability that the
individual is not in that group times the probability that the counting process produces

---

1. The betabin command is available by typing `net install st0337_1`,
   `from(http://www.stata-journal.com/software/sj18-1)`.
2. `glm survive adult i.class, family(binomial n)`

a zero; see Hilbe and Greene (2008). If we define $B(0)$ as the probability that the binary process results in a zero outcome and $\Pr(0)$ as the count process probability of a zero outcome, the probability of a zero outcome for the system is then given by

$$\Pr(y = 0) = B(0) + \{1 - B(0)\}\Pr(0) \tag{10.65}$$

and the probability of a nonzero outcome is

$$\Pr(y = k; k > 0) = \{1 - B(0)\}\Pr(k) \tag{10.66}$$

where $\Pr(k)$ is the Bernoulli (or binomial) process probability that the outcome is k. As is discussed in section 14.3, count models can also be characterized as having "excess" zeros. That is, after adjusting for other covariates, there remains a substantially higher observed proportion of zeros than can be explained by the model. In such cases, we can hypothesize that the data result from a mixture of two different processes. The first process always generates a zero outcome, and the second process is a usual Bernoulli (or binomial) process that sometimes generates a zero response.

Because these data do not have zero-inflation, we generate a simple dataset to illustrate the estimation of the parameters. To generate beta-binomial data with zero inflation, we use the following code:

```
. set seed 23942
. drop _all
. set obs 5000
number of observations (_N) was 0, now 5,000
. gen z1 = rbinomial(1,0.5)
. gen z2 = rbinomial(1,0.5)
. gen zg = -0.5 + 0.25*z1 - 0.25*z2
. gen z  = rbinomial(1, 1-exp(-exp(-zg)))    // link(log-log)
. gen x1 = rbinomial(1, 0.5)
. gen xb = -1+0.5*x1
. gen n  = floor(8*runiform()) + 1
. local sigma = 0.33
. gen p  = exp(xb)/(1+exp(xb))   // link(logit)
.
.           // y is beta-binomial, yo is zero-inflated beta-binomial
. gen y = rbinomial(n, rbeta((1/`sigma')*p, (1/`sigma')*(1-p)))
. gen yo = y*z
```

We fit the zero-inflated beta-binomial model with the community-contributed command `zibbin` (Hardin and Hilbe 2014a).[3] There are two sections for the model. The initial section is the beta-binomial model for the outcomes, and the second section is the Bernoulli process responsible for generating zero outcomes. The Bernoulli section is a model of the likelihood of a zero, so the interpretation should be a reflection of that focus. This is an important point for Stata users because nearly all other Bernoulli models estimated coefficients of the likelihood of an outcome of 1.

---

3. The `zibbin` command is available by typing `net install st0337_1`,
   `from(http://www.stata-journal.com/software/sj18-1)`.

```
. zibbin yo x1, inflate(z1 z2) n(n) link(logit) ilink(loglog) zib nolog
Zero-inflated beta-binomial regression              Number of obs    =        5000
Regression link: logit                              Nonzero obs      =        2509
Inflation link : loglog                             Zero obs         =        2491
                                                    LR chi2(1)       =       70.91
Log likelihood = -6741.874                          Prob > chi2      =      0.0000
```

yo	Coef.	Std. Err.	z	P>|z|	[95% Conf.	Interval]
yo						
x1	.4247326	.0498924	8.51	0.000	.3269452	.5225199
_cons	-1.008658	.0666281	-15.14	0.000	-1.139247	-.8780692
inflate						
z1	.198506	.0768891	2.58	0.010	.0478061	.3492059
z2	-.2669247	.0798883	-3.34	0.001	-.4235028	-.1103466
_cons	-.5258966	.116079	-4.53	0.000	-.7534073	-.298386
/lnsigma	-1.026614	.0839			-1.191055	-.8621728
sigma	.3582179	.0300545			.3039006	.4222436

```
Likelihood-ratio test of sigma=0:  chibar2(01) =  638.72 Prob>=chibar2 = 0.000
```

We specified the **zib** option in the **zibbin** command so that the output would include
a likelihood-ratio test comparing a zero-inflated binomial model with the zero-inflated
beta-binomial.

1   The problem of overdispersion

Contents

1.1   Overdispersion

The problem of overdispersion is central to all GLMs for discrete responses. We discuss what overdispersion is, how to identify it, and how to deal with it.

Overdispersion in discrete-response models occurs when the variance of the response is greater than the nominal variance. It is generally caused by positive correlation between responses or by an excess variation between response probabilities or counts. The problem with overdispersion is that it may cause underestimation of standard errors of the estimated coefficient vector. A variable may appear to be a significant predictor when it is not.

Overdispersion affects only the discrete models because continuous models fit the scale parameter ϕ, whereas discrete models do not have this extra parameter (it is theoretically equal to 1). Thus, continuous models still require the variance to be a function of the mean, but that relationship may be scaled. Discrete models, on the other hand, do not have the extra parameter to scale the relationship of the variance and mean.

We can recognize possible overdispersion by observing that the value of the Pearson χ^2 or deviance divided by the degrees of freedom $(n - p)$ is larger than 1. The quotient of either is called the dispersion. Small amounts of overdispersion are usually of little concern; however, if the dispersion statistic is greater than 2.0, then an adjustment to the standard errors may be required.

There is a distinct difference between true overdispersion and apparent overdispersion. Outward indicators such as a large dispersion statistic may be only a sign of apparent overdispersion. How do we tell?

Apparent overdispersion may arise from any of the following:

1. The model omits important explanatory predictors.
2. The data contain outliers.
3. The model fails to include enough interaction terms.
4. A predictor needs to be transformed (to the log or some other scale).
5. We have assumed a linear relationship, for example, between the logit transform and the response or predictors when the actual relationship is quadratic (or some higher order).

Any of the above situations may cause a model to appear inherently overdispersed. However, if we add the requisite predictor, discover the important interactions, adjust outliers, transform where necessary, and so forth, then the apparent overdispersion may be remedied. As a consequence, the model will be better fit.

As an example of how omitting an important explanatory predictor affects the model such that it appears to be overdispersed when in fact it is not, let us create a simulated model demonstrating how to accommodate apparent overdispersion.

```
. set seed 238947                    /* Set seed for reproducibility */
. set obs 10000                      /* Number of obs = 10,000        */
number of observations (_N) was 0, now 10,000
. generate x1 = rnormal()            /* Creation of random variates   */
. generate x2 = rnormal()
. generate x3 = rnormal()
. generate d  = 50+5*int((_n-1)/1000) /* Creation of 10 different denominators */
. tabulate d                         /* Tabulation of the denominator */

          d |      Freq.     Percent        Cum.
------------+-----------------------------------
         50 |      1,000       10.00       10.00
         55 |      1,000       10.00       20.00
         60 |      1,000       10.00       30.00
         65 |      1,000       10.00       40.00
         70 |      1,000       10.00       50.00
         75 |      1,000       10.00       60.00
         80 |      1,000       10.00       70.00
         85 |      1,000       10.00       80.00
         90 |      1,000       10.00       90.00
         95 |      1,000       10.00      100.00
------------+-----------------------------------
      Total |     10,000      100.00
```

Using these simulated denominators, we can now complete the generation of synthetic data and fit the binomial model. We use the Stata `rbinomial()` function for this purpose.

```
. set seed 214231
. generate xb = 1 + .5*x1 - .75*x2 + .25*x3    /* Create linear predictor */
. generate p  = exp(xb)/(1+exp(xb))
. generate y  = rbinomial(d,p)                 /* Create simulated dataset */
. glm y x1 x2 x3, family(binomial d) nolog     /* Model y on x1, x2, and x3 */
```

```
Generalized linear models                      No. of obs      =      10,000
Optimization      : ML                         Residual df     =       9,996
                                               Scale parameter =           1
Deviance          =   10319.15126              (1/df) Deviance =    1.032328
Pearson           =   10133.55075              (1/df) Pearson  =    1.013761

Variance function: V(u) = u*(1-u/d)            [Binomial]
Link function     : g(u) = ln(u/(d-u))         [Logit]
                                               AIC             =    5.312932
Log likelihood    = -26560.66203               BIC             =   -81747.41
```

y	Coef.	OIM Std. Err.	z	P>|z|	[95% Conf.	Interval]
x1	.5043512	.0028908	174.47	0.000	.4986854	.5100169
x2	-.7500798	.0030623	-244.94	0.000	-.7560819	-.7440777
x3	.2466322	.0028054	87.91	0.000	.2411337	.2521308
_cons	.9977827	.0029168	342.08	0.000	.9920659	1.003499

```
. save sim_data, replace
(note: file sim_data.dta not found)
file sim_data.dta saved
```

We set the random-number seed in the preceding dataset generation so that the results were replicable. Had we not set the seed, then each run of the code would result in slightly different results. To investigate the natural variability in the regression results for these (and other) simulated data, here we illustrate how to use the power of Stata to highlight certain properties. First, a program is defined to generate simulated predictors and refit the model.

```
. program GlogitSim, rclass
  1.      quietly {
  2.          drop _all
  3.          set obs 10000
  4.          generate x1 = rnormal()
  5.          generate x2 = rnormal()
  6.          generate x3 = rnormal()
  7.          generate d  = 50+5*int((_n-1)/1000)
  8.          generate xb = 1 + 0.5*x1 - .75*x2 + .25*x3
  9.          generate p  = exp(xb)/(1+exp(xb))
 10.          generate y  = rbinomial(d,p)
 11.          glm y x1 x2 x3, family(binomial d)
 12.      }
 13.      return scalar sx1 = _b[x1]
 14.      return scalar sx2 = _b[x2]
 15.      return scalar sx3 = _b[x3]
 16.      return scalar sxc = _b[_cons]
 17.      return scalar pds = e(dispers_p)
 18. end
```

Subsequently, the `simulate` command is used to run the program and the results summarized.

```
. set seed 2934288

. simulate x1=r(sx1) x2=r(sx2) x3=r(sx3) cons=r(sxc) disp=r(pds), reps(100):
> GlogitSim

      command:  GlogitSim
           x1:  r(sx1)
           x2:  r(sx2)
           x3:  r(sx3)
         cons:  r(sxc)
         disp:  r(pds)

Simulations (100)
  +---- 1 ----+---- 2 ----+---- 3 ----+---- 4 ----+---- 5
..................................................    50
..................................................   100

. summarize
    Variable |        Obs        Mean    Std. Dev.        Min         Max
-------------+--------------------------------------------------------------
          x1 |        100     .4997384    .0025868    .4937869     .506715
          x2 |        100    -.7503007    .0028862   -.7569348   -.7417647
          x3 |        100     .2496567    .0027887     .242906    .2556603
        cons |        100     1.000238    .0030107    .9927197     1.00844
        disp |        100     .9979798    .0142192    .9627941    1.029327
```

Other simulated datasets will have slightly different values but should have approximately the same means illustrated here. If we ran several hundred simulated models, however, we would discover that the parameter estimates would equal the assigned values and that the Pearson χ^2 statistic divided by the model degrees of freedom would center about 1.0, indicating a true grouped logistic model. The Pearson χ^2 dispersion statistic here is 1, and the parameter estimates are close to the values we specified. A Pearson statistic near 1.0 indicates that the model is not overdispersed; that is, it is well specified. Values greater than 1.0 indicate overdispersion. When the number of observations is high, as in this simulated dataset, values around 1.05 or more tell us that there may be greater correlation in the data than is allowed based on the distributional assumptions of the binomial probability distribution function. This situation biases the standard errors.

In the following, we use the original simulated data and consider what happens to the model fit when we omit the `x1` covariate.

```
. use sim_data
. glm y x2 x3, family(binomial d) nolog   /* Here we omit the x1 covariate */
Generalized linear models                      No. of obs      =      10,000
Optimization      : ML                         Residual df     =       9,997
                                               Scale parameter =           1
Deviance       =  42992.97762                  (1/df) Deviance =    4.300588
Pearson        =  42128.38902                  (1/df) Pearson  =    4.214103

Variance function: V(u) = u*(1-u/d)            [Binomial]
Link function    : g(u) = ln(u/(d-u))          [Logit]

                                               AIC             =    8.580115
Log likelihood   = -42897.57521                BIC             =    -49082.8
```

| | | OIM | | | | |
y	Coef.	Std. Err.	z	P>|z|	[95% Conf.	Interval]
x2	-.7141349	.0029691	-240.53	0.000	-.7199541	-.7083156
x3	.2309762	.0027352	84.45	0.000	.2256153	.236337
_cons	.944368	.0027959	337.77	0.000	.9388881	.9498478

The remaining parameter estimates are close to the original or true values. The Pearson dispersion is 4.21. Such a large value indicates an ill-fitted model. If this were our initial model, the value of the Pearson dispersion would indicate something was wrong and that some type of adjustment must be made. We know in this situation that the overdispersion is only apparent and that including predictor x1 eliminates the excess dispersion. Usually, we do not know this when modeling data at the outset.

Note also the difference in the AIC and BIC statistics between the two models. The values generated by the "correct" model are respectively, 5.31 and −81,747; the values for the apparently overdispersed model are 8.58 and −49,083. Because lower values of these GOF statistics indicate a better-fitted model, the latter values clearly represent a poorly fitted model. If we had not paid attention to the Pearson dispersion statistic, we may not have known that the model was poor, especially because the parameter estimate p-values all indicate that they contribute to the model.

Next, we will consider an example of a model that indicates overdispersion due to not having a requisite interaction term.

The interaction term, x23, is generated by multiplying x2 and x3. We give it a coefficient of 0.2 and model the data by using the main-effects terms and interaction. The main effects are the same as in the previous example.

```
. set seed 64321
. generate x23 = x2*x3                        /* create interaction */
. gen xb1 = 1 + .5*x1 - .75*x2 + .25*x3 + .2*x23 /* create linear predictor */
. gen p1  = exp(xb1)/(1+exp(xb1))             /* create probability of success */
. gen y1  = rbinomial(d, p1)                  /* create simulated dataset */
```

Here we model the main effects and interaction:

```
. glm y1 x1 x2 x3 x23, family(binomial d) nolog
Generalized linear models                       No. of obs        =     10,000
Optimization     : ML                           Residual df       =      9,995
                                                Scale parameter =          1
Deviance         =    9924.108197               (1/df) Deviance =   .9929073
Pearson          =    9774.81605                (1/df) Pearson  =   .9779706

Variance function: V(u) = u*(1-u/d)             [Binomial]
Link function    : g(u) = ln(u/(d-u))           [Logit]

                                                AIC               =   5.270965
Log likelihood   = -26349.82718                 BIC               =  -82133.24
```

		OIM			
y1	Coef.	Std. Err.	z	P>|z|	[95% Conf. Interval]
x1	.4983804	.002904	171.62	0.000	.4926886 .5040722
x2	-.7436916	.0030914	-240.57	0.000	-.7497506 -.7376326
x3	.2520277	.0028967	87.01	0.000	.2463504 .2577051
x23	.2031744	.0031319	64.87	0.000	.1970359 .2093129
_cons	.9991675	.0029164	342.61	0.000	.9934515 1.004884

Parameter estimates and the Pearson dispersion appear as we expect. We next model the data without the interaction term.

```
. glm y1 x1 x2 x3, family(binomial d) nolog
Generalized linear models                       No. of obs        =     10,000
Optimization     : ML                           Residual df       =      9,996
                                                Scale parameter =          1
Deviance         =   14231.48817                (1/df) Deviance =   1.423718
Pearson          =   14193.81173                (1/df) Pearson  =   1.419949

Variance function: V(u) = u*(1-u/d)             [Binomial]
Link function    : g(u) = ln(u/(d-u))           [Logit]

                                                AIC               =   5.701503
Log likelihood   = -28503.51717                 BIC               =  -77835.07
```

		OIM			
y1	Coef.	Std. Err.	z	P>|z|	[95% Conf. Interval]
x1	.4952235	.0028896	171.38	0.000	.4895601 .500887
x2	-.7451735	.0030621	-243.35	0.000	-.7511751 -.7391719
x3	.2905955	.0028214	103.00	0.000	.2850656 .2961253
_cons	1.002786	.0029224	343.14	0.000	.9970587 1.008514

The overdispersion statistic increased, the AIC statistic is 0.43 higher, and the BIC is about 4,300 higher. The model may appear to be well fit at first glance, but the Pearson dispersion statistic indicates that the model needs adjustment. If statistical output does not include a Pearson dispersion statistic, or if the user ignores it, a model having parameter estimate p-values of all 0.000 means nothing. If the model is overdispersed or underdispersed, it does not fit and we cannot trust the parameter estimates. On the other hand, if we observe overdispersion in the model by virtue of the Pearson χ^2 dispersion, then we need to determine whether the overdispersion is real. Here it is not. The model is only apparently overdispersed. Testing the criteria we listed will help ensure that we have genuine overdispersion. See Hilbe (2011) for an in-depth examination of overdispersion and for simulations of the remaining criteria.

Sometimes the overdispersion is indeed inherent in the dataset itself—no external remedy can be found to remove the overdispersion. Then we must use methods that are designed to deal with true overdispersion.

The following methods have been used with binomial models to adjust a model for real overdispersion:

1. Scaling of standard errors by the dispersion statistic

2. Using the Williams' procedure

3. Using robust variance estimators as well as bootstrapping or jackknifing standard errors

4. Using GEE or random-effects models, mixed models, or some other variety of binomial model that accounts for extra correlation in the data

We look at the first three of these alternatives in this chapter. We will relegate the fourth method to part VI of the text.

11.2 Scaling of standard errors

Standard errors may be adjusted post hoc by multiplying the model standard errors by the square root of the Pearson dispersion statistic. Most major software distributions allow such adjustment, including Stata.

The method of scaling is simple. First, the algorithm fits the full model. Second, weights are abstracted according to the specified scale where Stata allows users to scale by deviance dispersion, by Pearson dispersion, or by any user-specified scalar.

Third, the GLM estimation algorithm calculates one more iteration of the main weighted regression routine,

$$\boldsymbol{\beta} = (\mathbf{X'WX})^{-1}\mathbf{X'Wz} \tag{11.1}$$

with the new values of the scaled weight matrix, $\mathbf{W}$. The result is a scaling of the standard errors—a result that in effect adjusts the standard errors to a value that would have been obtained if the dispersion were 1. Thus, the greatest change in standard errors will be for those models having large dispersions.

Although Stata will allow different scaling specifications, it is usually performed using the Pearson-based dispersion (`scale(x2)`); see Hilbe (2009, 2011). The χ^2 dispersion statistic was used to fit the scale parameter for two-parameter continuous models. We noted then that the χ^2-based scale estimate was nearly identical to the scale or ancillary parameter produced using maximum likelihood methods.

The scaling of standard errors for binomial models is undertaken to adjust for overdispersion; however, there is a catch. Although scaling can be used to adjust for overdispersion as a result of clustering, for longitudinal effects, or for some other cause of correlated outcomes, Bernoulli (binary response) data are not overdispersed (or underdispersed). We can, however, describe such data as implicitly overdispersed if we can consider the grouped form of the model to be overdispersed. This method is not applicable to models with continuous predictors, although such predictors can be transformed to factors or leveled predictors, as is the case for `age2`, `age3`, and `age4` in `heart01.dta`. Information is lost to the model when continuous variables are categorized, but the loss may be offset by gaining greater interpretability and more checks for model fit.

Recall the table of estimates with data from `heart01.dta` produced by fitting the binary logistic model in chapter 9:

death	Odds Ratio	EIM Std. Err.	z	P>|z|	[95% Conf. Interval]	
anterior	1.901333	.3185755	3.83	0.000	1.369103	2.640464
hcabg	2.105275	.7430692	2.11	0.035	1.054076	4.204801
kk2	2.251732	.4064421	4.50	0.000	1.580786	3.207453
kk3	2.172105	.5844269	2.88	0.004	1.281907	3.680486
kk4	14.29137	5.087652	7.47	0.000	7.112966	28.71423
age2	1.63726	.5078572	1.59	0.112	.8914272	3.007112
age3	4.532029	1.20653	5.68	0.000	2.689572	7.636636
age4	8.893222	2.417514	8.04	0.000	5.219999	15.15123

Standard errors and scaled standard errors may be calculated for coefficients and odds ratios directly

```
. display _se[anterior]*sqrt(e(dispers_p))                  // Scaled SE for coef
.16273166

. display exp(_b[anterior])*_se[anterior]                   // SE for OR
.3185755

. display exp(_b[anterior])*_se[anterior]*sqrt(e(dispers_p)) // Scaled SE for OR
.30940707
```

Alternatively, as the output below displays, parameter results for `heart01.dta` with scaled standard errors may be requested from the `glm` command. We have omitted the header portion of the output. The z-values have all increased, hence reducing p-values down the table. The odds ratios remain the same; only standard errors and their effects have changed. Again, standard errors are displayed as if the deviance dispersion had originally been 1. Because this is a binary model, the scaling is not dealing with true overdispersion.

```
. use http://www.stata-press.com/data/hh4/heart01, clear

. glm death anterior hcabg kk2 kk3 kk4 age2-age4, family(binomial) irls
> scale(x2) eform noheader nolog
```

death	Odds Ratio	EIM Std. Err.	z	P>|z|	[95% Conf. Interval]	
anterior	1.901333	.3094071	3.95	0.000	1.382104	2.615626
hcabg	2.105275	.721684	2.17	0.030	1.075272	4.121914
kk2	2.251732	.3947449	4.63	0.000	1.596963	3.174962
kk3	2.172105	.5676073	2.97	0.003	1.30151	3.62505
kk4	14.29137	4.941232	7.69	0.000	7.257242	28.14338
age2	1.63726	.4932413	1.64	0.102	.9071615	2.954955
age3	4.532029	1.171807	5.84	0.000	2.730266	7.522815
age4	8.893222	2.347939	8.28	0.000	5.300657	14.92068
_cons	.0063961	.0016065	-20.11	0.000	.0039095	.0104642

```
Note: _cons estimates baseline odds.
(Standard errors scaled using square root of Pearson X2-based dispersion.)
```

Using the Stata programming language, one can manipulate binary-response models such as our example for true over- or underdispersion by converting the data to proportion format either using the collapse command or using the egen command as we illustrate here:

```
. egen grp=group(anterior-age4)    /* The covariate pattern is  */
(905 missing values generated)

.                                   /* assigned a group number   */
. drop if grp == .                  /* Discard missing values    */
(905 observations deleted)

. egen cases = count(grp), by(grp)  /* Size of covariate pattern */

. egen die = total(death), by(grp)  /* Deaths in pattern         */

. sort grp                          /* Sort by covariate pattern */

. quietly by grp: keep if _n==1     /* Keep one case per pattern */
```

The new model will have `die` as the binomial numerator and `cases` as the binomial denominator. It is fit using

```
. glm die anterior hcabg kk2 kk3 kk4 age2-age4, family(binomial cases) eform
> nolog
Generalized linear models                      No. of obs       =          53
Optimization      : ML                         Residual df      =          44
                                               Scale parameter  =           1
Deviance       =    54.5713578                 (1/df) Deviance  =    1.240258
Pearson        =    54.94033298                (1/df) Pearson   =    1.248644

Variance function: V(u) = u*(1-u/cases)        [Binomial]
Link function    : g(u) = ln(u/(cases-u))      [Logit]
                                               AIC              =    3.372005
Log likelihood    = -80.35813033               BIC              =   -120.1215
```

die	Odds Ratio	OIM Std. Err.	z	P>|z|	[95% Conf. Interval]	
anterior	1.901333	.3185757	3.83	0.000	1.369103	2.640464
hcabg	2.105275	.7430694	2.11	0.035	1.054076	4.204801
kk2	2.251732	.4064423	4.50	0.000	1.580786	3.207453
kk3	2.172105	.584427	2.88	0.004	1.281907	3.680487
kk4	14.29137	5.087654	7.47	0.000	7.112964	28.71423
age2	1.63726	.5078582	1.59	0.112	.8914261	3.007115
age3	4.532029	1.206534	5.68	0.000	2.689568	7.636647
age4	8.893222	2.41752	8.04	0.000	5.219991	15.15125
_cons	.0063961	.0016541	-19.54	0.000	.0038529	.010618

Note: _cons estimates baseline odds.

The odds ratios and standard errors are identical to the binary model. This is as it should be. However, there is a difference in value for the dispersion (the header for this model was included in chapter 9 where the Pearson dispersion had a value of 0.943). The dispersion is now 1.2486, which indicates considerable overdispersion in the data. Scaling, or some other type of adjustment, is necessary.

It must be remembered that when used with discrete response GLM models, the Pearson dispersion is not an estimate of a scale parameter as it was for continuous response models. The scale is 1.0 for discrete response models, unless a quasilikelihood model is being specified.

We scale the model to accommodate the overdispersion found in the binomial model.

```
. glm die anterior hcabg kk2 kk3 kk4 age2-age4, family(binomial cases)
> scale(x2) eform noheader nolog
```

die	Odds Ratio	OIM Std. Err.	z	P>|z|	[95% Conf. Interval]	
anterior	1.901333	.3559852	3.43	0.001	1.317311	2.744278
hcabg	2.105275	.8303261	1.89	0.059	.971835	4.560631
kk2	2.251732	.4541698	4.02	0.000	1.516461	3.343508
kk3	2.172105	.6530547	2.58	0.010	1.204932	3.915606
kk4	14.29137	5.685084	6.69	0.000	6.553411	31.16596
age2	1.63726	.5674947	1.42	0.155	.830005	3.229644
age3	4.532029	1.348214	5.08	0.000	2.529719	8.119194
age4	8.893222	2.701403	7.19	0.000	4.903412	16.12946
_cons	.0063961	.0018483	-17.48	0.000	.0036303	.0112691

```
Note: _cons estimates baseline odds.
(Standard errors scaled using square root of Pearson X2-based dispersion.)
```

Finally, we model the data by using the log-log link. We found in the last chapter that using the log-log link produced a smaller deviance statistic than using any of the other links. Scaling will also be applied with the scale(x2) option. However, the coefficients are not exponentiated. Noncanonical binomial links cannot be interpreted using odds ratios. Hence, probit, log-log, clog-log, and other noncanonical links are usually left in basic coefficient form. Exponentiated coefficients for the log-log and clog-log links do not admit an odds-ratio interpretation as explained in section 10.6.2, though some other noncanonical links do admit other interpretations for exponentiated coefficients as shown in section 10.6. Of course, this has no bearing on standard errors or p-values.

```
. glm die anterior hcabg kk2 kk3 kk4 age2-age4, family(binomial cases)
> link(loglog) scale(x2) eform nolog
Generalized linear models                        No. of obs       =          53
Optimization      : ML                           Residual df      =          44
                                                 Scale parameter  =           1
Deviance        =   47.26519568                  (1/df) Deviance  =    1.074209
Pearson         =   42.22306082                  (1/df) Pearson   =     .959615

Variance function: V(u) = u*(1-u/cases)          [Binomial]
Link function    : g(u) = -ln(-ln(u/cases))      [Log-log]

                                                 AIC              =    3.234153
Log likelihood  = -76.70504927                   BIC              =   -127.4276
```

die	exp(b)	OIM Std. Err.	z	P>|z|	[95% Conf. Interval]	
anterior	1.226474	.0624397	4.01	0.000	1.110002	1.355167
hcabg	1.260886	.1601373	1.83	0.068	.9830372	1.617266
kk2	1.287005	.0753592	4.31	0.000	1.147464	1.443516
kk3	1.370154	.1278584	3.37	0.001	1.141138	1.645133
kk4	3.257141	.6064738	6.34	0.000	2.261228	4.691685
age2	1.110362	.0851241	1.37	0.172	.9554513	1.290389
age3	1.516314	.1040671	6.07	0.000	1.32547	1.734637
age4	1.998016	.153782	8.99	0.000	1.718242	2.323345
_cons	.1827758	.011464	-27.10	0.000	.161633	.2066842

(Standard errors scaled using square root of Pearson X2-based dispersion.)

hcabg becomes a bit more suspect under this parameterization. Let us use a likelihood-ratio test to evaluate its worth to the model where the likelihood-ratio χ^2 statistic is given by

$$\chi^2_{(\nu)} = -2\left(\mathcal{L}_1 - \mathcal{L}_0\right) \tag{11.2}$$

where $\mathcal{L}_1$ is the log-likelihood value for fitting the full model and $\mathcal{L}_0$ is the log-likelihood value for the reduced model. Significance is based on a χ^2 distribution with the degrees of freedom ν equal to the number of predictors dropped from the full model. For our case, we have

$$\chi^2_{(1)} = -2\left\{(-78.27579023) - (-76.70504926)\right\} = 3.14 \tag{11.3}$$

With 1 degree of freedom, we can calculate the significance by using

```
. display chiprob(1, 3.14)
.07639381
```

which shows that deleting hcabg from the model does not significantly reduce the log-likelihood function. We should remove it in favor of the more parsimonious model. The likelihood-ratio test provides more evidence to the results obtained by the Wald test (displayed in the output).

The method of scaling is a post hoc method of adjustment and is fairly easy to implement. In the previous model, the measure of deviance dispersion does not indicate a problem of overdispersion. We included the scaled output in this model only for continuity of the illustration. The other methods we examine are more sophisticated but usually do not provide a better adjustment. An example of this is Williams' procedure presented in the next section.

All binary models need to be evaluated for the possibility of extra binomial dispersion. When it is significant, means need to be taken to adjust standard errors to take account for the excess correlation in the data. Ascertaining the source of the dispersion is helpful, though this is not always possible. Extra correlation is often a result of a clustering effect or longitudinal data taken on a group of individuals over time. Scaling and implementing robust variance estimators often addresses this issue. Researchers may, in fact, need more sophisticated methods to deal with the extra correlation they have found in their data. The Stata `glm` command incorporates several sophisticated variance adjustment options, including jackknife, bootstrap, Newey–West, and others.

The output below is obtained for the log-log model, assuming the data are in temporal order, where the variance matrix is a weighted sandwich estimate of variance of Newey–West type with Anderson quadratic weights for the first two lags. These data have no true temporal ordering, so the example is meant only to highlight the specification of the variance.

```
. generate t = _n
. tsset t
        time variable:  t, 1 to 53
                delta:  1 unit
. glm die anterior hcabg kk2 kk3 kk4 age2-age4, family(binomial cases)
> link(loglog) irls nwest(anderson 2) nolog
Generalized linear models                   No. of obs      =          53
Optimization     : MQL Fisher scoring       Residual df     =          44
                   (IRLS EIM)               Scale parameter =           1
Deviance         =  47.26519568             (1/df) Deviance =    1.074209
Pearson          =  42.22305114             (1/df) Pearson  =    .9596148

Variance function: V(u) = u*(1-u/cases)     [Binomial]
Link function    : g(u) = -ln(-ln(u/cases)) [Log-log]

HAC kernel (lags): Anderson (2)             BIC             =   -127.4276
```

die	Coef.	HAC Std. Err.	z	P>|z|	[95% Conf. Interval]	
anterior	.204143	.0499691	4.09	0.000	.1062054	.3020805
hcabg	.2318145	.0658359	3.52	0.000	.1027786	.3608505
kk2	.2523179	.0401534	6.28	0.000	.1736186	.3310171
kk3	.3149235	.1686791	1.87	0.062	-.0156815	.6455285
kk4	1.18085	.1204433	9.80	0.000	.9447852	1.416914
age2	.104686	.0578602	1.81	0.070	-.0087179	.2180899
age3	.4162827	.0707713	5.88	0.000	.2775734	.5549919
age4	.6921545	.0663559	10.43	0.000	.5620994	.8222097
_cons	-1.699495	.0687791	-24.71	0.000	-1.8343	-1.56469

GEE or random-effects models may be required. These are not post hoc methods. GEE methods incorporate an adjustment into the variance–covariance matrix, which is iteratively updated with new adjustment values. Adjustments are calculated for each cluster group and are summed across groups before entering the weight matrix.

Random-effects models can also be used but require specification of a distribution for the random element causing extra dispersion. The *Longitudinal-Data/Panel-Data Reference Manual* describes these possibilities.

11.3   Williams' procedure

Another method that has been widely used to deal with overdispersed grouped logit models is Williams' procedure. The Williams' procedure, introduced by Williams (1982), iteratively reduces the χ^2-based dispersion to approximately 1.0. The procedure uses an extra parameter, called ϕ, to scale the variance or weight function. It changes with each iteration. The value of ϕ, which results in a χ^2-based dispersion of 1.0, is then used to weight a standard grouped or proportion logistic regression. The weighting formula is $1/\{1 + (k - 1)\phi\}$, where k is the proportional denominator.

Pearson's χ^2 is defined as

$$\chi^2 = \sum_{i=1}^{n} \frac{(y_i - \mu_i)^2}{V(\mu_i)} = \sum_{i=1}^{n} \frac{(y_i - \mu_i)^2}{\mu_i(1 - \mu_i/k_i)} \tag{11.4}$$

Listing 11.1. Williams' procedure algorithm

```
1     φ = user input
2     else {φ = 1}
3     sc = 1 + (k − 1)φ
4     sc = 1   if  φ = 1
5     μ = k(y + 0.5)/(k + 1)
6     η = ln{μ/(k − μ)}
7     WHILE (abs(ΔDev) > tolerance)  {
8         w = {sc(μ)(k − μ)}/k
9         z = η + {k(y − μ)}/{sc(μ)(k − μ)}
10        β = (xᵀwx)⁻¹xᵀwz
11        η = xβ
12        μ = k/{1 + exp(−η)}
13        OldDev = Dev
14        Devᵢ = k ln{(k − y)/(k − μ)}     if (y = 0)
15        Devᵢ = y ln(y/μ)                 if (y = k)
16        Devᵢ = y ln(y/μ) + (k − y) ln{(k − y)/(k − μ)} if (y ≠ 0, y ≠ k)
17        Dev = 2∑((p)Devᵢ)/sc
18        ΔDev = Dev − OldDev
19    }
20    χ² = ∑(y − μ)²/(sc(μ){1 − μ/k})
21    dof = #observations − #predictors
22    dispersion = χ²/dof
```

```
23      * Note: Adjust φ so dispersion=1, then with final φ value:
24      w = 1/(1 + (k − 1)φ)
25      Rerun algorithm with prior weight equal to w and φ = 1
```

The user must input a value of ϕ in the above algorithm. To avoid delay in finding the optimal ϕ value to begin, we can use the algorithm below that iterates through the Williams' procedure, searching for the optimal scale value. We refer to the above algorithm simply as `williams`.

Listing 11.2. IRLS algorithm for Williams' procedure with optimal scale

```
1       Let ks = scale (sc above)
2       wt = weight
3       williams y k predictors
4       χ² = Σ(y − μ)²/{μ(1 − μ/k)}
5       dof = #observations - #predictors
6       Disp = χ²/dof
7       φ = 1/Disp
8       WHILE (abs(ΔDisp) > tolerance) {
9           OldDisp = Disp
10          williams y k predictors, ks=φ
11          χ² = Σ[(y − μ)²/{μ(1 − μ/k)}]
12          Disp = χ²/dof
13          φ = Dispφ
14          ΔDisp = Disp - OldDisp
15      }
16      wt = 1/{1 + (k − 1)φ}
17      williams y k predictors [w=wt]
```

You can download the `williams` command (which we use below) for this model by typing `net install williams, from(http://www.stata.com/users/jhilbe)`.

Running this version of the Williams' procedure yields the following output:

```
. williams die cases anterior hcabg kk2 kk3 kk4 age2-age4, eform nolog
Resid DF   =        44                         No obs.    =        53
Chi2       =  38.22372                         Deviance   =  41.0286
Dispersion =  .8687208                         Prob>chi2  =  .599722
                                               Dispersion =  .9324682
CHI2(8)    =  90.61774
Prob>CHI2  =  3.48e-16                          Pseudo R2  =  .1279288
logistic regression: William´s procedure
```

__00000I	OR	Std. Err.	t	P>|t|	[95% Conf. Interval]	
anterior	1.632604	.3784553	2.11	0.034	1.036479	2.571585
hcabg	1.969905	.7339105	1.82	0.069	.9491105	4.088594
kk2	2.11554	.5611586	2.82	0.005	1.257861	3.558032
kk3	2.239763	.7395005	2.44	0.015	1.172622	4.278051
kk4	13.52973	5.291466	6.66	0.000	6.286095	29.1204
age2	1.563919	.6988122	1.00	0.317	.6514241	3.754609
age3	3.715466	1.482642	3.29	0.001	1.699576	8.122429
age4	6.516559	2.578986	4.74	0.000	3.000155	14.15445
_cons	.0091257	.0037047	-11.57	0.000	.0041182	.0202223

```
Log Likelihood = -308.86356
Phi            =    .0080459
```

The estimates are different from those produced in the standard model. An extra parameter was put into the estimating algorithm itself, rather than being attached only to the standard errors as in the scaling method. And the final weighted logistic regression displays p-values based on t statistics per the recommendation of Collett (2003, 196).

The procedure changed the results only slightly, but this is not always the case. Standard errors and coefficients will change little when the standard model has a Pearson χ^2 statistic near 1. Here the standard model had a Pearson statistic of 1.25.

11.4   Robust standard errors

Parameter standard errors are derived from the variance–covariance matrix produced in the estimation process. As discussed in chapter 3, model standard errors are calculated as the square root of the matrix diagonal. If the model is fit using Fisher scoring, then the variance–covariance matrix from which standard errors are produced is the matrix of expected second derivatives. This is called the expected information matrix. If, on the other hand, standard errors are based on maximum likelihood using a variety of Newton–Raphson algorithms, then we have the matrix of observed second derivatives. As we discovered earlier, the more complex observed information matrix reduces to the expected information for canonical linked models.

In either case, the source of the variance–covariance matrix must not be forgotten. It is the matrix of observed second derivatives of the log-likelihood function, which itself is derived from an underlying probability distribution.

One of the prime requisites of a log-likelihood function is that the individual observations defined by it be independent of one another. We assume that there is no correlation between observations. When there is, the reliability of the model is in question.

We have discussed various types of overdispersion—some apparent, some genuine. One type of genuine overdispersion occurs when there is correlation between observations. Such correlation comes from a variety of sources. Commonly, correlation among response observations comes from the clustering effect due to grouping in the dataset itself. For instance, suppose that we are attempting to model the probability of death on the basis of patient characteristics as well as certain types of medical procedures. Data come from several different facilities. Perhaps there are differences in how facilities tend to treat patients. Facilities may even prefer treating certain types of patients over other types. If there are treatment differences between hospitals, then there may well be extra correlation of treatment within hospitals, thus violating the independence-of-observations criterion. Treatment across facilities may be independent, but not within facilities. When this type of situation occurs, as it does so often in practice, then standard errors need to be adjusted.

Methods have been designed to adjust standard errors by a means that does not depend directly on the underlying likelihood of the model. One such method is redefining the weight such that "robust" standard errors are produced. As we will see, robust standard errors still have a likelihood basis, but the assumption of independence is relaxed. The method we discuss was first detailed by Huber (1967) and White (1980). It was introduced into Stata software and discussed by (a former student of Huber) Rogers (1992) and others. It plays an important role in GEE and survey-specific algorithms. Robust variance estimators have only recently found favor in the biological and health sciences, but they are now popular in those areas. In fact, they are probably overused.

Robust standard errors, and there are several versions of them (see section 3.6), can be understood as a measurement of the standard error of the calculated parameter if we indefinitely resample and refit the model. For instance, if we use a 95% confidence interval, we can interpret robust standard errors as follows:

> If the sampling of data were continually repeated and estimated, we would expect the collection of robust constructed confidence intervals to contain the "true" coefficient 95% of the time.

We place the quotes around "true" for the same reason that we mentioned that the robust variance estimate is probably overused. Although the robust variance estimate is robust to several violations of assumptions in the linear regression model, the same cannot be said in other circumstances. For example, the robust variance estimate applied to a probit model does not have the same properties. If the probit model is correctly specified, then the issue of robustness is not relevant, because either the usual Hessian matrix or the sandwich estimate may be used. The probit estimator itself is not consistent in the presence of heteroskedasticity, for example, and the sandwich estimate of variance is appropriate when the estimator converges in probability. With our example, the probit model may not converge to anything interesting. Using the robust variance estimate does not change this, and so the utility of the estimate is limited to inferring coverage of true parameters (parameters of the probit model) for the infinite population from which we obtained our sample. As Bill Sribney, one of the authors of Stata's original survey commands, likes to say, "There are no free lunches!"

Part IV

Count Response Models

12 The Poisson family

Contents

12.1 Count response regression models

In previous chapters, we discussed regression models having either binary or grouped response data. We now turn to models having counts as a response. Included are the Poisson, geometric, generalized Poisson, and negative binomial regression models. We discuss multinomial and ordinal models, which entail multiple levels of a response, in later chapters.

Counts refer to a simple counting of events, whether they are classified in tables or found as raw data. The model is named for the study presented in Poisson (1837). Examples of count response data include hospital length of stay (LOS), number of rats that died in an drug experiment, or the classic count of Prussian soldiers who died as a result of being kicked by horses described in von Bortkiewicz (1898).

Count data may also take the form of a rate. In a manner analogous to the grouped or proportional binomial model, we may think of rate data as counts per unit of population (for example, the comparative incidence rate of death for people exposed to a particular virus in Arizona). Incidence rates will be computed for each county, and a model will be developed to help understand comparative risks.

We begin the discussion of both the count and rate data with the Poisson model. It is a comparatively simple one-parameter model and has been extensively used in epidemiological studies. We discover, however, that it is too simple (restrictive) for many real-life data situations.

12.2   Derivation of the Poisson algorithm

The Poisson probability distribution function is formulated as

$$f(y; \mu) = e^{-\mu} \mu^y / y! \tag{12.1}$$

or in exponential-family form as

$$f(y; \mu) = \exp\left\{ y \ln(\mu) - \mu - \ln \Gamma(y + 1) \right\} \tag{12.2}$$

The link and cumulant functions are then derived as

$$\theta = \ln(\mu) \tag{12.3}$$

$$b(\theta) = \mu \tag{12.4}$$

The canonical link is found to be the log link. Hence, the inverse link is $\exp(\eta)$, where η is the linear predictor. The mean and variance functions are calculated as the first and second derivatives with respect to θ.

$$b'(\theta) = \frac{\partial b}{\partial \mu} \frac{\partial \mu}{\partial \theta} \tag{12.5}$$

$$= (1)(\mu) \tag{12.6}$$

$$= \mu \tag{12.7}$$

$$b''(\theta) = \frac{\partial^2 b}{\partial \mu^2} \left(\frac{\partial \mu}{\partial \theta}\right)^2 + \frac{\partial b}{\partial \mu} \frac{\partial^2 \mu}{\partial \theta^2} \tag{12.8}$$

$$= (0)(1)^2 + (\mu)(1) \tag{12.9}$$

$$= \mu \tag{12.10}$$

The mean and variance functions of the Poisson distribution are identical. As a prelude to future discussion, models having greater values for the variance than the mean are said to be overdispersed.

The derivative of the link is essential to the Poisson algorithm. It is calculated as

$$\frac{\partial \theta}{\partial \mu} = \frac{\partial \{\ln(\mu)\}}{\partial \mu} = \frac{1}{\mu} \tag{12.11}$$

The log-likelihood function can be abstracted from the exponential form of the distribution by removing the initial exponential and associated braces.

$$\mathcal{L}(\mu; y) = \sum_{i=1}^{n} \left\{ y_i \ln(\mu_i) - \mu_i - \ln \Gamma(y_i + 1) \right\} \tag{12.12}$$

When the response has the value of zero, the individual log-likelihood functions reduce to

$$\mathcal{L}_i(\mu_i; 0) = -\mu_i \tag{12.13}$$

The log-likelihood function can also be parameterized in terms of $\mathbf{x}\boldsymbol{\beta}$. This is required for all maximum likelihood methods. Substituting the inverse link, $\exp(\mathbf{x}\boldsymbol{\beta})$, for each instance of μ in the GLM parameterization above,

$$\mathcal{L}(\mathbf{x}\boldsymbol{\beta};y) \;=\; \sum_{i=1}^{n} [y_i \ln\{\exp(\mathbf{x}_i\boldsymbol{\beta})\} - \exp(\mathbf{x}_i\boldsymbol{\beta}) - \ln\Gamma(y_i+1)] \quad (12.14)$$

$$=\; \sum_{i=1}^{n} \{y_i(\mathbf{x}_i\boldsymbol{\beta}) - \exp(\mathbf{x}_i\boldsymbol{\beta}) - \ln\Gamma(y_i+1)\} \quad (12.15)$$

$$\mathcal{L}_i(\mathbf{x}\boldsymbol{\beta};y_i=0) \;=\; -\exp(\mathbf{x}_i\boldsymbol{\beta}) \quad (12.16)$$

The deviance function, required for the traditional GLM algorithm, is defined as $2\{\mathcal{L}(y;y) - \mathcal{L}(\mu;y)\}$. It is derived as

$$D \;=\; 2\sum_{i=1}^{n} \{y_i \ln(y_i) - y_i - y_i \ln(\mu_i) + \mu_i\} \quad (12.17)$$

$$=\; 2\sum_{i=1}^{n} \left\{ y_i \ln\left(\frac{y_i}{\mu_i}\right) - (y_i - \mu_i) \right\} \quad (12.18)$$

Again when the response is zero, the deviance function reduces to

$$D_i(y_i = 0) = 2\mu_i \quad (12.19)$$

We can now enter the necessary functions into the basic iteratively reweighted least-squares (IRLS) GLM algorithm.

Listing 12.1. IRLS algorithm for Poisson regression

```
1     Dev = 0
2     μ = {y − mean(y)}/2           /* initialization of μ */
3     η = ln(μ)                     /* initialization of η */
4     WHILE (Abs(ΔDev) > tolerance )  {
5          W = μ
6          z = η + (y − μ)/μ − offset
7          β = (XᵀWX)⁻¹XᵀWz
8          η = Xβ + offset
9          μ = exp(η)
10         OldDev = Dev
11         Dev = 2∑{y ln(y/μ) − (y − μ)} if (y ≥ 1)
12         Dev = 2∑μ if (y = 0)
13         ΔDev = Dev − OldDev
14    }
15    χ² = ∑(y − μ)²/μ
```

The log-likelihood function entails using the log-gamma function. Most commercial software packages have the log-gamma as a user-accessible function. However, the traditional GLM algorithm favors excluding the log likelihood except when used to calculate certain fit statistics. Some software packages have the algorithm iterate on the basis of the log-likelihood function, just as we do here on the deviance. For example, Stata's `poisson` command iterates on the basis of log-likelihood difference. The results of iteration based on the deviance or the log likelihood will be identical. However, some programs (not Stata), when using the log likelihood to iterate, drop the log-gamma normalization term from the function. Usually, results vary only slightly, though the reported value of the log likelihood will differ.

When the deviance is used as the basis of iteration, the log-likelihood functions can be used to construct a (likelihood ratio) χ^2 GOF statistic and pseudo-R^2 given in (4.50) defined by

$$\chi^2 = -2\left(\mathcal{L}_0 - \mathcal{L}_F\right) \tag{12.20}$$

$$R^2 = 1 - \frac{\mathcal{L}_F}{\mathcal{L}_0} \tag{12.21}$$

where $\mathcal{L}_0$ is the value of the log likelihood for a constant-only model and $\mathcal{L}_F$ is the final or full value function. Different versions exist for both χ^2 and R^2 functions. One should always check the software manual to determine which formulas are being implemented. As we outlined in section 4.6.1, there are several definitions for R^2. From the list in that section, the Cameron–Windmeijer definition is the preferred one. Calculating

$$G^2 = 2\sum_{i=1}^{n} y_i \ln(y_i/\widehat{\mu}_i) \tag{12.22}$$

as a GOF statistic is also common. This statistic is zero for a perfect fit.

The above algorithm varies from the standard algorithm we have been using thus far. We have added an offset to the algorithm. In fact, previous GLM algorithms could well have had offsets. We chose not to put them in until now because they are rarely used for anything but Poisson and negative binomial algorithms. Sometimes offsets are used with clog-log data, but this is not the ordinary case.

John Nelder first conceived of offsets as an afterthought to the IRLS algorithm that he and Wedderburn designed in 1972. The idea began as a method to put a constant term directly into the linear predictor without that term being estimated. It affects the algorithm directly before and after regression estimation. Only later did Nelder discover that the notion of an offset could be useful for modeling rate data.

An offset must be put into the same metric as the linear predictor, η. Thus, an offset enters into a (canonical link) Poisson model as the natural log of the variable. We must log-transform the variable to be used as an offset before inserting it into the estimation algorithm unless the software does this automatically or unless the software has an option for this task, like Stata (see the documentation on `exposure()`).

Epidemiologists interpret the notion of offset as exposure (for example, the number of people who are exposed to a disease). The incidence, or number of people who suffer some consequence, enter as the response. Hence, we can think of the Poisson model, parameterized as a rate, as modeling incidence over exposure.

Poisson data (counts of events) are generated by a process that conforms to the following four properties:

1. The probability of a single event over a small interval (small amount of exposure) is approximately proportional to the size of the interval (amount of the exposure).

2. The probability of two events occurring in the same narrow interval (the same period of exposure) is negligible.

3. The probability of an event within a certain interval length (certain amount of exposure) does not change over different intervals of the same length (different periods of the same amount of exposure).

4. The probability of an event in one interval (in one certain period of exposure) is independent of the probability of an event in any other nonoverlapping interval (period of exposure).

When the latter two properties do not hold, the data can have variance that exceeds the mean; this is called overdispersion. For our hospital discharge example, if the likelihood of discharge on a Friday is different than it is on a Monday, then the third property does not hold and the LOS could be more variable than would be modeled by Poisson regression.

Also, we find that there is no possibility that 0 can be a value for the response hospital LOS. Day 1, representing a count of 1, begins the day that a patient is admitted to the hospital. To be included in the dataset, the patient must have a count of 1 or greater.

The Poisson model assumes that there is the possibility of zero counts even if there are no such records in the data. When the possibility of a zero count is precluded, as in our example, then the Poisson model may not be appropriate. A zero-truncated Poisson model could be used, which adjusts the Poisson probability appropriately to exclude zero counts. To do this, subtract the probability of a Poisson zero count, $\exp(-\mu)$, from 1, and condition this event out of the model. The resulting log likelihood is given by

$$\mathcal{L}(\mu; y|y > 0) = \sum_{i=1}^{n} \left(y_i(\mathbf{x}_i\boldsymbol{\beta}) - \exp(\mathbf{x}_i\boldsymbol{\beta}) - \ln\Gamma(y_i + 1) - \ln\left[1 - \exp\{-\exp(\mathbf{x}_i\boldsymbol{\beta})\}\right] \right)$$
(12.23)

This expression amounts to apportioning the probability of the zero count to the remaining counts (weighted by their original probability).

We mention this non-GLM model because such data are often inappropriately modeled using unadjusted Poisson regression. Even so, experience has shown us that estimates and standard errors do not differ greatly between the two models, particularly when there are many observations in the data. Stata has built-in commands for such models; see the documentation for `tpoisson` and `tnbreg` commands. We discuss this model more and give examples in chapter 14.

12.3   Poisson regression: Examples

Let us use an example from the health field. We estimate LOS, adjusted by several predictors, including

hmo	binary variable coding membership in a health maintenance organization (HMO)
white	binary variable coding (self-reported) white race
type1	binary variable coding elective admission
type2	binary variable coding urgent admission
type3	binary variable coding emergency admission

`type` was a factor variable with three levels. It was separated into three factors before estimation, resulting in `type1`, `type2`, and `type3`. We will have `type1` serve as the reference.

All hospitals in this dataset are in Arizona and have the same diagnostic related group (DRG). Patient data have been randomly selected to be part of this dataset.

```
. use http://www.stata-press.com/data/hh4/medpar

. glm los hmo white type2 type3, family(poisson) eform nolog
Generalized linear models                     No. of obs       =      1,495
Optimization     : ML                         Residual df      =      1,490
                                              Scale parameter  =          1
Deviance       =  8142.666001                 (1/df) Deviance  =   5.464877
Pearson        =  9327.983215                 (1/df) Pearson   =   6.260391

Variance function: V(u) = u                   [Poisson]
Link function    : g(u) = ln(u)               [Log]

                                              AIC              =   9.276131
Log likelihood   = -6928.907786               BIC              =  -2749.057
```

los	IRR	OIM Std. Err.	z	P>|z|	[95% Conf. Interval]	
hmo	.9309504	.0222906	-2.99	0.003	.8882708	.9756806
white	.8573826	.0235032	-5.61	0.000	.8125327	.904708
type2	1.248137	.0262756	10.53	0.000	1.197685	1.300713
type3	2.032927	.0531325	27.15	0.000	1.931412	2.139778
_cons	10.30813	.2804654	85.74	0.000	9.77283	10.87275

Note: _cons estimates baseline incidence rate.

We requested that the software provide incidence-rate ratios rather than coefficients. Incidence-rate ratios are interpreted similarly to odds ratios, except that the response is a count. We can interpret the IRR for `type2` and `type3` as the following:

> Urgent admissions are expected to have a LOS that is 25% longer than elective admissions. Emergency admissions stay twice as long as elective admissions.

We can also see that HMO patients stay in the hospital for slightly less time than do non-HMO patients, 7% less time, and nonwhite patients stay in the hospital about 14% longer than do white patients.

At first glance, each predictor appears to have a statistically significant relationship to the response, adjusted by other predictors. However, by checking the Pearson-based dispersion, we find substantial overdispersion in the data. The dispersion statistic is 6.26, much higher than it should be for a viable Poisson model.

There are two approaches to handing the data at this point. We can simply scale the standard errors by the dispersion, adjusting standard errors to a value that would be the case if there were no dispersion in the data, that is, a dispersion of 1. Not restricting the dispersion to one specifies a quasi-Poisson model. We may also seek to discover the source of the overdispersion and take more directed measures to adjust for it. First, we scale standard errors.

```
. glm los hmo white type2 type3, family(poisson) irls scale(x2) eform noheader
> nolog
```

		EIM				
los	IRR	Std. Err.	z	P>|z|	[95% Conf.	Interval]
hmo	.9309504	.0557729	-1.19	0.232	.8278113	1.04694
white	.8573826	.0588069	-2.24	0.025	.7495346	.9807484
type2	1.248137	.0657437	4.21	0.000	1.12571	1.383878
type3	2.032927	.1329416	10.85	0.000	1.788373	2.310923
_cons	10.30813	.701746	34.27	0.000	9.020545	11.77951

Note: _cons estimates baseline incidence rate.
(Standard errors scaled using square root of Pearson X2-based dispersion.)

The standard errors have markedly changed. `hmo` now appears to be noncontributory to the model.

Provider type may have bearing on LOS. Some facilities may tend to keep patients longer than others. We now attempt to see whether overdispersion is caused by extra correlation due to provider effect.

```
. glm los hmo white type2 type3, family(poisson) cluster(provnum) eform
> noheader nolog
```

                              (Std. Err. adjusted for 54 clusters in provnum)

| | | Robust | | | | |
los	IRR	Std. Err.	z	P>|z|	[95% Conf.	Interval]
hmo	.9309504	.0490889	-1.36	0.175	.8395427	1.03231
white	.8573826	.0625888	-2.11	0.035	.7430825	.9892642
type2	1.248137	.0760289	3.64	0.000	1.107674	1.406411
type3	2.032927	.4126821	3.49	0.000	1.365617	3.02632
_cons	10.30813	.6898133	34.86	0.000	9.041034	11.75281

Note: _cons estimates baseline incidence rate.

Again we see somewhat similar results—hmo does not contribute. There indeed appears to be an excess correlation of responses within providers. That is, LOS is more highly correlated within providers, given particular patterns of covariates, than between providers. This correlation accounts for overdispersion, but we know nothing more about the nature of the overdispersion.

Finally, we show an example of a rate model. The dataset used in this example is also used in an example in the poisson entry of the *Stata Base Reference Manual*, and it comes from Doll and Hill (1966). The data deal with coronary deaths among British physicians due to smoking background. Person-years is entered as an offset. We let the software log pyears prior to estimation. The variables a1–a5 represent 10-year-age categories, starting with age 35 and ending at age 84.

```
. use http://www.stata-press.com/data/hh4/doll
. glm deaths smokes a2-a5, family(poisson) exposure(pyears) eform nolog
```

Generalized linear models   No. of obs   =   10
Optimization   : ML   Residual df   =   4
                                                   Scale parameter =   1
Deviance   =   12.1323664   (1/df) Deviance = 3.033092
Pearson   =   11.15533338   (1/df) Pearson = 2.788833

Variance function: V(u) = u   [Poisson]
Link function   : g(u) = ln(u)   [Log]

                                                   AIC   = 7.920031
Log likelihood   = -33.60015344   BIC   = 2.922026

| | | OIM | | | | |
deaths	IRR	Std. Err.	z	P>|z|	[95% Conf.	Interval]
smokes	1.425519	.1530638	3.30	0.001	1.154984	1.759421
a2	4.410584	.8605198	7.61	0.000	3.009011	6.464998
a3	13.8392	2.542639	14.30	0.000	9.654328	19.83809
a4	28.51678	5.269878	18.13	0.000	19.85177	40.96396
a5	40.45121	7.775511	19.25	0.000	27.75326	58.95885
_cons	.0003636	.0000697	-41.30	0.000	.0002497	.0005296
ln(pyears)	1	(exposure)				

Note: _cons estimates baseline incidence rate.

We model only a main-effects model. The *Stata Base Reference Manual* explores interactions, finding several significant contributions to the model.

Smoking and the age categories all appear to significantly contribute to the model. In fact, we see a rather abrupt increase in the number of deaths as the age category increases. However, once again we find overdispersion. Note the Pearson-based dispersion statistic of 2.79. We may then attempt to deal with the overdispersion by scaling standard errors.

```
. glm deaths smokes a2-a5, family(poisson) irls scale(x2) exposure(pyears)
> eform noheader nolog
```

deaths	IRR	EIM Std. Err.	z	P>|z|	[95% Conf. Interval]	
smokes	1.425519	.255613	1.98	0.048	1.003095	2.025834
a2	4.410584	1.437046	4.55	0.000	2.328947	8.352807
a3	13.8392	4.246139	8.56	0.000	7.584826	25.25087
a4	28.51678	8.800557	10.86	0.000	15.5744	52.21432
a5	40.45121	12.9849	11.53	0.000	21.56227	75.88719
_cons	.0003636	.0001165	-24.73	0.000	.0001941	.0006812
ln(pyears)	1	(exposure)				

```
Note: _cons estimates baseline incidence rate.
(Standard errors scaled using square root of Pearson X2-based dispersion.)
```

We find that smoking and the age categories are still significant at a significance level of 0.05.

In chapter 11, we detailed the various components to overdispersion. We pointed out the difference between apparent and inherent overdispersion. This is a case of the former. By creating various interaction terms based on smoking and age category, the model loses its apparent overdispersion. Smoking risk is associated with different levels of age. Understanding a as age-category and s as smoking, we can create the variables for the following interactions. The variable sa12 indicates smokers in age groups a1 and a2, sa34 indicates smokers in age groups a3 and a4, and sa5 indicates smokers in age group a5. We can model the data with these covariates by typing

```
. generate sa12 = smokes*(a1+a2)

. generate sa34 = smokes*(a3+a4)

. generate sa5  = smokes*(a5)

. glm deaths sa12 sa34 sa5 a2-a5, family(poisson) irls scale(x2)
> exposure(pyears) eform noheader nolog
```

deaths	IRR	EIM Std. Err.	z	P>|z|	[95% Conf. Interval]	
sa12	2.636259	.6921945	3.69	0.000	1.575763	4.410476
sa34	1.412229	.1885011	2.59	0.010	1.087148	1.834517
sa5	.9047304	.1733675	-0.52	0.601	.6214541	1.317132
a2	4.294559	.7834725	7.99	0.000	3.003521	6.140539
a3	23.42263	7.276351	10.15	0.000	12.74107	43.05915
a4	48.2631	15.01423	12.46	0.000	26.23094	88.80073
a5	97.87965	32.05599	14.00	0.000	51.51342	185.9792
_cons	.0002166	.0000609	-30.00	0.000	.0001248	.0003759
ln(pyears)	1	(exposure)				

```
Note: _cons estimates baseline incidence rate.
(Standard errors scaled using square root of Pearson X2-based dispersion.)
```

Scaling produces substantially different standard errors, indicating that the dispersion is real and not just apparent.

The moral of our discussion is that we must be constantly aware of overdispersion when dealing with count data. If the model appears to be overdispersed, we must seek out the source of overdispersion. Maybe interactions are required or predictors need to be transformed. If the model still shows evidence of overdispersion, then scaling or applying robust variance estimators may be required. Perhaps other types of models need to be applied. The negative binomial would be a logical alternative.

12.4   Example: Testing overdispersion in the Poisson model

As we have mentioned in this chapter, a shortcoming of the Poisson model is its assumption that the variance is equal to the mean. We can use a regression-based test for this assumption by following the procedure shown below:

1. Obtain the fitted values $\widehat{\mu}_i$.

2. Calculate

$$z = \frac{(y_i - \widehat{\mu}_i)^2 - y_i}{\widehat{\mu}_i \sqrt{2}}$$

3. Regress z as a constant-only model.

The test of the hypotheses

$$H_0: V(y) = E(y)$$
$$H_1: V(y) = E(y) + \alpha g\{E(y)\}$$

is carried out by the t test of the constant in the regression; see Cameron and Trivedi (2013).

Returning to our example, we may assess the overdispersion in our model by using the following procedure:

```
. use http://www.stata-press.com/data/hh4/medpar, clear
. glm los hmo white type2 type3, family(poisson) eform noheader nolog
```

		OIM				
los	IRR	Std. Err.	z	P>|z|	[95% Conf. Interval]	
hmo	.9309504	.0222906	-2.99	0.003	.8882708	.9756806
white	.8573826	.0235032	-5.61	0.000	.8125327	.904708
type2	1.248137	.0262756	10.53	0.000	1.197685	1.300713
type3	2.032927	.0531325	27.15	0.000	1.931412	2.139778
_cons	10.30813	.2804654	85.74	0.000	9.77283	10.87275

```
Note: _cons estimates baseline incidence rate.
. predict double mu, mu
. generate z = ((los-mu)^2-los)/(mu*sqrt(2))
. regress z
```

Source	SS	df	MS	Number of obs	=	1,495
				F(0, 1494)	=	0.00
Model	0	0	.	Prob > F	=	.
Residual	348013.947	1,494	232.941062	R-squared	=	0.0000
				Adj R-squared	=	0.0000
Total	348013.947	1,494	232.941062	Root MSE	=	15.262

z	Coef.	Std. Err.	t	P>|t|	[95% Conf. Interval]	
_cons	3.704561	.3947321	9.39	0.000	2.930273	4.478849

The hypothesis of no overdispersion is rejected. This is not surprising because we have already seen evidence that the overdispersion exists.

There are other tests for overdispersion in the Poisson model, including a Lagrange multiplier test given by

$$\chi^2 = \frac{(\sum_i \widehat{\mu}_i^2 - n\overline{y})^2}{2\sum_i \widehat{\mu}_i^2} \tag{12.24}$$

This test is distributed χ^2 with 1 degree of freedom; see Cameron and Trivedi (2013). We can continue our use of Stata to compute the test:

```
. quietly summarize los if e(sample), meanonly
. quietly scalar nybar = r(sum)
. quietly generate double musq = mu*mu
. quietly summarize musq if e(sample), meanonly
. quietly scalar mu2 = r(sum)
. quietly scalar chival = (mu2-nybar)^2/(2*mu2)
```

```
. display as txt "LM value = " as res chival _n as txt "P-value  = " as res
> %6.4f chi2tail(1, chival)
LM value = 62987.844
P-value  = 0.0000
```

This approach yields the same conclusions.

12.5 Using the Poisson model for survival analysis

In section 6.6, we used the gamma model for survival analysis. Although the gamma
model cannot incorporate censoring into its routine, a right-censored exponential regression model may be duplicated with GLM by using the Poisson family. We simply
log-transform the time response, incorporating it into the model as an offset. The
new response is then the 0/1 censor variable. Using cancer.dta, we compare the two
methods.

```
. sysuse cancer, clear
(Patient Survival in Drug Trial)
. glm died age i.drug, exposure(studytime) family(poisson) nolog
Generalized linear models                    No. of obs      =         48
Optimization     : ML                        Residual df     =         44
                                             Scale parameter =          1
Deviance        =   34.33673608              (1/df) Deviance =   .7803804
Pearson         =   42.74908562              (1/df) Pearson  =   .9715701

Variance function: V(u) = u                  [Poisson]
Link function     : g(u) = ln(u)             [Log]

                                             AIC             =   2.173682
Log likelihood   = -48.16836804              BIC             =  -135.9961
```

died	Coef.	OIM Std. Err.	z	P>|z|	[95% Conf. Interval]	
age	.0854187	.0334969	2.55	0.011	.0197659	.1510716
drug						
2	-1.461711	.4761143	-3.07	0.002	-2.394878	-.5285437
3	-1.857458	.4683077	-3.97	0.000	-2.775324	-.9395916
_cons	-6.962982	1.899393	-3.67	0.000	-10.68572	-3.240241
ln(studyt~e)	1	(exposure)				

```
. stset studytime, failure(died)

     failure event:  died != 0 & died < .
obs. time interval:  (0, studytime]
 exit on or before:  failure
```

```
      48  total observations
       0  exclusions

      48  observations remaining, representing
      31  failures in single-record/single-failure data
     744  total analysis time at risk and under observation
                                     at risk from t =          0
                                earliest observed entry t =    0
                                 last observed exit t =        39
```

```
. streg age i.drug, nohr distribution(exponential) nolog

        failure _d:  died
   analysis time _t:  studytime
```

Exponential PH regression

No. of subjects =	48		Number of obs =		48
No. of failures =	31				
Time at risk =	744				
			LR chi2(3) =		26.35
Log likelihood =	-48.168368		Prob > chi2 =		0.0000

_t	Coef.	Std. Err.	z	P>|z|	[95% Conf. Interval]
age	.0854187	.0334969	2.55	0.011	.0197659 .1510716
drug					
2	-1.461711	.4761143	-3.07	0.002	-2.394878 -.5285437
3	-1.857458	.4683077	-3.97	0.000	-2.775324 -.9395916
_cons	-6.962982	1.899393	-3.67	0.000	-10.68572 -3.240241

The scale value produced by the Poisson model is nearly 1, which is to be expected. Theoretically, the exponential distribution has a scale of 1. In that, it is a subset of the gamma distribution. The **streg** model displayed above is parameterized in terms of a log hazard-rate (as opposed to log-time parameterization), hence the **nohr** option. Log-time rates would reverse the signs of the parameter estimates.

2.6   Using offsets to compare models

If we wish to test a subset of coefficients, we can fit a model and obtain a Wald test. Stata provides this functionality through the **test** command. For example, if we wish to test that H_0: $\beta_{age} = 0$ and $\beta_{drug2} = -1.5$, for the survival model presented in section 12.5, we first run the model

```
. glm died age i.drug, exposure(studytime) family(poisson) noheader nolog
```

	Coef.	OIM Std. Err.	z	P>|z|	[95% Conf.	Interval]
age	.0854187	.0334969	2.55	0.011	.0197659	.1510716
drug						
2	-1.461711	.4761143	-3.07	0.002	-2.394878	-.5285437
3	-1.857458	.4683077	-3.97	0.000	-2.775324	-.9395916
_cons	-6.962982	1.899393	-3.67	0.000	-10.68572	-3.240241
ln(studyt~e)	1	(exposure)				

To test our hypothesis using a Wald test, we use the `test` command.

```
. test age=0, notest
 ( 1)  [died]age = 0
. test 2.drug=-1.5, accumulate
 ( 1)  [died]age = 0
 ( 2)  [died]2.drug = -1.5
           chi2(  2) =    6.80
         Prob > chi2 =    0.0333
```

If we prefer a likelihood-ratio test, we can generate an offset that reflects our constraints and fits the model. Because our model already includes an offset on the log scale, we must combine all our constraints into one variable. First, we tell Stata to save the log likelihood from the full model.

```
. estimates store Unconstrained
```

Next, we create an offset reflecting the constraints that we wish to impose on the model.

```
. generate double offvar = 0*age - 1.5*2.drug + log(studytime)
```

Finally, we can fit the constrained model by specifying our new variable as the offset and then asking for the likelihood-ratio test of our constraints.

```
. glm died 3.drug, offset(offvar) family(poisson) nolog
Generalized linear models                    No. of obs      =          48
Optimization     : ML                        Residual df     =          46
                                             Scale parameter =           1
Deviance         =  40.83400585              (1/df) Deviance =    .8876958
Pearson          =  57.71533028              (1/df) Pearson  =    1.254681
Variance function: V(u) = u                  [Poisson]
Link function    : g(u) = ln(u)              [Log]
                                             AIC             =    2.225708
Log likelihood   = -51.41700292              BIC             =   -137.2412
```

died	Coef.	OIM Std. Err.	z	P>|z|	[95% Conf. Interval]
3.drug	-1.875897	.4546061	-4.13	0.000	-2.766908 -.9848854
_cons	-2.204461	.2	-11.02	0.000	-2.596454 -1.812469
offvar	1	(offset)			

```
. lrtest Unconstrained
Likelihood-ratio test                        LR chi2(2)   =        6.50
(Assumption: . nested in Unconstrained)      Prob > chi2  =      0.0388
```

Both tests yield the conclusion that the full model is significantly different from the constrained model. The likelihood-ratio test can be verified, by hand, using the listed values of the log likelihood from each model $\chi^2_{\mathrm{LR}}(2) = 6.5 = -2\{(-51.417)-(-48.168)\}$.

The example highlighted how to use the offset to test constrained models, but Stata can more directly implement constraints. The same test can be obtained using

```
. constraint 1 age=0

. constraint 2 2.drug=-1.5

. glm died age i.drug, exposure(studytime) family(poisson) constraints(1 2)
> nolog
```

```
Generalized linear models                    No. of obs      =          48
Optimization     : ML                        Residual df     =          46
                                             Scale parameter =           1
Deviance         =    40.83400585            (1/df) Deviance =    .8876958
Pearson          =    57.71533028            (1/df) Pearson  =    1.254681

Variance function: V(u) = u                  [Poisson]
Link function    : g(u) = ln(u)              [Log]

                                             AIC             =    2.225708
Log likelihood   = -51.41700292              BIC             =   -137.2412

 ( 1)   [died]age = 0
 ( 2)   [died]2.drug = -1.5
```

died	Coef.	OIM Std. Err.	z	P>|z|	[95% Conf. Interval]
age	0	(omitted)			
drug					
2	-1.5	(constrained)			
3	-1.875897	.4546061	-4.13	0.000	-2.766908 -.9848854
_cons	-2.204461	.2	-11.02	0.000	-2.596454 -1.812469
ln(studyt~e)	1	(exposure)			

```
. lrtest Unconstrained

Likelihood-ratio test                        LR chi2(2)   =        6.50
(Assumption: . nested in Unconstrained)      Prob > chi2 =      0.0388
```

12.7   Interpretation of coefficients

Interpreting the coefficients under the log link function can be difficult. Earlier, we illustrated a model for the LOS in a hospital where the duration of the stay was measured in days. The identity-linked Poisson model fits coefficients that are interpreted as the rate difference. However, as is customary, our previous example specified the canonical log link to relate the mean and linear predictor.

```
. use http://www.stata-press.com/data/hh4/medpar, clear
. glm los hmo white type2 type3, family(poisson) noheader nolog
```

los	Coef.	OIM Std. Err.	z	P>|z|	[95% Conf. Interval]	
hmo	-.0715493	.023944	-2.99	0.003	-.1184786	-.02462
white	-.153871	.0274128	-5.61	0.000	-.2075991	-.100143
type2	.2216518	.0210519	10.53	0.000	.1803908	.2629127
type3	.7094767	.026136	27.15	0.000	.6582512	.7607022
_cons	2.332933	.0272082	85.74	0.000	2.279606	2.38626

Under this standard link, the coefficients are interpreted as the difference in the log of the expected response. For example, this output illustrates that with other variables held constant, there is an expected 0.07 reduction in the log-days of stay for HMO patients compared with the LOS for private pay patients. Also with other variables held constant, there is a 0.22 increase in the log-days of stay for urgent admissions over elective admissions.

Although one can interpret the coefficients as the change in the log of the outcome, such an explanation does not satisfy most researchers. Now we motivate an alternative metric for a transformation of the coefficients that leads to easier interpretation for the canonical log link.

For algebraic illustration, we assume a model with two covariates x_1 and x_2 along with a constant. We focus on the interpretation of β_1.

Because the inverse canonical link is nonlinear, interpreting the coefficient is difficult.

$$\Delta y_i = \exp\{\beta_0 + (x_{1i} + 1)\beta_1 + x_{2i}\beta_2\} - \exp(\beta_0 + x_{1i}\beta_1 + x_{2i}\beta_2) \quad (12.25)$$

The difficulty arises because the difference is not constant and depends on the values of the covariates.

Instead of focusing on the difference in the outcome, we define a measure of the incidence-rate ratio as the rate of change in the outcome (incidence)

$$\text{Incidence-rate ratio for } x_1 = \frac{\exp\{\beta_0 + (x_{1i} + 1)\beta_1 + x_{2i}\beta_2\}}{\exp(\beta_0 + x_{1i}\beta_1 + x_{2i}\beta_2)} \quad (12.26)$$

$$= \exp(\beta_1) \quad (12.27)$$

The calculation of the incidence-rate ratio simplifies such that there is no dependence on a particular observation. The incidence-rate ratio is therefore constant—it is independent of the particular values of the covariates. This interpretation does not hold for other inverse links.

Interpretation for canonical Poisson-family models is straightforward for exponentiated coefficients. These exponentiated coefficients represent incidence-rate ratios. An incidence-rate ratio of 2 indicates that there is twice the incidence (the outcome is twice as prevalent) if the associated covariate is increased by one.

13 The negative binomial family

Contents

The negative binomial is the only discrete GLM distribution supported by Stata's `glm` command that has an ancillary parameter, τ. Unlike the exponential family scale parameter associated with Gaussian, gamma, and inverse Gaussian distributions, τ is included in the conditional variance $v(\mu)$. When τ is equal to 1, the negative binomial distribution is known as the geometric distribution. The geometric is the discrete analog to the continuous negative exponential distribution, and the negative binomial is the discrete analog to the continuous gamma distribution. The exponential model can be fit using log-gamma regression with the scale constrained to 1. Likewise, the geometric model can be fit using negative binomial regression with the ancillary scale parameter τ set to 1.

In the previous chapter, we illustrated the Poisson distribution in a graph with counts entering cells at a uniform rate. We mentioned the situation where counts may enter each cell with a predefined gamma shape. This was the Poisson–gamma mixture or contagion distribution model or negative binomial regression.

The negative binomial has long been recognized as a full member of the exponential family of distributions. Likewise, the geometric distribution is listed in nearly every source as an exponential family member. However, until recently neither the geometric nor the negative binomial was considered suitable for entry into the mainstream family of GLMs because of the complexity of the ancillary parameter. The negative binomial distribution is a member of the two-parameter exponential family, and we allow it in this discussion by assuming that the ancillary parameter is fixed rather than stochastic. That said, the parameter can be estimated by extending the usual GLM estimation algorithms to update estimation of this parameter at each step. Stata allows this estimation through the family specification, `family(nbreg ml)`.

The negative binomial regression model seems to have been first discussed in 1949 by Anscombe. Others have mentioned it, pointing out its usefulness for dealing with overdispersed count data. Lawless (1987) detailed the mixture model parameterization of the negative binomial, providing formulas for its log likelihood, mean, variance, and moments. Later, Breslow (1990) cited Lawless' work while manipulating the Poisson model to fit negative binomial parameters. From its inception to the late 1980s, the negative binomial regression model has been construed as a mixture model that is useful for accommodating otherwise overdispersed Poisson data. The models were fit using maximum likelihood or as an adjusted Poisson model.

McCullagh and Nelder (1989) gave brief attention to the negative binomial but did recognize that, with a fixed value for τ, it could be considered a GLM. The authors mentioned the existence of a canonical link, breaking away from the strict mixture model concept, but they did not proceed beyond a paragraph explanation.

Not until the mid-1990s was the negative binomial construed to be a full-fledged member of the GLM family. Hilbe (1993a) detailed how both the GLM-based negative binomial and geometric regression models could be derived from their respective probability distributions. The idea behind the arguments is that the model that is traditionally called the negative binomial is, within the GLM framework, a log-linked negative binomial. The canonical link, although usable, has properties that often result in nonconvergence. This is particularly the case when the data are highly unbalanced and occurs because the ancillary parameter is a term in both the link and inverse link functions. This problem, as well as how the canonical model is to be interpreted, spurred the creation of a SAS log-negative binomial (Hilbe 1994) and geometric regression macro and the inclusion of the negative binomial within the Stata `glm` command. It was also made part of the XploRe statistical package (Hilbe 1992; Hilbe and Turlach 1995). The negative binomial is now widely available in GLM programs and is used to model overdispersed count data in many disciplines.

The negative binomial, when fit using maximum likelihood (ML), is almost always based on a Poisson–gamma mixture model. This is the log-linked version of the GLM negative binomial and is the default parameterization for the `nbreg` command. The negative binomial, as mentioned above, may have other links, including the canonical and the identity. Except for demonstrating how these links are constructed, though, we spend little time with them here.

There are two methods for motivating the negative binomial regression model, and we present them both for completeness. Only one derivation, however, fits within the GLM framework. Both methods are parameterizations of a Poisson–gamma mixture and yield what the literature has begun to distinguish as the NB-1 (constant overdispersion) and the NB-2 (variable overdispersion) regression models. See Hilbe (2011) for a survey of count models and the negative binomial model in particular.

13.1   **Constant overdispersion**

In the first derivation, we consider a Poisson–gamma mixture in which

$$f(y_i|\lambda_i) = e^{-\lambda_i}\lambda_i^{y_i}/y_i! \tag{13.1}$$

where the Poisson parameter λ_i is a random variable. The λ_i are distributed $\Gamma(\delta, \mu_i)$, where

$$\mu_i = e^{\mathbf{x}_i\boldsymbol{\beta}+\text{offset}_i} \tag{13.2}$$

As such, the expected value of the Poisson parameter is

$$E(\lambda_i) = e^{\mathbf{x}_i\boldsymbol{\beta}+\text{offset}_i}/\delta \tag{13.3}$$

and the variance is

$$V(\lambda_i) = e^{\mathbf{x}_i\boldsymbol{\beta}+\text{offset}_i}/\delta^2 \tag{13.4}$$

The resulting mixture distribution is then derived as

$$f(y_i|x_i) = \int_0^\infty \frac{e^{-(\lambda_i)}(\lambda_i)^{y_i}}{y_i!}\frac{\delta^{\mu_i}}{\Gamma(\mu_i)}\lambda_i^{\mu_i-1}e^{-\lambda_i\delta}d\lambda_i \tag{13.5}$$

$$= \frac{\delta^{\mu_i}}{\Gamma(y_i+1)\Gamma(\mu_i)}\int_0^\infty \lambda_i^{(y_i+\mu_i)-1}e^{-\lambda_i(\delta+1)}d\lambda_i \tag{13.6}$$

$$= \frac{\delta^{\mu_i}}{\Gamma(y_i+1)\Gamma(\mu_i)}\frac{\Gamma(y_i+\mu_i)}{(\delta+1)^{y_i+\mu_i}}$$
$$\int_0^\infty \frac{(\delta+1)^{y_i+\mu_i}}{\Gamma(y_i+\mu_i)}\lambda_i^{(y_i+\mu_i)-1}e^{-\lambda_i(\delta+1)}d\lambda_i \tag{13.7}$$

$$= \frac{\delta^{\mu_i}}{\Gamma(y_i+1)\Gamma(\mu_i)}\frac{\Gamma(y_i+\mu_i)}{(\delta+1)^{y_i+\mu_i}} \tag{13.8}$$

$$= \frac{\Gamma(y_i+\mu_i)}{\Gamma(\mu_i)\Gamma(y_i+1)}\left(\frac{\delta}{1+\delta}\right)^{\mu_i}\left(\frac{1}{1+\delta}\right)^{y_i} \tag{13.9}$$

The moments of this distribution are given by

$$E(y_i) = e^{\mathbf{x}_i\boldsymbol{\beta}+\text{offset}_i}/\delta \tag{13.10}$$
$$V(y_i) = e^{\mathbf{x}_i\boldsymbol{\beta}+\text{offset}_i}(1+\delta)/\delta^2 \tag{13.11}$$

such that the variance-to-mean ratio is given by $(1+\delta)/\delta$ (constant for all observations). The variance is a scalar multiple of the mean and is thus referred to as NB-1 in the literature. For a complete treatment of this and other models, see Hilbe (2011).

We may rewrite the model as

$$\frac{\Gamma(y_i + \mu_i)}{\Gamma(\mu_i)\Gamma(y_i + 1)}\left(\frac{1}{1+\alpha}\right)^{\mu_i}\left(\frac{\alpha}{1+\alpha}\right)^{y_i} \tag{13.12}$$

where $\alpha = 1/\delta$.

We now show that the Poisson is the limiting case when $\alpha = 0$.

$$\alpha = \exp(\tau) \qquad \mu_i = \exp\{\mathbf{x}_i\boldsymbol{\beta} + \text{offset}_i + \ln(\alpha)\} \tag{13.13}$$

$$\mathcal{L} = \sum_{i=1}^{n}\left[\ln\{\Gamma(\mu_i + y_i)\} - \ln\{\Gamma(y_i + 1)\} - \ln\{\Gamma(\mu_i)\}\right.$$

$$\left. + y_i\ln(\alpha) - (y_i + \mu_i)\ln(1+\alpha)\right] \tag{13.14}$$

In the above, we allowed that

$$\mu_i = \exp\{\mathbf{x}_i\boldsymbol{\beta} + \text{offset}_i + \ln(\alpha)\} \tag{13.15}$$

where the inclusion of the $+\ln(\alpha)$ term standardizes the $\boldsymbol{\beta}$ coefficients such that the expected value of the outcome is then equal to the exponentiated linear predictor $\exp(\mathbf{X}\boldsymbol{\beta})$; to obtain the expected outcome without this term, we would have to multiply the exponentiated linear predictor by α. This specification parameterizes $\alpha > 0$ as $\exp(\tau)$ to enforce the boundary condition that $\alpha > 0$. The model is fit using Stata's `nbreg` command along with specification of the `dispersion(constant)` option. However, this distribution cannot be written as a member of the exponential family of distributions.

13.2   Variable overdispersion

One can illustrate the derivation of the negative binomial, within the GLM framework, in two ways. The first is to conceive the negative binomial distribution in the traditional sense—as a Poisson model with gamma heterogeneity where the gamma noise has a mean of 1. The second way of looking at the negative binomial is as a probability function in its own right (independent of the Poisson). In this sense, one can think of the negative binomial probability function as the probability of observing y failures before the rth success in a series of Bernoulli trials. Each of these two ways of looking at the negative binomial converges to the same log-likelihood function. Though the approaches are different, reviewing them is educational. Regardless of the approach, the variance is a quadratic function of the mean and is thus referred to as NB-2 in the literature.

13.2.1   Derivation in terms of a Poisson–gamma mixture

The derivation of the negative binomial (NB-2) is usually illustrated as a Poisson model with gamma heterogeneity where the gamma noise has a mean of 1. First, we introduce an individual unobserved effect into the conditional Poisson mean:

$$\ln \mu_i \;=\; \mathbf{x}_i \boldsymbol{\beta} + \epsilon_i \tag{13.16}$$

$$=\; \ln \lambda_i + \ln u_i \tag{13.17}$$

The distribution of y conditioned on x_i and u_i remains Poisson with the conditional mean and variance given by μ_i.

$$f(y_i|u_i) = \frac{e^{-\lambda_i u_i}(\lambda_i u_i)^{y_i}}{y_i!} \tag{13.18}$$

The conditional mean of y_i under the gamma heterogeneity is then given by $\lambda_i u_i$ instead of just λ_i. The unconditional distribution of y is then derived as

$$f(y_i|x_i) = \int_0^\infty \frac{e^{-(\lambda_i u_i)}(\lambda_i u_i)^{y_i}}{y_i!} g(u_i) du_i \tag{13.19}$$

where the choice of $g(\cdot)$ defines the unconditional distribution. Here we specify a gamma distribution for $u_i = \exp(\epsilon_i)$. Like other models of heterogeneity, the mean of the distribution is unidentified if the gamma specification contains a constant term (because the disturbance enters multiplicatively). Therefore, the gamma mean is assumed to be one. Using this normalization, we have that

$$f(y_i|x_i) \;=\; \int_0^\infty \frac{e^{-(\lambda_i u_i)}(\lambda_i u_i)^{y_i}}{y_i!} \frac{\nu^\nu}{\Gamma(\nu)} u_i^{\nu-1} e^{-\nu u_i} du_i \tag{13.20}$$

$$=\; \frac{\lambda_i^{y_i}}{\Gamma(y_i+1)} \frac{\nu^\nu}{\Gamma(\nu)} \int_0^\infty e^{-(\lambda_i+\nu)u_i} u_i^{(y_i+\nu)-1} du_i \tag{13.21}$$

$$=\; \frac{\lambda_i^{y_i}}{\Gamma(y_i+1)} \frac{\nu^\nu}{\Gamma(\nu)} \frac{\Gamma(y_i+\nu)}{(\lambda_i+\nu)^{y_i+\nu}}$$
$$\int_0^\infty \frac{(\lambda_i+\nu)^{y_i+\nu}}{\Gamma(y_i+\nu)} e^{-(\lambda_i+\nu)u_i} u_i^{(y_i+\nu)-1} du_i \tag{13.22}$$

$$=\; \frac{\lambda_i^{y_i}}{\Gamma(y_i+1)} \frac{\nu^\nu}{\Gamma(\nu)} \frac{\Gamma(y_i+\nu)}{(\lambda_i+\nu)^{y_i+\nu}} \tag{13.23}$$

$$=\; \frac{\lambda_i^{y_i}}{\Gamma(y_i+1)} \frac{\nu^\nu}{\Gamma(\nu)} \Gamma(y_i+\nu) \left(\frac{\nu}{\lambda_i+\nu}\right)^\nu \frac{1}{\nu^\nu} \left(\frac{\lambda_i}{\lambda_i+\nu}\right)^{y_i} \frac{1}{\lambda_i^{y_i}} \tag{13.24}$$

$$=\; \frac{\Gamma(y_i+\nu)}{\Gamma(y_i+1)\Gamma(\nu)} \left(\frac{\nu}{\lambda_i+\nu}\right)^\nu \left(\frac{\lambda_i}{\lambda_i+\nu}\right)^{y_i} \tag{13.25}$$

$$=\; \frac{\Gamma(y_i+\nu)}{\Gamma(y_i+1)\Gamma(\nu)} \left(\frac{1}{1+\lambda_i/\nu}\right)^\nu \left(1 - \frac{1}{1+\lambda_i/\nu}\right)^{y_i} \tag{13.26}$$

We may then rewrite the result as

$$\frac{\Gamma(y_i+1/\alpha)}{\Gamma(y_i+1)\Gamma(1/\alpha)} \left(\frac{1}{1+\alpha\mu_i}\right)^{1/\alpha} \left(1 - \frac{1}{1+\alpha\mu_i}\right)^{y_i} \tag{13.27}$$

such that

$$\mu_i = \lambda_i = \exp(\mathbf{x}_i\boldsymbol{\beta} + \text{offset}_i) \tag{13.28}$$

which allows that the mean of the negative binomial is given by $\exp(\mathbf{x}_i\boldsymbol{\beta} + \text{offset}_i)$ (parameterized using the exponential function to restrict the outcomes to positive counts) and the overdispersion (variance divided by the mean) is given by $1 + \alpha\mu_i$.

In this parameterization, α is positive. The boundary case, $\alpha = 0$, corresponds to the nested Poisson model, and as α increases, so does the assumed overdispersion of the model. The likelihood is sometimes coded in terms of $\tau = \ln(\alpha)$, so the interesting boundary condition for the nested Poisson model corresponds to $\tau = -\infty$.

Inverting the scale parameter from that appearing in McCullagh and Nelder (1989) allows a direct relationship between the mean and scale. Estimation results will not differ between parameterizations, as long as each is used consistently for all related formulas. However, we think that the direct relationship approach allows enhanced interpretation of excess model dispersion. John Nelder first introduced this approach to one of the authors of this book. Moreover, Nelder used this parameterization in his GENSTAT macro, called k-system, and he preferred this specification from soon after the publication of McCullagh and Nelder (1989) until his death in 2010.

In exponential-family notation, we have that

$$f(y; \mu, \alpha) = \exp\left\{y \ln\left(\frac{\alpha\mu}{1 + \alpha\mu}\right) + \frac{1}{\alpha}\ln\left(\frac{1}{1 + \alpha\mu}\right) + \ln\Gamma\left(y + \frac{1}{\alpha}\right)\right.$$
$$\left. - \ln\Gamma(y + 1) - \ln\Gamma\left(\frac{1}{\alpha}\right)\right\} \tag{13.29}$$

so that we now have the major formulas that are required for the GLM negative binomial algorithm.

$$\theta = \ln\left(\frac{\alpha\mu}{1+\alpha\mu}\right) \tag{13.30}$$

$$b(\theta) = -\frac{1}{\alpha}\ln\left(\frac{1}{1+\alpha\mu}\right) \tag{13.31}$$

$$b'(\theta) = \frac{\partial b}{\partial\mu}\frac{\partial\mu}{\partial\theta} \tag{13.32}$$

$$= \left(\frac{1}{1+\alpha\mu}\right)\{\mu(1+\alpha\mu)\} \tag{13.33}$$

$$= \mu \tag{13.34}$$

$$b''(\theta) = \frac{\partial^2 b}{\partial\mu^2}\left(\frac{\partial\mu}{\partial\theta}\right)^2 + \frac{\partial b}{\partial\mu}\frac{\partial^2\mu}{\partial\theta^2} \tag{13.35}$$

$$= \left\{-\frac{\alpha}{(1+\alpha\mu)^2}\right\}\{\mu^2(1+\alpha\mu)^2\} + \left(\frac{1}{1+\alpha\mu}\right)(1+\alpha\mu)(\mu+2\alpha\mu^2) \tag{13.36}$$

$$= -\alpha\mu^2 + \mu + 2\alpha\mu^2 \tag{13.37}$$

$$= \mu + \alpha\mu^2 \tag{13.38}$$

$$V(\mu) = \mu + \alpha\mu^2 \tag{13.39}$$

$$\frac{\partial V(\mu)}{\partial\mu} = 1 + 2\alpha\mu \tag{13.40}$$

The full log-likelihood and deviance functions are given by

$$\mathcal{L}(\mu;y,\alpha) = \sum_{i=1}^{n}\left\{y_i\ln\left(\frac{\alpha\mu_i}{1+\alpha\mu_i}\right) - \frac{1}{\alpha}\ln(1+\alpha\mu_i) + \ln\Gamma\left(y_i + \frac{1}{\alpha}\right)\right.$$
$$\left. - \ln\Gamma(y_i+1) - \ln\Gamma\left(\frac{1}{\alpha}\right)\right\} \tag{13.41}$$

The deviance function, defined as usual, $2\{\mathcal{L}(y;y) - \mathcal{L}(\mu;y)\}$, is calculated as

$$D = 2\sum_{i=1}^{n}\left\{y_i\ln\left(\frac{y_i}{\mu_i}\right) - \left(y_i + \frac{1}{\alpha}\right)\ln\left(\frac{1+\alpha y_i}{1+\alpha\mu_i}\right)\right\} \tag{13.42}$$

13.2.2 Derivation in terms of the negative binomial probability function

The second approach to understanding the negative binomial within the framework of GLM is to begin with the negative binomial probability function. As stated earlier, we can conceive of the distribution as the probability of observing y failures before the rth success in a series of Bernoulli trials. The negative binomial probability mass function is written

$$f(y;r,p) = \binom{y+r-1}{r-1}p^r(1-p)^y \tag{13.43}$$

In exponential-family form, the above translates to

$$f(y; r, p) = \exp\left\{ y \ln(1 - p) + r \ln(p) + \ln \binom{y + r - 1}{r - 1} \right\} \tag{13.44}$$

Thus,

$$\theta = \ln(1 - p) \tag{13.45}$$
$$e^\theta = 1 - p \tag{13.46}$$
$$p = 1 - e^\theta \tag{13.47}$$
$$b(\theta) = -r \ln(p) = -r \ln\left(1 - e^\theta\right) \tag{13.48}$$

where the scale parameter $a(\phi) = 1$. The mean and variance may be obtained by taking the derivatives of $b(\theta)$ with respect to θ:

$$b'(\theta) = \frac{\partial b}{\partial p} \frac{\partial p}{\partial \theta} \tag{13.49}$$

$$= \left(-\frac{r}{p}\right)\{-(1 - p)\} = \frac{r(1 - p)}{p} \tag{13.50}$$

$$= \mu \tag{13.51}$$

$$b''(\theta) = \frac{\partial^2 b}{\partial p^2}\left(\frac{\partial p}{\partial \theta}\right)^2 + \frac{\partial b}{\partial p}\frac{\partial^2 p}{\partial \theta^2} \tag{13.52}$$

$$= \left(\frac{r}{p^2}\right)(1 - p)^2 + \frac{r}{p}(1 - p) \tag{13.53}$$

$$= \frac{r(1 - p)^2 + rp(1 - p)}{p^2} = \frac{r(1 - p)}{p^2} \tag{13.54}$$

In terms of the mean, the variance may be written

$$V(\mu) = b''(\theta) = \mu + \frac{\mu^2}{r} \tag{13.55}$$

Reparameterizing $\alpha = 1/r$ so that the variance is directly (instead of inversely) proportional to the mean yields

$$p = 1 - e^{\theta} = \frac{1}{1 + \alpha\mu} \tag{13.56}$$

$$\theta = \ln(1 - p) = \ln\left(\frac{\alpha\mu}{1 + \alpha\mu}\right) \tag{13.57}$$

$$b(\theta) = -\frac{1}{\alpha}\ln(p) = \frac{1}{\alpha}\ln(1 + \alpha\mu) \tag{13.58}$$

$$b'(\theta) = \frac{1 - p}{\alpha p} = \mu = \frac{1}{\alpha(e^{-\theta} - 1)} \tag{13.59}$$

$$b''(\theta) = \frac{1 - p}{\alpha p^2} = \mu + \alpha\mu^2 \tag{13.60}$$

$$g'(\theta) = \frac{\partial}{\partial\mu}\ln\left(\frac{\alpha\mu}{1 + \alpha\mu}\right) = \frac{1}{\mu + \alpha\mu^2} \tag{13.61}$$

where the variance in terms of the mean is given by

$$V(\mu) = b''(\theta) = \mu + \alpha\mu^2 \tag{13.62}$$

The full log-likelihood and deviance functions are then found by substituting (13.56) and the reparameterization, $\alpha = 1/r$, into (13.44).

$$\mathcal{L}(\mu; y, \alpha) = \sum_{i=1}^{n}\left\{ y_i \ln\left(\frac{\alpha\mu_i}{1 + \alpha\mu_i}\right) - \frac{1}{\alpha}\ln(1 + \alpha\mu_i) + \ln\Gamma\left(y_i + \frac{1}{\alpha}\right)\right.$$
$$\left. - \ln\Gamma(y_i + 1) - \ln\Gamma\left(\frac{1}{\alpha}\right)\right\} \tag{13.63}$$

The deviance function, defined as usual, $2\{\mathcal{L}(y; y) - \mathcal{L}(\mu; y)\}$, is calculated as

$$D = 2\sum_{i=1}^{n}\left\{ y_i \ln\left(\frac{y_i}{\mu_i}\right) - \left(y_i + \frac{1}{\alpha}\right)\ln\left(\frac{1 + \alpha y_i}{1 + \alpha\mu_i}\right)\right\} \tag{13.64}$$

where these results coincide with the results of the previous subsection. Thus, we may view the negative binomial as a mixture distribution of the Poisson and gamma, or we may view it as a probability function on its own.

13.2.3 The canonical link negative binomial parameterization

We construct a canonical link negative binomial algorithm by inserting all the required formulas into the standard GLM algorithm. This algorithm is the result of either derivation given in the last two subsections.

As in previous algorithm construction, we obtain the relevant terms for our algorithm by substituting the inverse link of the linear predictor for μ. For the canonical case, we have that $\eta = \theta$, and we can derive the relationships between μ and η by using the results from either of the previous two subsections.

$$\eta \ = \ \theta \tag{13.65}$$

$$= \ \ln\left(\frac{\alpha\mu}{1+\alpha\mu}\right) \tag{13.66}$$

$$= \ -\ln\left(\frac{1}{\alpha\mu}+1\right) \tag{13.67}$$

$$\mu \ = \ \frac{1}{\alpha(e^{-\eta}-1)} \tag{13.68}$$

In algorithm construction, we also use the derivative of the link function that was given before as

$$g'(\theta) \ = \ \frac{\partial}{\partial\mu}\ln\left(\frac{\alpha\mu}{1+\alpha\mu}\right) \tag{13.69}$$

$$= \ \frac{1}{\mu+\alpha\mu^2} \tag{13.70}$$

so that the log likelihood for the canonical link substitutes $\exp(\eta)/[\alpha\{1-\exp(\eta)\}]$ for μ in (13.63) and then calculate

$$\mathcal{L}(\mu;y,\alpha) \ = \ \sum_{i=1}^{n}\left[y_i\eta_i - \frac{1}{\alpha}\ln\left\{1+\frac{\exp(\eta_i)}{1+\exp(\eta_i)}\right\} \right. \tag{13.71}$$

$$\left. +\ln\Gamma\left(y_i+\frac{1}{\alpha}\right)-\ln\Gamma(y_i+1)+\ln\Gamma(\alpha)\right] \tag{13.72}$$

The relevant algorithm is then given by

Listing 13.1. IRLS algorithm for negative binomial regression

```
1      μ = {y + mean(y)}/2
2      η = − ln{1/(αμ) + 1}
3      WHILE (abs(ΔDev) > tolerance) {
4          W = μ + αμ²
5          z = {η + (y − μ)/W} − offset
6          β = (XᵀWX)⁻¹XᵀWz
7          η = Xβ + offset
8          μ = 1/{α(e⁻ⁿ − 1)}
9          OldDev = Dev
10         Dev  = 2∑[y ln(y/μ) − (y + 1/α) ln{(1 + αy)/(1 + αμ)}]
11         Dev  = 2∑[ln{1/(1 + αμ)}/α]  if y = 0
12         ΔDev = Dev - OldDev
13     }
14     χ² = ∑{(y − μ)²/(μ + αμ²)}
```

In addition to the `glm` command, this model is also available through Stata's `nbreg` command by specifying the `dispersion(mean)` option. The option is not required, because this is the default model. Our emphasis on the option, in this case, serves to differentiate this case from the `dispersion(constant)` option associated with the model in section 13.1.

13.3   The log-negative binomial parameterization

The negative binomial is rarely used in canonical form. Its primary use is to serve as an overdispersed Poisson regression model. That is, because Poisson models are mostly overdispersed in real data situations, the negative binomial is used to model such data in their place. To facilitate the use of the Poisson to overdispersion, the negative binomial is nearly always parameterized using a log link given in (13.28).

To convert the canonical form to log form, we substitute the log and inverse link into the algorithm. We then amend the weight function, w, and the working response, z. Recalling that $w = 1/(vg'^2)$ and $z = \eta + (y - \mu)g'$, we see that using the log link results in

$$w \quad = \quad \frac{1}{(\mu + \alpha\mu^2)/\mu^2} \tag{13.73}$$

$$= \quad \frac{\mu}{1 + \alpha\mu} \tag{13.74}$$

$$z \quad = \quad \eta + \frac{y - \mu}{\mu} \tag{13.75}$$

so we may specify the log-negative binomial algorithm as

Listing 13.2. IRLS algorithm for log-negative binomial regression

```
1    μ = {y + mean(y)}/2
2    η = ln(μ)
3    WHILE (abs(ΔDev) > tolerance) {
4        W = μ/(1 + αμ)
5        z = {η + (y − μ)/μ} − offset
6        β = (XᵀWX)⁻¹XᵀWz
7        η = Xβ + offset
8        μ = exp(η)
9        OldDev = Dev
10       Dev  = 2∑[ln(y/μ) − (y + 1/α) ln{(1 + αy)/(1 + αμ)}]
11       Dev  = 2∑[ln{1/(1 + αμ)}/α]  if y = 0
12       ΔDev = Dev − OldDev
13   }
14   χ² = ∑{(y − μ)²/(μ + αμ²)}
```

We have emphasized throughout this book that, when possible, GLMs should be recast to use the observed rather than the traditional expected information matrix to abstract standard errors. The above algorithm uses the expected information matrix. We can change the above algorithm to use the observed information matrix by converting the weight and working response.

Formulas common to using the observed information matrix and expected information matrix are the following:

$$\eta = g(\mu) = \ln(\mu) \tag{13.76}$$
$$g'(\mu) = g' = 1/\mu \tag{13.77}$$
$$g''(\mu) = g'' = -1/\mu^2 \tag{13.78}$$
$$V(\mu) = V = \mu + \alpha\mu^2 \tag{13.79}$$
$$V^2 = (\mu + \alpha\mu^2)^2 \tag{13.80}$$
$$V'(\mu) = V' = 1 + 2\alpha\mu \tag{13.81}$$
$$w = \mu/(1 + \alpha\mu) \tag{13.82}$$

The observed information matrix adjusts the weights such that

$$w_o = w + (y - \mu)(Vg'' + V'g')/\{V^2(g')^3\} \tag{13.83}$$
$$z = \eta + (y - \mu)/\{w_o(1 + \alpha\mu)\} \tag{13.84}$$

Substituting these results into the GLM negative binomial algorithm:

Listing 13.3. IRLS algorithm for log-negative binomial regression using OIM

```
1      μ = {y + mean(y)}/2
2      η = ln(μ)
3      WHILE (abs(ΔDev) > tolerance) {
4          W = μ/(1 + αμ) + (y − μ){αμ/(1 + 2αμ + α²μ²)}
5          z = {η + (y − μ)/(Wαμ)} − offset
6          β = (XᵀWX)⁻¹XᵀWz
7          η = Xβ + offset
8          μ = exp(η)
9          OldDev = Dev
10         Dev  = 2∑ln(y/μ) − (y + 1/α)ln{(1 + αy)/(1 + αμ)}
11         Dev  = 2∑ln{1/(1 + αμ)}/α  if y = 0
12         ΔDev = Dev − OldDev
13     }
14     χ² = ∑{(y − μ)²/(μ + αμ²)}
```

ML methods require the log likelihood to be parameterized in terms of $\mathbf{x}\boldsymbol{\beta}$. We provide them here, letting C_i represent the normalization terms.

$$C_i = \ln\Gamma(y_i + 1/\alpha) - \ln\Gamma(y_i + 1) - \ln\Gamma(1/\alpha) \tag{13.85}$$

$$\mathcal{L}(\mathbf{x}\boldsymbol{\beta}; y, \alpha) = \sum_{i=1}^{n}\left[y_i\ln\{\alpha\exp(\mathbf{x}_i\boldsymbol{\beta})\} - \left(y_i + \frac{1}{\alpha}\right)\ln\{1 + \alpha\exp(\mathbf{x}_i\boldsymbol{\beta})\} + C_i\right] \tag{13.86}$$

$$= \sum_{i=1}^{n}\left[y_i\ln\left\{\frac{\alpha\exp(\mathbf{x}_i\boldsymbol{\beta})}{1 + \alpha\exp(\mathbf{x}_i\boldsymbol{\beta})}\right\} - \frac{\ln\{1 + \alpha\exp(\mathbf{x}_i\boldsymbol{\beta})\}}{\alpha} + C_i\right] \tag{13.87}$$

These log-likelihood functions are the ones usually found in ML programs that provide negative binomial modeling. The canonical link, as well as identity and other possible links, are usually unavailable.

One can fit a negative binomial model with an estimate of the scale parameter by iteratively forcing the χ^2 dispersion to unity. The following algorithm can be used for this purpose. It is the basis for the original SAS log-negative binomial macro. Hilbe's `negbin` command implements this approach. You can download this command by typing `net install negbin, from(http://www.stata.com/users/jhardin)`. We show results from this command in the next section.

Listing 13.4. IRLS algorithm for log-negative binomial regression with estimation of α

```
1      Poisson estimation: y <predictors>
2      χ² = Σ(y - μ)²/μ
3      Disp  = χ²/df
4      α = 1/disp
5      WHILE (abs(ΔDisp) > tolerance) {
6          OldDisp = Disp
7          estimation: y <predictors>, alpha≈ α
8          χ² = Σ{(y - μ)²/(μ + αμ²)}
9          Disp = χ²/df
10         α = disp(α)
11         ΔDisp = Disp - OldDisp
12     }
```

The above algorithm provides accurate estimates of α and coefficients of β. One can obtain estimates of both parameters by using the above method if ML software is unavailable, that is, if the software requires specification of α while estimating only β. We illustrate these issues in the following section.

13.4   Negative binomial examples

The development of negative binomial models is typically motivated from overdispersed Poisson models. We examine how the negative binomial model parameters, standard errors, and dispersion statistics differ from those of the Poisson. The limiting case (that which results when $\alpha = 0$) is identical to the Poisson model. In fact, when α is close to zero, we must consider whether the Poisson model might be preferred.

Stata's `glm` command includes support for the user to specify that the dispersion parameter should be estimated using ML. In previous versions of the command, there was no support and the user had to specify the value. Technically, with no specification of α, the model is not a GLM because the distribution would be a member of the two-parameter exponential family of distributions. The GLM algorithm is customized so that between each estimation iteration step, an additional step estimates α using the current estimate of β. Whether the user first uses the `nbreg` command to obtain an estimate of α to specify or the user simply requests the `glm` command to obtain the estimate, the results will be the same.

The added value of fitting the model with `glm` rather than `nbreg` is in the host of associated residuals and fit statistics available with the `glm` command. The one advantage of the `nbreg` command is the inclusion of a formal assessment of whether the α parameter is different from zero. Here we show results using both approaches.

```
. use http://www.stata-press.com/data/hh4/medpar

. nbreg los hmo white type2 type3, irr nolog
```

Negative binomial regression			Number of obs	=	1,495
			LR chi2(4)	=	118.03
Dispersion	= mean		Prob > chi2	=	0.0000
Log likelihood = -4797.4766			Pseudo R2	=	0.0122

los	IRR	Std. Err.	z	P>|z|	[95% Conf. Interval]	
hmo	.9343023	.0497622	-1.28	0.202	.8416883	1.037107
white	.8789164	.0602425	-1.88	0.060	.7684307	1.005288
type2	1.247634	.063121	4.37	0.000	1.129855	1.37769
type3	2.026193	.1542563	9.28	0.000	1.745332	2.352252
_cons	10.07724	.6847215	34.00	0.000	8.820729	11.51273
/lnalpha	-.807982	.0444542			-.8951107	-.7208533
alpha	.4457567	.0198158			.4085624	.4863371

```
Note: Estimates are transformed only in the first equation.
Note: _cons estimates baseline incidence rate.
LR test of alpha=0: chibar2(01) = 4262.86                Prob >= chibar2 = 0.000

. glm los hmo white type2 type3, family(nbinomial ml) eform nolog
```

Generalized linear models			No. of obs	=	1,495
Optimization	: ML		Residual df	=	1,490
			Scale parameter =		1
Deviance	=	1568.14286	(1/df) Deviance =		1.052445
Pearson	=	1624.538251	(1/df) Pearson	=	1.090294
Variance function: V(u) = u+(.4458)u^2			[Neg. Binomial]		
Link function	: g(u) = ln(u)		[Log]		
			AIC	=	6.424718
Log likelihood	= -4797.476603		BIC	=	-9323.581

los	IRR	OIM Std. Err.	z	P>|z|	[95% Conf. Interval]	
hmo	.9343023	.0497622	-1.28	0.202	.8416883	1.037107
white	.8789164	.0602424	-1.88	0.060	.768431	1.005288
type2	1.247634	.063121	4.37	0.000	1.129855	1.37769
type3	2.026193	.1542563	9.28	0.000	1.745332	2.352252
_cons	10.07724	.6847202	34.00	0.000	8.820732	11.51273

```
Note: _cons estimates baseline incidence rate.
Note: Negative binomial parameter estimated via ML and treated as fixed once
```

Note the default log-link parameterization for the GLM model and compare the incidence-rate ratios and standard errors for the two regression tables. When using Stata, the canonical link or others must be specifically requested. The ml specification is not supported for any of the other families. The default value for α is 1; when using the glm command having specified the nbinomial family, the value of α is not estimated if it is specified by the user.

The model results demonstrate that most of the overdispersion in the Poisson model has been accommodated in the negative binomial. However, the negative binomial model itself is overdispersed (1.09). The likelihood-ratio test from the **nbreg** command indicates that the model is significantly different from a Poisson specification. However, the negative binomial model itself has not completely addressed the overdispersion. The key to successfully addressing the shortcoming is the identification of the cause of the overdispersion. We need to evaluate the model via residual analysis. We always recommend plotting Anscombe residuals or standardized residuals versus predictors. We also observe here that the mean of μ, the predicted fit, and length of stay (LOS) are 9.85.

We mentioned that the scale can be estimated within the GLM approach independent of ML techniques. In this approach, we iterate estimating a negative binomial in a loop in which we update the estimate of α at each iteration. Using this method, we search for and find the estimate that has a dispersion statistic as close to 1.0 as possible.

```
. negbin los hmo white type2 type3, eform
1:    Deviance:     3110   Alpha (k):   .1597  Disp:    2.298
2:    Deviance:     1812   Alpha (k):    .367  Disp:    1.274
3:    Deviance:     1512   Alpha (k):   .4677  Disp:    1.048
4:    Deviance:     1458   Alpha (k):   .4902  Disp:    1.008
5:    Deviance:     1449   Alpha (k):   .4942  Disp:    1.001
6:    Deviance:     1447   Alpha (k):   .4949  Disp:        1
7:    Deviance:     1447   Alpha (k):    .495  Disp:        1
8:    Deviance:     1447   Alpha (k):    .495  Disp:        1

                                        No obs.      =       1495
   Poisson Dp =      6.26                Poisson Dev  =       8143
   Alpha (k)  =      .495                Neg Bin Dev  =       1447
                                        Prob > Chi2  =          0

   Chi2       =   1490.01               Deviance     =   1447.139
   Prob>chi2  =  .4950578               Prob>chi2    =  .7824694
   Dispersion =  1.000006               Dispersion   =  .9712345

Log negative binomial regression
```

los	IRR	Std. Err.	z	P>|z|	[95% Conf. Interval]	
hmo	.9343301	.0519053	-1.22	0.221	.8379405	1.041808
white	.8792076	.0629791	-1.80	0.072	.7640442	1.011729
type2	1.247625	.0659398	4.19	0.000	1.124854	1.383795
type3	2.026132	.1615912	8.85	0.000	1.732932	2.36894

```
Loglikelihood = -4800.2524
```

The estimate of α uses the Pearson dispersion for convergence. This method provides a close approximation to the true scale by iteratively reducing the χ^2 dispersion to 1.

We next use **doll.dta**. After modeling the data using the **negbin** command (as was done in the previous example), we found that the value of the ancillary parameter was 0.09414; that value resulted in the dispersion statistic being close to one. Specifying this as the value of α in the **glm** command, with the **family(nbinomial 0.09414)** and **eform** options, produces results similar to those from the **negbin** command. Note that users who try and replicate this need to generate a variable equal to the log of **pyears** because the **negbin** command supports only the **offset()** option.

```
. use http://www.stata-press.com/data/hh4/doll

. glm deaths smokes a2-a5, family(nbinomial 0.09414) exposure(pyears) eform
> nolog
Generalized linear models                No. of obs     =         10
Optimization      : ML                   Residual df    =          4
                                         Scale parameter =         1
Deviance          =  4.564400632         (1/df) Deviance =    1.1411
Pearson           =  3.999657112         (1/df) Pearson  =  .9999143

Variance function: V(u) = u+(.0941)u^2   [Neg. Binomial]
Link function     : g(u) = ln(u)         [Log]
                                         AIC            =   8.868605
Log likelihood    = -38.34302481         BIC            =   -4.64594
```

		OIM				
deaths	IRR	Std. Err.	z	P>|z|	[95% Conf. Interval]	
smokes	1.584135	.3674839	1.98	0.047	1.005385	2.496042
a2	4.645222	1.783327	4.00	0.000	2.188897	9.857973
a3	16.00471	6.058747	7.32	0.000	7.621081	33.6108
a4	33.84958	12.85897	9.27	0.000	16.0766	71.27095
a5	54.43735	21.05176	10.34	0.000	25.51081	116.1635
_cons	.0002936	.0000998	-23.93	0.000	.0001508	.0005715
ln(pyears)	1	(exposure)				

```
Note: _cons estimates baseline incidence rate.
```

What does this tell us? The model still appears to be overdispersed. We could try scaling, but what happened?

A negative binomial having an ancillary parameter of 0 is a Poisson model. Recall that the overdispersion we found in the last chapter was due to apparent overdispersion, which could be corrected by appropriate interaction terms. The overdispersion was not inherent in the data.

We learned that we can make a first-round test of apparent versus inherent overdispersion by modeling the data using both Poisson and negative binomial. If the negative binomial does not clear up the problem and has an ancillary parameter estimate near 0, then the model is likely Poisson. You must look for interactions, and so forth. Adding the interactions to the negative binomial model does not help. We can scale if we wish, but it is better to revert to the Poisson model.

The exponentiated parameter estimates of the log-link negative binomial model are to be understood as incidence-rate ratios, IRRs. This is not the case for other links.

A final point should be made prior to leaving our discussion of the GLM-based negative binomial. In the previous chapter, we mentioned that the zero-truncated Poisson model may be preferable when modeling count data situations in which there is no possibility of zero counts. That is, when the response is such that zero counts are excluded, we should consider the zero-truncated Poisson. We used the hospital LOS data as an example. The same situation may exist with an otherwise negative binomial model. If the response is not capable of having a zero count, then one should consider the zero-truncated negative binomial.

The zero-truncated negative binomial log-likelihood function is a modification of the standard form. The modification is based on an adjustment to the probability function. The adjustment excludes 0 in the probability range and adjusts the probability function so that it still sums to 1. If we label the log-link negative binomial likelihood as L_{NB} and the log-link negative binomial log likelihood as $\mathcal{L}_{\text{NB}}$, then the zero-truncated log likelihood for a log-link negative binomial is given by

$$\mathcal{L} = \sum_{i=1}^{n} \ln\left\{ \frac{L_{\text{NB}\,i}}{1 - P_{\text{NB}}(y_i = 0)} \right\} \tag{13.88}$$

$$= \sum_{i=1}^{n} \left(\mathcal{L}_{\text{NB}\,i} - \ln\left[1 - \{1 + \exp(\mathbf{x}\boldsymbol{\beta})\}^{-1/\alpha} \right] \right) \tag{13.89}$$

If we find that there is little difference in parameter and scale estimates between the standard and the zero-truncated form, then using the standard form is wiser. We can interpret this situation to indicate that the lack of zero counts had little impact on the probability of the response. On the other hand, if large differences do exist, then we must maintain the zero-truncated model. Convergence can be a problem for some types of data situations.

Stata also includes commands for fitting truncated models; see the `tpoisson` and `tnbreg` manual entries in the *Stata Base Reference Manual*. We cover these and other models in chapter 14.

As demonstrated throughout the text, we now illustrate how to synthesize data and fit appropriate models. Interested readers should experiment by changing specified coefficient values (and other parameters such as sample size) to see the effect on outcomes. Here we illustrate generating data and then fitting a negative binomial model.

```
. clear
. set obs 10000
number of observations (_N) was 0, now 10,000
. set seed 4321
. generate x1 = rnormal()
. generate x2 = rnormal()
. generate expxb = exp(0.75*x1 - 1.25*x2 + 2)
. generate nby = rpoisson(rgamma(1/.5, .5) * expxb)
```

```
. glm nby x1 x2, family(nbinomial ml) nolog
Generalized linear models                    No. of obs      =      10,000
Optimization     : ML                        Residual df     =       9,997
                                             Scale parameter =           1
Deviance       =   10829.21032               (1/df) Deviance =    1.083246
Pearson        =   9965.766594               (1/df) Pearson  =    .9968757

Variance function: V(u) = u+(.4957)u^2       [Neg. Binomial]
Link function    : g(u) = ln(u)              [Log]

                                             AIC             =     6.13562
Log likelihood  = -30675.09925               BIC             =  -81246.56
```

nby	Coef.	OIM Std. Err.	z	P>|z|	[95% Conf. Interval]
x1	.7505527	.0085902	87.37	0.000	.7337163 .7673892
x2	-1.228814	.0089378	-137.49	0.000	-1.246332 -1.211296
_cons	2.005219	.008764	228.80	0.000	1.988042 2.022396

```
Note: Negative binomial parameter estimated via ML and treated as fixed once
```

Note how the Pearson dispersion statistic is close to 1, whereas the deviance dispersion seems somewhat larger. We can investigate whether these results are anomalous with a Monte Carlo simulation. First, we write an r-class program that stores results we want to access.

```
. program NegbinSim, rclass
  1.      quietly {
  2.          drop _all
  3.          set obs 10000
  4.          generate x1 = rnormal()
  5.          generate x2 = rnormal()
  6.          generate nby = rpoisson(rgamma(1/.5,.5)*exp(0.75*x1-1.25*x2+2))
  7.          glm nby x1 x2, family(nbinomial ml)
  8.      }
  9.      return scalar pearson  = e(dispers_p)
 10.      return scalar deviance = e(dispers_s)
 11. end
```

We then use `simulate` to execute the `NegbinSim` command, and we then use the `summarize` command.

```
. simulate pearson=r(pearson) deviance=r(deviance), reps(100) nodots: NegbinSim
      command:  NegbinSim
      pearson:  r(pearson)
     deviance:  r(deviance)

. summarize
```

Variable	Obs	Mean	Std. Dev.	Min	Max
pearson	100	.9995479	.0115009	.9705971	1.037193
deviance	100	1.084003	.0036461	1.075858	1.0956

The value of the Pearson and deviance dispersion statistics are summarized in the simulation. The Pearson statistic is close to the theoretical value of 1, but the deviance statistic is somewhat higher. The deviance dispersion is substantially greater than the Pearson dispersion for true Poisson data. This is indicative of the usefulness of the Pearson dispersion as the measure for assessing evidence of overdispersion.

With Poisson and negative binomial regression models, scaling of standard errors, using the sandwich variance estimator, bootstrapping, and jackknifing can all help ensure proper fit to the data. The negative binomial model can overcorrect for overdispersion that is indicated by a Poisson model. Such results are more common with small datasets.

13.5   The geometric family

We mentioned at the beginning of this chapter that the geometric regression model is simply the negative binomial with the ancillary or scale parameter set at 1. It is the discrete correlate to the negative exponential distribution, taking a smooth downward slope from an initial high point at zero counts. If the data are distributed in such a manner, then the geometric model may be appropriate. The log-link form has the same gradual downward smooth slope as the canonical link.

For example, if we wanted to look at the distribution of the number of patients as a function of the length of stay, example data might look like

```
. drop _all
. input LOS npatients
            LOS  npatients
  1.  1   50
  2.  2   40
  3.  3   30
  4.  4   20
  5.  5   15
  6.  6   12
  7.  7   10
  8.  8    7
  9.  9    6
 10. 10    6
 11. 11    5
 12. 12    5
 13. 13    4
 14. 14    3
 15. 15    2
 16. 16    0
 17. end
```

```
. twoway (scatter npatients LOS), ylabel(0 2 4 6 8 10 20 30 40 50)
> xlabel(1(1)16)
```

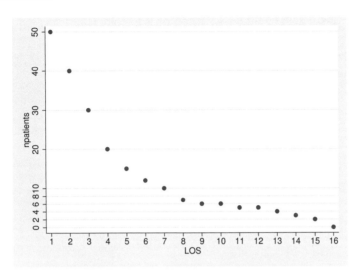

Figure 13.1. Frequency of occurrence versus LOS

The canonical geometric regression algorithm may be specified as

Listing 13.5. IRLS algorithm for geometric regression

1    $\mu = \{y + \text{mean}(y)\}/2$
2    $\eta = -\ln(1 + 1/\mu)$
3   WHILE (abs(ΔDev) > tolerance) {
4        $W = \mu + \mu^2$
5        $z = \{\eta + (y - \mu)/W\} - \text{offset}$
6        $\beta = (X^T W X)^{-1} X^T W z$
7        $\eta = X\beta + \text{offset}$
8        $\mu = 1/\{\exp(-\eta) - 1\}$
9   OldDev = Dev
10   Dev = $2\sum[\ln(y/\mu) - (1 + y)\ln\{(1 + y)/(1 + \mu)\}]$
11   Dev = $2\sum -\ln(1 + \mu)$ if $y = 0$
12   ΔDev = Dev - OldDev
13   }
14   $\chi^2 = \sum\{(y - \mu)^2/(\mu + \mu^2)\}$
```
```

The log geometric may be parameterized so that standard errors are produced from the observed information matrix and may be specified as

Listing 13.6. IRLS algorithm for log-geometric regression

```
1    μ = {y + mean(y)}/2
2    η = ln(μ)
3    WHILE (abs(ΔDev) > tolerance) {
4        W = μ/(1 + μ) + (y − μ){μ/(1 + 2μ + μ²)}
5        z = [η + (y − μ)/{W(μ)}] − offset
6        β = (XᵀWX)⁻¹XᵀWz
7        η = Xβ + offset
8        μ = exp(η)
9        OldDev = Dev
10       Dev = 2 ∑[ln(y/μ) − (1 + y) ln{(1 + y)/(1 + μ)}]
11       Dev = 2 ∑ ln{1/(1 + μ)}  if y = 0
12       ΔDev = Dev − OldDev
13   }
14   χ² = ∑{(y − μ)²/(μ + μ²)}
```

$$1 \quad \mu = \{y + \text{mean}(y)\}/2$$
$$2 \quad \eta = \ln(\mu)$$
$$3 \quad \text{WHILE (abs}(\Delta\text{Dev) > tolerance) \{}$$
$$4 \quad W = \mu/(1 + \mu) + (y - \mu)\{\mu/(1 + 2\mu + \mu^2)\}$$
$$5 \quad z = [\eta + (y - \mu)/\{W(\mu)\}] - \text{offset}$$
$$6 \quad \beta = (X^T W X)^{-1} X^T W z$$
$$7 \quad \eta = X\beta + \text{offset}$$
$$8 \quad \mu = \exp(\eta)$$
$$9 \quad \text{OldDev = Dev}$$
$$10 \quad \text{Dev} = 2 \sum [\ln(y/\mu) - (1 + y) \ln\{(1 + y)/(1 + \mu)\}]$$
$$11 \quad \text{Dev} = 2 \sum \ln\{1/(1 + \mu)\} \text{ if } y = 0$$
$$12 \quad \Delta\text{Dev} = \text{Dev} - \text{OldDev}$$
$$13 \quad \}$$
$$14 \quad \chi^2 = \sum\{(y - \mu)^2/(\mu + \mu^2)\}$$

If there is no evidence to reject the null hypothesis $H_0 : \alpha = 1$, then the geometric distribution yields a better model than does the negative binomial because the geometric model is more parsimonious. We spent considerable time with the LOS data in the previous section. We fit a negative binomial model with the data that worked well. Now let's constrain the ancillary parameter to that of a geometric model and compare it with alternatives. We know from graphing the LOS response that it shows a high initial start and a steady monotonic decline. Such characteristics can be indicative of a geometric distribution.

```
. use http://www.stata-press.com/data/hh4/medpar, clear
. glm los white type2 type3, family(nbinomial 1) link(nbinomial) eform nolog
Generalized linear models                      No. of obs      =       1,495
Optimization      : ML                         Residual df     =       1,491
                                               Scale parameter =           1
Deviance          =   812.0438886              (1/df) Deviance =    .5446304
Pearson           =   806.6148763              (1/df) Pearson  =    .5409892
Variance function: V(u) = u+(1)u^2             [Neg. Binomial]
Link function     : g(u) = ln(u/(u+(1/1)))     [Neg. Binomial]
                                               AIC             =    6.639057
Log likelihood    = -4958.695274               BIC             =   -10086.99
```

los	exp(b)	OIM Std. Err.	z	P>|z|	[95% Conf. Interval]	
white	.9875833	.0073704	-1.67	0.094	.9732427	1.002135
type2	1.020941	.0063076	3.35	0.001	1.008653	1.033379
type3	1.054697	.0066635	8.43	0.000	1.041718	1.067839
_cons	.9086739	.0068259	-12.75	0.000	.8953934	.9221513

The Pearson dispersion shown above is half the value of the NB-2 model, shown on page 239, indicating substantial underdispersion. The AIC statistic is only slightly higher (6.640 to 6.425), but the Raftery BIC is lower for the canonical geometric (-10087) than for the NB-2 (-9323). The incidence-rate ratios are only loosely similar.

Scaling is typically a first-round strategy for adjusting the standard errors for an underdispersed model. The results of scaling the canonical geometric model are displayed below. The scaled model is followed by the same model scaled by provider, which may be the source of the underdispersion.

```
. glm los white type2 type3, family(nbinomial 1) link(nbinomial) scale(x2)
> eform nolog noheader
```

los	exp(b)	OIM Std. Err.	z	P>|z|	[95% Conf. Interval]	
white	.9875833	.0054211	-2.28	0.023	.9770151	.9982658
type2	1.020941	.0046394	4.56	0.000	1.011888	1.030075
type3	1.054697	.0049011	11.46	0.000	1.045135	1.064347
_cons	.9086739	.0050206	-17.33	0.000	.8988868	.9185674

```
(Standard errors scaled using square root of Pearson X2-based dispersion.)
. glm los white type2 type3, family(nbinomial 1) link(nbinomial)
> vce(cluster provnum) eform nolog noheader
```

(Std. Err. adjusted for 54 clusters in provnum)

los	exp(b)	Robust Std. Err.	z	P>|z|	[95% Conf. Interval]	
white	.9875833	.0049677	-2.48	0.013	.9778946	.997368
type2	1.020941	.0052132	4.06	0.000	1.010774	1.03121
type3	1.054697	.0113407	4.95	0.000	1.032702	1.07716
_cons	.9086739	.0043548	-19.98	0.000	.9001785	.9172493

In adjusting the standard errors, scaling provides results similar to those of the modified sandwich estimate of variance. This is an indication that the underdispersion in the original model is accounted for by addressing the correlation of multiple observations from each provider.

Note that the predictor `white` is significant after scaling. Because this model has a significantly lower BIC statistic (although similar to the AIC statistic), it may be argued that it is preferred to the traditional negative binomial model. However, this geometric model could also be compared with other GLMs such as the negative binomial GLM with the canonical link.

```
. glm los white type2 type3, family(nbinomial ml) link(nbinomial) eform nolog
Generalized linear models                        No. of obs      =       1,495
Optimization      : ML                           Residual df     =       1,491
                                                 Scale parameter =           1
Deviance         =   1566.850816                 (1/df) Deviance =    1.050872
Pearson          =   1623.879743                 (1/df) Pearson  =    1.089121

Variance function: V(u) = u+(.4463)u^2                [Neg. Binomial]
Link function     : g(u) = ln(u/(u+(1/.4463318088585395)))[Neg. Binomial]
                                                 AIC             =    6.423541
Log likelihood    = -4797.597197                 BIC             =   -9332.182
```

| | | OIM | | | | |
	exp(b)	Std. Err.	z	P>|z|	[95% Conf. Interval]	
los						
white	.9748169	.0104275	-2.38	0.017	.9545921	.9954701
type2	1.04255	.0091588	4.74	0.000	1.024752	1.060656
type3	1.115134	.010149	11.97	0.000	1.095419	1.135204
_cons	.8165925	.0087719	-18.86	0.000	.7995796	.8339674

Note: Negative binomial parameter estimated via ML and treated as fixed once

The canonical-link negative binomial GLM is clearly preferred to the geometric model and appears to be preferred to the traditional NB-2 model as well. We could scale the above model to determine what may be more accurate standard errors (not displayed), but there are no appreciable changes to the standard errors and the above model is preferred. Consideration must be made, however, regarding interpretability.

We printed the model results with exponentiated coefficients, which Stata has correctly labeled. For the canonical link, the exponentiated coefficients are not interpreted as incidence-rate ratios.

13.6   Interpretation of coefficients

Interpreting the coefficients can be difficult depending on the link function. The canonical link is seldom used because of the difficulty of interpreting results under the nonlinear relationship of the mean and linear predictor. The most common reason for investigating a negative binomial model is to compare it with a Poisson model. The two models (variance functions) are compared under a common link, and that link function is almost always the log link.

Although the two models are compared under a common link, if a researcher determines the negative binomial variance is preferred, then a different link can be used. This includes the negative binomial link function; Hilbe (2011) explores this in greater detail.

Because interpretation of the coefficients under the log link can be difficult, we motivate an alternative metric depending on a transformation of the coefficients for easier interpretation.

For illustration, we assume a model with two covariates $\mathbf{x}_1$ and $\mathbf{x}_2$ along with a constant. We focus on the interpretation of β_1. Because the inverse canonical link is nonlinear, interpretation of the coefficient is difficult.

$$\Delta y_i = \exp\{\beta_0 + (x_{1i} + 1)\beta_1 + x_{2i}\beta_2\} - \exp(\beta_0 + x_{1i}\beta_1 + x_{2i}\beta_2) \quad (13.90)$$

The difficulty arises because the difference is not constant and depends on the other covariate values.

Instead of focusing on the difference in the outcome, we define a measure of the incidence-rate ratio based on the log link function as the ratio of change in the outcome (incidence)

$$\text{Incidence-rate ratio for } \mathbf{x}_1 = \frac{\exp\{\beta_0 + (x_{1i} + 1)\beta_1 + x_{2i}\beta_2\}}{\exp(\beta_0 + x_{1i}\beta_1 + x_{2i}\beta_2)} \quad (13.91)$$

$$= \exp(\beta_1) \quad (13.92)$$

The calculation of the incidence-rate ratio simplifies such that there is no dependence on a particular observation. The incidence-rate ratio is therefore constant—it is independent of the particular values of the covariates. This interpretation does not hold for other inverse links.

Interpretation for log-linked negative binomial or log-linked geometric family models is straightforward for exponentiated coefficients. These exponentiated coefficients represent incidence-rate ratios. An incidence-rate ratio of 2 indicates that there is twice the incidence (the outcome is twice as prevalent) if the associated covariate is increased by one.

14   Other count-data models

Contents

14.1   Count response regression models

We observed in the last several chapters that both Poisson and negative binomial models, as count response regression models, can succumb to ill effects of overdispersion. Overdispersion exists in discrete-response models when the observed model variance exceeds the expected model variance. A fundamental assumption of the Poisson model is that the mean and variance functions are equal. When the variance exceeds the mean, the data are overdispersed.

The negative binomial model can also be affected by excess dispersion when the variance exceeds the calculated value of $\mu + \alpha\mu^2$. For both the Poisson and negative binomial, overdispersion is indicated by the value of the Pearson χ^2 dispersion statistic. Values greater than 1.0 indicate overdispersion.

We have also learned that indicated overdispersion is not always real (that is, data that are centered)—it may be only apparent overdispersion (or a model misspecification). Apparent overdispersion occurs when we can externally adjust the model to reduce the dispersion statistic to near 1.0. We indicated how to do so in chapter 11. Apparent overdispersion may occur because there is a missing explanatory predictor, the data contain outliers, the model requires an interaction term, a predictor needs to be transformed to another scale, or the link function is misspecified. We demonstrated how apparent overdispersion disappears when the appropriate remedy is provided.

Once we determine that the model overdispersion is real, we next should, if possible, attempt to identify the cause or source for the overdispersion. What causes the extra variation observed in the model? Scaling and applying robust variance estimators to adjust model standard errors may provide a post hoc remedy to accommodate bias in the standard errors, but it does not address the underlying cause of the overdispersion.

There are two general causes for real overdispersion in binomial and count response models, as outlined in Hilbe (2011). The first is a violation of the distributional assumptions upon which the model is based. For instance, with count models, the distributional mean specifies an expected number of zero counts in the response. If there are far more zeros than expected under distributional assumptions, the data may be overdispersed. If the response has no zero counts, the data may be underdispersed. Zero-inflated Poisson (ZIP) and zero-inflated negative binomial (ZINB) models adjust for excessive zeros in the response. One can also use hurdle models to deal with such a situation. On the other hand, when there are no zero counts, one can use a zero-truncated Poisson (ZTP) or zero-truncated negative binomial (ZTNB) model to fit the data. Usually, these types of models fit better than standard models, but not always. We list these models in table 14.1, together with several others that have also been used to overcome distributional problems with the respective base model.

Table 14.1. Other count-data models

Zero-truncated Poisson and zero-truncated negative binomial (ZTP, ZTNB)
Truncated Poisson and truncated negative binomial
Zero-inflated Poisson and zero-inflated negative binomial (ZIP, ZINB)
Censored Poisson and censored negative binomial
Poisson and negative binomial selection models
Negative binomial with endogenous stratification
Generalized Poisson regression
Poisson inverse Gaussian (PIG) regression
Generalized negative binomial(P) regression (NB-P)
Generalized negative binomial(Famoye) regression (NB-F)
Generalized negative binomial(Waring) regression (NB-W)
Generalized binomial regression
Finite mixture models
Quantile count models
Flexible polynomial count models
Bivariate count models
Panel models for count outcomes (GEE, RE, FE, multilevel)
Bayesian count models

Other similar types of models have been developed, but biases and other difficulties have been identified with their use; for example, certain early parameterizations of generalized negative binomial regression. Most of the above listed models are either official commands in the Stata software or community-contributed commands.

Another cause of overdispersion relates to misspecified variance. The GLM variance may be reparameterized to take the quasilikelihood form, $\mu^2(1 - \mu)^2$, rather than the basic Bernoulli variance $\mu(1 - \mu)$. For count models, we produce table 14.2, which is similar to information summarized in Hilbe (2011), detailing the various count variance functions.

Table 14.2. Variance functions V for count-data models; ϕ, α, δ, p, θ, and ρ are constants

Poisson	μ
QL	$\phi\mu$
NB-1	$\mu(1 + \alpha) = \mu + \alpha\mu$
Gen. Poisson	$\mu/(1 - \delta)^2$
Dbl. Poisson	μ/ϕ
Geometric	$\mu(1 + \mu) = \mu + \mu^2$
NB-2	$\mu(1 + \alpha\mu) = \mu + \alpha\mu^2$
NB-H	$\mu + (\alpha_i)\mu^2$ (where $\alpha_i = z_i\boldsymbol{\gamma}$)
NB-P	$\mu + \alpha\mu^p$
NB-F	$\theta\mu(1 - \mu)(1 - \phi\mu)^{-3}$
NB-W	$\mu + \mu\left(\frac{k+1}{\rho-2}\right) + \mu^2\left(\frac{k+\rho-1}{k(\rho-2)}\right)$
GEE	$V_iR_iV_i$ where R_i is a correlation matrix

NB-H represents the heterogeneous negative binomial model, called generalized negative binomial in Stata. The term "generalized negative binomial" has a history in the literature of the area and refers to an entirely different model. We prefer to use the term heterogeneous for this model, following the usage of Greene (2002) and LIMDEP software.

NB-P refers to the power negative binomial, that is, the NB-2 variance function for which the exponent parameter is to be estimated rather than specified. The NB-P model is therefore a three-parameter model with μ or $\boldsymbol{\beta}$ estimated, as well as the heterogeneity parameter, α, and the exponent, p. We include code to fit this model and illustrate its use.

Panel models constitute another model type that relates to the distributional aspects of the Poisson and negative binomial probability distribution functions (PDFs). A collection of standard approaches is listed in table 14.3.

Table 14.3. Poisson and negative binomial panel-data models

Fixed-effects models
  (Unconditional) fixed effects
  Conditional fixed effects
Random-effects models
  Poisson with gamma-distributed random effects
  Poisson with Gaussian-distributed random effects
  Negative binomial with beta-distributed random effects
  Negative binomial with Gaussian-distributed random effects
GEE
Mixed linear models
  Random intercepts (multiple levels)
  Random coefficients

Each of the models in table 14.3 was developed to adjust an overdispersed Poisson or negative binomial model by enhancing the variance function itself or by addressing the distributional assumptions of the Poisson and negative binomial models. A statistician has a variety of count response models available that, if selected properly, will yield a well-fitted model to the data.

In this chapter, we will look at only a few of the above count models. For another examination of each of the above listed models, see Hilbe (2011).

14.2   Zero-truncated models

We often find that our count response data fail to have zero counts. That is, we discover not only that the count has no zeros but also that it structurally excludes having zero counts. Hospital length of stay data, common in the health analysis industry, are a prime example of this type of data situation. When a patient enters the hospital, the count begins with one. There is no zero length of hospital stay. Zero counts are structurally excluded from lengths of stay. A patient stays in the hospital 1, 2, 3, or more days before being discharged.

The problem with modeling zero-excluded counts relates to the distributional assumptions of count models such as Poisson, negative binomial, generalized Poisson, and Poisson inverse Gaussian. Their probability functions require the possibility of zeros. For a specified value of the mean of the distribution, there is an expected percentage of zeros, ones, twos, and so forth in the data. Figure 14.1 displays the probabilities of outcomes given a mean of 4. The three distributions include the Poisson, negative binomial with $\alpha = 0.5$, and a negative binomial with $\alpha = 1.5$. Because a negative binomial with $\alpha = 0$ is a Poisson distribution, we have labeled the curve for Poisson as NB-0. The probability of 0 is about 2% for a Poisson distribution with a mean of 4, about 11% for a negative binomial with $\alpha = 0.5$, and about 27% for a negative binomial with an $\alpha = 1.5$. For a Poisson model having 100 observations with a count response mean of 4, we would expect that there will be two zeros.

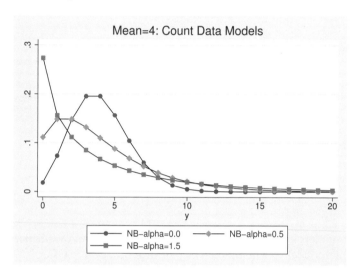

Figure 14.1. Probability mass functions for negative binomial models

A standard method of dealing with count models such as Poisson and negative binomial models for which the data exclude the possibility of zero counts is to use a model usually called a zero-truncated model. The model is not adjusted by using some type of scaling or robust variance estimator because the parameter estimates are biased as well as standard errors. The solution is to adjust the probability function itself so that zeros are excluded. We do this by determining the probability of a zero count from the usual pdf, subtracting it from one, and then dividing the count probability function by the resultant value. We focus on the zero-truncated Poisson in this section, leaving discussion of more general truncated models to section 14.4.

The Poisson PDF can be expressed as

$$f(y_i; \mu) = e^{-\mu}\mu^{y_i}/y_i! \tag{14.1}$$

The probability of a zero count is given as $\exp(-\mu)$. The probability of not having a zero count is then $1 - \exp(-\mu)$. Dividing the Poisson PDF for positive outcomes by the total probability of a positive outcome rescales each term so that the result will sum to unity.

$$f(y_i; \mu|y_i > 0) = \frac{\exp(-\mu)\mu^{y_i}}{\{1 - \exp(-\mu)\}y_i!} \tag{14.2}$$

With $\mu = \exp(x\beta)$, the log-likelihood form of the above is expressed as

$$\mathcal{L}(\mu_i; y_i | y_i > 0) = \sum \Big(y_i \left(\mathbf{x}_i\boldsymbol{\beta}\right) - \exp\left(\mathbf{x}_i\boldsymbol{\beta}\right) - \ln \Gamma(y_i + 1) - \ln \left[1 - \exp\left\{ -\exp(\mathbf{x}_i\boldsymbol{\beta}) \right\} \right] \Big)$$
$$(14.3)$$

We use the same logic to determine the log-likelihood function of the zero-truncated negative binomial. The probability of a zero negative binomial count is $(1 + \alpha\mu) - 1/\alpha$ or, in terms of the linear predictor $x\beta$, $\{1 + \alpha \exp(x\beta)\} - 1/\alpha$.

After subtracting from 1, we use the resultant formula to rescale probabilities for strictly positive outcomes for the negative binomial log-likelihood function. The rescaled function is then given by

$$\mathcal{L}(\mu; y | y > 0) = \sum [\mathcal{L}_{\mathrm{NB}} - \ln\{-\alpha \exp(x\beta) + 1/\alpha\}] \qquad (14.4)$$

where $\mathcal{L}_{\mathrm{NB}}$ is the (unscaled) negative binomial log likelihood.

The iteratively reweighted least-squares estimating algorithm used for standard GLM models is not appropriate for this type of model. The rescaled log-likelihood functions are not members of the exponential family of distributions and therefore cannot take advantage of the simpler Fisher scoring method of estimation. We must use a Newton–Raphson type of maximum likelihood method to estimate parameters, standard errors, and other model statistics.

To demonstrate the difference between models, we will create a 50,000-observation synthetic Poisson data with the following format: y = -1.0 + 0.5*x1 + 1.5*x2.

```
. set seed 2349
. set obs 50000
number of observations (_N) was 0, now 50,000
. generate x1 = rnormal()
. generate x2 = rnormal()
. generate xb = -1.0 + 0.5*x1 + 1.5*x2
. generate lambda = exp(xb)
. generate yp = rpoisson(lambda)
```

We next model the data by using a Poisson model. We expect that the parameter estimates will be close to the values we specified in the construction of the data.

```
. poisson yp x1 x2
Iteration 0:   log likelihood = -46461.854
Iteration 1:   log likelihood = -46461.844
Iteration 2:   log likelihood = -46461.844
Poisson regression                              Number of obs    =      50,000
                                                LR chi2(2)       =   154187.55
                                                Prob > chi2      =      0.0000
Log likelihood = -46461.844                     Pseudo R2        =      0.6240
```

yp	Coef.	Std. Err.	z	P>|z|	[95% Conf. Interval]	
x1	.5025049	.0039275	127.94	0.000	.494807	.5102027
x2	1.497894	.0040731	367.75	0.000	1.489911	1.505877
_cons	-.9953207	.0074831	-133.01	0.000	-1.009987	-.9806541

We find our expectations met. Next, for illustrative purposes, we model only the positive outcomes; we treat the data as if there are no zero outcomes.

```
. poisson yp x1 x2 if y>0, nolog
Poisson regression                              Number of obs    =      20,033
                                                LR chi2(2)       =    62939.63
                                                Prob > chi2      =      0.0000
Log likelihood = -33777.371                     Pseudo R2        =      0.4823
```

yp	Coef.	Std. Err.	z	P>|z|	[95% Conf. Interval]	
x1	.3941379	.0040124	98.23	0.000	.3862737	.4020021
x2	1.178548	.0047894	246.08	0.000	1.169161	1.187935
_cons	-.2739272	.0085306	-32.11	0.000	-.2906469	-.2572075

The parameter estimates differ substantially from those we specified in constructing the data. We now investigate a zero-truncated model on the positive outcome subset of data.

```
. tpoisson yp x1 x2 if y>0, nolog
Truncated Poisson regression                    Number of obs    =      20,033
Limits:   lower =          0                     LR chi2(2)       =    77879.10
          upper =       +inf                     Prob > chi2      =      0.0000
Log likelihood = -25458.683                     Pseudo R2        =      0.6047
```

yp	Coef.	Std. Err.	z	P>|z|	[95% Conf. Interval]	
x1	.501717	.004571	109.76	0.000	.492758	.510676
x2	1.495111	.0055686	268.49	0.000	1.484197	1.506025
_cons	-.9901485	.0115713	-85.57	0.000	-1.012828	-.9674691

We again obtain parameter estimates that closely match those values used in the construction of the data. If we do have data not including zero outcomes, we see that using a zero-truncated model provides a more superior fit than simply using the standard count model and ignoring the distributional violation involved. The next two sections address the occasions where there is an excess frequency of zero outcomes in the data;

that is, the number of zero outcomes exceeds what would otherwise be expected for the given distribution.

14.3 Zero-inflated models

Using the same logic as we did with the zero-truncated models, count response models having far more zeros than expected by the distributional assumptions of the Poisson and negative binomial models result in incorrect parameter estimates as well as biased standard errors. We will briefly discuss the nature of the zero-inflated count models and how they adjust the likelihood function to produce appropriate model statistics. Of course, this statement hinges on excess zeros being the only problem in the data. There are often multiple sources resulting in correlated data and therefore overdispersion. For zero-inflated binomial models, see section 10.9.

Zero-inflated models consider two distinct sources of zero outcomes. One source is generated from individuals who do not enter the counting process, the other from those who do enter the counting process but result in a zero outcome. Suppose that we have data related to physician visits. We intend to model the number of visits made to the office of the patient's physician over a given period. However, suppose that our data also have information on individuals who do not have a physician during the specified period. Such patients are given a count of zero.

When there are far more observed outcomes of zero (individuals not visiting a physician) than prescribed by the standard Poisson or negative binomial distribution upon which the respective models are based, we may consider breaking the model into a part, indicating whether an individual has a physician ($y = 1$ or $y = 0$); that is, one part of the model focuses on whether the individual entered into the counting process, and another part models (if the individual enters the counting process) the number of visits made to the physician ($y \geq 0$).

Lambert (1992) first described this type of mixture model in the context of process control in manufacturing. It has since been used in many applications and is now discussed in nearly every book or article dealing with count response models.

For the zero-inflated model, the probability of observing a zero outcome equals the probability that an individual is in the always-zero group plus the probability that the individual is not in that group times the probability that the counting process produces a zero; see Hilbe and Greene (2008). If we define $B(0)$ as the probability that the binary process results in a zero outcome and $\Pr(0)$ as the counting process probability of a zero outcome, the probability of a zero outcome for the system is then given by

$$\Pr(y = 0) = B(0) + \{1 - B(0)\}\Pr(0) \tag{14.5}$$

and the probability of a nonzero count is

$$\Pr(y = k; k > 0) = \{1 - B(0)\}\Pr(k) \tag{14.6}$$

where $\Pr(k)$ is the counting process probability that the outcome is k.

Logistic and probit are the most commonly used link functions for the Bernoulli part of the log-likelihood function. Mixing the binary logistic distribution for zero counts with the negative binomial results in the following zero-inflated log-likelihood function in terms of the linear predictor $x\beta$. In the parameterization below, we indicate the Bernoulli linear predictor as $z\gamma$ and the negative binomial as $x\beta$.

In the following, $F(z\gamma)$ may be in terms of probit or logit. This model has also been called the "with-zeros" or WZ model (for example, the "with-zeros Poisson model"). Another estimator is available by allowing that $v_i = z_i\gamma = \tau(x_i\beta)$, which introduces only one new parameter, τ, rather than a new covariate list. In the formulas below, we denote the set of zero outcomes as S. We include both weights and offsets in the presentation of the log likelihood and estimating equations for the Poisson version.

$$\xi_i^\beta = \mathbf{x}_i\boldsymbol{\beta} + \text{offset}_i^\beta \tag{14.7}$$

$$\xi_i^\gamma = \mathbf{z}_i\boldsymbol{\gamma} + \text{offset}_i^\gamma \tag{14.8}$$

$$\lambda_i = \exp(\xi_i^\beta) \tag{14.9}$$

$$\Pr(Y_i = 0) = F(\xi_i^\gamma) + \{1 - F(\xi_i^\gamma)\}\frac{e^{-\lambda_i}\lambda_i^{y_i}}{y_i!} \tag{14.10}$$

$$\Pr(Y_i > 0) = \{1 - F(\xi_i^\gamma)\}\frac{e^{-\lambda}\lambda^{y_i}}{y_i!} \tag{14.11}$$

$$\mathcal{L} = \sum_{i \in S} w_i \ln\left[F(\xi_i^\gamma) + \{1 - F(\xi_i^\gamma)\}\exp(-\lambda_i)\right]$$

$$+ \sum_{i \notin S} w_i \left[\ln\{1 - F(\xi_i^\gamma)\} - \lambda_i + \xi_i^\beta y_i - \ln(y_i!)\right] \tag{14.12}$$

These formulas have associated derivatives

$$\frac{\partial\mathcal{L}}{\partial\beta_j} = -\sum_{i \in S} w_i x_{ji}\frac{\{1 - F(\xi_i^\gamma)\}\lambda_i\exp(-\lambda_i)}{F(\xi_i^\gamma) + \{1 - F(\xi_i^\gamma)\}\exp(-\lambda_i)} + \sum_{i \notin S} w_i x_{ji}(y_i - \lambda_i) \tag{14.13}$$

$$\frac{\partial\mathcal{L}}{\partial\gamma_j} = \sum_{i \in S} w_i z_{ji}\frac{f(\xi_i^\gamma)\{1 - \exp(-\lambda_i)\}}{F(\xi_i^\gamma) + \{1 - F(\xi_i^\gamma)\}\exp(-\lambda_i)} - \sum_{i \notin S} w_i z_{ji}\frac{f(\xi_i^\gamma)}{1 - F(\xi_i^\gamma)} \tag{14.14}$$

Similarly, for the negative binomial, $F(z\gamma)$ may be in terms of probit or logit. Another estimator is available through assuming that $v_i = z_i\gamma = \tau(x_i\beta)$ (introducing only one new parameter τ). We include both weights and offsets in the presentation of the log likelihood and estimating equations for the negative binomial version of zero-inflated models.

$$
\begin{align}
\psi(z) &= \text{digamma function evaluated at } z & (14.15)\\
m &= 1/\alpha & (14.16)\\
p_i &= 1/(1+\alpha\mu_i) & (14.17)\\
\mu_i &= \exp(x_i\beta + \text{offset}_i) & (14.18)\\
\Pr(Y=0) &= F(z_i\gamma) + \{1 - F(z_i\gamma)\}p_i^m & (14.19)\\
\Pr(Y>0) &= \{1 - F(z_i\gamma)\}\frac{\Gamma(m+y_i)}{\Gamma(y_i+1)\Gamma(m)}p_i^m(1-p_i)^{y_i} & (14.20)
\end{align}
$$

$$
\begin{align}
\mathcal{L} &= \sum_{i\in S} w_i \ln\left[F(z_i\gamma) + \{1 - F(z_i\gamma)\}p_i^m\right] + \sum_{i\notin S} w_i \ln\{1 - F(z_i\gamma)\}\\
&\quad + w_i \ln\Gamma(m+y_i) - w_i\Gamma(y_i+1) - w_i \ln\Gamma(m) + w_i m \ln p_i\\
&\quad + w_i y_i \ln(1-p_i) & (14.21)
\end{align}
$$

with associated estimating equations

$$
\frac{\partial\mathcal{L}}{\partial\beta_j} = \sum_{i\in S} w_i x_{ji}\frac{-\{1 - F(z_i\gamma)\}\mu_i p_i^{m+1}}{F(z_i\gamma) + \{1 - F(z_i\gamma)\}p_i^m} + \sum_{i\notin S} w_i x_{ji}p_i(y_i - \mu_i) \tag{14.22}
$$

$$
\frac{\partial\mathcal{L}}{\partial\gamma_j} = \sum_{i\in S} w_i z_{ji}\frac{f(z_i\gamma)(1 - p_i^m)}{F(z_i\gamma) + \{1 - F(z_i\gamma)\}p_i^m} + \sum_{i\notin S} w_i z_{ji}\frac{-f(z_i\gamma)}{1 - F(z_i\gamma)} \tag{14.23}
$$

$$
\begin{align}
\frac{\partial\mathcal{L}}{\partial\alpha} &= -\sum_{i\in S} w_i\frac{m^2 p_i^m \ln p_i - m\mu_i p_i^{m-1}}{F(z_i\gamma) + \{1 - F(z_i\gamma)\}p_i^m} - \sum_{i\notin S} w_i\alpha^{-2}\left\{\frac{\alpha(\mu_i - y_i)}{1+\alpha\mu_i} - \ln(1+\alpha\mu_i)\right.\\
&\quad \left. + \psi(y_i + 1/\alpha) - \psi(1/\alpha)\right\} \tag{14.24}
\end{align}
$$

In the following examples, we use the same data for fitting zero-inflated models as we did for fitting the zero-truncated models, except that we will include the zero outcomes. A tabulation of the smallest 15 outcomes is given by

```
. tabulate yp if yp <= 15
         yp |      Freq.     Percent        Cum.
------------+-----------------------------------
          0 |     29,967       60.52       60.52
          1 |      9,520       19.23       79.75
          2 |      3,997        8.07       87.82
          3 |      1,994        4.03       91.85
          4 |      1,171        2.36       94.21
          5 |        775        1.57       95.78
          6 |        561        1.13       96.91
          7 |        390        0.79       97.70
          8 |        301        0.61       98.30
          9 |        209        0.42       98.73
         10 |        164        0.33       99.06
         11 |        142        0.29       99.34
         12 |        102        0.21       99.55
         13 |         93        0.19       99.74
         14 |         61        0.12       99.86
         15 |         69        0.14      100.00
------------+-----------------------------------
      Total |     49,516      100.00
```

The corresponding model is fit by

```
. zip yp x1 x2, inflate(x1 x2) iterate(20) nolog
Zero-inflated Poisson regression                  Number of obs    =      50,000
                                                  Nonzero obs      =      20,033
                                                  Zero obs         =      29,967

Inflation model = logit                           LR chi2(2)       =    78662.62
Log likelihood  = -46460.54                       Prob > chi2      =      0.0000

------------------------------------------------------------------------------
          yp |      Coef.   Std. Err.       z    P>|z|     [95% Conf. Interval]
-------------+----------------------------------------------------------------
yp           |
          x1 |   .5016162   .0040405    124.15   0.000     .493697    .5095354
          x2 |   1.497917   .0040853    366.66   0.000     1.48991    1.505924
       _cons |  -.9944515   .0076147   -130.60   0.000    -1.009376   -.979527
-------------+----------------------------------------------------------------
inflate      |
          x1 |  -1.878906   .6580896     -2.86   0.004    -3.168738   -.5890744
          x2 |   .4106878   1.292919      0.32   0.751    -2.123386    2.944762
       _cons |  -9.011493   3.549654     -2.54   0.011    -15.96869   -2.054299
------------------------------------------------------------------------------
```

The inflated frame provides parameter estimates for the binary part of the mixture model. We find here that x1 and x2 do not contribute to a Bernoulli process generating extra zero outcomes. The upper frame is interpreted just like a standard Poisson model. See Long and Freese (2014) and Hilbe (2011) for more discussion, examples, and interpretation of the zero-inflated models. Both references also include discussion of marginal effects.

To highlight a pathway for illustrating these models, here we show how to synthesize data. This can be an instructive method for observing how various alterations of model parameters affect estimation. However, our purposes here are simply to highlight the initial data synthesis.

Here we illustrate generation of data for the ZIP model for which the count-data model includes two covariates x1 and x2. The Bernoulli will use the same covariates.

```
. clear
. set obs 50000
number of observations (_N) was 0, now 50,000
. set seed 1234
. generate x1 = rnormal()        /* x1 ~ N(0,1) */
. generate x2 = rnormal()        /* x2 ~ N(0,1) */
. generate yp = rpoisson(exp(2.0+0.75*x1-1.25*x2)) /* yp ~ Poisson            */
. generate p  = 1/(1+exp((0.2+0.9*x1+0.1*x2)))     /* p=P(B=0) for Bernoulli */
. generate yb = runiform() < p                     /* yb ~ Bernoulli(1-p)    */
. generate yz = yb*yp                              /* yz ~ ZIP               */
. generate yzero = yz == 0                         /* yzero = 1 if yz=0      */
```

The important point is to remember that the Bernoulli process is modeling the inflation of zero outcomes; that is, it models the probability of a zero outcome. In the usual (Stata) approach for modeling Bernoulli outcomes, the focus is on the probability of positive (nonzero) outcomes. The two approaches result in opposite signs on coefficients. To illustrate the results of our data, the estimation results are given by

```
. zip yz x1 x2, inflate(x1 x2) nolog
Zero-inflated Poisson regression                   Number of obs    =     50,000
                                                   Nonzero obs      =     20,517
                                                   Zero obs         =     29,483

Inflation model = logit                            LR chi2(2)       = 650531.06
Log likelihood  = -77991.97                        Prob > chi2      =     0.0000
```

yz	Coef.	Std. Err.	z	P>|z|	[95% Conf. Interval]	
yz						
x1	.746065	.0018011	414.23	0.000	.742535	.7495951
x2	-1.248262	.0017667	-706.55	0.000	-1.251725	-1.2448
_cons	2.00479	.0028616	700.59	0.000	1.999181	2.010399
inflate						
x1	.8943852	.0128306	69.71	0.000	.8692377	.9195328
x2	.0915107	.0115939	7.89	0.000	.0687871	.1142343
_cons	.2042994	.0107085	19.08	0.000	.1833111	.2252878

Over half (59%) the synthesized data are zero, which far exceeds that which would be true for the Poisson half of the mixture of distributions.

Construction of zero-inflated count-data models is not limited to models for which the counting process is either a Poisson or a negative binomial. We could just as

easily consider a zero-inflated generalized Poisson model as a zero-inflated negative
binomial model. Either of these approaches assumes that there are two distinct sources
of overdispersion: the excess zero outcomes resulting from mixing in outcomes from the
Bernoulli process, and the possible overdispersion in the counting process. Here we load
data on doctor visits and then fit a zero-inflated generalized Poisson (ZIGP) model with
the `zigp` command.[1]

```
. use http://www.stata-press.com/data/hh4/rwm1984, clear
(German health data for 1984; Hardin & Hilbe, GLM and Extensions, 4th ed)
. zigp docvis outwork female age married, inflate(female age) zip nolog
```

```
Zero-inflated generalized Poisson regression     Number of obs    =        3874
Regression link:                                 Nonzero obs      =        2263
Inflation link : logit                           Zero obs         =        1611
                                                 LR chi2(4)       =       89.82
Log likelihood = -8246.647                       Prob > chi2      =      0.0000
```

docvis	Coef.	Std. Err.	z	P>|z|	[95% Conf. Interval]	
docvis						
outwork	.2201645	.0523103	4.21	0.000	.1176383	.3226908
female	.1875331	.0711449	2.64	0.008	.0480918	.3269745
age	.0139482	.0026954	5.17	0.000	.0086653	.0192311
married	-.0590289	.0535342	-1.10	0.270	-.1639539	.0458962
_cons	.5973603	.1506705	3.96	0.000	.3020515	.8926691
inflate						
female	-.5893982	.2034934	-2.90	0.004	-.988238	-.1905583
age	-.0237719	.0073725	-3.22	0.001	-.0382218	-.0093221
_cons	-.0425846	.3715118	-0.11	0.909	-.7707343	.685565
/atanhdelta	.7718012	.0170026	45.39	0.000	.7384767	.8051257
delta	.6479756	.0098637			.628224	.6668926

```
LR test of delta=0 (zigp versus zip):            X2 =   7749.32    Pr>X2= 0.0000
. estat ic
Akaike's information criterion and Bayesian information criterion
```

Model	Obs	ll(null)	ll(model)	df	AIC	BIC
.	3,874	-8291.556	-8246.647	9	16511.29	16567.65

```
        Note: N=Obs used in calculating BIC; see [R] BIC note.
```

The likelihood-ratio test of $\delta = 0$ at the bottom of the output assesses the overdispersion parameter of the counting process. In effect, this test assesses whether there is
a significant preference of the ZIGP model over the ZIP model for these data.

---

1. The `zipg` command can be installed by typing `net install zigp`,
   from(http://www.stata.com/users/jhardin/).

We can fit the generalized Poisson without zero inflation using the community-contributed `gpoisson` command (Harris, Yang, and Hardin 2012).[2]

```
. quietly gpoisson docvis outwork female age married

. estat ic
```

Akaike's information criterion and Bayesian information criterion

Model	Obs	ll(null)	ll(model)	df	AIC	BIC
.	3,874	−8426.955	−8276.891	6	16565.78	16603.35

Note: N=Obs used in calculating BIC; see [R] BIC note.

We can see a preference for the ZIGP model over the generalized Poisson model by comparing the AIC and BIC statistics from the two models.

Note that zero-inflated models estimate coefficients of the two mixing distributions. If the analyst is interested in the interpretation of the coefficients of the marginal count distribution (across all observations), then the zero-inflated model may not be of interest. One could simply fit the count model, or one could fit a "marginalized" version of the zero-inflated model.

To begin, we generate synthetic data for which we can fit and compare models.

```
. clear
. set seed 2323423
. set obs 1000
number of observations (_N) was 0, now 1,000
. generate x1 = runiform() < .5
. generate x2 = runiform() < .5
. generate x3 = runiform() < .5
. generate yc = rpoisson(exp(1+x1-x2))
. generate yz = rbinomial(1, 1/(1+exp(1+x1-x3)))
. generate y  = yc * (yz!=0)
```

---

2. The `gpoisson` command can be installed by typing `net install st0279_2`, `from(http://www.stata-journal.com/software/sj18-1)`.

The zero-inflated Poisson model will estimate the coefficients of the component processes. The estimated coefficients should be close to the parameters used in the data
synthesis.

```
. program zipllf
  1.      args lnf beta1 beta2
  2.      quietly {
  3.        tempvar pi mu gamma
  4.        generate double `pi´ = exp(`beta2´)
  5.        generate double `gamma´ = `pi´/(1+`pi´)
  6.        generate double `mu´ = exp(`beta1´)
  7.        replace `lnf´ = cond($ML_y1==0,
  >            ln(`pi´/(1+`pi´)+1/(1+`pi´)*exp(-`mu´)),
  >            ln(1/(1+`pi´)) -`mu´ + $ML_y1 * log(`mu´)
  >            - lngamma($ML_y1 + 1))
  8.      }
  9. end
. ml model lf zipllf (y:y=x1 x2) (logit: x1 x3)
. ml maximize, search(on) nolog title(Zero-inflated Poisson)
Zero-inflated Poisson                          Number of obs   =      1,000
                                               Wald chi2(2)    =     344.16
Log likelihood = -1011.4802                    Prob > chi2     =     0.0000
```

y	Coef.	Std. Err.	z	P>|z|	[95% Conf.	Interval]
y						
x1	.978296	.0740531	13.21	0.000	.8331545	1.123438
x2	-1.159871	.0889903	-13.03	0.000	-1.334289	-.9854537
_cons	1.028199	.0623722	16.48	0.000	.9059513	1.150446
logit						
x1	.9936171	.1650329	6.02	0.000	.6701585	1.317076
x3	-.76528	.1607036	-4.76	0.000	-1.080253	-.4503067
_cons	.677723	.1428233	4.75	0.000	.3977945	.9576515

As can be seen, the estimated coefficients are close to the values that were used to
generate the data. However, they are not close to an "overall" model of the counts. Analysts, in this case, are interested in the estimation of population-averaged parameters;
see also section 18.6.

Long et al. (2014) described a marginalized ZIP regression model that directly models
the population mean count, therefore giving the ability to interpret population-wide
parameters. In the straightforward approach, the mean of the count process, μ_i, is
rescaled as a function of the model parameters but through the use of a nonlinear
optimization. This results in a relatively straightforward change to the definition of the
mean in the log-likelihood function. We illustrate this in the following output:

```
. capture program drop zimpllf
. program zimpllf
  1.      args lnf beta1 beta2
  2.      quietly {
  3.          tempvar pi mu gamma
  4.          generate double `pi´ = exp(`beta2´)
  5.          generate double `gamma´ = `pi´/(1+`pi´)
  6.          generate double `mu´ = exp(`beta1´)/(1-`gamma´)
  7.          replace `lnf´ = cond($ML_y1==0,
  >          ln(`pi´/(1+`pi´)+1/(1+`pi´)*exp(-`mu´)),
  >          ln(1/(1+`pi´)) -`mu´ + $ML_y1 * log(`mu´)
  >          - lngamma($ML_y1 + 1))
  8.      }
  9. end
. ml model lf zimpllf (margin:y=x1 x2) (logit: x1 x3)
. ml maximize, search(on) nolog title(Zero-inflated marginalized Poisson)
```

```
Zero-inflated marginalized Poisson          Number of obs   =      1,000
                                            Wald chi2(2)    =     174.79
Log likelihood = -1019.7728                 Prob > chi2     =     0.0000
```

y	Coef.	Std. Err.	z	P>|z|	[95% Conf. Interval]	
margin						
x1	.3482449	.1206852	2.89	0.004	.1117062	.5847835
x2	-1.140214	.0893833	-12.76	0.000	-1.315402	-.9650257
_cons	.1676297	.0816992	2.05	0.040	.0075022	.3277572
logit						
x1	.9406856	.1604287	5.86	0.000	.6262513	1.25512
x3	-.2317826	.0889378	-2.61	0.009	-.4060974	-.0574677
_cons	.4321513	.125734	3.44	0.001	.1857171	.6785854

14.4 General truncated models

In Hardin and Hilbe (2015), we presented a new command to fit count regression models based on truncated distributions. To install the command, type `net install st0378_1, from(http://www.stata-journal.com/software/sj18-1)`.

Earlier, we illustrated the use and estimation of zero-truncation. However, one might be faced with modeling data that are the result of an even more restricted range of outcomes. Though one could model data using models based on usual distributions (supporting a range from 0 to $+\infty$), it is more reasonable to use a truncated distribution if we believe that the outcomes really are limited.

Regression modeling of truncated Poisson and negative binomial count outcomes is supported in Stata's `tpoisson` and `tnbreg` commands. These commands allow users to fit models for left-truncated ($y \in \{L+1, L+2, \dots\}$) distributions; in Stata 15, the commands also support right-truncated distributions. The user can either specify a common truncation value or a truncation variable so that each observation has its own truncation value and thus an observation-specific truncated distribution. Though left-truncation is far more commonly considered in regression models, the `trncregress` command considers right-truncation ($y \in \{0, 1, \ldots, R-1\}$) and even truncation on both sides ($y \in \{L+1, L+2, \ldots, R-2, R-1\}$).

Prior to the development of the `trncregress` command, support for estimation of truncated regression models was given only for the specific zero-truncated models that are now deprecated. Missing from these former and current commands is support for count distributions aside from Poisson and negative binomial, as well as support for truncation on the right and on both sides.

`trncregress` evaluates count-data regression models for truncated distributions including Poisson, negative binomial, generalized Poisson, Poisson-inverse Gaussian, negative binomial(P), three-parameter negative binomial(Waring), and three-parameter negative binomial(Famoye).

Equivalent in syntax to the basic count-data commands, the syntax for the truncated regression command is given by

> `trncregress` *depvar* [*indepvars*] [*if*] [*in*] [*weight*] [, `ltrunc(`#` | `*varname*`)`
>
>   `rtrunc(`#` | `*varname*`)` `dist(`*distname*`)` `offset(`*varname*`)` *display_options*
>
>   *maximization_options*]

In the above syntax, the allowable distribution names (*distnames*) are given as follows: `poisson`, `negbin`, `gpoisson`, `pig`, `nbp`, `nbf`, or `nbw`. Help files are included for the estimation and postestimation specifications of these models. The help files include the complete syntax and further examples.

The output header includes the usual summary information for the model. Also, we include a short description of the support for the outcome by the designated truncated distribution. This description is of the form {#$_1$, ..., #$_2$}, where #$_1$ is the minimum and #$_2$ is the maximum. Thus, for a zero-truncated model, the support is given by #$_1$ = 1 and #$_2$ = . (or positive infinity).

Model predictions are available through Stata's `predict` command. Specifically, there is support for linear predictions, predictions of the mean, and standard errors of the linear prediction.

To highlight, we use data collected from the 1991 Arizona MedPar database; these data consist of the inpatient records for Medicare patients. In this study, all patients are over 65 years of age. The diagnostic related group classification is confidential for privacy concerns.

The response variable is the patient length of stay (`los`), which commences with a count of 1. There are no length of stay records of 0, which in other circumstances could indicate that a patient was not admitted to the hospital.

hospital Length of Stay	Freq.	Percent	Cum.
1	126	8.43	8.43
2	71	4.75	13.18
3	75	5.02	18.19
4	104	6.96	25.15
5	123	8.23	33.38
6	97	6.49	39.87
.	. . .		
70	1	0.07	99.80
74	1	0.07	99.87
91	1	0.07	99.93
116	1	0.07	100.00
Total	1,495	100.00	

The mean of `los` is 9.85. In fact, using a zero-truncated model will make little difference in the estimates. However, if the mean of the response is low, under 3 or 4, then there will be a substantial difference in coefficient values. The closer the mean is to zero, the greater the difference will be in coefficient values. Closeness of coefficients notwithstanding for this example, the foremost point is the use of the appropriate count model for a given data situation. The explanatory predictors for our example model include an indicator of white race (`white`), an indicator of HMO (`hmo`), an indicator of elective admittance (`type1`, used as the referent group for admittance types), an indicator of urgent admittance (`type2`), and an indicator of emergency admittance (`type3`); all the indicators are combined to form the classification variable `type`.

We first fit the model using a zero-truncated Poisson (ZTP) model. As illustrated by this command, we include the `nolog` option to inhibit the display of the iteration log, the `eform` option to display model coefficients in exponentiated form, and automatic generation of indicator variables from categorical variable names through the `i.` prefix.

```
. use http://www.stata-press.com/data/hh4/medpar, clear
. generate byte type = type1 + 2*type2 + 3*type3
. label define tlab 1 "elective" 2 "urgent" 3 "emergency"
. label values type tlab
```

```
. trncregress los white hmo i.type, dist(poisson) ltrunc(0) nolog eform
```

Truncated Poisson regression   Number of obs   =   1495
Dist. support on {1, ..., .}   LR chi2(4)   =   758.68
Log likelihood = -6928.723   Prob > chi2   =   0.0000

los	exp(b)	Std. Err.	z	P>|z|	[95% Conf. Interval]	
white	.8573203	.0235048	-5.61	0.000	.8124676	.9046491
hmo	.930858	.0223067	-2.99	0.003	.8881484	.9756214
type						
urgent	1.248297	.0262846	10.53	0.000	1.197829	1.300892
emergency	2.033211	.053145	27.15	0.000	1.931672	2.140087
_cons	10.30738	.2804854	85.73	0.000	9.772044	10.87205

```
. estimates store trnc_poi
```

For comparison and to highlight functionality, we also fit a zero-truncated generalized Poisson model and a zero-truncated Poisson inverse Gaussian (PIG) model.

```
. trncregress los white hmo i.type, dist(gpoisson) ltrunc(0) nolog eform
```

Truncated gen. Poisson regression   Number of obs   =   1495
Dist. support on {1, ..., .}   LR chi2(4)   =   36.92
Log likelihood = -4781.699   Prob > chi2   =   0.0000

los	exp(b)	Std. Err.	z	P>|z|	[95% Conf. Interval]	
white	.8670813	.0574607	-2.15	0.031	.7614677	.9873434
hmo	.9498428	.0542766	-0.90	0.368	.8492035	1.062409
type						
urgent	1.191689	.0602171	3.47	0.001	1.079322	1.315755
emergency	1.457548	.1064875	5.16	0.000	1.263091	1.681942
_cons	10.34951	.6813722	35.50	0.000	9.096614	11.77496
/atanhdelta	.7175918	.0164439			.6853623	.7498213
delta	.6154154	.010216			.5949944	.6350423

Note: Estimates are transformed only in the first equation.

```
. estimates store trnc_gpo
```

```
. trncregress los white hmo i.type, dist(pig) nolog ltrunc(0) vce(robust)
```

Truncated Poisson IG regression				Number of obs	=	1495
Dist. support on {1, ..., .}				LR chi2(4)	=	63.10
Log pseudolikelihood = -4778.698				Prob > chi2	=	0.0000

los	Coef.	Robust Std. Err.	z	P>|z|	[95% Conf. Interval]	
white	-.147871	.0742113	-1.99	0.046	-.2933226	-.0024195
hmo	-.0643421	.0574163	-1.12	0.262	-.1768759	.0481917
type						
urgent	.2098073	.0559506	3.75	0.000	.1001462	.3194684
emergency	.5427813	.1047398	5.18	0.000	.337495	.7480676
_cons	2.324182	.0722792	32.16	0.000	2.182517	2.465847
/lnalpha	-.5109055	.065811			-.6398927	-.3819183
alpha	.5999521	.0394835			.527349	.6825508

```
. estimates store trnc_pig
```

To highlight a more general application, we isolate the observations for which the outcome variable is in {1, 2, ..., 18} and model the variable using a truncated distribution.

```
. trncregress los white hmo i.type if los>0 & los<19, dist(gpoisson) ltrunc(0)
> rtrunc(19) nolog eform vce(robust)
Truncated gen. Poisson regression              Number of obs   =      1335
Dist. support on {1, ..., 18}                  LR chi2(4)      =     11.46
Log pseudolikelihood = -3775.192               Prob > chi2     =    0.0218
```

los	exp(b)	Robust Std. Err.	z	P>|z|	[95% Conf. Interval]	
white	.8898304	.070152	-1.48	0.139	.7624312	1.038518
hmo	.9676663	.06314	-0.50	0.614	.8515004	1.09968
type						
urgent	1.106379	.0697724	1.60	0.109	.9777414	1.251941
emergency	1.051078	.1268398	0.41	0.680	.8296895	1.331541
_cons	10.21698	.7983777	29.74	0.000	8.766123	11.90796
/atanhdelta	.6811347	.0175962			.6466467	.7156226
delta	.5922566	.011424			.5694082	.6141905

Note: Estimates are transformed only in the first equation.

```
. estimates store trnc_poirt
```

```
. estimates stats trnc*
```

Akaike's information criterion and Bayesian information criterion

Model	Obs	ll(null)	ll(model)	df	AIC	BIC
trnc_poi	1,495	-7308.063	-6928.723	5	13867.45	13894
trnc_gpo	1,495	-4800.157	-4781.699	6	9575.398	9607.257
trnc_pig	1,495	-4810.25	-4778.698	6	9569.395	9601.255
trnc_poirt	1,335	-3780.924	-3775.192	6	7562.383	7593.563

Note: N=Obs used in calculating BIC; see [R] BIC note.

Note that you cannot compare AIC or BIC values for models that are not on the same analytic sample. Thus, we cannot compare the final model in the table with the others because of the discrepancies in the number of observations. For the other three models, there is a preference for the truncated PIG model. Interested readers should try fitting additional models, because we have found an even better fit by the three-parameter negative binomial(Waring) model for zero-truncated data.

14.5   Hurdle models

Like zero-inflated models, hurdle models were developed (Mullahy 1986) to deal with count responses having more zeros than allowed by the distributional assumptions of Poisson and negative binomial regression. (Also see McDowell [2003].) Unlike zero-inflated models, however, hurdle models, which have at times been referred to as zero-altered (ZAP and ZANB) models, clearly partition the data into two groupings.

The first part consists of a binary process that generates zeros. The second part consists of a count process that generates positive counts (it does not generate zero counts). In effect, we imagine that the data-generation process has to cross the hurdle of generating zeros before it starts to generate positive counts.

This process continues until an unspecified event occurs that begins the generation of a nonzero counting process. In effect, the hurdle to be crossed is the generation of zeros.

The primary difference that the hurdle models have from the zero-inflated models is that the count aspect of the model does not incorporate zero counts in its structure; the counting process is incorporated using a zero-truncated model. The binary aspect is modeled using a Bernoulli model with the logit, probit, or clog-log link.

The binary and count aspects of the Poisson hurdle log-likelihood function are represented by

$$\xi_i^\beta = \mathbf{x}_i\boldsymbol{\beta} + \text{offset}_i^\beta \tag{14.25}$$

$$\xi_i^\gamma = \mathbf{z}_i\boldsymbol{\gamma} + \text{offset}_i^\gamma \tag{14.26}$$

$$\lambda_i = \exp(\xi_i^\beta) \tag{14.27}$$

$$\Pr(Y_i = 0) = F(\xi_i^\gamma) \tag{14.28}$$

$$\Pr(Y_i > 0) = \{1 - F(\xi_i^\gamma)\}\frac{e^{-\lambda_i}\lambda_i^{y_i}}{y_i!\{1 - \exp(\lambda_i)\}} \tag{14.29}$$

$$\mathcal{L} = \sum_{i \in S} w_i \ln\{F(\xi_i^\gamma)\} + \sum_{i \notin S} w_i \left[\ln\{1 - F(\xi_i^\gamma)\} - \lambda_i \right.$$
$$\left. + \xi_i^\beta y_i - \ln(y_i!) - \ln\{1 - \exp(\lambda_i)\}\right] \tag{14.30}$$

Given the structure of the model, the estimating equations for $\boldsymbol{\beta}$ and $\boldsymbol{\gamma}$ from the hurdle model are going to be the same as those obtained separately from the binary and zero-truncated count models.

Here we use the Poisson-logit hurdle model on the same data we used with the zero-inflated Poisson (with logit inflation). Unlike the zero-inflated Poisson model, the Poisson-logit hurdle model requires generation of zero-truncated Poisson outcomes for when the outcome is not zero. Without a built-in command for generating zero-truncated values, we discard zero observations from the Poisson and replace them with nonzero random values (in this and the next example of data synthesis).

```
. clear

. set obs 50000
number of observations (_N) was 0, now 50,000

. set seed 1234

. generate x1 = rnormal()

. generate x2 = rnormal()

. // Generate Poissons in a loop; keep replacing
. // values until there are no zeros
. generate yp = 0

. count if yp==0
  50,000

. while r(N) {
  2.          quietly {
  3.                  replace yp = rpoisson(exp(2.0+0.75*x1-1.25*x2)) if yp==0
  4.                  count if yp==0
  5.          }
  6. }

. generate p   = 1/(1+exp((0.2+0.9*x1+0.1*x2)))   // Binary p

. generate yb = runiform() < p   // Zero outcomes

. generate yz = yb*yp            // Hurdle outcome to model

. generate yzero = (yz == 0)     // Indicator of zeros

. program HurdPoissLogitLL
  1.      args lnf beta1 beta2
  2.      quietly {
  3.          tempvar pi mu
  4.          generate double `mu´ = exp(`beta1´)
  5.          generate double `pi´ = exp(`beta2´)
  6.          replace `lnf´ = cond($ML_y1==0,
>           ln(`pi´/(1+`pi´)),
>           ln(1/(1+`pi´)) -`mu´ + $ML_y1 * `beta1´
>           - lngamma($ML_y1 + 1) - ln(1 - exp(-`mu´)))
  7.      }
  8. end

. ml model lf HurdPoissLogitLL (yz:yz=x1 x2) (logit:yzero=x1 x2)

. ml maximize, search(on) nolog title(Poisson Logistic Hurdle Regression)
```

Poisson Logistic Hurdle Regression   Number of obs   =   50,000
                                                  Wald chi2(2)   =   745317.68
Log likelihood = -80637.231   Prob > chi2   =   0.0000

		Coef.	Std. Err.	z	P>|z|	[95% Conf.	Interval]
yz							
	x1	.7506657	.0017678	424.62	0.000	.7472008	.7541306
	x2	-1.250193	.0016833	-742.70	0.000	-1.253492	-1.246894
	_cons	2.002522	.0028048	713.96	0.000	1.997024	2.008019
logit							
	x1	.9015221	.0113286	79.58	0.000	.8793185	.9237256
	x2	.0872327	.0097722	8.93	0.000	.0680795	.106386
	_cons	.1915	.0097549	19.63	0.000	.1723808	.2106192

To check the viability of the above hurdle model, we model the two parts separately. First, we must create a new binary response variable for the logit model. We generate a variable called yp0 that takes the value of 1 if yp is greater than zero and 0 if yp equals zero. The results are

```
. logit yzero x1 x2, nolog
Logistic regression                          Number of obs    =      50,000
                                             LR chi2(2)       =     7976.72
                                             Prob > chi2      =      0.0000
Log likelihood = -30524.305                  Pseudo R2        =      0.1156
```

yzero	Coef.	Std. Err.	z	P>|z|	[95% Conf. Interval]	
x1	.9015217	.0113286	79.58	0.000	.8793182	.9237253
x2	.0872327	.0097722	8.93	0.000	.0680794	.1063859
_cons	.1914999	.0097549	19.63	0.000	.1723807	.2106191

The independent logit model is identical to the hurdle model logit. We next model the positive outcomes in the data by using a zero-truncated Poisson. This time we illustrate Stata's **tpoisson** command instead of the **trncregress** command highlighted in the previous section.

```
. tpoisson yz x1 x2 if yz>0, nolog
Truncated Poisson regression                 Number of obs    =      23,099
Limits:  lower =            0                 LR chi2(2)       =   770694.57
         upper =         +inf                 Prob > chi2      =      0.0000
Log likelihood = -50112.926                  Pseudo R2        =      0.8849
```

yz	Coef.	Std. Err.	z	P>|z|	[95% Conf. Interval]	
x1	.7506657	.0017678	424.62	0.000	.7472008	.7541306
x2	-1.250193	.0016833	-742.70	0.000	-1.253492	-1.246894
_cons	2.002522	.0028048	713.96	0.000	1.997024	2.008019

You may obtain predicted values for each equation in the hurdle model by specifying (you must run these commands immediately after fitting the hurdle model)

```
. predict xb1 if yz>0, eq(#1)        /* zero-truncated Poisson prediction */
(26,901 missing values generated)
. predict xb2, eq(#2)               /* logit prediction                 */
. generate mu1 = exp(xb1)           /* predicted ZTP count              */
(26,901 missing values generated)
. generate mu2 = exp(xb2)/(1+exp(xb2))  /* predicted Pr(yz=0)           */
```

To summarize the model fit, we can compare the predicted probabilities of zero for yzero, and we can compare the predicted counts with the observed counts for the nonzero outcomes in yz. As we would expect, the predicted probabilities for a zero outcome are higher for the zero outcomes, and the averages of the predicted and observed counts are very close.

```
. label define zlab 0 "yz != zero" 1 "yz = zero"
. label values yzero zlab
```

```
. table yzero, c(freq mean mu2)          // Show mean predicted probabilities
```

yzero	Freq.	mean(mu2)
yz != zero	23,099	.4576602
yz = zero	26,901	.6070223

```
. summarize mu1 yz if yz > 0     // Show mean predicted and observed counts
```

Variable	Obs	Mean	Std. Dev.	Min	Max
mu1	23,099	16.20896	41.98424	.0081356	1905.152
yz	23,099	16.38235	42.08849	1	1875

In the preceding approach to fitting the hurdle model, we illustrated how to specify the log likelihood and use Stata's `ml` commands. For completeness, here we illustrate the output of the community-contributed `hplogit` command (Hilbe 2005). This command can be installed by typing `ssc install hplogit`.

```
. hplogit yz x1 x2, nolog
Poisson-Logit Hurdle Regression                Number of obs   =      50,000
                                               Wald chi2(2)    =     6369.55
Log likelihood = -80637.231                    Prob > chi2     =      0.0000
```

	Coef.	Std. Err.	z	P>|z|	[95% Conf. Interval]
logit					
x1	.9015221	.0113286	79.58	0.000	.8793185 .9237256
x2	.0872327	.0097722	8.93	0.000	.0680795 .106386
_cons	.1915	.0097549	19.63	0.000	.1723808 .2106192
poisson					
x1	.7506657	.0017678	424.62	0.000	.7472008 .7541306
x2	-1.250193	.0016833	-742.70	0.000	-1.253492 -1.246894
_cons	2.002522	.0028048	713.96	0.000	1.997024 2.008019

```
AIC Statistic =      3.226
```

As illustrated in the previous hurdle model, we again highlight the generation of data in another example. This time, the two processes making up the hurdle model each have their own specific covariate list.

```
. clear
. set obs 50000
number of observations (_N) was 0, now 50,000
. set seed 1234
. generate x1 = rnormal()        /* x1 ~ N(0,1)   */
. generate x2 = rnormal()        /* x2 ~ N(0,1)   */
. generate z1 = runiform() < .2 /* z1 ~ Ber(0.20) */
. generate z2 = runiform() < .4 /* z2 ~ Ber(0.40) */
. generate yp = 0
. count if yp==0
  50,000
```

```
. while r(N) {
  2.          quietly {
  3.                  replace yp = rpoisson(exp(2.0+0.75*x1-1.25*x2)) if yp==0
  4.                  count if yp==0
  5.          }
  6. }                                           /* yp ~ ZTP */
. generate p  = 1/(1+exp((-1.0+0.9*z1+0.1*z2)))  /* p=P(B=0) for Bernoulli */
. generate yb = runiform() < p                   /* yb ~ Bernoulli(p) */
. generate yh = yp*yb                            /* yh ~ Poisson-logit hurdle */
```

Fitting the model, we can see that the design specifications are represented in the estimates.

```
. generate yhzero = yh==0
. ml model lf HurdPoissLogitLL (yh:yh=x1 x2) (logit:yhzero=z1 z2)
. ml maximize, search(on) nolog title(Poisson Logistic Hurdle Regression)
Poisson Logistic Hurdle Regression          Number of obs    =      50,000
                                            Wald chi2(2)     = 1567916.87
Log likelihood = -109095.01                 Prob > chi2      =      0.0000
```

		Coef.	Std. Err.	z	P>|z|	[95% Conf.	Interval]
yh							
	x1	.7522847	.0011775	638.88	0.000	.7499769	.7545926
	x2	-1.25043	.0011753	-1063.95	0.000	-1.252733	-1.248126
	_cons	1.997747	.0020967	952.82	0.000	1.993638	2.001856
logit							
	z1	.9351966	.0228822	40.87	0.000	.8903484	.9800448
	z2	.1074629	.0198617	5.41	0.000	.0685347	.1463912
	_cons	-1.008222	.013845	-72.82	0.000	-1.035358	-.9810864

14.6 Negative binomial(P) models

Over the past thirty years, researchers have developed several models that have each been termed a generalized negative binomial model. Many of these models were later discovered to produce biased estimates.

Three viable types of generalizations to the negative binomial model have recently been developed where each is a three-parameter distribution based in some way on the negative binomial distribution. Australian Malcolm Faddy has worked on such a model over the past twenty years with a sophisticated working model emerging in 2010 with co-developer Michael Smith.

Another approach developed from the transportation and insurance industry. The model is generally referred to as generalized Waring regression, or the Waring generalized negative binomial model. Fostered by Irwin (1968) and twenty years later by Xekalaki (1983), the model has recently been resurrected in Rodríguez-Avi et al. (2009), who posted an R package called GWRM to CRAN. Additional discussion of this model can be found in Hilbe (2011).

The third approach and the one we discuss in more detail is the one in Greene (2008). The author based his model on the prior work of Winkelmann and Zimmermann (1995) and Cameron and Trivedi (2013), who advanced what they termed a generalized event count [GEC(k)] model. Essentially, Greene's model, called NB-P, was created as a generalization to both the NB-1 and NB-2 negative binomial models. He specifically developed the model as a more flexible version of the negative binomial model such that the variance function has a parameterized power. This generalized the specific powers for the NB-1 and NB-2 models.

$$P(Y = y_i | \mathbf{x}_i) = \frac{\Gamma\left(\theta\lambda_i^{2-P} + y_i\right)}{\Gamma\left(y_i + 1\right)\Gamma\left(\theta\lambda_i^{2-P}\right)} \left(\frac{\lambda_i}{\lambda_i + \theta\lambda_i^{2-P}}\right)^{\theta\lambda_i^{2-P}} \left(\frac{\theta\lambda_i^{2-P}}{\lambda_i + \theta\lambda_i^{2-P}}\right)^{y_i} \quad (14.31)$$

To fit the model, we again rely on the utility of Stata's `ml` commands. Further, we synthesize data to validate our specifications. We generate a large dataset with two outcomes: **ynb1**, which follows a NB-1 distribution, and **ynb2**, which follows a NB-2 distribution.

```
. clear
. set seed 2942
. set obs   50000
number of observations (_N) was 0, now 50,000
. generate x1 = rnormal()
. generate x2 = rnormal()
. generate expxb1 = exp(1.00 - 1.00*x1 + 1.25*x2)
. generate ap1     = 0.33*expxb1
. generate expxb2 = exp(1.00 + 1.00*x1 + 1.00*x2)
. generate ap2     = 0.25
. generate ynb1 = rpoisson(rgamma(ap1, 1/ap1) * expxb1) // ynb1 ~ NB(1)
(2,714 missing values generated)
. generate ynb2 = rpoisson(rgamma(1/ap2, ap2) * expxb2) // ynb2 ~ NB(2)
```

Next, we specify the program to calculate the log likelihood:

```
. program NegbinpLL
  1.      args lnf xb P lntheta
  2.      quietly {
  3.          tempvar g r lambda theta
  4.          generate double `theta´ = exp(`lntheta´)
  5.          generate double `lambda´ = exp(`xb´)
  6.          generate double `g´ = `theta´*`lambda´^(2-`P´)
  7.          generate double `r´ = `g´/(`g´+`lambda´)
  8.          replace `lnf´ = lngamma($ML_y1+`g´) - lngamma(`g´)
>               - lngamma($ML_y1+1) + `g´*ln(`r´) + $ML_y1*ln(1-`r´)
  9.      }
 10. end
```

Finally, we illustrate using Stata's `ml` commands to obtain and use starting values to fit the model for `ynb1`:

```
. quietly nbreg ynb1 x1 x2, disp(cons)  // Run NB(1) model
. matrix b = e(b)                       // Get coefficient vector: beta,lntheta
. local nc = colsof(b)
. local ncp1 = `nc´+1                    // Start vals NB(P=1): beta,1,1/lntheta
. local lntheta = 1/b[1,`nc´]
. matrix b[1,`nc´] = 1                   // Insert 1 for power
. matrix b = b,`lntheta´                 // Add lntheta to end of vector
. ml model lf NegbinpLL (ynb1:ynb1 = x1 x2) (P:) (lntheta:)
. ml init b, copy
. ml maximize, search(off) nolog title(Negative Binomial(p) Regression)
> diparm(lntheta, exp label("theta"))
```

Negative Binomial(p) Regression				Number of obs	=	47,286
				Wald chi2(2)	=	291897.72
Log likelihood = -107265.23				Prob > chi2	=	0.0000

ynb1	Coef.	Std. Err.	z	P>|z|	[95% Conf. Interval]	
ynb1						
x1	-.9980349	.0028407	-351.33	0.000	-1.003603	-.9924672
x2	1.245551	.002889	431.14	0.000	1.239889	1.251213
_cons	1.008698	.0054453	185.24	0.000	.9980257	1.019371
P						
_cons	1.024039	.0078517	130.42	0.000	1.00865	1.039428
lntheta						
_cons	-1.059144	.0175932	-60.20	0.000	-1.093626	-1.024662
theta	.3467526	.0061005			.3349996	.3589179

Now, we illustrate how we can obtain and use starting values to fit the model for `ynb2`:

```
. quietly nbreg ynb2 x1 x2        // Run NB(2) model
. matrix b = e(b)                 // Get coefficient vector: beta,lntheta
. local nc = colsof(b)
. local ncp1 = `nc'+1             // Start vals NB(P=2): beta,2,lntheta
. local lntheta = b[1,`nc']
. matrix b[1,`nc'] = 2            // Insert 2 for power
. matrix b = b,`lntheta'          // Add lntheta to end of vector
. ml model lf NegbinpLL (ynb2:ynb2 = x1 x2) (P:) (lntheta:)
. ml init b, copy
. ml maximize, search(off) nolog title(Negative Binomial(p) Regression)
> diparm(lntheta, exp label("theta"))
```

```
Negative Binomial(p) Regression          Number of obs    =     50,000
                                          Wald chi2(2)     = 124259.58
Log likelihood = -108125.38               Prob > chi2      =     0.0000
```

ynb2	Coef.	Std. Err.	z	P>|z|	[95% Conf. Interval]
ynb2					
x1	.9979473	.0037405	266.79	0.000	.990616	1.005279
x2	.9977913	.0037162	268.50	0.000	.9905076	1.005075
_cons	1.002792	.004173	240.31	0.000	.9946134	1.010971
P					
_cons	1.986373	.0118249	167.98	0.000	1.963197	2.00955
lntheta					
_cons	1.338105	.0305591	43.79	0.000	1.27821	1.397999
theta	3.811812	.1164855			3.590207	4.047095

Next, we model German health data from 1984 through 1988. The outcome variable `docvis` contains the number of doctor visits for each patient; `outwork` is an indicator that the person is unemployed, `age` is the age in years, and `female` is an indicator of female. We will also use the indicators of the two highest levels of education `edlevel3` and `edlevel4`. We quietly fit both the NB-1 and NB-2 models so that we may subsequently compare the estimated NB-P results. We use the fitted coefficients of the NB-2 model for starting values.

```
. use http://www.stata-press.com/data/hh4/rwm5panel, clear
(German health data for 1984-1988: ID=5 obs each panel)
. quietly nbreg docvis outwork age female edlevel3 edlevel4,
> dispersion(constant)
. estimates store A1
. quietly nbreg docvis outwork age female edlevel3 edlevel4
. estimates store A2
. matrix b = e(b)
. local nc = colsof(b)
. local ncp1 = `nc'+1
```

```
. local lntheta = 1/b[1,`nc´]
. matrix b[1,`nc´] = 2
. matrix b = b,`lntheta´
. ml model lf NegbinpLL (docvis:docvis = outwork age female edlevel3 edlevel4)
> (P:) (lntheta:)
. ml init b, copy
. ml maximize, search(off) nolog title(Negative Binomial(p) Regression)
> diparm(lntheta, exp label("theta")) irr
```

```
Negative Binomial(p) Regression          Number of obs     =       8,000
                                          Wald chi2(5)      =      370.83
Log likelihood = -17423.768               Prob > chi2       =      0.0000
```

docvis	IRR	Std. Err.	z	P>|z|	[95% Conf. Interval]	
docvis						
outwork	1.252267	.0459953	6.12	0.000	1.165286	1.34574
age	1.019847	.0017304	11.58	0.000	1.016461	1.023244
female	1.275654	.044214	7.02	0.000	1.191874	1.365323
edlevel3	.7250289	.0522886	-4.46	0.000	.6294587	.8351094
edlevel4	.5740902	.0541758	-5.88	0.000	.4771488	.690727
_cons	1.079703	.0879133	0.94	0.346	.9204425	1.266521
P						
_cons	1.300363	.0792371	16.41	0.000	1.145061	1.455665
lntheta						
_cons	-1.494214	.0953078	-15.68	0.000	-1.681014	-1.307415
theta	.2244248	.0213894			.186185	.2705185

```
Note: Estimates are transformed only in the first equation.
Note: _cons estimates baseline incidence rate.
. estimates store B
```

For comparison, we show the coefficient table for the NB-1 model and the results of a likelihood-ratio test of the two models. This test assesses whether there is significant evidence to support rejecting $\alpha = 1$.

```
. estimates restore A1
(results A1 are active now)

. ml display, irr noheader
```

docvis	IRR	Std. Err.	z	P>|z|	[95% Conf. Interval]	
outwork	1.222652	.0400691	6.13	0.000	1.146587	1.303763
age	1.017032	.0014133	12.15	0.000	1.014266	1.019806
female	1.259288	.039425	7.36	0.000	1.184339	1.338979
edlevel3	.7700581	.0479431	-4.20	0.000	.6815982	.8699986
edlevel4	.6460549	.049588	-5.69	0.000	.5558216	.7509369
_cons	1.244645	.0831925	3.27	0.001	1.091819	1.418861
/lndelta	1.848	.027307			1.794479	1.901521
delta	6.347114	.1733206			6.016342	6.696071

```
Note: Estimates are transformed only in the first equation.
Note: _cons estimates baseline incidence rate.

. lrtest A1 B, force
```

Likelihood-ratio test	LR chi2(1) =	11.07
(Assumption: A1 nested in B)	Prob > chi2 =	0.0009

To complete the analysis, we also compare results to the NB-2 model. Again, there is evidence that the NB-P model is preferred.

```
. estimates restore A2
(results A2 are active now)

. ml display, irr noheader
```

docvis	IRR	Std. Err.	z	P>|z|	[95% Conf. Interval]	
outwork	1.241635	.0506979	5.30	0.000	1.146141	1.345085
age	1.021475	.0017674	12.28	0.000	1.018017	1.024945
female	1.241917	.0467307	5.76	0.000	1.153622	1.33697
edlevel3	.6858095	.0505458	-5.12	0.000	.5935645	.7923902
edlevel4	.4852264	.0424805	-8.26	0.000	.4087179	.5760567
_cons	1.027431	.0836272	0.33	0.740	.8759302	1.205135
/lnalpha	.6876398	.0214354			.6456271	.7296524
alpha	1.989015	.0426354			1.907183	2.074359

```
Note: Estimates are transformed only in the first equation.
Note: _cons estimates baseline incidence rate.

. lrtest A2 B, force
```

Likelihood-ratio test	LR chi2(1) =	114.64
(Assumption: A2 nested in B)	Prob > chi2 =	0.0000

Of course, one need not fit all the models if this is not desirable. The evaluation of whether the NB-P model is preferred over the more standard versions of negative binomial regression may be carried out with Wald tests of the parameter.

```
. estimates restore B
(results B are active now)
. test [P]_cons=1
 ( 1)  [P]_cons = 1
           chi2( 1) =    14.37
         Prob > chi2 =    0.0002
. test [P]_cons=2
 ( 1)  [P]_cons = 2
           chi2( 1) =    77.96
         Prob > chi2 =    0.0000
```

We now illustrate the community-contributed `nbregp` command (Hardin and Hilbe 2014b) for fitting NB-P models. This command offers all the support of those commands described thus far, including alternate variance estimates and weighting. `nbregp` can be installed by typing `net install st0336_1, from(http://www.stata-journal.com/software/sj18-1)`.

```
. use http://www.stata-press.com/data/hh4/rwm5panel, clear
(German health data for 1984-1988: ID=5 obs each panel)
. nbregp docvis outwork age female edlevel3 edlevel4, irr nolog
Negative binomial-P regression                  Number of obs   =       8000
                                                Wald chi2(5)    =     370.83
Log likelihood = -17423.77                      Prob > chi2     =     0.0000
```

docvis	IRR	Std. Err.	z	P>|z|	[95% Conf. Interval]	
outwork	1.252267	.0459953	6.12	0.000	1.165286	1.34574
age	1.019847	.0017304	11.58	0.000	1.016461	1.023244
female	1.275654	.044214	7.02	0.000	1.191874	1.365323
edlevel3	.7250289	.0522886	-4.46	0.000	.6294587	.8351094
edlevel4	.5740902	.0541758	-5.88	0.000	.4771488	.690727
_cons	1.079704	.0879133	0.94	0.346	.9204425	1.266521
/P	1.300363	.0792371	16.41	0.000	1.145061	1.455665
/lntheta	1.494215	.0953078			1.307415	1.681014
theta	4.455835	.4246757			3.696605	5.371001

```
Note: Estimates are transformed only in the first equation.
Note: _cons estimates baseline incidence rate.
Likelihood-ratio test of P=1:    chi2 =     11.07 Prob > chi2    =    0.0009
Likelihood-ratio test of P=2:    chi2 =    114.64 Prob > chi2    =    0.0000
```

This command produces the same results as the `ml` solution already presented. We have presented both solutions to maintain as many `ml` examples in this text as possible.

14.7   Negative binomial(Famoye)

As implemented in the community-contributed `nbregf` command (Harris, Hilbe, and Hardin 2014),[3] the NB-F model assumes that θ is a scalar unknown parameter. Thus, the probability mass function, mean, and variance are given by

$$P(Y = y) = \frac{\theta}{\theta + \phi y} \binom{\theta + \phi y}{y} \mu^y (1 - \mu)^{\theta - y + \phi y} \tag{14.32}$$

where $0 < \mu < 1$, $1 \le \phi < \mu^{-1}$, for $\theta > 0$ and nonnegative outcomes $y_i \in \{0, 1, 2, \ldots\}$.

$$
\begin{align}
E(Y) &= \theta\mu(1 - \phi\mu)^{-1} \tag{14.33} \\
V(Y) &= \theta\mu(1 - \mu)(1 - \phi\mu)^{-3} \tag{14.34}
\end{align}
$$

The variance is equal to that of the negative binomial distribution when $\phi = 1$. Thus, because the $\phi \ge 1$, it generalizes the negative binomial distribution in the NB-F model to have greater variance than allowed for in a negative binomial regression model. To construct a regression model, we use log link, $\log(\mu) = x\beta$, to make results comparable with Poisson and negative binomial models.

```
. use http://www.stata-press.com/data/hh4/rwm1984, clear
(German health data for 1984; Hardin & Hilbe, GLM and Extensions, 4th ed)
. nbregf docvis edlevel4 age hh, eform nolog
Generalized negative binomial-F regression        Number of obs    =      3874
                                                  LR chi2(3)       =    166.51
Log likelihood = -8337.884                        Prob > chi2      =    0.0000
```

docvis	IRR	Std. Err.	z	P>|z|	[95% Conf. Interval]	
edlevel4	.7205452	.0831698	-2.84	0.005	.5746591	.9034669
age	1.025957	.0024634	10.67	0.000	1.02114	1.030796
hhninc	.9252175	.0149634	-4.81	0.000	.8963496	.955015
_cons	2.366462	.3349416	6.09	0.000	1.793177	3.123028
/lnphim1	-3.252403	.4280259			-4.091318	-2.413488
/lntheta	-.6445887	.0760764			-.7936957	-.4954816
phi	1.038681	.0165565			1.016717	1.089503
theta	.5248784	.0399309			.4521706	.6092774

```
Note: Estimates are transformed only in the first equation.
Note: _cons estimates baseline incidence rate.
```

---

3. The `nbregf` command is available by typing `net install st0351_1`, from(http://www.stata-journal.com/software/sj18-1)).

14.8 Negative binomial(Waring)

As illustrated in Irwin (1968), the generalized Waring distribution can be constructed under the following specifications:

1. $Y|x, \lambda_x, v \sim \text{Poisson}(\lambda_x)$

2. $\lambda_x|v \sim \text{Gamma}(a_x, v)$

3. $v \sim \text{Beta}(\rho, k)$

In the author's presentation for accident data, the author specifies $\lambda|v$ as "accident liability" and v to indicate "accident proneness". The probability mass function is ultimately given by

$$P(Y = y) = \frac{\Gamma(a_x + \rho)\Gamma(k + \rho)}{\Gamma(\rho)\Gamma(a_x + k + \rho)} \frac{(a_x)_y (k)_y}{(a_x + k + \rho)_y} \frac{1}{y!}$$

where $k, \rho, a_x > 0$ and $(a)_w$ is the Pochhammer notation for $\Gamma(a + w)/\Gamma(w)$ if $a > 0$. Thus, the pmf could be given just using gamma functions as

$$P(Y = y) = \frac{\Gamma(a_x + \rho)\Gamma(k + \rho)\Gamma(a_x + y)\Gamma(k + y)}{\Gamma(\rho)\Gamma(a_x + k + \rho)\Gamma(y)\Gamma(a_x + k + \rho + y)\Gamma(y + 1)}$$

The expected value and variance of the distribution are

$$\begin{aligned} E(Y) &= \mu \\ V(Y) &= \mu + \mu\left(\frac{k+1}{\rho-2}\right) + \mu^2\left\{\frac{k+\rho-1}{k(\rho-2)}\right\} \end{aligned}$$ (14.35)

where a_x is $\mu(\rho - 1)/k$ (simplifies the expression and ensures that the mean is given by μ), $k > 0$ and $\rho > 2$ (to ensure nonnegative variance). To construct a regression model, we use log link, $\log(\mu) = x\beta$, to make results comparable with Poisson and negative binomial models. A unique characteristic of this model occurs when the data are from a different underlying distribution. For instance, when the data are from a Poisson distribution [with $V(Y) = \mu$] indicates $(k + 1)/(\rho - 2) \to 0$ and $(k + \rho - 1)/\{k(\rho - 2)\} \to 0$, then $k, \rho \to \infty$, and $\rho \to \infty$ faster than does k. Also, if the data have an underlying NB-2 distribution with $V(Y) = \mu + \alpha\mu^2$ (where α is the dispersion parameter), indicates $(k + 1)/(\rho - 2) \to 0$ and $(k + \rho - 1)/\{k(\rho - 2)\} \to \alpha$, where $k \to 1/\alpha$ and $\rho \to \infty$.

We implement this model using the community-contributed `nbregw` command (Harris, Hilbe, and Hardin 2014), which was downloaded with the `nbregf` command in the previous section.

```
. nbregw docvis edlevel4 age hh, eform nolog
Generalized negative binomial-W regression          Number of obs    =      3874
                                                    LR chi2(3)       =    163.80
Log likelihood = -8315.421                          Prob > chi2      =    0.0000
```

docvis	IRR	Std. Err.	z	P>|z|	[95% Conf. Interval]	
edlevel4	.6909839	.0865305	-2.95	0.003	.5405971	.8832062
age	1.027734	.0024924	11.28	0.000	1.02286	1.032631
hhninc	.9271367	.0171783	-4.08	0.000	.8940719	.9614243
_cons	1.14262	.1430973	1.06	0.287	.8939239	1.460505
/lnrhom2	.9047973	.1993192			.5141388	1.295456
/lnk	-.6112581	.0522087			-.7135853	-.5089308
rho	4.471431	.4926037			3.672198	5.652661
k	.5426677	.028332			.4898846	.601138

```
Note: Estimates are transformed only in the first equation.
Note: _cons estimates baseline incidence rate.
```

14.9   Heterogeneous negative binomial models

We next consider the heterogeneous negative binomial model. Stata refers to this model as a generalized negative binomial, which is fit using the `gnbreg` command, but generalized negative binomial has long been used in the literature to refer to an entirely different model. The term heterogeneous, however, has been used to refer to this model, and by referring to it as such we follow common usage.

The heterogeneous negative binomial model not only estimates the standard list of parameters and their associated standard errors but also allows parameterization of the ancillary or overdispersion parameter α. The dispersion parameters for the NB-2 model may incorporate a linear predictor so that the dispersion parameter is allowed to vary across observations. The log-likelihood function of the heterogeneous NB-2 model can be derived, allowing for an offset in the covariate list for the mean parameter and for the dispersion parameter, as

$$\alpha_i = \exp(\mathbf{z}_i\boldsymbol{\gamma} + \text{offset}_i^\gamma) \tag{14.36}$$

$$m_i = 1/\alpha_i \qquad p_i = 1/(1+\alpha_i\mu_i) \qquad \mu_i = \exp(\mathbf{x}_i\boldsymbol{\beta} + \text{offset}_i^\beta) \tag{14.37}$$

$$\mathcal{L} = \sum_{i=1}^{n}\left[\ln\{\Gamma(m_i+y_i)\} - \ln\{\Gamma(y_i+1)\} - \ln\{\Gamma(m_i)\}\right.$$

$$\left. + m_i\ln(p_i) + y_i\ln(1-p_i)\right] \tag{14.38}$$

For an example of this model, we return to the previously used `medpar.dta`. We fit a negative binomial (NB-2) before using `gnbreg` for comparison.

```
. use http://www.stata-press.com/data/hh4/medpar, clear
. nbreg los hmo white type2 type3 died, nolog
Negative binomial regression                    Number of obs   =      1,495
                                                LR chi2(5)      =     151.08
Dispersion      = mean                          Prob > chi2     =     0.0000
Log likelihood = -4780.9543                     Pseudo R2       =     0.0156
```

los	Coef.	Std. Err.	z	P>|z|	[95% Conf. Interval]	
hmo	-.065878	.0527626	-1.25	0.212	-.1692908	.0375348
white	-.1218214	.0678367	-1.80	0.073	-.2547789	.011136
type2	.2418841	.0502247	4.82	0.000	.1434454	.3403227
type3	.7286303	.0754246	9.66	0.000	.5808008	.8764597
died	-.2355563	.040457	-5.82	0.000	-.3148506	-.1562619
_cons	2.372991	.0682086	34.79	0.000	2.239305	2.506678
/lnalpha	-.8347201	.0448234			-.9225724	-.7468679
alpha	.4339959	.0194532			.3974952	.4738484

```
LR test of alpha=0: chibar2(01) = 4108.03         Prob >= chibar2 = 0.000
```

In this NB-2 model, the dispersion parameter is a scalar. We now fit a model using `gnbreg` for which the dispersion is allowed to vary across levels of `type` and `died`.

```
. gnbreg los hmo white type2 type3 died, lnalpha(type2 type3 died) nolog
Generalized negative binomial regression        Number of obs   =      1,495
                                                LR chi2(5)      =     107.03
                                                Prob > chi2     =     0.0000
Log likelihood = -4731.6185                     Pseudo R2       =     0.0112
```

los	Coef.	Std. Err.	z	P>|z|	[95% Conf. Interval]	
los						
hmo	-.0630361	.0492268	-1.28	0.200	-.1595188	.0334466
white	-.0756874	.0642429	-1.18	0.239	-.2016013	.0502264
type2	.2349034	.049132	4.78	0.000	.1386065	.3312003
type3	.7701078	.0971028	7.93	0.000	.5797897	.9604258
died	-.2283633	.0444074	-5.14	0.000	-.3154003	-.1413263
_cons	2.327675	.063542	36.63	0.000	2.203135	2.452215
lnalpha						
type2	.1660303	.1165805	1.42	0.154	-.0624631	.3945238
type3	.74406	.1517917	4.90	0.000	.4465538	1.041566
died	.77198	.0932257	8.28	0.000	.5892611	.9546989
_cons	-1.258687	.0657166	-19.15	0.000	-1.387489	-1.129885

The AIC statistic is lower for the heterogeneous model than for the basic NB-2. We also find that `type3`—that is, emergency admissions—and `died` significantly contribute to the dispersion α. That is, emergency admissions and the death of a patient contribute to the overdispersion in the data. Displaying only the header output from the `glm` command negative binomial output, with the ancillary or heterogeneity parameter given as the constant of the full-maximum likelihood model, we see

```
. glm los hmo white type2 type3 died, family(nbinomial .4339959) notable nolog
Generalized linear models                    No. of obs     =       1,495
Optimization     : ML                        Residual df    =       1,489
                                             Scale parameter =          1
Deviance        =  1566.760836               (1/df) Deviance =   1.052224
Pearson         =  1687.716966               (1/df) Pearson  =   1.133457

Variance function: V(u) = u+(.434)u^2        [Neg. Binomial]
Link function    : g(u) = ln(u)              [Log]

                                             AIC            =    6.403952
Log likelihood   = -4780.954342              BIC            =  -9317.653
```

This output shows that there is some negative binomial overdispersion in the data, 13%. This overdispersion is derived largely from patients who were admitted with more serious conditions, including those who died. Other patients do not seem to contribute to the extra variation in the data. This information is considerably superior to simply knowing that the Poisson model of the same data is overdispersed (as seen in the header output of a Poisson model where you should especially note the Pearson dispersion statistic).

```
. glm los hmo white type2 type3 died, family(poisson) notable nolog
Generalized linear models                    No. of obs     =       1,495
Optimization     : ML                        Residual df    =       1,489
                                             Scale parameter =          1
Deviance        =  7954.791152               (1/df) Deviance =   5.342371
Pearson         =  9251.713475               (1/df) Pearson  =   6.213374

Variance function: V(u) = u                  [Poisson]
Link function    : g(u) = ln(u)              [Log]

                                             AIC            =      9.1518
Log likelihood   = -6834.970362              BIC            =  -2929.622
```

We will show another method of interpreting the ancillary or heterogeneity parameter. We use Stata's gnbreg command to fit a heterogeneous (or Stata's generalized) negative binomial regression model on our data, but this time allowing only the scale parameter α to depend on whether the patient died.

```
. gnbreg los hmo white type2 type3, lnalpha(died) irr nolog
```

```
Generalized negative binomial regression          Number of obs   =      1,495
                                                  LR chi2(4)      =     131.39
                                                  Prob > chi2     =     0.0000
Log likelihood = -4757.8933                       Pseudo R2       =     0.0136
```

los	IRR	Std. Err.	z	P>|z|	[95% Conf. Interval]	
hmo	.9350806	.0479923	-1.31	0.191	.8455937	1.034038
white	.9107966	.0592616	-1.44	0.151	.8017471	1.034678
type2	1.245232	.0612273	4.46	0.000	1.13083	1.371208
type3	2.13208	.1611669	10.02	0.000	1.838485	2.472561
_cons	10.0167	.6431153	35.89	0.000	8.832306	11.35993
lnalpha	(type gnbreg to see ln(alpha) coefficient estimates)					

Note: _cons estimates baseline incidence rate.

The outcome shows similar incidence-rate ratios to what we obtained with previous models. To see the coefficients in the parameterization of α, we type

```
. gnbreg
```

```
Generalized negative binomial regression          Number of obs   =      1,495
                                                  LR chi2(4)      =     131.39
                                                  Prob > chi2     =     0.0000
Log likelihood = -4757.8933                       Pseudo R2       =     0.0136
```

los	Coef.	Std. Err.	z	P>|z|	[95% Conf. Interval]	
los						
hmo	-.0671226	.0513243	-1.31	0.191	-.1677163	.0334712
white	-.0934357	.0650656	-1.44	0.151	-.2209621	.0340906
type2	.219322	.0491694	4.46	0.000	.1229518	.3156923
type3	.7570981	.0755914	10.02	0.000	.6089417	.9052545
_cons	2.304254	.0642043	35.89	0.000	2.178416	2.430092
lnalpha						
died	.8171296	.0917266	8.91	0.000	.6373487	.9969105
_cons	-1.135773	.0575386	-19.74	0.000	-1.248547	-1.023

From the output, we can determine that

$$\alpha_{\text{died}} = \exp\{-1.135773 + 1.0(0.8171296)\} \qquad (14.39)$$
$$= \exp(-0.3186434) \qquad (14.40)$$
$$= 0.7271348 \qquad (14.41)$$
$$\alpha_{\text{survived}} = \exp\{-1.135773 + 0.0(0.8171296)\} \qquad (14.42)$$
$$= \exp(-1.135773) \qquad (14.43)$$
$$= 0.32117376 \qquad (14.44)$$

compared with $\alpha = 0.4457567$ when we do not account for the survival of the patient. The coefficient of the died variable is significant, and we conclude that this is a better model for the data than the previous models. There is clearly even greater overdispersion for the subset of data in which patients died, and accounting for that variation results in a better model.

The next model we will consider in this chapter is the generalized Poisson. This model is similar to the negative binomial model but is not based on the negative binomial PDF. Moreover, as will be discussed, one can use it to model both overdispersion and underdispersion.

14.10   Generalized Poisson regression models

The most commonly used regression model for count data is the Poisson regression model with a log-link function described by the probability mass function:

$$f(y_i; \theta_i, \mathbf{x}_i) = \frac{\theta_i^{y_i} e^{-\theta_i}}{y_i!} \quad y_i = 0, 1, 2, \dots \quad \theta_i > 0 \tag{14.45}$$

Covariates and regression parameters are then introduced with the expected outcome by the inverse link function $\theta_i = \exp(\mathbf{x}_i \boldsymbol{\beta})$.

We see that $\mathbf{x}_i$ is a covariate vector and $\boldsymbol{\beta}$ is a vector of unknown coefficients to be estimated. When there is overdispersion due to latent heterogeneity, people often assume a gamma mixture of Poisson variables. Suppose that ϑ_i is an unobserved individual heterogeneity factor, with $\exp(\vartheta_i)$ following a gamma distribution with mean 1 and variance κ. Now assume the response vector follows a modification of the Poisson model with mean $\theta_i^\star = \exp(\mathbf{x}_i \boldsymbol{\beta} + \vartheta_i)$. The result is the well-known negative binomial regression model with mass function

$$f(y_i; \theta_i, \mathbf{x}_i, \kappa) = \frac{\Gamma(y_i + 1/\kappa)}{\Gamma(y_i + 1)\Gamma(1/\kappa)} \frac{(\kappa\theta_i)^{y_i}}{(1 + \kappa\theta_i)^{y_i + 1/\kappa}} \tag{14.46}$$

where κ is a nonnegative parameter, indicating the degree of overdispersion. The negative binomial model reduces to a Poisson model in the limit as $\kappa \to 0$. We emphasize this material throughout chapter 13.

As an alternative to the negative binomial, we may assume that the response variable follows a generalized Poisson (GP) distribution with mass function given by

$$f(y_i; \theta_i, \delta) = \frac{\theta_i(\theta_i + \delta y_i)^{y_i - 1} e^{-\theta_i - \delta y_i}}{y_i!} \tag{14.47}$$

for $y_i = 0, 1, 2, \dots$, $\theta_i > 0$, $0 \le \delta < 1$. Following Joe and Zhu (2005), we see that

$$E(Y_i) \;=\; \mu_i = \frac{\theta_i}{1 - \delta} \tag{14.48}$$

$$V(Y_i) \;=\; \frac{\theta_i}{(1 - \delta)^3} = \frac{1}{(1 - \delta)^2} E(Y_i) = \phi E(Y_i) \tag{14.49}$$

More information on the GP regression model is also presented in Consul and Famoye (1992).

As presented in Yang et al. (2007), the term $\phi = 1/(1 - \delta)^2$ plays the role of a dispersion factor in the GP mass function. Clearly, when $\delta = 0$, the GP distribution reduces to the usual Poisson distribution with parameter θ_i. Further, when $\delta < 0$ underdispersion is assumed in the model, and if $\delta > 0$, we get overdispersion. In this discussion, we are concerned with overdispersion, $\delta > 0$.

In chapter 13, we investigated the negative binomial model as an alternative to the Poisson for the length of stay data. Here we consider the GP as an alternative.

```
. gpoisson los hmo white type2 type3, irr nolog
Generalized Poisson regression                    Number of obs   =        1495
                                                  LR chi2(4)      =       36.18
Dispersion     = .5956076                         Prob > chi2     =      0.0000
Log likelihood = -4816.9845                       Pseudo R2       =      0.0037
```

los	IRR	Std. Err.	z	P>|z|	[95% Conf. Interval]	
hmo	.9578067	.0470583	-0.88	0.380	.8698758	1.054626
white	.8838994	.0533205	-2.05	0.041	.7853348	.9948345
type2	1.16412	.0528339	3.35	0.001	1.065039	1.272417
type3	1.411021	.09608	5.06	0.000	1.234732	1.612478
_cons	10.51186	.6322697	39.11	0.000	9.342892	11.82709
/atanhdelta	.686312	.0151052			.6567064	.7159177
delta	.5956076	.0097466			.5761674	.6143742

```
Note: Estimates are transformed only in the first equation.
Likelihood-ratio test of delta=0:  chi2(1) = 4223.85      Prob>=chi2 = 0.0000
```

Again, we find that there is evidence of overdispersion such that the Poisson model is not flexible enough for these data. Now we are faced with choosing the GP or the negative binomial.

We investigated a Poisson model in chapter 12 for the number of deaths as a function of whether patients were smokers and their age category.

```
. use http://www.stata-press.com/data/hh4/doll, clear

. gpoisson deaths smokes a2-a5, exposure(pyears) irr nolog
```

Generalized Poisson regression   Number of obs   =   10
                                                   LR chi2(5)   =   53.13
Dispersion   = .0789024   Prob > chi2   =   0.0000
Log likelihood = -33.53767   Pseudo R2   =   0.4420

deaths	IRR	Std. Err.	z	P>|z|	[95% Conf. Interval]	
smokes	1.430444	.1675984	3.06	0.002	1.136945	1.799708
a2	4.438658	.9479536	6.98	0.000	2.920537	6.745914
a3	13.9343	2.807428	13.08	0.000	9.388334	20.6815
a4	28.71523	5.819276	16.57	0.000	19.30244	42.71813
a5	40.64951	8.548174	17.62	0.000	26.91884	61.38386
_cons	.0003604	.0000756	-37.80	0.000	.0002389	.0005436
ln(pyears)	1	(exposure)				
/atanhdelta	.0790667	.221058			-.3541991	.5123325
delta	.0789024	.2196818			-.3400942	.4717606

Note: Estimates are transformed only in the first equation.
Likelihood-ratio test of delta=0: chi2(1) =   0.12   Prob>=chi2 = 0.3619

The likelihood-ratio test of the hypothesis that the dispersion parameter is zero is not rejected, indicating a preference for the more constrained Poisson model over the more general GP model.

Here we discuss an example again using the Arizona diagnostic dataset (azdrg112.dta). The length of stay (los measured in number of days) is the response variable of interest, and here we model the outcome based on associations with sex (male), being at least 75 years old (age75), and whether the hospital admission was an emergency or urgent admission (emergency).

```
. use http://www.stata-press.com/data/hh4/azdrg112, clear
(AZ 1991: DRG112, no deaths)

. gpoisson los male age75 emergency, nolog
```

Generalized Poisson regression   Number of obs   =   1798
                                                   LR chi2(3)   =   334.75
Dispersion   = .2811441   Prob > chi2   =   0.0000
Log likelihood = -4287.833   Pseudo R2   =   0.0376

los	Coef.	Std. Err.	z	P>|z|	[95% Conf. Interval]	
male	-.1248703	.0296043	-4.22	0.000	-.1828936	-.066847
age75	.1274631	.0313672	4.06	0.000	.0659844	.1889417
emergency	.5695026	.0338531	16.82	0.000	.5031518	.6358535
_cons	1.214749	.0367966	33.01	0.000	1.142629	1.286869
/atanhdelta	.2889239	.0134865			.2624909	.3153569
delta	.2811441	.0124205			.2566239	.3053026

Likelihood-ratio test of delta=0: chi2(1) = 594.88   Prob>=chi2 = 0.0000

```
. estat ic
Akaike´s information criterion and Bayesian information criterion
```

Model	Obs	ll(null)	ll(model)	df	AIC	BIC
.	1,798	-4455.209	-4287.833	5	8585.666	8613.138

Note: N=Obs used in calculating BIC; see [R] BIC note.

For comparison, we can list out the AIC and BIC values for the Poisson model.

```
. quietly poisson los male age75 emergency
. estat ic
Akaike´s information criterion and Bayesian information criterion
```

Model	Obs	ll(null)	ll(model)	df	AIC	BIC
.	1,798	-4975.852	-4585.273	4	9178.545	9200.523

Note: N=Obs used in calculating BIC; see [R] BIC note.

We used the GP command, `gpoisson`, to model underdispersed count data in section 22.2.6. That section illustrated how to generate such data and illustrated one of the foremost uses of this modeling command.

14.11   Poisson inverse Gaussian models

The Poisson inverse Gaussian (PIG) model is derived similar to the techniques and assumptions underlying the negative binomial (NB-2) model. That is, a multiplicative random effect with unit mean following the inverse Gaussian (rather than the gamma) distribution is introduced to the Poisson distribution. The PIG model is recognized as a two-parameter version of the Sichel distribution.

The log likelihood of the PIG regression model involves the Bessel function of the second kind. This function is not included in most statistics packages, so the regression model is not typically supported. However, we have developed a command called `pigreg` for this reason.[4] Without access to a built-in function for the Bessel function of the second kind, the computation of the log likelihood and its derivatives is carried out with an iterative algorithm. This can be computationally burdensome, and so the command can take much longer to complete than it takes to complete alternative model commands. The PIG model is most applicable for outcomes that are characterized by distributions with an initial peak followed by a long tail. Using the diagnostic-related group data, we illustrate the estimation of the PIG model.

---

4. You can download `pigreg` by typing `net install pigreg`, `from(http://www.stata.com/users/jhardin)`.

```
. use http://www.stata-press.com/data/hh4/azdrg112, clear
(AZ 1991: DRG112, no deaths)

. pigreg los male age75 emergency, nolog
```

Poisson-Inverse Gaussian regression

			Number of obs	=	1798
			LR chi2(3)	=	364.93
			Prob > chi2	=	0.0000
Log likelihood = -4249.6504			Pseudo R2	=	0.0412

los	Coef.	Std. Err.	z	P>|z|	[95% Conf. Interval]	
male	-.1405327	.0313484	-4.48	0.000	-.2019745	-.0790909
age75	.1220959	.033536	3.64	0.000	.0563666	.1878252
emergency	.6208627	.0341523	18.18	0.000	.5539254	.6877999
_cons	1.184932	.0367502	32.24	0.000	1.112903	1.256961
/lnalpha	-1.578208	.072001			-1.719327	-1.437088
alpha	.2063446	.014857			.1791867	.2376187

Likelihood-ratio test of alpha=0:   chibar2(01) =   671.24 Prob>=chibar2 = 0.000

```
. estat ic
```

Akaike's information criterion and Bayesian information criterion

Model	Obs	ll(null)	ll(model)	df	AIC	BIC
.	1,798	-4432.118	-4249.65	5	8509.301	8536.773

Note: N=Obs used in calculating BIC; see [R] BIC note.

Comparing the AIC statistics with those obtained for the generalized Poisson model, it appears that the PIG model is preferred. To view the IRRs, we redisplay the results

```
. pigreg, irr
```

Poisson-Inverse Gaussian regression

			Number of obs	=	1798
			LR chi2(3)	=	364.93
			Prob > chi2	=	0.0000
Log likelihood = -4249.6504			Pseudo R2	=	0.0412

los	IRR	Std. Err.	z	P>|z|	[95% Conf. Interval]	
male	.8688952	.0272385	-4.48	0.000	.8171157	.9239559
age75	1.129862	.037891	3.64	0.000	1.057985	1.206623
emergency	1.860532	.0635414	18.18	0.000	1.74007	1.989334
_cons	3.270464	.1201901	32.24	0.000	3.043179	3.514723
/lnalpha	-1.578208	.072001			-1.719327	-1.437088
alpha	.2063446	.014857			.1791867	.2376187

Note: Estimates are transformed only in the first equation.
Likelihood-ratio test of alpha=0:   chibar2(01) =   671.24 Prob>=chibar2 = 0.000

In either format, the likelihood-ratio test at the bottom of the output assesses whether there is significant preference for the PIG model over the Poisson model. In this case, there is strong evidence with $p < 0.001$.

14.12 Censored count response models

One can amend both the Poisson and negative binomial models to allow censored observations. Censoring occurs when a case or observation is not known to an exact value. In a truncated model, a new distribution is defined without support for certain values or ranges of values. Any observed values not in the support of the truncated distribution are not modeled. With censoring, the observed value is known to either an exact value or only a range of values. In either case, relevant probabilities based on the model's distribution may be calculated.

Censoring, with respect to count response models, is often found in the domains of econometrics and social science. Data are almost always collected as some exact measurement. Censoring describes those occasions where data are inexact. Essentially, censoring defines data in terms of intervals for which the endpoints might be defined for the entire dataset or might be observation specific. Examples of censored counts include knowing an outcome y_i is no larger than some value y_R; that is, only knowing that $y_i \leq y_R$. One might only know that an outcome y_i is at least equal to some cutoff value y_L; that is, only knowing that $y_i \geq y_L$. Finally, we might only know that an observation y_i is between two cutoffs; that is, only knowing that $y_L \leq y_i \leq y_R$. Adding observation-specific subscripts i to the cutpoints, we can see that one general specification for possibly censored outcomes is given by y_L and y_R.

Table 14.4. Types of censoring for outcome y_i

$0 \leq y_{Li} = y_{Ri} < \infty$	The outcome is exactly known (uncensored); $y_i = y_{Li} = y_{Ri}$
$0 \leq y_{Li} < \infty; y_{Ri} = \infty$	The outcome is at least y_{Li} (right-censored); $y_i \in \{y_{Li}, y_{Li} + 1, \ldots\}$
$0 = y_{Li} < y_{Ri} < \infty$	The outcome is at most y_{Ri} (left-censored); $y_i \in \{0, \ldots, y_{Ri} - 1, y_{Ri}\}$
$0 < y_{Li} < y_{Ri} < \infty$	The outcome is known to an interval (interval-censored); $y \in \{y_{Li}, y_{Li} + 1, \ldots, y_{Ri} - 1, y_{Ri}\}$

In essence, every observation y_i is known to be in the interval $[y_{Li}, y_{R_i}]$ where the end of the endpoint is open when $Y_{Ri} = \infty$. This description mimics the Stata documentation for the `intreg` command.

There are two basic approaches to communicate the intervals for each observation. One approach is as we have just described. Another would be to have a single observation `y` and two options for the lower limit `ll` and the upper limit `ul`. As we pointed out, the former description applies to the `intreg` command, and the latter description applies to the `cpoisson` command. If we were to create a complete command, we could invoke the `intreg` syntax because we would be able to interpret the missing values of the second outcome variable according to the table. In the following examples, however, we will be using calls to the `ml` command, which will interpret those missing values as incomplete observations and will thus exclude those observations from the estimation.

Thus, we must make a substitute value for $y_{Ri} = \infty$. Fortunately, all the count models are nonnegative, so we use $y_{Ri} = -1$ in the examples below to indicate left-censored observations.

Additional discussion of censoring can be found in Cameron and Trivedi (2013), Greene (2012), and Hilbe (2011), where these presentations use additional terms to describe datasets and the nature of censoring that is present in the observations.

Censoring is an essential attribute of all survival models, with left-censored observations referring to the event of interest happening to a case before entry into the study. Admittedly, left-censoring is relatively rare. Right-censored observations, on the other hand, are those that are lost to the study before it is concluded. Observations can be lost by participants simply leaving the study, by their being disqualified from continuing in the study, or by some other means. Cleves, Gould, and Marchenko (2016) provide an excellent introduction to survival models using Stata; see this source for more discussion of this type of survival modeling.

Given the need to utilize outcome variables that are not missing, we will utilize a syntax such that:

Censor type	y1	y2
Uncensored	y_{Li}	y_{Ri}
Right censored	y_{Li}	-1
Left censored	0	y_{Ri}
Interval censored	y_{Li}	y_{Ri}

The log-likelihood function of a censored Poisson is then given by

$$p_{1i} = \Gamma_I \{y_{Li}, \exp(\mathbf{x}_i\boldsymbol{\beta})\} \tag{14.50}$$

$$p_{2i} = \texttt{cond}(\texttt{y2} == -1, 0, \Gamma_I \{y_{Ri} + 1, \exp(\mathbf{x}_i\boldsymbol{\beta})\}) \tag{14.51}$$

where Γ_I is the incomplete gamma function such that $\Gamma_I(y, \mu) = P(Y \geq y)$ for $Y \sim$ Poisson(μ). Thus, $P(Y \in \{y_{Li}, y_{Li} + 1, \ldots, y_{Ri} - 1, y_{Ri}\}) = p_{1i} - p_{2i}$ such that the log likelihood is then given by

$$\mathcal{L}(\boldsymbol{\beta}; y) = \sum_{i=1}^{n} \ln(p_{1i} - p_{2i}) \tag{14.52}$$

Similarly, the log likelihood for a censored negative binomial is given by

$$p_{1i} = B_I [y_{Li}, \alpha, 1/\{1 + \exp(\mathbf{x}_i\boldsymbol{\beta})/\alpha\}] \tag{14.53}$$

$$p_{2i} = \texttt{cond}(\texttt{y2} == -1, 0, B_I [y_{Ri} + 1, \alpha, 1/\{1 + \exp(\mathbf{x}_i\boldsymbol{\beta})/\alpha\}]) \tag{14.54}$$

$$\mathcal{L}(\boldsymbol{\beta}; y) = \sum_{i=1}^{n} \ln(p_{1i} - p_{2i}) \tag{14.55}$$

where $B_I(\cdot)$ is the three-parameter incomplete beta function such that $B_I(y, \alpha, \mu) = P(Y \geq y)$ for $Y \sim$ negative binomial(μ, α).

Here we illustrate how one can fit these censored models to data. First, we load a modification of medpar.dta used earlier.

```
. use http://www.stata-press.com/data/hh4/medparc, clear
. describe
Contains data from data/medparc.dta
  obs:         1,495
  vars:           12                          11 Dec 2006 10:10
  size:       44,850

              storage   display    value
variable name   type    format     label      variable label

provnum         str6    %9s                    Provider number
died            float   %9.0g                  1=Died; 0=Alive
white           float   %9.0g                  1=White
hmo             byte    %9.0g                  HMO/readmit´
los             int     %9.0g                  Length of Stay
age80           float   %9.0g                  1=age>=80
age             byte    %9.0g                  9 age groups
type1           byte    %8.0g                  Elective Admit
type2           byte    %8.0g                  Urgent Admit
type3           byte    %8.0g                  Emergency Admit
cen1            float   %9.0g                  censor=1:  no censoring
cenx            byte    %8.0g                  censor: -1 lost to study

Sorted by: provnum
```

We will use cenx as the censor variable. Here we illustrate those observations that are right-censored.

```
. tabulate cenx
  censor: -1 |
     lost to |
       study |      Freq.     Percent        Cum.

          -1 |          8        0.54        0.54
           1 |      1,487       99.46      100.00

       Total |      1,495      100.00
```

There are eight observations lost to study. To see which ones they are, we list the length of stay for those observations that are right-censored.

. list los cenx if cenx == -1

	los	cenx
943.	24	-1
970.	24	-1
1066.	24	-1
1127.	24	-1
1450.	24	-1
1452.	116	-1
1466.	91	-1
1481.	74	-1

These eight censored observations all have long lengths of stay. The associated patients exited the study after being in the hospital for (at least) 24 days. We will next tabulate those observations censored after lengths of stay of more than 60 days.

. tabulate los if los > 60

Length of Stay	Freq.	Percent	Cum.
63	1	16.67	16.67
65	1	16.67	33.33
70	1	16.67	50.00
74	1	16.67	66.67
91	1	16.67	83.33
116	1	16.67	100.00
Total	6	100.00	

All patients staying longer than 70 days were lost to the study (censored). We know that those patients stayed for the days indicated but not what happened thereafter. Many outcome registries provide information on patient disposition—that is, whether patients died, were transferred to another facility, or developed symptoms that disqualified them from further participation in the study. We do not have this type of information here. Nevertheless, let us look at the observations from 20 days to 30.

. tabulate los if los > 20 & los < 30

Length of Stay	Freq.	Percent	Cum.
21	18	22.50	22.50
22	15	18.75	41.25
23	10	12.50	53.75
24	11	13.75	67.50
25	4	5.00	72.50
26	7	8.75	81.25
27	7	8.75	90.00
28	5	6.25	96.25
29	3	3.75	100.00
Total	80	100.00	

Five of 11 patients staying 24 days are lost to study. This result is common in real outcomes data. It probably indicates that patients having a given diagnostic-related group are transferred to a long-term facility after 24 days. Patients having other diagnostic-related groups remain.

We can now model the data. We model length of stay (los) by whether the patient is a member of an HMO (hmo), ethnic identity (white = 1; all others = 0), and type of admission (1 = elective, 2 = urgent, 3 = emergency). For this example, all patients are alive throughout the term of the study.

```
. generate cens = cond(cenx==-1,-1,los)
. program CensPoisson
  1.      args lnf xb
  2.      quietly {
  3.        tempvar p1 p2
  4.        generate double `p1´ = cond($ML_y1==0,1,gammap($ML_y1,exp(`xb´)))
  5.        generate double `p2´ = cond($ML_y2==-1,0,gammap($ML_y2+1,exp(`xb´)))
  6.        replace `lnf´ = ln(`p1´-`p2´) if $ML_samp
  7.      }
  8. end
. ml model lf CensPoisson (los:los cens=hmo white type2 type3)
. ml maximize, search(on) nolog title(Censored Poisson Regression) irr
```

```
Censored Poisson Regression                  Number of obs   =     1,495
                                             Wald chi2(4)    =    868.24
Log likelihood = -6925.2494                  Prob > chi2     =    0.0000
```

	IRR	Std. Err.	z	P>|z|	[95% Conf. Interval]	
hmo	.9307211	.0222852	-3.00	0.003	.8880519	.9754405
white	.8575596	.0235092	-5.61	0.000	.8126985	.9048971
type2	1.248249	.0262797	10.53	0.000	1.19779	1.300834
type3	2.035556	.0532321	27.18	0.000	1.933852	2.142609
_cons	10.3082	.280477	85.74	0.000	9.772879	10.87285

```
Note: _cons estimates baseline incidence rate.
. estat ic
Akaike's information criterion and Bayesian information criterion
```

Model	Obs	ll(null)	ll(model)	df	AIC	BIC
.	1,495	.	-6925.249	5	13860.5	13887.05

```
Note: N=Obs used in calculating BIC; see [R] BIC note.
```

All predictors are significant. However, even by including censored measurements in a tabulation of the mean and standard deviation of the los variable for each covariate pattern, there is apparent overdispersion (the standard deviation should be approximately equal to the square root of the associated mean).

```
. egen covpattern = group(hmo white type2 type3)
. table covpattern if e(sample), contents(freq mean los sd los)
```

group(hmo white type2 type3)	Freq.	mean(los)	sd(los)
1	72	9.013889	6.038592
2	10	25.5	29.14809
3	33	14.72727	10.25083
4	857	8.94049	6.615082
5	83	17.6988	19.69693
6	201	10.63682	8.590835
7	8	9.25	4.559135
8	4	9.25	8.655441
9	197	8.269035	5.956792
10	3	9	6.928203
11	27	11.33333	8.138228

Modeling the data by using a censored negative binomial may be more appropriate.

```
. program CensNegBin
  1.      args todo b lnf
  2.      quietly {
  3.          tempvar xb lnalpha r p ll
  4.          mleval `xb´     = `b´, eq(1)
  5.          mleval `lnalpha´ = `b´, eq(2)
  6.          generate double `r´ = exp(-`lnalpha´)
  7.          generate double `p´ = 1/(1+exp(-`xb´-`lnalpha´))
  8.          tempvar p1 p2
  9.          generate double `p1´ = cond($ML_y1==0,1,ibeta($ML_y1,`r´,`p´))
 10.          generate double `p2´ = cond($ML_y2==-1,0,ibeta($ML_y2+1,`r´,`p´))
 11.          replace `lnf´ = ln(`p1´-`p2´)
 12.      }
 13. end
```

```
. ml model gf0 CensNegBin (los:los cens=hmo white type2 type3) (lnalpha:)
. ml maximize, search(on) nolog ti(Censored Negative Binomial Regression)
> diparm(lnalpha, exp label("alpha")) irr
```

```
Censored Negative Binomial Regression        Number of obs    =       1,495
                                             Wald chi2(4)     =      112.27
Log likelihood = -4780.4545                  Prob > chi2      =      0.0000
```

	IRR	Std. Err.	z	P>|z|	[95% Conf.	Interval]
los						
hmo	.9323925	.0498796	-1.31	0.191	.839581	1.035464
white	.8785736	.060604	-1.88	0.061	.7674716	1.005759
type2	1.248278	.0634928	4.36	0.000	1.129836	1.379136
type3	2.068304	.1598223	9.40	0.000	1.777625	2.406515
_cons	10.09924	.6904923	33.82	0.000	8.832659	11.54745
lnalpha						
_cons	-.797411	.0444701	-17.93	0.000	-.8845708	-.7102513
alpha	.4504938	.0200335			.4128914	.4915207

```
Note: Estimates are transformed only in the first equation.
Note: _cons estimates baseline incidence rate.
```

```
. estat ic
```

Akaike´s information criterion and Bayesian information criterion

Model	Obs	ll(null)	ll(model)	df	AIC	BIC
.	1,495	.	-4780.455	6	9572.909	9604.768

Note: N=Obs used in calculating BIC; see [R] BIC note.

The `hmo` and `white` variables no longer appear to be significant predictors of length of stay. Given the substantial reduction in the value of the AIC statistic over that of the censored Poisson, we favor the censored negative binomial.

We next see what difference the censored observations make to the model. To illustrate this effect, we will highlight a fitted model that does not account for censoring versus a fitted model that does account for censoring. Specifying the same outcome variable twice in the `ml model` statement is equivalent to a declaration that all observations are uncensored (making the model equivalent to the NB-2). Here none of the observations are assumed to be censored:

```
. ml model gf0 CensNegBin (los:los los=hmo white type2 type3) (lnalpha:)
. ml maximize, search(on) nolog ti(Censored Negative Binomial Regression)
> diparm(lnalpha, exp label("alpha")) irr
Censored Negative Binomial Regression          Number of obs    =      1,495
                                               Wald chi2(4)     =     109.82
Log likelihood = -4797.4766                    Prob > chi2      =     0.0000
```

	IRR	Std. Err.	z	P>|z|	[95% Conf.	Interval]
los						
hmo	.9343023	.0497622	-1.28	0.202	.8416883	1.037107
white	.8789166	.0602425	-1.88	0.060	.7684309	1.005288
type2	1.247634	.063121	4.37	0.000	1.129855	1.37769
type3	2.026193	.1542563	9.28	0.000	1.745332	2.352252
_cons	10.07723	.6847214	34.00	0.000	8.820728	11.51273
lnalpha						
_cons	-.8079821	.0444542	-18.18	0.000	-.8951108	-.7208534
alpha	.4457566	.0198158			.4085623	.486337

Note: Estimates are transformed only in the first equation.
Note: _cons estimates baseline incidence rate.

```
. estat ic
Akaike's information criterion and Bayesian information criterion
```

Model	Obs	ll(null)	ll(model)	df	AIC	BIC
.	1,495	.	-4797.477	6	9606.953	9638.812

Note: N=Obs used in calculating BIC; see [R] BIC note.

There is little difference in inference for these two models because there are only 8 censored observations out of the 1,495 total observations. However, we favor the censored model for its faithful treatment of the outcomes.

Here we investigate specification of a robust variance estimator with the censored negative binomial model to see whether there is evidence for a provider clustering effect. That is, perhaps length of stay depends partly on the provider; los is more highly correlated within providers than between them.

```
. ml model gf0 CensNegBin (los:los cens=hmo white type2 type3) (lnalpha:),
> vce(cluster provnum)
. ml maximize, search(on) nolog ti(Censored Negative Binomial Regression)
> diparm(lnalpha, exp label("alpha")) irr
```

```
Censored Negative Binomial Regression              Number of obs   =      1,495
                                                   Wald chi2(4)    =      26.30
Log pseudolikelihood = -4780.4545                  Prob > chi2     =     0.0000
```

(Std. Err. adjusted for 54 clusters in provnum)

	IRR	Robust Std. Err.	z	P>|z|	[95% Conf. Interval]
los						
hmo	.9323925	.0478629	-1.36	0.173	.8431478	1.031083
white	.8785736	.0639788	-1.78	0.075	.7617151	1.01336
type2	1.248278	.0769924	3.60	0.000	1.10614	1.40868
type3	2.068304	.4511879	3.33	0.001	1.348745	3.171749
_cons	10.09924	.6823747	34.22	0.000	8.846584	11.52927
lnalpha						
_cons	-.797411	.0691505	-11.53	0.000	-.9329435	-.6618786
alpha	.4504938	.0311519			.3933941	.5158813

Note: Estimates are transformed only in the first equation.
Note: _cons estimates baseline incidence rate.

```
. estat ic
```

Akaike´s information criterion and Bayesian information criterion

Model	Obs	ll(null)	ll(model)	df	AIC	BIC
.	1,495	.	-4780.455	6	9572.909	9604.768

Note: N=Obs used in calculating BIC; see [R] BIC note.

There does appear to be a clustering effect. Though most of the standard errors change very little, the standard error for the **type3** indicator changes substantially. The interested reader could consider alternative approaches to this model. For example, one could assess whether this model's allowance for unspecified correlation among outcomes within provider is superior to a random-effects specification; see chapter 18 for details.

To further highlight applications to censored data, we redefine the outcome variables by imposing a dataset-specific right-censoring point of 30. That is, we define $y_{Li} = 30$ if the length of stay is at least 30, and we define $y_{Ri} = -1$ as an indicator that the observation is censored. Fitting this model, we see that

```
. replace cens  = -1 if los>=30  // Censored at 30
(34 real changes made)

. generate los30 = min(los,30)   // Dataset-censor value

. ml model gf0 CensNegBin (los:los30 cens=hmo white type2 type3) (lnalpha:),
> vce(cluster provnum)

. ml maximize, search(on) nolog ti(Censored Negative Binomial Regression)
> diparm(lnalpha, exp label("alpha")) irr
```

```
Censored Negative Binomial Regression          Number of obs    =     1,495
                                               Wald chi2(4)     =     24.65
Log pseudolikelihood = -4636.4278              Prob > chi2      =    0.0001
```

```
                            (Std. Err. adjusted for 54 clusters in provnum)
```

	IRR	Robust Std. Err.	z	P>|z|	[95% Conf. Interval]
los					
hmo	.9403067	.0484507	-1.19	0.232	.8499827 1.040229
white	.8958405	.0631204	-1.56	0.119	.7802892 1.028504
type2	1.244742	.0731647	3.72	0.000	1.109294 1.396729
type3	1.686614	.2939181	3.00	0.003	1.198618 2.373288
_cons	9.863324	.6556857	34.43	0.000	8.658404 11.23592
lnalpha					
_cons	-.8704337	.0523524	-16.63	0.000	-.9730425 -.7678249
alpha					
alpha	.4187699	.0219236			.3779314 .4640212

```
Note: Estimates are transformed only in the first equation.
Note: _cons estimates baseline incidence rate.
```

In addition to the `ml` commands highlighted above, we have discussed in other sections of this text estimation commands that have been developed for censored (and other) models. We presented the models above using Stata's `ml` commands to highlight the relative ease with which one can specify such models and the power and ease of the `ml` command. We emphasize again that censor levels may be specified for the entire sample or may be specified for each observation. Both approaches are handled in the discussed commands.

14.13   Finite mixture models

Finite mixture models (FMMs) are often found in econometric research, though their application is just as appropriate with data from health, political science, astrostatistics, and other disciplines. We have already seen specific applications of FMMs that we called hurdle and zero-inflated models. In such analyses, the researcher believes that the outcome of interest is best analyzed as a mixture of component distributions. In hurdle models, the component distributions were nonoverlapping, and in zero-inflated models, the components did overlap.

FMMs are a type of latent class model for which the distribution of an outcome variable is assumed to consist of a finite mixture of component data generating processes. Studies differ on the optimum number of components and depend on the nature of the outcome under analysis; Aitkin et al. (2009) include illustration of a study in which there are six different components, whereas Cameron and Trivedi (2010) advise that there are rarely applications needing more than two components.

FMMs are warranted when we believe that there is more than one process generating the outcome of interest. The `fmm` prefix command can be used with estimation commands to allow components to follow the normal (Gaussian), Poisson, NB-1, NB-2, and many other distributions. See `help fmm estimation` for a full list of commands that support the `fmm` prefix.

Consider an example for which a count response is generated from two processes where each of the processes has a different mean. We could hypothesize a mixture of two different Poisson processes: one with mean $\lambda_1 = 1.25$ and the other with mean $\lambda_2 = 3.75$. Moreover, each of the component distributions is assumed to yield a given proportion of counts π_i where over the total number of components C, $\sum \pi_i = 1$. Thus, the mixture of densities is given by $f(x) = \sum \pi_i f_i(x)$.

To highlight estimation and interpretation of FMMs, we use the Arizona Medicare diagnostic-related group data (`azdrg112.dta`) and model the length of stay (in days) as our outcome. The model will assume that there are three different Poisson distributions (each depending on the same list of covariates). The AIC and BIC statistics are also displayed.

```
. use http://www.stata-press.com/data/hh4/azdrg112, clear
(AZ 1991: DRG112, no deaths)

. fmm 3, noheader nolog: poisson los male emergency age75
```

	Coef.	Std. Err.	z	P>|z|	[95% Conf.	Interval]
1.Class	(base outcome)					
2.Class						
_cons	-.8477609	.2102199	-4.03	0.000	-1.259784	-.4357376
3.Class						
_cons	-3.253356	.2448453	-13.29	0.000	-3.733244	-2.773468

Class   : 1

	Coef.	Std. Err.	z	P>|z|	[95% Conf.	Interval]
los						
male	-.1151784	.0379308	-3.04	0.002	-.1895215	-.0408354
emergency	.4139503	.0583021	7.10	0.000	.2996803	.5282203
age75	.127431	.0398688	3.20	0.001	.0492895	.2055724
_cons	1.043817	.0551014	18.94	0.000	.9358206	1.151814

Class   : 2

	Coef.	Std. Err.	z	P>|z|	[95% Conf.	Interval]
los						
male	-.1856325	.0577096	-3.22	0.001	-.2987412	-.0725238
emergency	1.139434	.1013195	11.25	0.000	.9408516	1.338017
age75	.1352596	.0608519	2.22	0.026	.0159919	.2545272
_cons	1.093901	.1006497	10.87	0.000	.8966313	1.291171

Class   : 3

	Coef.	Std. Err.	z	P>|z|	[95% Conf.	Interval]
los						
male	-.2627711	.135278	-1.94	0.052	-.527911	.0023688
emergency	.6982351	.1035648	6.74	0.000	.4952518	.9012184
age75	-.0118023	.1757276	-0.07	0.946	-.356222	.3326174
_cons	2.540302	.1046631	24.27	0.000	2.335166	2.745438

```
. estat ic
```

Akaike's information criterion and Bayesian information criterion

Model	Obs	ll(null)	ll(model)	df	AIC	BIC
.	1,798	.	-4198.957	14	8425.914	8502.836

Note: N=Obs used in calculating BIC; see [R] BIC note.

The output reflects the results of our specification. The best three Poisson models are displayed.

We use `estat lcprob` to see that the data are fit by 68% in the first component, 29% in the second component, and 3% in the third component. Being older than 75 years is significant to the first and second components but not to the third.

```
. estat lcprob
Latent class marginal probabilities           Number of obs    =    1,798
```

	Margin	Delta-method Std. Err.	[95% Conf. Interval]	
Class				
1	.6816552	.0441761	.5896186	.7614034
2	.2920027	.0425636	.2159883	.3817434
3	.0263421	.0056605	.0172515	.0400278

We obtain estimates of the outcome of interest by using `predict` with the `marginal` option.

```
. predict fmmhat, marginal
(option mu assumed)
. summarize fmmhat
```

Variable	Obs	Mean	Std. Dev.	Min	Max
fmmhat	1,798	4.894109	1.534629	2.706389	7.109684

Following the synthetic data examples highlighted throughout this chapter, we provide example code to illustrate data generation for the FMM. For this example, we generate data from a mixture of three NB-2 components.

```
. clear
. set obs 75000
number of observations (_N) was 0, now 75,000
. set seed 2342
. generate n1 = rnormal()
. generate n2 = rnormal()
. generate b1 = runiform() < .4
. generate b2 = runiform() < .6
. generate expxb1 = exp(2.00 + 1.50*n1 + -0.5*n2 + 3.00*b1 - 4.0*b2)
. generate ap1     = 0.66
. generate expxb2 = exp(1.00 + 1.00*n1 + 1.00*b1 - 2.0*b2)
. generate ap2     = 0.25
. generate expxb3 = exp(1.00 - 1.00*n1 + 1.25*n2)
. generate ap3     = 0.33
. generate ynb1 = rpoisson(rgamma(1/ap1, ap1) * expxb1)   // ynb1 ~ NB(2)
. generate ynb2 = rpoisson(rgamma(1/ap2, ap2) * expxb2)   // ynb2 ~ NB(2)
. generate ynb3 = rpoisson(rgamma(1/ap3, ap3) * expxb3)   // ynb3 ~ NB(2)
. generate mix = runiform()
. generate yfmm = (mix>0.40)*ynb1 + (mix<0.25)*ynb2 +
> (mix>0.25 & mix<0.40)*ynb3
```

We fit the model by typing

```
. fmm 3, noheader nolog: nbreg yfmm n1 n2 b1 b2
```

	Coef.	Std. Err.	z	P>|z|	[95% Conf.	Interval]
1.Class	(base outcome)					
2.Class						
_cons	.4794876	.0265268	18.08	0.000	.427496	.5314791
3.Class						
_cons	1.357791	.0167562	81.03	0.000	1.324949	1.390632

Class   : 1

	Coef.	Std. Err.	z	P>|z|	[95% Conf.	Interval]
yfmm						
n1	-.997569	.0107877	-92.47	0.000	-1.018712	-.9764256
n2	1.261432	.0116659	108.13	0.000	1.238567	1.284297
b1	.0655935	.0212628	3.08	0.002	.0239191	.1072679
b2	.010755	.022177	0.48	0.628	-.0327112	.0542212
_cons	.9353676	.0223128	41.92	0.000	.8916353	.9791
/yfmm						
lnalpha	-1.125046	.0353439			-1.194319	-1.055773

Class   : 2

	Coef.	Std. Err.	z	P>|z|	[95% Conf.	Interval]
yfmm						
n1	1.010655	.0123645	81.74	0.000	.9864213	1.034889
n2	.0118766	.0115705	1.03	0.305	-.0108011	.0345543
b1	1.005114	.0257906	38.97	0.000	.9545654	1.055663
b2	-1.983701	.0297979	-66.57	0.000	-2.042104	-1.925298
_cons	.9911895	.0231664	42.79	0.000	.9457841	1.036595
/yfmm						
lnalpha	-1.303205	.0513584			-1.403866	-1.202545

Class   : 3

	Coef.	Std. Err.	z	P>|z|	[95% Conf.	Interval]
yfmm						
n1	1.501873	.006946	216.22	0.000	1.488259	1.515487
n2	-.5104871	.0062599	-81.55	0.000	-.5227563	-.4982179
b1	2.997435	.0136221	220.04	0.000	2.970737	3.024134
b2	-4.014121	.0149299	-268.86	0.000	-4.043383	-3.984859
_cons	2.00439	.0115614	173.37	0.000	1.98173	2.02705
/yfmm						
lnalpha	-.4265563	.018129			-.4620885	-.3910242

```
. estat ic
Akaike's information criterion and Bayesian information criterion
```

Model	Obs	ll(null)	ll(model)	df	AIC	BIC
.	75,000	.	-189684.4	20	379408.7	379593.2

```
                Note: N=Obs used in calculating BIC; see [R] BIC note.
```

Note how the model used the complete set of covariates for each component even though the synthesized data use only subsets.

fmm also has a hybrid syntax that allows us to specify different sets of covariates. We could have fit the mode we previously simulated by typing

```
fmm, noheader nolog: (nbreg yfmm n1 n2) (nbreg yfmm n1 b1 b2)   ///
     (nbreg yfmm n1 n2 b1 b2)
```

14.14   Quantile regression for count outcomes

Efron (1992) proposed the asymmetric maximum-likelihood estimator to estimate conditional location functions for count data similar to conditional expectiles of a linear model; expectiles are not required to be integers. The asymmetric maximum-likelihood estimator is given by

$$\widehat{\beta}_w = \operatorname*{argmax} \sum_{i=1}^{n} \{y_i x_i \beta - \exp(x_i \beta) - \ln \Gamma(y_i + 1)\} w^{I\{y_i > \exp(x_i \beta)\}} \tag{14.56}$$

where $w > 0$. Efron defined the 100αth asymmetric maximum-likelihood regression percentile is obtained for w such that $\alpha = \sum I\{y_i \leq \exp(x_i \widehat{\beta}_w)\}$.

The main challenge with defining quantile regression for count outcomes is that the outcomes are discrete integers. To overcome the difficulty associated with the step function appearance of the distribution, Machado and Santos Silva (2005) describe the use of "jittering" to smooth out the distribution function and allow application of usual quantile methods.

The basic idea is that estimation is bootstrapped r times. In each bootstrap iteration, there is a pseudo-outcome variable

$$y_b = \log(y + u) \tag{14.57}$$

where $u \sim \text{Uniform}(0, 1)$ is defined. In this way, the outcome is jittered or smeared into a continuous representation. Quantile linear regression is carried out at the specified quantile. The estimated parameter vector is collected and stored so that the collection of r estimates can be used as an empirical distribution. Estimates and standard errors are calculated from the r estimates.

We use the `qcount` command (Miranda 2006) to implement these methods. This command can be installed by typing `ssc install qcount`. The model specification is standard, and the particular quantile is specified in an option. Using the data already specified in this chapter, a quantile regression model may be fit:

```
. use http://www.stata-press.com/data/hh4/medpar, clear
. quietly qcount los hmo white type2 type3 died, quantile(.50) repetition(1000)
. qcount
```

Count Data Quantile Regression
(Quantile 0.50)

			Number of obs	=	1495	
			No. jittered samples	=	1000	

los	Coef.	Std. Err.	z	P>|z|	[95% Conf. Interval]	
hmo	-.0896129	.079142	-1.13	0.258	-.2447284	.0655025
white	-.0707864	.1132315	-0.63	0.532	-.2927161	.1511433
type2	.2379269	.0958676	2.48	0.013	.0500299	.4258238
type3	.4566922	.3934314	1.16	0.246	-.3144193	1.227804
died	-.3470832	.1089957	-3.18	0.001	-.5607109	-.1334556
_cons	2.172156	.1111166	19.55	0.000	1.954372	2.389941

We can also fit models on either side of the median to investigate whether the "low responders" and "high responders" have different associations.

```
. quietly qcount los hmo white type2 type3 died, quantile(.25) repetition(1000)
. qcount
```

Count Data Quantile Regression
(Quantile 0.25)

			Number of obs	=	1495	
			No. jittered samples	=	1000	

los	Coef.	Std. Err.	z	P>|z|	[95% Conf. Interval]	
hmo	-.0490511	.0621588	-0.79	0.430	-.17088	.0727778
white	-.1726938	.103235	-1.67	0.094	-.3750307	.0296431
type2	.2524764	.0846606	2.98	0.003	.0865447	.4184081
type3	.4891432	.1490266	3.28	0.001	.1970563	.78123
died	-.865719	.1062624	-8.15	0.000	-1.07399	-.6574485
_cons	1.833929	.1012713	18.11	0.000	1.635441	2.032417

```
. quietly qcount los hmo white type2 type3 died, quantile(.75) repetition(1000)
. qcount

Count Data Quantile Regression
( Quantile 0.75 )
                                                Number of obs      =        1495
                                                No. jittered samples =        1000
```

los	Coef.	Std. Err.	z	P>|z|	[95% Conf. Interval]	
hmo	-.0264222	.0707646	-0.37	0.709	-.1651183	.112274
white	-.2105584	.0881462	-2.39	0.017	-.3833217	-.037795
type2	.2442494	.0468144	5.22	0.000	.1524949	.336004
type3	.6224817	.1635853	3.81	0.000	.3018605	.9431029
died	-.1062508	.0524463	-2.03	0.043	-.2090437	-.0034579
_cons	2.687726	.0872895	30.79	0.000	2.516642	2.858811

There is no option in the software (as of this writing) to eliminate preliminary printing of dots on the screen as the bootstrap process executes, so we elected to run the command quietly to eliminate all output, and then we reissue the command without arguments to "redisplay" the model estimates.

14.15   Heaped data models

When count data are collected by asking people to recall a number of events over a specific time period, there is a chance that the respondent will think about the count for a shorter time period and then multiply; for example, a person reporting a number of events over six months might think of a typical month estimate that count and then report six times their estimated one-month count. When count data for (naturally) grouped events are collected, as when smokers are asked to report the number of cigarettes smoked in a day, they tend to report numbers in half-packs (multiples of 10) and full packs (multiples of 20). Both of these examples lead to reports of "heaped data" or "digit preference". Basically, the reported events tend to "heap" (be overrepresented) at multiples of the heaping interval. For reported cigarettes smoked, the reported counts tend to be overrepresented for the number 20 and underrepresented at 19 and 21.

Cummings et al. (2015) describe terminology to describe mixtures of heaped data. They term nonheaped data as being heaped on 1; that is, the data are heaped on multiples of 1 or they are exact for the time period requested. Data that are heaped on 7 are assumed to be the result of considering a time period of 1/7th the requested time period. Before multiplication by 7, the conditional mean on the shorter time period is 1/7th the size of the conditional mean of the requested time period. The authors deem reporting behavior in terms of the multiples on which data are reported. Given a reasonable covariate list, one could simultaneously estimate the propensity to report on a given scale and the incidence rate ratios common to all heaping intervals.

For the proposed mixture of rescaled distributions, we consider two behaviors that subjects would choose. Behavior 1 consists of subjects reporting an exact count of the requested frequency. Whereas those subjects who remember the requested frequency over a $1/k$th of a specified period of time and then reporting k times that amount exhibit behavior 2. Under behavior 1, we consider covariates and parameters β are associated with the mean $\mu^{[1]}$ under the log-link function and $\ln(\mu^{[1]}) = X\beta$. In the same way, for behavior 2, we use the same covariates and parameters for the associated mean, $\mu^{[2]}$, as $1/k$ times $\mu^{[1]}$ (mean of behavior 1) under the log-link function $\ln(k\mu^{[2]}) = X\beta$, which is $\ln(\mu^{[2]}) = X\beta - \ln(k)$. Here, notice the only difference in the means for the two behaviors is the offset term $\ln(k)$. We exponentiate coefficients show rates that will remain constant over all time periods because of the reparameterization of the covariates and parameters.

For a two behavior mixture, Cummings et al. (2015) denote a binary model B to represent whether a subject chooses behavior 2 as a function of covariates J and coefficients ϕ. The likelihood of behavior 2 multiplies the likelihood of the unscaled outcome (the reported outcome divided by the heaping number) where the reciprocal of the heaping number is then used as the exposure $[\ln(k)]$ as part of the linear predictor. Thus, the mixture probability model for a reported outcome is given by

$$
\begin{aligned}
P(Y = y) \quad = \quad & P_B(b = 2|J, \phi) P_{[2]}\Big[Y = \frac{y}{k}\Big|\mu = \exp\{X\beta - \ln(k)\}\Big] I_{(y \bmod k = 0)} \\
& + P_B(b = 1|J, \phi) P_{[1]}\{Y = y|\mu = \exp(X\beta)\} \qquad (14.58)
\end{aligned}
$$

where $P_B(b = 2|J, \phi) = \exp(J\phi)/\{1 + \exp(J\phi)\}$ and $P_B(b = 1|J, \phi) = 1 - P_B(b = 2|J, \phi) = 1/\{1 + \exp(J\phi)\}$.

To generalize, consider s heaping numbers $k_2, \ldots, k_s, k_{s+1}$, where each response is the result of one of $s+1$ behaviors that match the s heaping numbers and one behavior of exact reporting on the specified time period. A multinomial model S is then used to estimate whether a subject chooses $s + 1, s, \ldots, 2$ as a function of covariates J and coefficients ϕ using behavior 1 (exact reporting) as the reference. The probability model for a particular outcome is then given by

$$
\begin{aligned}
P(Y = y|x, \beta, \phi) = \ & P_S(b = s + 1|J, \phi_2, \ldots, \phi_{s+1}) \\
& \times P_{[s+1]}\Big[Y = \frac{y}{k_{s+1}}\Big|\mu = \exp\{X\beta - \ln(k_{s+1})\}\Big] I_{(y \bmod k_{s+1}=0)} \\
& + P_S(b = s|J, \phi_2, \ldots, \phi_{s+1}) \\
& \times P_{[s]}\Big[Y = \frac{y}{k_s}\Big|\mu = \exp\{X\beta - \ln(k_s)\}\Big] I_{(y \bmod k_s=0)} \\
& + \cdots + P_S(b = 2|J, \phi_2, \ldots, \phi_{s+1}) \\
& \times P_{[2]}\Big[Y = \frac{y}{k_2}\Big|\mu = \exp\{X\beta - \ln(k_2)\}\Big] I_{(y \bmod k_2=0)} \\
& + P_S(b = 1|J, \phi_2, \ldots, \phi_{s+1}) P_{[1]}\Big\{Y = y|\mu = \exp(X\beta)\Big\} \qquad (14.59)
\end{aligned}
$$

where $P_S(b = k|J, \phi_2, \ldots, \phi_{s+1}) = \exp(J\phi_k)/\{1 + \exp(J\phi_2) + \cdots + \exp(J\phi_{s+1})\}$ for $k = 2, \ldots, s+1$, and $P_S(b = 1|J, \phi_2, \ldots, \phi_{s+1}) = 1/\{1 + \exp(J\phi_2) + \cdots + \exp(J\phi_{s+1})\}$. The log likelihood for a count-data model, in a general form, is the sum of the logs of the probabilities of observed outcomes given by

$$\mathcal{L} = \sum_{i=1}^{n} \ln \{P(Y = y|x, \beta, \phi)\} \tag{14.60}$$

where $P(Y = y|x, \beta, \phi)$ is from (14.59). When only one heaping multiple k exists (2 behavior mixture) and the counting process is given by the Poisson distribution, the log likelihood is given by

$$\mathcal{L} = \sum_{i=1}^{n} \ln \left[\left\{ \frac{\exp(J\phi)}{1 + \exp(J\phi)} \right\} \exp\left[-\exp\{x\beta - \ln(k)\} + \frac{y}{k}\{x\beta - \ln(k)\} \right. \right.$$
$$\left. - \ln \Gamma\left(\frac{y}{k} + 1\right) \right] I_{(y \bmod k=0)} + \left\{ \frac{1}{1 + \exp(J\phi)} \right\} \exp\left\{ -\exp(x\beta) \right.$$
$$\left. \left. + y(x\beta) - \ln \Gamma(y+1) \right\} \right] \tag{14.61}$$

Similarly, where the counting process is given by the GP distribution, the log likelihood is given by

$$\mathcal{L} = \sum_{i=1}^{n} \ln \left(\left\{ \frac{\exp(J\phi)}{1 + \exp(J\phi)} \right\} \exp\left[-\exp\{x\beta - \ln(k)\} + \frac{y}{k}\{x\beta - \ln(k)\} \right. \right.$$
$$\left. - \ln \Gamma\left(\frac{y}{k} + 1\right) \right] I_{(y \bmod k=0)} + \left\{ \frac{1}{1 + \exp(J\phi)} \right\} \exp\left[-(1-\alpha)x\beta + \alpha y \right.$$
$$\left. \left. + (y-1)\ln\{(1-\alpha)x\beta + \alpha y\} + \ln(x\beta) + \ln(1-\alpha) - \ln \Gamma(y+1) \right] \right) \tag{14.62}$$

Finally, for an NB distribution counting process, the log likelihood is given by

$$\mathcal{L} = \sum_{i=1}^{n} \ln \left[\left\{ \frac{\exp(J\phi)}{1 + \exp(J\phi)} \right\} \exp\left[-\exp\{x\beta - \ln(k)\} + \frac{y}{k}\{x\beta - \ln(k)\} \right. \right.$$
$$\left. - \ln \Gamma\left(\frac{y}{k} + 1\right) \right] I_{(y \bmod k=0)} + \left(\frac{1}{1 + \exp(J\phi)} \right) \exp\left\{ \ln \Gamma\left(\frac{1}{\alpha} + y\right) - \ln \Gamma\left(\frac{1}{\alpha}\right) \right.$$
$$\left. \left. + \frac{1}{\alpha}\ln\left(\frac{1}{1+\alpha\mu}\right) + y\ln\left(\frac{\alpha\mu}{1+\alpha\mu}\right) - \ln \Gamma(y+1) \right\} \right] \tag{14.63}$$

Because we assume that the mixture distributions all have the same regression covariates and that those covariates have the same associations in terms of incidence rate ratios. Here we synthesize 1,000 observations from a mixture of data from two Poisson distributions Y_1 and $3 \times Y_2$, where $E(Y_1) = 3E(Y_2)$. To generate data for a particular regression model, we assume that $Y_1 \sim \text{Poisson}\{\exp(1 + x_1 + 2x_2)\}$ and $Y_2 \sim \text{Poisson}[\exp\{1 + x_1 + 2x_2 - \ln(3)\}]$. We mix the two outcomes into our analysis set where Y_2 (the heaping behavior) is selected with probability $\exp(-x_1 + 2x_2)/(1 + \exp(-x_1 + 2x_2))$. Finally, we assume that x_1 and x_2 are both distributed Bernoulli(0.5).

```
clear
set seed 29342
set obs 1000
generate x1 = runiform() < .5
generate x2 = runiform() < .5
generate p  = exp(-x1+2*x2) / (1+exp(-x1+2*x2))
generate y1 = rpoisson(exp(1+x1+2x2))
generate y2 = 3*rpoisson(exp(1+x1+2x2-ln(3)))
generate y  = cond(runiform()<p,y2,y1)
```

This example uses only two binary predictors, so there are only four covariate patterns. As such, we can easily illustrate the observed and predicted outcomes. The observed data are illustrated in figure 14.2. Note the characteristic regularly appearing spikes in the histogram. These spikes are a result of the rescaled distribution, which causes outcomes to heap on these regular intervals.

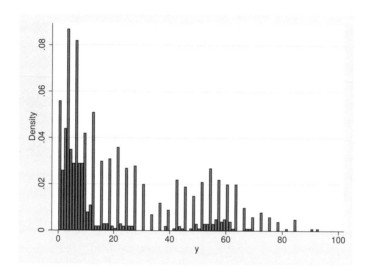

Figure 14.2. Histogram of response variables created as a mixture of scaled Poissons

We obtain predicted values from the Poisson regression model:

```
poisson y x1 x2, nolog
forvalues k=0/100 {
        predict predp`k', pr(`k')
}
```

Here, we obtain predicted values from the heaped regression model:

```
heapr y x1 x2, heap(3) hvars(x1 x2) nolog
forvalues k=0/100 {
        predict predh`k', pr(`k')
}
```

Note that we used the community-contributed `heapr` command (Cummings et al. 2015), which can be installed by typing `net install st0388, from(http://www.stata-journal.com/software/sj15-2)`.

Now, to illustrate the differences in predicted values, we will write a small program that (for each covariate pattern) will estimate a suitable range of outcomes the proportion of observed values equal to that outcome, the predicted probability of that value from the Poisson model, and the predicted probability of that value for the heaped model. Because there are only four possible covariate patterns, we will save each graph of the results and then combine the graphs into one.

Our program looks like the following:

```
capture program drop GetHeapPred
program define GetHeapPred
        args x1 x2 ll uu
        * The arguments are:
        * x1,x2 = the specific values of the covariates
        * ll,uu = the specific range (min,max) for which to
        *         to generate predictions.
        * assumed but not enforced: ll>=0 and uu<=100
        quietly {
                count if x1==`x1´ & x2==`x2´
                local Nden = `r(N)´
                tempvar xx yd yp yh
                foreach var in xx yd yp yh {
                        generate double ``var´´ = .
                        replace `xx´ = _n-1
                }
                forvalues k=0/100 {
                        count if `yp´==`k´ & x1==`x1´ & x2==`x2´
                        replace `yd´ = `r(N)´/`Nden´ if _n==`k´+1
                        summ predp`k´ if x1==`x1´ & x2==`x2´
                        replace `yp´ = `r(mean)´ if _n==`k´+1
                        summ predh`k´ if x1==`x1´ & x2==`x2´
                        replace `yh´ = `r(mean)´ if _n==`k´+1
                }
                label var `yd´ "Data"
                label var `yp´ "Poisson"
                label var `yh´ "Heap"
                label var `xx´ "x"
                twoway (bar `yd´ `xx´, bstyle(outline)) ///
                    (connected `yp´ `xx´, msymbol(circle_hollow) msize(small))   ///
                    (scatter `yh´ `xx´, msymbol(smtriangle_hollow) msize(small)) ///
                    in `ll´/`uu´, scheme(s1mono) legend(rows(1))
        }
end
```

Our program makes assumptions about the passed in arguments that are okay for our specific application. For a similar application with other data, several changes may be required. In any event, we call the program by typing

```
GetHeapPred 0 0  1 13
graph save pred00, replace
GetHeapPred 0 1  1 41
graph save pred01, replace
```

```
GetHeapPred 1 0  1 21
graph save pred10, replace
GetHeapPred 1 1 40 81
graph save pred11, replace
```

The program considers each of the four covariates and calculates $P(Y = k)$ for the specified range given by the final two arguments. You should appreciate that for x1=0, x2=0, we investigated the range $\{1, 2, \ldots, 13\}$ because this is the most likely portion of the underlying population; it is mean $\exp(1 + \text{x1} + 2\text{x2}) = \exp(1) \approx 2.7$. The model predictions for the Poisson regression model and the heaped Poisson regression model are calculated along with the observed proportion of responses. All are plotted together for comparison, where it is clear that the heaped Poisson model predictions are much closer to the observed proportions than are the Poisson model predictions.

To combine the above four graphs into one, we type

```
graph combine pred00.gph pred01.gph pred10.gph pred11.gph, rows(2)
```

The final combined graph is illustrated in figure 14.3.

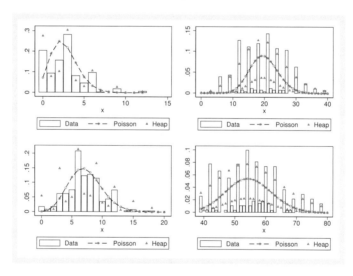

Figure 14.3. Graphs are organized for the conditional distribution of the outcome conditional on the covariates (x_1, x_2). The values of the covariates are $(0,0)$ in the upper left, $(0,1)$ in the upper right, $(1,0)$ in the lower left, and $(1,1)$ in the lower right. Bars represent the empirical distribution of the outcome variable. Circles represent the estimated probabilities $\widehat{P}(Y = y|x, \beta_{\text{Poisson}})$ evaluated at $\beta_{\text{Poisson}} = \widehat{\beta}_{\text{Poisson}}$ generated by the fitted Poisson regression model. Triangles represent the estimated probabilities $\widehat{P}(Y = y|x, \beta_{\text{Heap}}, \gamma_{\text{Heap}})$ evaluated at $\beta_{\text{Heap}} = \widehat{\beta}_{\text{Heap}}, \gamma_{\text{Heap}} = \widehat{\gamma}_{\text{Heap}}$ of the fitted heaped Poisson regression model.

These model specifications are similar to those considered in zero-inflation models. In fact, one can consider zero-inflated versions of these mixture models. As in (14.64), we use the same general definition of zero inflation. However, for our mixture of rescaled distributions approach, the zero-inflated heaped count-data log likelihoods for Poisson, generalized Poisson, and negative binomial distributions can be shown as

$$
\begin{aligned}
\mathcal{L} &= \sum_{i \in Z} \ln \left\{ w_i + (1 - w_i) f(0) \right\} \\
&+ \sum_{i \notin Z} \ln \left((1 - w_i) \left[P_B(b = 1 | J, \phi) P_{[1]} \{ Y = y | \mu = \exp(X\beta) \} \right] \right) (14.64)
\end{aligned}
$$

where Z, again, is the set of zero outcomes and $P_B(b = 1 | J, \phi) P_{[1]} \{ Y = y | \mu = \exp(X\beta) \}$ is from (14.58) above. Zero-inflation models for heaped data are supported under the `ziheapr` command using options `poisson`, `gpoisson`, and `nbreg` for zero-inflated Poisson, zero-inflated generalized Poisson, and zero-inflated negative binomial distributions, respectively.

In the next chapter, we discuss polytomous response models, which include models having both ordered and unordered levels of response. Like the models we have addressed in this section, ordered and unordered response models are a variety of discrete response regression model. Ordered response models in particular have undergone some interesting enhancements in recent years. We will address many of these in the next chapter.

Part V

Multinomial Response Models

15 Unordered-response family

Contents

This chapter addresses data in which we record an integer response. Our responses are limited to a finite choice set, and there is no meaning to the values or to the relative magnitude of the outcomes that are assigned.

A useful tool for analyzing such data may be introduced using a utility argument. Assume that we collect data from respondents who provide one choice from a list of alternatives. We code these alternatives $1, 2, \ldots, r$ for convenience. Here we assume that the alternatives are unordered—that there is no indication of one alternative's being better or higher. For our $i = 1, \ldots, n$ respondents, we denote the selection

$$y_{ij}^* = \mathbf{x}_i \boldsymbol{\beta}_j + \epsilon_i \tag{15.1}$$

if respondent i makes choice j. Therefore, y_{ij} is the maximum among the r utilities $y_{i1}^*, y_{i2}^*, \ldots, y_{ir}^*$ such that

$$\Pr(y_{ij} > y_{ij}^*) \tag{15.2}$$

To use a parametric model, we must assume a distribution for ϵ. In the next two sections, we present the results of assuming logit and then probit.

15.1   The multinomial logit model

Sometimes called the polytomous logistic regression or the nominal logit model, the multinomial logit model assumes r equations for the r outcomes. One equation sets $\boldsymbol{\beta}$ to zero so that the problem is identifiable. The associated outcome is the base reference group. Below we assume that the outcomes are coded using the set $\{1, 2, \ldots, r\}$ and that the outcome of one is used as the base reference group. Our choice for which outcome to use as the base reference group will affect the estimated coefficients but not the predicted probabilities.

We may derive the log likelihood from

$$\xi_{ik} = \mathbf{x}_i \boldsymbol{\beta}_k + \text{offset}_{ki} \tag{15.3}$$

$$\Pr(y_i = k) = \frac{\exp(\xi_{ik})}{1 + \sum_{m=2}^{r} \exp(\xi_{im})} \qquad k = 2, 3, \ldots, r \tag{15.4}$$

$$\Pr(y_i = 1) = \frac{1}{1 + \sum_{m=2}^{r} \exp(\xi_{im})} \tag{15.5}$$

$$d_{ij} = \begin{cases} 1 & \text{if observation } i \text{ has outcome } j \text{ (if } y_i = j) \\ 0 & \text{otherwise} \end{cases} \tag{15.6}$$

$$\mathcal{L} = \sum_{i=1}^{n} \sum_{j=1}^{r} d_{ij} \ln \Pr(y_i = j) \tag{15.7}$$

$$\lambda_{ik} = d_{ik} - \Pr(y_i = k) \tag{15.8}$$

First and second derivatives of the log likelihood with respect to the coefficient vectors are given by

$$\frac{\partial \mathcal{L}}{\partial \beta_{kt}} = \sum_{i=1}^{n} w_i x_{ti} \lambda_{ik} \tag{15.9}$$

$$\frac{\partial^2 \mathcal{L}}{\partial \beta_{kt} \partial \beta_{mu}} = -\sum_{i=1}^{n} \Pr(y_i = k) \left\{ I(k = m) - \Pr(y_i = m) \right\} w_i x_{ti} x_{ui} \tag{15.10}$$

15.1.1   Interpretation of coefficients: Single binary predictor

To illustrate the assumptions of the model and the interpretation of the predictors, we investigate the relationships between an outcome with three levels and one binary predictor; see section 16.1 for similar description of the ordered model. In the simple design with only one binary predictor, there are six means because we allow the association of the predictor to be different for each of the outcome levels. Table 15.1 presents the layout of the means. Given the number of parameters and the constraints that the probabilities sum to unity, the parameters associated with the reference level are set to zero. Table 15.2 shows the layout of the means with the parameters of the reference level set to zero.

Table 15.1. Multinomial (three-levels) logistic regression with one binary predictor

Outcome	Binary predictor 0	Binary predictor 1
1	$\dfrac{e^{\beta_{10}}}{e^{\beta_{10}} + e^{\beta_{20}} + e^{\beta_{30}}}$	$\dfrac{e^{\beta_{10}+\beta_{11}}}{e^{\beta_{10}+\beta_{11}} + e^{\beta_{20}+\beta_{21}} + e^{\beta_{30}+\beta_{31}}}$
2	$\dfrac{e^{\beta_{20}}}{e^{\beta_{10}} + e^{\beta_{20}} + e^{\beta_{30}}}$	$\dfrac{e^{\beta_{20}+\beta_{21}}}{e^{\beta_{10}+\beta_{11}} + e^{\beta_{20}+\beta_{21}} + e^{\beta_{30}+\beta_{31}}}$
3	$\dfrac{e^{\beta_{30}}}{e^{\beta_{10}} + e^{\beta_{20}} + e^{\beta_{30}}}$	$\dfrac{e^{\beta_{30}+\beta_{31}}}{e^{\beta_{10}+\beta_{11}} + e^{\beta_{20}+\beta_{21}} + e^{\beta_{30}+\beta_{31}}}$
Total	1	1

Table 15.2. Multinomial (three-levels) logistic regression with one binary predictor where the coefficients of the reference outcome (β_{10}, β_{11}) are set to zero

Outcome	Binary predictor 0	Binary predictor 1
1	$\dfrac{1}{1 + e^{\beta_{20}} + e^{\beta_{30}}}$	$\dfrac{1}{1 + e^{\beta_{20}+\beta_{21}} + e^{\beta_{30}+\beta_{31}}}$
2	$\dfrac{e^{\beta_{20}}}{1 + e^{\beta_{20}} + e^{\beta_{30}}}$	$\dfrac{e^{\beta_{20}+\beta_{21}}}{1 + e^{\beta_{20}+\beta_{21}} + e^{\beta_{30}+\beta_{31}}}$
3	$\dfrac{e^{\beta_{30}}}{1 + e^{\beta_{20}} + e^{\beta_{30}}}$	$\dfrac{e^{\beta_{30}+\beta_{31}}}{1 + e^{\beta_{20}+\beta_{21}} + e^{\beta_{30}+\beta_{31}}}$
Total	1	1

The relative risk of outcome 2 versus the reference level (outcome 1) for the binary predictor having outcome 1 is then

$$\left(\frac{e^{\beta_{20}+\beta_{21}}}{1 + e^{\beta_{20}+\beta_{21}} + e^{\beta_{30}+\beta_{31}}}\right) \Big/ \left(\frac{1}{1 + e^{\beta_{20}+\beta_{21}} + e^{\beta_{30}+\beta_{31}}}\right) = e^{\beta_{20}+\beta_{21}} \qquad (15.11)$$

whereas the relative risk of outcome 2 versus the reference level (outcome 1) for the binary predictor having outcome 0 is then

$$\left(\frac{e^{\beta_{20}}}{1 + e^{\beta_{20}} + e^{\beta_{30}}}\right) \bigg/ \left(\frac{1}{1 + e^{\beta_{20}} + e^{\beta_{30}}}\right) = e^{\beta_{20}} \qquad (15.12)$$

Thus, the relative-risk ratio for outcome 2 versus the reference level of the binary predictor is $\exp(\beta_{21})$. Similarly, the interpretation of $\exp(\beta_{31})$ is the relative-risk ratio for outcome 3 versus the reference level of the binary predictor.

Note that the exponentiated coefficients in this model always have an interpretation as the relative-risk ratio. Although the subsequent section illustrates an equivalence to logistic regression when there are only two outcomes, exponentiated coefficients should not be interpreted as odds ratios when there are more than two outcome levels.

15.1.2 Example: Relation to logistic regression

The multinomial logit model is a generalization of logistic regression to more than two outcomes. It is equivalent to logistic regression when there are only two outcomes.

Recall our example from chapter 9 in which we investigated `heart.dta`. Here we show the equivalence of multinomial logit to logistic regression for two outcomes:

```
. use http://www.stata-press.com/data/hh4/heart
. set seed 21392
. generate sortu = runiform()
. sort sortu
. mlogit death anterior hcabg kk2 kk3 age2-age4, nolog
```

Multinomial logistic regression		Number of obs	=	4,483
		LR chi2(7)	=	211.35
		Prob > chi2	=	0.0000
Log likelihood = −636.6339		Pseudo R2	=	0.1424

death	Coef.	Std. Err.	z	P>|z|	[95% Conf. Interval]	
0	(base outcome)					
1						
anterior	.6411581	.1672199	3.83	0.000	.3134131	.9689031
hcabg	.7450945	.3528945	2.11	0.035	.053434	1.436755
kk2	.80282	.1671932	4.80	0.000	.4751274	1.130513
kk3	2.659937	.3559227	7.47	0.000	1.962341	3.357532
age2	.4923474	.3101561	1.59	0.112	−.1155474	1.100242
age3	1.509629	.2659662	5.68	0.000	.9883452	2.030913
age4	2.182796	.2711773	8.05	0.000	1.651298	2.714293
_cons	−5.049838	.257972	−19.58	0.000	−5.555454	−4.544222

```
. glm death anterior hcabg kk2 kk3 age2-age4, family(binomial) noheader nolog
```

death	Coef.	OIM Std. Err.	z	P>|z|	[95% Conf. Interval]	
anterior	.6411581	.1672199	3.83	0.000	.3134131	.9689031
hcabg	.7450945	.3528945	2.11	0.035	.053434	1.436755
kk2	.80282	.1671932	4.80	0.000	.4751274	1.130513
kk3	2.659936	.3559226	7.47	0.000	1.962341	3.357532
age2	.4923457	.310156	1.59	0.112	-.1155489	1.10024
age3	1.509628	.265966	5.68	0.000	.9883438	2.030911
age4	2.182794	.2711772	8.05	0.000	1.651296	2.714291
_cons	-5.049836	.2579718	-19.58	0.000	-5.555451	-4.54422

15.1.3   Example: Relation to conditional logistic regression

McFadden (1974) describes the analysis of multinomial data in which each respondent can choose only one from a common list of alternatives. The data may be organized such that there is one observation for each respondent for each possible choice. One choice is marked with 1, and $r - 1$ choices are marked with 0. In conditional logistic regression, we assume that we have $i = 1, \ldots, n$ clusters (panels) of data where each cluster has n_i observations. In conditional logistic regression, any number of observations in a cluster may be marked with a 1, so that the McFadden's choice model is a special case of conditional logistic regression.

Greene (2012) includes a 210-respondent dataset from Greene (2002) in which a choice model is applied to travel mode between Sydney and Melbourne, Australia. The modes of travel are air, train, bus, and car. There are also two person-specific constants in the data: hinc, the household income, and psize, the party size in the chosen mode. The hinc and psize variables are constant for all choices for a given person (they vary between persons). The relation of the multinomial logit model to the conditional logistic model exists when we limit our covariates to person-specific constants.

For this example, we have 210 respondents and four choices per respondent, for a total of 840 observations. The variable choice is set to a member of the set $\{1, 2, 3, 4\}$, meaning $\{$air, train, bus, car$\}$. Indicator variables for the modes of travel are also created, as well as the interactions of the hinc and psize variables with the mode indicator variables air, bus, and train. Here we show the equivalence of the two commands:

First, we generate the necessary data:

```
. use http://www.stata-press.com/data/hh4/tbl19-2, clear
. generate cc      = mod(_n-1,4)+1
. generate g       = int((_n-1)/4)+1
. generate air     = cc==1
. generate bus     = cc==3
. generate train   = cc==2
. generate airinc  = air*hinc
. generate choice  = cc*mode
. generate hair    = hinc*air
. generate htrain  = hinc*train
. generate hbus    = hinc*bus
. generate psair   = psize*air
. generate pstrain = psize*train
. generate psbus   = psize*bus
. label define mlab 1 "air" 2 "train" 3 "bus" 4 "car"
. label values choice mlab
```

Then we fit the alternative model:

```
. clogit mode hair psair air htrain pstrain train hbus psbus bus, group(g)
> nolog
Conditional (fixed-effects) logistic regression
                                          Number of obs    =        840
                                          LR chi2(9)       =      75.56
                                          Prob > chi2      =     0.0000
Log likelihood = -253.34085               Pseudo R2        =     0.1298
```

mode	Coef.	Std. Err.	z	P>|z|	[95% Conf. Interval]	
hair	.0035438	.0103047	0.34	0.731	-.0166531	.0237407
psair	-.6005541	.1992005	-3.01	0.003	-.9909798	-.2101284
air	.9434924	.549847	1.72	0.086	-.1341881	2.021173
htrain	-.0573078	.0118416	-4.84	0.000	-.0805169	-.0340987
pstrain	-.3098126	.1955598	-1.58	0.113	-.6931027	.0734775
train	2.493848	.5357211	4.66	0.000	1.443854	3.543842
hbus	-.0303253	.0132228	-2.29	0.022	-.0562415	-.004409
psbus	-.9404139	.3244532	-2.90	0.004	-1.57633	-.3044974
bus	1.977971	.671715	2.94	0.003	.6614336	3.294508

```
. mlogit choice hinc psize if choice>0, base(4) nolog
Multinomial logistic regression                    Number of obs   =        210
                                                   LR chi2(6)      =      60.84
                                                   Prob > chi2     =     0.0000
Log likelihood = -253.34085                        Pseudo R2       =     0.1072
```

choice	Coef.	Std. Err.	z	P>|z|	[95% Conf. Interval]
air						
hinc	.0035438	.0103047	0.34	0.731	-.0166531 .0237407
psize	-.6005541	.1992005	-3.01	0.003	-.9909798 -.2101284
_cons	.9434923	.549847	1.72	0.086	-.1341881 2.021173
train						
hinc	-.0573078	.0118416	-4.84	0.000	-.0805169 -.0340987
psize	-.3098126	.1955598	-1.58	0.113	-.6931028 .0734775
_cons	2.493848	.5357211	4.66	0.000	1.443854 3.543842
bus						
hinc	-.0303253	.0132228	-2.29	0.022	-.0562415 -.004409
psize	-.940414	.3244532	-2.90	0.004	-1.576331 -.3044974
_cons	1.977971	.671715	2.94	0.003	.6614334 3.294508
car	(base outcome)					

15.1.4   Example: Extensions with conditional logistic regression

In the previous section, we presented a pedagogical example illustrating the conditions under which the multinomial logit model and the conditional logistic regression model are the same. The conditions were restrictive and the example was contrived.

In the analysis presented in Greene (2012), the independent variables actually used in the data analysis are not respondent-specific constants. In fact, they are choice-specific covariates that cannot be used in the multinomial logit model—the dataset may not be collapsed by respondent because the values of the choice-specific covariates are not constant for the respondent. Here the conditional logit model is preferred. In the reference cited, the illustrated example is another model entirely (nested logit model). Although the tools to fit the nested logit model are now available in Stata, we will limit our discussion to a construction of the conditional logit model (called the unconditional model in the cited text).

We have choice-specific covariates: `gc`, the generalized cost constructed from measures of the in-vehicle cost and a product of a wage measure with the time spent traveling; `ttme`, reflecting the terminal time (there is zero waiting time for a car); `hinc`, the household income; and `airinc`, the interaction of the household income with the `fly` mode of travel. A conditional logistic model is preferred because of the presence of the choice-specific constants. The data are fit below.

```
. clogit mode air bus train gc ttme airinc, group(g) nolog
Conditional (fixed-effects) logistic regression
                                            Number of obs    =        840
                                            LR chi2(6)       =     183.99
                                            Prob > chi2      =     0.0000
Log likelihood = -199.12837                 Pseudo R2        =     0.3160
```

mode	Coef.	Std. Err.	z	P>|z|	[95% Conf. Interval]	
air	5.207443	.7790551	6.68	0.000	3.680523	6.734363
bus	3.163194	.4502659	7.03	0.000	2.280689	4.045699
train	3.869043	.4431269	8.73	0.000	3.00053	4.737555
gc	-.0155015	.004408	-3.52	0.000	-.024141	-.006862
ttme	-.0961248	.0104398	-9.21	0.000	-.1165865	-.0756631
airinc	.013287	.0102624	1.29	0.195	-.0068269	.033401

15.1.5 The independence of irrelevant alternatives

The multinomial logit model carries with it the assumption of the independence of irrelevant alternatives (IIA). A humorous example of violating this assumption is illustrated in a story about Groucho Marx, a famous American comedian. Marx was dining in a posh restaurant when the waiter informed him that the specials for the evening were steak, fish, and chicken. Groucho ordered the steak. The waiter returned later and apologized that there was no fish that evening. Groucho replied, "In that case, I'll have the chicken."[1]

A common application of the multinomial logit model is to model choices in available modes of transportation for commuters. Let us assume, for the sake of emphasizing the role of the IIA assumption, that commuters may choose from the following:

- Driver of a car

- Passenger in a car

- Passenger on a train

- Bicyclist

- Pedestrian

In fitting a model, let us further assume that we use car drivers as our reference group. We obtain a risk ratio of 5 in comparing the propensity to ride the train with the propensity to drive a car. The IIA assumption says that this preference is unaffected by the presence of the other choices (not involved in the ratio). If all other modes of commuting vanished and the associated people had to choose between driving a car and riding a train, they would do so in such a way that the risk ratio of 5 would hold constant.

---

1. Unfortunately, we have no reference for this story and fear that it may be urban legend. In any case, we hope that it is a memorable way to introduce the topic at hand.

Because this is a rather strong assumption, researchers using this model are interested in available tests to evaluate the assumption. Hausman (1978) introduced such a test. As we might guess, the test is built from fitting two multinomial logit models. The first estimation is the full model, and the second estimation deletes all observations for one of the choices in the outcome variable. Equivalently, the second model eliminates one choice from the outcome set. If the IIA assumption holds, then the full model is more efficient than the second model, but both models are consistent and we expect to see no systematic change in the coefficients that are common to both models. The test statistic is given by

$$\chi_\nu^2 = (\boldsymbol{\beta}_s - \boldsymbol{\beta}_f)^{\mathrm{T}} (V_s - V_f)^{-1} (\boldsymbol{\beta}_s - \boldsymbol{\beta}_f) \tag{15.13}$$

where the s subscript is the subset (second) model, the f subscript is the full (first) model, and the ν subscript denotes the degrees of freedom of the test statistic.

There may be numeric problems in calculating the inverse of the difference in the variance matrices, so the inverse term in the middle of the equation is a generalized inverse. The degrees of freedom of the test statistic are taken to be the number of common coefficients in the two estimation models. When a generalized inverse is needed, the degrees of freedom are taken to be the rank of the difference matrix.

15.1.6 Example: Assessing the IIA

Let us return to our first analysis of the travel mode data and assess the IIA assumption. Stata provides the `hausman` test. To use this command, we must fit two models. The first model is the full model, and the second model is a subset where we remove one of the choices.

```
. mlogit choice hinc psize if choice>0, base(4) nolog
Multinomial logistic regression                    Number of obs    =         210
                                                   LR chi2(6)       =       60.84
                                                   Prob > chi2      =      0.0000
Log likelihood = -253.34085                        Pseudo R2        =      0.1072
```

choice	Coef.	Std. Err.	z	P>|z|	[95% Conf. Interval]
air						
hinc	.0035438	.0103047	0.34	0.731	-.0166531	.0237407
psize	-.6005541	.1992005	-3.01	0.003	-.9909798	-.2101284
_cons	.9434923	.549847	1.72	0.086	-.1341881	2.021173
train						
hinc	-.0573078	.0118416	-4.84	0.000	-.0805169	-.0340987
psize	-.3098126	.1955598	-1.58	0.113	-.6931028	.0734775
_cons	2.493848	.5357211	4.66	0.000	1.443854	3.543842
bus						
hinc	-.0303253	.0132228	-2.29	0.022	-.0562415	-.004409
psize	-.940414	.3244532	-2.90	0.004	-1.576331	-.3044974
_cons	1.977971	.671715	2.94	0.003	.6614334	3.294508
car	(base outcome)					

```
. estimates store all
. mlogit choice hinc psize if choice!=0 & choice!=2, base(4) nolog
Multinomial logistic regression                    Number of obs    =         147
                                                   LR chi2(4)       =       27.02
                                                   Prob > chi2      =      0.0000
Log likelihood =  -141.9678                        Pseudo R2        =      0.0869
```

choice	Coef.	Std. Err.	z	P>|z|	[95% Conf. Interval]
air						
hinc	.0037738	.0106795	0.35	0.724	-.0171577	.0247053
psize	-.5861604	.19851	-2.95	0.003	-.9752327	-.197088
_cons	.9074443	.5592262	1.62	0.105	-.1886189	2.003507
bus						
hinc	-.03247	.0138722	-2.34	0.019	-.059659	-.005281
psize	-.9604402	.3423731	-2.81	0.005	-1.631479	-.2894012
_cons	2.074974	.7159861	2.90	0.004	.6716666	3.47828
car	(base outcome)					

```
. estimates store partial
```

```
. hausman partial all, alleqs constant
                       ──── Coefficients ────
                        (b)         (B)           (b-B)      sqrt(diag(V_b-V_B))
                       partial      all         Difference         S.E.

air
       hinc          .0037738     .0035438        .00023          .0028044
       psize        -.5861604    -.6005541       .0143938             .
       _cons         .9074443     .9434923      -.0360481         .1019909

bus
       hinc          -.03247     -.0303253      -.0021447         .0041946
       psize        -.9604402    -.940414       -.0200262         .1093137
       _cons        2.074974     1.977971        .097003          .2478609

                    b = consistent under Ho and Ha; obtained from mlogit
            B = inconsistent under Ha, efficient under Ho; obtained from mlogit
      Test:  Ho:  difference in coefficients not systematic
                  chi2(6) = (b-B)´[(V_b-V_B)^(-1)](b-B)
                          =         0.26
                Prob>chi2 =       0.9997
                (V_b-V_B is not positive definite)
```

The results show that the IIA assumption is not violated for our data.

One cannot apply the Hausman test with blind faith. The test does not require us
to choose any particular outcome to remove from the nested model and, annoyingly,
one can see conflicting results from the test depending on which outcome we remove.
For the variance matrix used in the Hausman test to be singular is also common. These
problems are usually associated with models that are ill-fitted or based on small samples
(or both). In these cases, a more comprehensive review of the model is in order, for
which a reliable application of this test may not be possible.

15.1.7   Interpreting coefficients

Interpreting the coefficients is hard because of the nonlinearity of the link function and
the incorporation of a base reference group. As in section 10.6, we can motivate an
alternative metric that admits a transformation of the coefficients for easier interpretation.

Because the model is fit using one of the outcomes as a base reference group, the
probabilities that we calculate are relative to that base group. We can define a relative-
risk measure for an observation i as the probability of outcome k over the probability
of the reference outcome; see (15.4) and (15.5).

$$\text{Relative risk for outcome } k \;=\; \exp(\xi_{ik}) \qquad (15.14)$$

This measure can be calculated for each outcome and each covariate.

For illustration of the interpretation of a given coefficient, we will assume a model with two covariates $\mathbf{x}_1$ and $\mathbf{x}_2$ along with a constant. The relative-risk ratio for $\mathbf{x}_1$ and outcome k is then calculated as the ratio of relative-risk measures. The numerator relative risk is calculated such that the specific covariate is incremented by one relative to the value used in the denominator. The relative-risk ratio for $\mathbf{x}_1$ and outcome k is calculated as

$$\text{Relative-risk ratio for } \mathbf{x}_1 = \frac{\exp\{\beta_0 + (x_{1ki} + 1)\beta_1 + x_{2ki}\beta_2\}}{\exp(\beta_0 + x_{1ki}\beta_1 + x_{2ki}\beta_2)} \tag{15.15}$$

$$= \exp(\beta_1) \tag{15.16}$$

The calculation of the relative-risk ratio simplifies such that there is no dependence on a particular observation. The relative-risk ratio is therefore constant—it is independent of the particular values of the covariates.

Interpreting exponentiated coefficients is straightforward for multinomial logit models. These exponentiated coefficients represent relative-risk ratios. A relative-risk ratio of 2 indicates that an outcome is twice as likely relative to the base reference group if the associated covariate is increased by one.

15.1.8   Example: Medical admissions—introduction

We now wish to demonstrate the development of a multinomial model from the medical data used earlier in the text. We must first convert the three `type` variables to one response variable called `admit`. We then tabulate `admit`, showing the number of cases in each type of admission.

```
. use http://www.stata-press.com/data/hh4/medpar, clear
. generate byte admit = type1 + 2*type2 + 3*type3
. label define admitlab 1 "Elective" 2 "Urgent" 3 "Emergency"
. label values admit admitlab
. label define whitelab 1 "White" 0 "Other"
. label values white whitelab
. tabulate admit
```

admit	Freq.	Percent	Cum.
Elective	1,134	75.85	75.85
Urgent	265	17.73	93.58
Emergency	96	6.42	100.00
Total	1,495	100.00	

As expected, emergency admissions are less common than elective or urgent admissions. We next summarize variables `white` (1 = white patient; 0 = otherwise) and `los` (hospital length of stay in days) within each level or category of `admit`. The mean value of `white` indicates the percentage of white patients in each level of `admit`. Of the patients being admitted as elective, 93% are white and 7% are nonwhite.

```
. tabulate admit white
```

| | | white | |
admit	Other	White	Total
Elective	80	1,054	1,134
Urgent	37	228	265
Emergency	10	86	96
Total	127	1,368	1,495

The odds ratios for the urgent and emergency levels, with level one (elective) as the reference, can be determined directly:

$$\frac{(80)(228)}{(37)(1054)} \quad = \quad 0.46771629 \tag{15.17}$$

$$\frac{(80)(86)}{(10)(1054)} \quad = \quad 0.65275142 \tag{15.18}$$

Directly modeling the relationship between the measures by using multinomial logistic regression confirms the direct calculations.

```
. mlogit admit white, rrr nolog
Multinomial logistic regression              Number of obs   =      1,495
                                             LR chi2(2)      =      12.30
                                             Prob > chi2     =     0.0021
Log likelihood = -1029.3204                  Pseudo R2       =     0.0059
```

admit	RRR	Std. Err.	z	P>|z|	[95% Conf.	Interval]
Elective	(base outcome)					
Urgent						
white	.4677163	.0990651	-3.59	0.000	.3088111	.7083894
_cons	.4625	.0919515	-3.88	0.000	.3132427	.6828771
Emergency						
white	.6527514	.2308532	-1.21	0.228	.3263701	1.305525
_cons	.125	.0419263	-6.20	0.000	.0647751	.2412191

Note: _cons estimates baseline relative risk for each outcome.

The printed relative-risk ratios agree with the direct calculations. Moreover, for `admit=3` versus `admit=2`, we can calculate

$$\frac{(37)(86)}{(10)(228)} = 1.395614 \tag{15.19}$$

The coefficient of `white` for `admit=3` versus `admit=2` is given by $\log(1.395614) = 0.33333446$. You can also obtain this coefficient by subtracting the coefficients for `white` in the model using `admit=1` as the reference; $(0.7598934 - 0.4265589 = 0.3333345)$.

We run `mlogit` again but this time using `admit=2` as the reference category.

```
. mlogit admit white, base(2) nolog
Multinomial logistic regression                    Number of obs   =      1,495
                                                   LR chi2(2)      =      12.30
                                                   Prob > chi2     =     0.0021
Log likelihood = -1029.3204                        Pseudo R2       =     0.0059
```

admit	Coef.	Std. Err.	z	P>|z|	[95% Conf. Interval]	
Elective						
white	.7598934	.2118059	3.59	0.000	.3447614	1.175025
_cons	.7711087	.1988141	3.88	0.000	.3814403	1.160777
Urgent	(base outcome)					
Emergency						
white	.3333345	.3782075	0.88	0.378	-.4079386	1.074608
_cons	-1.308333	.3564085	-3.67	0.000	-2.006881	-.609785

In the next section, we illustrate interpreting the relative-risk ratios and coefficients.

15.1.9   Example: Medical admissions—summary

We will next develop a more complex model by adding an additional predictor with the aim of interpreting and testing the resulting model. We use the data introduced in the previous section with the same `admit` response variable. We begin by summarizing the mean, standard deviation, minimum value, and maximum value for each predictor, `white` and `los`, for all three levels of the outcome measure `admit`.

```
. by admit, sort: summarize white los
```

-> admit = Elective

Variable	Obs	Mean	Std. Dev.	Min	Max
white	1,134	.9294533	.2561792	0	1
los	1,134	8.830688	6.456009	1	60

-> admit = Urgent

Variable	Obs	Mean	Std. Dev.	Min	Max
white	265	.8603774	.3472509	0	1
los	265	11.19623	8.824852	1	63

-> admit = Emergency

Variable	Obs	Mean	Std. Dev.	Min	Max
white	96	.8958333	.3070802	0	1
los	96	18.23958	20.61259	1	116

We next model `admit` on `white` and `los` by using a multinomial model. There is a loose order to the levels (qualitative rather than quantitative). We will evaluate the ordinality of the model afterward.

```
. mlogit admit white los, nolog
Multinomial logistic regression                 Number of obs   =       1,495
                                                LR chi2(4)      =       87.02
                                                Prob > chi2     =      0.0000
Log likelihood = -991.96238                     Pseudo R2       =      0.0420
```

admit	Coef.	Std. Err.	z	P>|z|	[95% Conf. Interval]
Elective	(base outcome)					
Urgent						
white	-.7086604	.2140645	-3.31	0.001	-1.128219	-.2891017
los	.0375103	.0084408	4.44	0.000	.0209666	.054054
_cons	-1.187453	.2227772	-5.33	0.000	-1.624088	-.7508174
Emergency						
white	-.2267416	.377415	-0.60	0.548	-.9664613	.5129782
los	.0780083	.0097884	7.97	0.000	.0588233	.0971932
_cons	-3.188991	.3917186	-8.14	0.000	-3.956746	-2.421237

```
. estat ic
Akaike's information criterion and Bayesian information criterion
```

Model	Obs	ll(null)	ll(model)	df	AIC	BIC
.	1,495	-1035.471	-991.9624	6	1995.925	2027.784

Note: N=Obs used in calculating BIC; see [R] BIC note.

Interpretation is facilitated by expressing the coefficients in exponential form, using relative-risk ratios.

```
. mlogit admit white los, rrr nolog
Multinomial logistic regression                 Number of obs    =        1,495
                                                LR chi2(4)       =        87.02
                                                Prob > chi2      =       0.0000
Log likelihood = -991.96238                     Pseudo R2        =       0.0420
```

admit	RRR	Std. Err.	z	P>|z|	[95% Conf. Interval]	
Elective	(base outcome)					
Urgent						
white	.4923032	.1053847	-3.31	0.001	.323609	.7489361
los	1.038223	.0087634	4.44	0.000	1.021188	1.055542
_cons	.3049972	.0679464	-5.33	0.000	.1970914	.4719806
Emergency						
white	.7971268	.3008476	-0.60	0.548	.3804269	1.670258
los	1.081132	.0105826	7.97	0.000	1.060588	1.102073
_cons	.0412134	.0161441	-8.14	0.000	.0191253	.0888117

```
Note: _cons estimates baseline relative risk for each outcome.
. predict p1 p2 p3
(option pr assumed; predicted probabilities)
```

Elective admission is the reference level; both of the other levels are interpreted relative to the reference level. Therefore, we may state, on the basis of the model, that whites are half as likely to be admitted as an urgent admission than as an elective admission. Whites are approximately 20% less likely to be admitted as an emergency admission than as elective. We can also say that for a one-unit increase in the variable los, the relative risk of having an urgent admission rather than an elective admission increases by about 4%. Likewise, for a one-unit change in the variable los, the relative risk of having an emergency admission rather than an elective admission increases by about 8%.

We can graph the relationship of the length of stay and race variables for each level by using the specifications below. The three respective graphs are given in figures 15.1, 15.2, and 15.3.

```
. sort white los
. line p1 los if white==0 || line p1 los if white==1,
> legend(order(1 "nonwhite" 2 "white")) title(Medicare LOS vs Admission Type)
> l1title("Admit=1: Elective Admission") clpattern("-###")
```

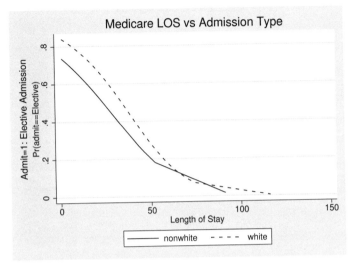

Figure 15.1. Length of stay versus admission type for elective admissions

```
. line p2 los if white==0 || line p2 los if white==1,
> legend(order(1 "nonwhite" 2 "white")) title(Medicare LOS vs Admission Type)
> l1title("Admit=2: Urgent Admission") clpattern("-###")
```

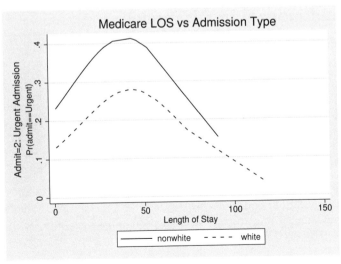

Figure 15.2. Length of stay versus admission type for urgent admissions

```
. line p3 los if white==0 || line p3 los if white==1,
> legend(order(1 "nonwhite" 2 "white")) title(Medicare LOS vs Admission Type)
> l1title("Admit=3: Emergency Admission") clpattern("-###")
```

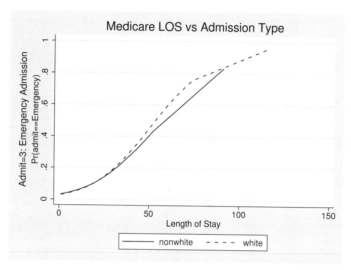

Figure 15.3. Length of stay versus admission type for emergency admissions

In particular, figure 15.2 illustrates the quadratic relationship for each race related to hospital stay for urgent admissions.

15.2   The multinomial probit model

The multinomial probit model was first described by Albright, Lerman, and Manski (1977) and examined in full by Daganzo (1979). It is a computationally difficult model to fit. It became an official part of Stata with version 9.

The multinomial probit model is to the binary probit as the multinomial logit is to the binary logit model. The multinomial probit model is often referred to as the random utility model in that an individual's choice among r alternatives is that choice having the maximum utility. The utility U_{ji} is the utility of alternative j for individual i and is assumed to be structured as a linear combination of observed alternative choices.

The multinomial probit model relaxes the assumption of irrelevant alternatives; it does not assume r different equations. However, it is a much more difficult model to fit, requiring computation of multivariate normal probabilities.

To use a parametric multinomial probit model, we assume that $y_i = j$ if $U_{ji} = \max\{U_{1i}, \ldots, U_{ri}\}$; that is, the outcome of a particular observation is determined by the maximum utility. Thus,

$$\Pr(y_i = j) = \Pr(U_{ji} > U_{j'i}) \tag{15.20}$$

for all $j' \neq j$. In the multinomial probit model, the utility for a choice j given by U_{ji} is parameterized as a linear predictor from a matrix of covariates and coefficients introduced as $\mathbf{x}_i\boldsymbol{\beta}$.

A generalization of this approach allows the covariates to differ for each choice $j = 1, \ldots, r$ such that the linear predictor used to parameterize the utility is given by $\mathbf{x}_{ji}\boldsymbol{\beta}$. This model may be fit in Stata by using the `asmprobit` command.

Under the usual multinomial probit specification, we assume that the outcome may take one of r values with r unobserved errors $\epsilon_{1i}, \ldots, \epsilon_{ri}$. For the multinomial probit model, we assume that these errors are distributed as multivariate normal

$$[\epsilon_{1i}, \epsilon_{2i}, \ldots, \epsilon_{ri}] \sim N(\mathbf{0}_{r\times 1}, \boldsymbol{\Sigma}_{r\times r}) \tag{15.21}$$

Instead of normalizing one of the coefficient vectors as we did for the multinomial logit model, we must normalize the covariance matrix entries. We normalize the variances (diagonal elements of $\boldsymbol{\Sigma}$) to one because we are making comparisons of the variances. We must normalize $r - 1$ of the covariances to zero because we are making comparisons of the outcomes. The size of $\boldsymbol{\Sigma}$ is $r \times r$, leaving $(r - 1)(r - 2)/2$ unrestricted covariances. Thus, for a model with three choices ($r = 3$), all three probabilities are based on the cumulative bivariate normal distribution where the variance matrix has ones on the diagonal and $(3 - 1)(3 - 2)/2 = 1$ covariance parameter.

As mentioned previously, the difficulty in implementing this model is the need to calculate the multivariate normal probabilities. Lerman and Manski (1981) demonstrate how one can estimate these probabilities with simulation. When the first edition of this text was written, there were no programs to fit these models in Stata. This is no longer a limitation because users now have the `mprobit` command at their disposal. Furthermore, an extension allowing choice-specific covariates may also be fit using the `asmprobit` command.

```
. mprobit admit white los if admit > 0, base(1) nolog
Multinomial probit regression                    Number of obs    =       1,495
                                                 Wald chi2(4)     =       73.36
Log likelihood = -993.34447                      Prob > chi2      =      0.0000
```

admit	Coef.	Std. Err.	z	P>|z|	[95% Conf. Interval]	
Elective	(base outcome)					
Urgent						
white	-.5706817	.1771321	-3.22	0.001	-.9178542	-.2235091
los	.0302881	.0064955	4.66	0.000	.0175572	.0430189
_cons	-.9956175	.183421	-5.43	0.000	-1.355116	-.636119
Emergency						
white	-.2238527	.2454722	-0.91	0.362	-.7049694	.257264
los	.0549894	.0070543	7.80	0.000	.0411632	.0688157
_cons	-2.283496	.2523319	-9.05	0.000	-2.778057	-1.788935

The final example in the next subsection illustrates that we can compare the multinomial probit fit with the earlier model fit by using multinomial logit. There is a similarity in the coefficients from these models that we will now explore.

15.2.1   Example: A comparison of the models

We repeat a model of admit on white and los for comparing results with mprobit. Note the comparison of the log-likelihood functions, AIC GOF statistics, predicted probabilities for each level of admit from observations 1–10, and correlations of the probabilities for each level. Also compare the respective parameter estimates and their *p*-values. All the values are similar, indicating that neither model is preferred over the other. Because the multinomial logit has many accepted fit statistics associated with the model, we would recommend it. However, if the theory underlying the data prefers normality, then multinomial probit should be selected.

The estimates for the multinomial logit model are given by

```
. set seed 29233
. generate sortu = runiform() // so we can sort data
. mlogit admit white los, nolog
```

Multinomial logistic regression   Number of obs   =   1,495
                                                   LR chi2(4)   =   87.02
                                                   Prob > chi2   =   0.0000
Log likelihood = -991.96238   Pseudo R2   =   0.0420

admit	Coef.	Std. Err.	z	P>|z|	[95% Conf. Interval]	
Elective	(base outcome)					
Urgent						
white	-.7086604	.2140645	-3.31	0.001	-1.128219	-.2891017
los	.0375103	.0084408	4.44	0.000	.0209666	.054054
_cons	-1.187453	.2227772	-5.33	0.000	-1.624088	-.7508174
Emergency						
white	-.2267416	.377415	-0.60	0.548	-.9664613	.5129782
los	.0780083	.0097884	7.97	0.000	.0588233	.0971932
_cons	-3.188991	.3917186	-8.14	0.000	-3.956746	-2.421237

```
. estat ic
```
Akaike's information criterion and Bayesian information criterion

Model	Obs	ll(null)	ll(model)	df	AIC	BIC
.	1,495	-1035.471	-991.9624	6	1995.925	2027.784

Note: N=Obs used in calculating BIC; see [R] BIC note.

```
. predict pl1 pl2 pl3
(option pr assumed; predicted probabilities)
```

For comparison, the estimates for the multinomial probit are given by

```
. mprobit admit white los, nolog
```

Multinomial probit regression

Log likelihood = -993.34447

admit	Coef.	Std. Err.	z	P>|z|	[95% Conf. Interval]	
Elective	(base outcome)					
Urgent						
white	-.5706817	.1771321	-3.22	0.001	-.9178542	-.2235091
los	.0302881	.0064955	4.66	0.000	.0175572	.0430189
_cons	-.9956175	.183421	-5.43	0.000	-1.355116	-.636119
Emergency						
white	-.2238527	.2454722	-0.91	0.362	-.7049694	.257264
los	.0549894	.0070543	7.80	0.000	.0411632	.0688157
_cons	-2.283496	.2523319	-9.05	0.000	-2.778057	-1.788935

Number of obs = 1,495
Wald chi2(4) = 73.36
Prob > chi2 = 0.0000

```
. estat ic
```

Akaike's information criterion and Bayesian information criterion

Model	Obs	ll(null)	ll(model)	df	AIC	BIC
.	1,495	.	-993.3445	6	1998.689	2030.548

Note: N=Obs used in calculating BIC; see [R] BIC note.

```
. predict pp1 pp2 pp3
(option pr assumed; predicted probabilities)
```

The predicted probabilities are similar, as we would expect.

```
. sort sortu
```

```
. list pp1 pl1 pp2 pl2 pp3 pl3 in 1/10
```

	pp1	pl1	pp2	pl2	pp3	pl3
1.	.8269817	.8267539	.1396609	.1389237	.0333574	.0343224
2.	.7282639	.7302323	.1913027	.192463	.0804334	.0773047
3.	.805451	.8061198	.1522757	.1515902	.0422733	.0422899
4.	.718506	.7204143	.195627	.1971329	.0858669	.0824528
5.	.840225	.8393431	.1314712	.1308454	.0283038	.0298115
6.	.6982979	.6999208	.2041922	.2064461	.09751	.0936331
7.	.8269817	.8267539	.1396609	.1389237	.0333574	.0343224
8.	.7377875	.7397655	.1869595	.1877975	.075253	.0724369
9.	.7649376	.7666773	.1738808	.1739149	.0611816	.0594077
10.	.6878543	.6892461	.2084193	.2110681	.1037263	.0996858

The pairwise correlations of the predicted probabilities (for level 1, 2, and 3 admission types) for the multinomial logit and multinomial probit models are given below. Not surprisingly, the correlations are very high.

```
. correlate pp1 pl1
(obs=1,495)
                 |      pp1       pl1
      -----------+------------------
             pp1 |   1.0000
             pl1 |   0.9997    1.0000

. correlate pp2 pl2
(obs=1,495)
                 |      pp2       pl2
      -----------+------------------
             pp2 |   1.0000
             pl2 |   0.9955    1.0000

. correlate pp3 pl3
(obs=1,495)
                 |      pp3       pl3
      -----------+------------------
             pp3 |   1.0000
             pl3 |   0.9976    1.0000
```

15.2.2 Example: Comparing probit and multinomial probit

We may suspect that if we were to combine two categories of a three-category response variable and then model the two responses on predictors using both probit and multinomial probit, the results would be identical. However, as described by Long (1997) and Long and Freese (2014), this is not the case. The multinomial probit parameter estimates are greater—by a scale factor of $\sqrt{2} \approx 1.414$.

To illustrate this relationship, we combine categories two and three, defining them as outcome zero, but we keep category one the same.

```
. tabulate admit, nolabel
      admit |      Freq.     Percent        Cum.
------------+-----------------------------------
          1 |      1,134       75.85       75.85
          2 |        265       17.73       93.58
          3 |         96        6.42      100.00
------------+-----------------------------------
      Total |      1,495      100.00

. generate byte admitbin = (admit==1)   /* Create a binary variable */
. tabulate admitbin
   admitbin |      Freq.     Percent        Cum.
------------+-----------------------------------
          0 |        361       24.15       24.15
          1 |      1,134       75.85      100.00
------------+-----------------------------------
      Total |      1,495      100.00
```

We next model the binary outcome `admitbin` by using a probit model.

```
. probit admitbin white los, nolog
Probit regression                                 Number of obs    =      1,495
                                                  LR chi2(2)       =      65.65
                                                  Prob > chi2      =     0.0000
Log likelihood = -793.56443                       Pseudo R2        =     0.0397

   admitbin |      Coef.   Std. Err.      z    P>|z|     [95% Conf. Interval]
------------+----------------------------------------------------------------
      white |   .3588006   .1223274     2.93   0.003     .1190433    .5985579
        los |  -.0300373   .0043505    -6.90   0.000    -.0385641   -.0215105
      _cons |   .6863979   .1261822     5.44   0.000     .4390854    .9337105
```

Then, we model the same data by using multinomial probit, making sure to constrain the reference category or level to zero, just as probit does automatically.

```
. mprobit admitbin white los, baseoutcome(0) nolog
Multinomial probit regression                     Number of obs    =      1,495
                                                  Wald chi2(2)     =      58.02
Log likelihood = -793.56443                       Prob > chi2      =     0.0000

   admitbin |      Coef.   Std. Err.      z    P>|z|     [95% Conf. Interval]
------------+----------------------------------------------------------------
0           |  (base outcome)
------------+----------------------------------------------------------------
1           |
      white |   .5074207   .1729971     2.93   0.003     .1683526    .8464887
        los |  -.0424792   .0061525    -6.90   0.000    -.0545379   -.0304204
      _cons |   .9707133   .1784486     5.44   0.000     .6209605    1.320466
```

Now, we illustrate that dividing the multinomial probit coefficient by the square root of 2 produces the original probit coefficient.

```
. display "white: " _b[white]/sqrt(2) _n "los  : " _b[los]/sqrt(2) _n
> "_cons: " _b[_cons]/sqrt(2)
white: 0
los  : 0
_cons: 0
```

Thus, the relationship holds as stated. The label above the list of coefficient names in the output table says _outcome_2. This is a simple shift in the frame of reference. _outcome_1 is the reference level, which is 0 for the binary and multinomial probit models, but with 0 meaning a combined `admit=2` or `admit=3`.

We now look at the relationship of predicted probabilities. We again model `admitbin` on `white` and `los` by using probit regression, calculating probabilities for level 1, that is, the probability of success, which here is the probability of an elective admission. How the previous model has been reparameterized should be clear.

```
. probit admitbin white los, nolog
```

Probit regression				Number of obs	=	1,495
				LR chi2(2)	=	65.65
				Prob > chi2	=	0.0000
Log likelihood = -793.56443				Pseudo R2	=	0.0397

admitbin	Coef.	Std. Err.	z	P>|z|	[95% Conf. Interval]	
white	.3588006	.1223274	2.93	0.003	.1190433	.5985579
los	-.0300373	.0043505	-6.90	0.000	-.0385641	-.0215105
_cons	.6863979	.1261822	5.44	0.000	.4390854	.9337105

```
. predict prob1
(option pr assumed; Pr(admitbin))
```

We then model the same data by using **mprobit** and forcing the reference level to be zero. We specify the **probitparam** option to adjust the model by the square root of 2. The coefficients are now the same as those obtained from the **probit** command shown above. After fitting the model, we obtain the predicted probabilities.

```
. mprobit admitbin white los, probitparam baseoutcome(0) nolog
```

Multinomial probit regression				Number of obs	=	1,495
				Wald chi2(2)	=	58.02
Log likelihood = -793.56443				Prob > chi2	=	0.0000

admitbin	Coef.	Std. Err.	z	P>|z|	[95% Conf. Interval]	
0	(base outcome)					
1						
white	.3588006	.1223274	2.93	0.003	.1190433	.5985579
los	-.0300373	.0043505	-6.90	0.000	-.0385641	-.0215105
_cons	.6863979	.1261822	5.44	0.000	.4390854	.9337105

```
. predict mprob1 mprob2
(option pr assumed; predicted probabilities)
```

A quick check of the probabilities confirms that the probabilities from the multinomial model match those from the binary probit model.

```
. compare prob1 mprob2
```

		difference		
	count	minimum	average	maximum
prob1=mprob2	1493			
prob1>mprob2	2	4.66e-10	7.68e-09	1.49e-08
jointly defined	1495	0	1.03e-11	1.49e-08
total	1495			

A listing illustrates how the multinomial predicted probabilities relate to the predicted probability from the binary model.

```
. sort sortu
. list prob1 mprob1 mprob2 in 1/10
```

	prob1	mprob1	mprob2
1.	.830233	.1697669	.830233
2.	.7239576	.2760424	.7239576
3.	.8064737	.1935263	.8064737
4.	.7138277	.2861723	.7138277
5.	.8449855	.1550145	.8449855
6.	.6930545	.3069455	.6930545
7.	.830233	.1697669	.830233
8.	.7339082	.2660917	.7339082
9.	.76263	.2373699	.76263
10.	.6824247	.3175753	.6824247

The `mprob1` probabilities are calculated as `1.0 - mprob2`, which makes sense when we recall that for a given variable the probability of 1 and the probability of 0 must sum to 1.0. The `mprob1` probabilities correlate perfectly, but inversely, with `mprob2` and with `prob1`.

Of ancillary interest, the probabilities appear the same regardless of whether we use the `probitparam` option. The coefficients, however, differ.

```
. sort sortu
. quietly mprobit admitbin white los, baseoutcome(0) nolog
. predict pm1 pm2
(option pr assumed; predicted probabilities)
. correlate prob1 mprob1 mprob2 pm1 pm2
(obs=1,495)
```

	prob1	mprob1	mprob2	pm1	pm2
prob1	1.0000				
mprob1	-1.0000	1.0000			
mprob2	1.0000	-1.0000	1.0000		
pm1	-1.0000	1.0000	-1.0000	1.0000	
pm2	1.0000	-1.0000	1.0000	-1.0000	1.0000

```
. list prob1 mprob1 mprob2 pm1 pm2 in 1/5
```

	prob1	mprob1	mprob2	pm1	pm2
1.	.830233	.1697669	.830233	.1697669	.830233
2.	.7239576	.2760424	.7239576	.2760424	.7239576
3.	.8064737	.1935263	.8064737	.1935263	.8064737
4.	.7138277	.2861723	.7138277	.2861723	.7138277
5.	.8449855	.1550145	.8449855	.1550145	.8449855

15.2.3   Example: Concluding remarks

Many think that the advantage of multinomial probit over multinomial logit is that the former is immune to violation of the IIA. However, as Long and Freese (2014) discuss, the parameter estimates and probabilities from the two models are nearly always very close.

We have not discussed fit statistics here, but many of the tests of multinomial logit models can be used to evaluate multinomial probit models. Exponentiated probit coefficients are not odds ratios, even though prima facie they appear to be. For probit models, exponentiated coefficients have no statistically interesting interpretation. Therefore, the multinomial tests related to odds ratios are not to be used for evaluating multinomial probit models. However, commands such as `prvalue`, which is an extremely useful utility, and several others that are found in Long and Freese (2014) can be used to evaluate these models.

You can find many other categorical response models in the statistical literature. Stata has several more, including stereotype logistic regression, first discussed by Anderson (1984); alternative-specific multinomial probit; conditional logistic regression, which we briefly looked at when comparing it with multinomial models; and rank-ordered logistic regression. Long and Freese (2014) and Hilbe (2009) also discuss each of the above-mentioned models.

16 The ordered-response family

This chapter addresses data in which we record an integer response. Our responses are limited to a finite choice set, and there is some meaning to the order of the values assigned. The actual values are irrelevant but are interpreted to have meaning in that a larger value is considered higher in some sense. There is no interpretation regarding the difference between the outcomes (the labels may be equally spaced but may represent different measures). Because the actual values are irrelevant, without loss of generality we may present a discussion of the ordered outcome model, assuming that the outcomes are defined by the set $\{1, 2, \ldots, r\}$.

We begin by considering a model given by

$$y^* = \mathbf{X}\boldsymbol{\beta} + \epsilon \tag{16.1}$$

where y^* is unobserved. Instead, we observe an outcome y defined by

$$y \; = \; 1 \quad \text{if } \kappa_0 < y^* \le \kappa_1 \tag{16.2}$$
$$= \; 2 \quad \text{if } \kappa_1 < y^* \le \kappa_2 \tag{16.3}$$
$$= \; 3 \quad \text{if } \kappa_2 < y^* \le \kappa_3 \tag{16.4}$$
$$\vdots$$
$$= \; r \quad \text{if } \kappa_{r-1} < y^* < \kappa_r \tag{16.5}$$

$\kappa_0, \kappa_1, \ldots, \kappa_r$ are cutpoints that satisfy

$$-\infty = \kappa_0 < \kappa_1 < \kappa_2 < \cdots < \kappa_r = \infty \tag{16.6}$$

such that the cutpoints represent $r-1$ additional unknowns in the model. The probability of outcome i corresponds to the probability that the linear function plus the error is within the range of the associated cutpoints (κ_{i-1}, κ_i). For more reading, you can find an excellent review of models for ordered responses in Boes and Winkelmann (2006).

Note that if the outcome has only two levels, then an unordered logistic regression model and an ordered logistic regression model will produce the same estimates for $\beta_1, \ldots, \beta_p$ and that $\beta_0 = -\kappa_1$.

16.1   Interpretation of coefficients: Single binary predictor

To illustrate the assumptions of the model and the interpretation of the predictors, we investigate the relationships between an outcome with three levels and one binary predictor; see section 15.1.1 for similar description of the unordered model. In this example, we will focus on the logit link. In this simple design, there are six means subject to constraints. Table 16.1 presents the layout of the means. Given the number of parameters and the constraints that the probabilities sum to unity, the parameters associated with the reference level are set to zero. Table 16.2 shows the layout of the means with the parameters of the reference level set to zero.

Table 16.1. Ordered (three-levels) logistic regression with one binary predictor

Outcome	Binary predictor			
	0		1	
1	$\dfrac{e^{\kappa_1}}{1+e^{\kappa_1}}$	$-\dfrac{e^{\kappa_0}}{1+e^{\kappa_0}}$	$\dfrac{e^{\kappa_1-\beta_1}}{1+e^{\kappa_1-\beta_1}}$	$-\dfrac{e^{\kappa_0-\beta_1}}{1+e^{\kappa_0-\beta_1}}$
2	$\dfrac{e^{\kappa_2}}{1+e^{\kappa_2}}$	$-\dfrac{e^{\kappa_1}}{1+e^{\kappa_1}}$	$\dfrac{e^{\kappa_2-\beta_1}}{1+e^{\kappa_2-\beta_1}}$	$-\dfrac{e^{\kappa_1-\beta_1}}{1+e^{\kappa_1-\beta_1}}$
3	$\dfrac{e^{\kappa_3}}{1+e^{\kappa_3}}$	$-\dfrac{e^{\kappa_2}}{1+e^{\kappa_2}}$	$\dfrac{e^{\kappa_3-\beta_1}}{1+e^{\kappa_3-\beta_1}}$	$-\dfrac{e^{\kappa_2-\beta_1}}{1+e^{\kappa_2-\beta_1}}$
Total	1		1	

Table 16.2. Ordered (three-levels) logistic regression with one binary predictor where $\kappa_0 = -\infty$ and $\kappa_3 = \infty$

Outcome	Binary predictor			
	0		1	
1	$\dfrac{e^{\kappa_1}}{1+e^{\kappa_1}}$		$\dfrac{e^{\kappa_1-\beta_1}}{1+e^{\kappa_1-\beta_1}}$	
2	$\dfrac{e^{\kappa_2}}{1+e^{\kappa_2}}$	$-\dfrac{e^{\kappa_1}}{1+e^{\kappa_1}}$	$\dfrac{e^{\kappa_2-\beta_1}}{1+e^{\kappa_2-\beta_1}}$	$-\dfrac{e^{\kappa_1-\beta_1}}{1+e^{\kappa_1-\beta_1}}$
3	$1-\dfrac{e^{\kappa_2}}{1+e^{\kappa_2}}$		$1-\dfrac{e^{\kappa_2-\beta_1}}{1+e^{\kappa_2-\beta_1}}$	
Total	1		1	

Finally, we create tables 16.3 and 16.4 for the two respective models defined by outcome={1} versus outcome={2,3} and outcome={1,2} versus outcome={3}. Because we add up probabilities for various outcome levels to compare a given split, we call these cumulative logit models.

Table 16.3. Cumulative logits outcome {1} versus outcomes {2, 3}

Outcome	Binary predictor	
	0	1
1	$\dfrac{e^{\kappa_1}}{1 + e^{\kappa_1}}$	$\dfrac{e^{\kappa_1 - \beta_1}}{1 + e^{\kappa_1 - \beta_1}}$
2,3	$1 - \dfrac{e^{\kappa_1}}{1 + e^{\kappa_1}}$	$1 - \dfrac{e^{\kappa_1 - \beta_1}}{1 + e^{\kappa_1 - \beta_1}}$
Total	1	1

Table 16.4. Cumulative logits outcomes {1, 2} versus outcome {3}

Outcome	Binary predictor	
	0	1
1,2	$\dfrac{e^{\kappa_2}}{1 + e^{\kappa_2}}$	$\dfrac{e^{\kappa_2 - \beta_1}}{1 + e^{\kappa_2 - \beta_1}}$
3	$1 - \dfrac{e^{\kappa_2}}{1 + e^{\kappa_2}}$	$1 - \dfrac{e^{\kappa_2 - \beta_1}}{1 + e^{\kappa_2 - \beta_1}}$
Total	1	1

Note that using either of the two collapsed tables (tables 16.3 or 16.4), the odds ratio for the binary predictor of being in the higher set of outcomes is given by $\exp(\beta_1)$.

Thus, the exponentiated coefficients for the ordered model can be interpreted as the odds ratio for higher outcomes. Note that the model assumes that the odds ratio is the same no matter where we split the outcomes.

16.2 Ordered outcomes for general link

We assume that there are r total outcomes possible and that there are $r + 1$ cutpoints, κ_i for $i = 0, \ldots, r$, where $\kappa_0 = -\infty$ and $\kappa_r = \infty$. We must therefore estimate $r - 1$ cutpoints in addition to obtaining an estimate of $\boldsymbol{\beta}$. This presentation does not include a constant in the model, because the cutpoints take the place of the constant. Other presentations fit a constant and one fewer cutpoint. To emphasize: if we fit $r - 1$ cutpoints, then the constant is not identifiable, but if we fit $r - 2$ cutpoints, we may include a constant term. Regardless of this parameterization choice, the two approaches are equivalent and result in the same inference.

To present the ordered outcome likelihood in the most general case, we let f denote the probability density function and F denote the cumulative distribution function. We may then derive the properties of the general model and substitute the logistic, probit, clog-log, or other links where appropriate.

Below we list the quantities of interest (including analytic derivatives) to derive the log likelihood. Because the model of interest includes cutpoints in addition to the coefficient vector, we will examine this model only for maximum likelihood and include output from Stata's collection of model-specific commands.

$$I(a = b) = \begin{cases} 1 & \text{if } a = b \\ 0 & \text{otherwise} \end{cases} \tag{16.7}$$

$$\xi_i = \mathbf{x}_i\boldsymbol{\beta} + \text{offset}_i \tag{16.8}$$

$$\Pr(y_i = k) = \Pr(\kappa_{k-1} < \xi_i + \epsilon_i \le \kappa_k) \tag{16.9}$$

$$\Pr(\xi_i + \epsilon_i < \kappa) = F(\kappa - \xi_i) \tag{16.10}$$

$$\Pr(\xi_i + \epsilon_i > \kappa) = 1 - F(\kappa - \xi_i) \tag{16.11}$$

$$\Pr(\kappa_{k-1} < \xi_i + \epsilon_i < \kappa_k) = F(\kappa_k - \xi_i) - F(\kappa_{k-1} - \xi_i) \tag{16.12}$$

The log likelihood is given by

$$\mathcal{L} = \sum_{k=1}^{r} \sum_{i=1}^{n} \ln\left\{ F(\kappa_k - \xi_i) - F(\kappa_{k-1} - \xi_i) \right\} I(y_i = k) \tag{16.13}$$

First derivatives are given by

$$\frac{\partial \mathcal{L}}{\partial \beta_t} = \sum_{i=1}^{n} x_{ti} \sum_{k=1}^{r} \left\{ \frac{-f(\kappa_k - \xi_i) + f(\kappa_{k-1} - \xi_i)}{F(\kappa_k - \xi_i) - F(\kappa_{k-1} - \xi_i)} \right\} I(y_i = k) \tag{16.14}$$

$$\frac{\partial \mathcal{L}}{\partial \kappa_t} = \sum_{i=1}^{n} \left\{ \frac{f(\kappa_t - \xi_i)}{F(\kappa_t - \xi_i) - F(\kappa_{t-1} - \xi_i)} I(y_i = t) \right.$$
$$\left. - \frac{f(\kappa_t - \xi_i)}{F(\kappa_{t+1} - \xi_i) - F(\kappa_t - \xi_i)} I(y_i = t+1) \right\} \tag{16.15}$$

and second derivatives are given by

$$
\frac{\partial^2 \mathcal{L}}{\partial \beta_t \partial \beta_u} = \sum_{i=1}^{n} x_{ti} x_{ui} \sum_{k=1}^{r} \left[\frac{f'(\kappa_k - \xi_i) - f'(\kappa_{k-1} - \xi_i)}{F(\kappa_k - \xi_i) - F(\kappa_{k-1} - \xi_i)} \right.
$$
$$
\left. - \frac{\{-f(\kappa_k - \xi_i) + f(\kappa_{k-1} - \xi_i)\}^2}{\{F(\kappa_k - \xi_i) - F(\kappa_{k-1} - \xi_i)\}^2} \right] I(y_i = k) \tag{16.16}
$$

$$
\frac{\partial^2 \mathcal{L}}{\partial \kappa_t \partial \kappa_u} = - \sum_{i=1}^{n} \frac{f(\kappa_t - \xi_i) f(\kappa_u - \xi_i)}{\{F(\kappa_t - \xi_i) - F(\kappa_u - \xi_i)\}^2}
$$
$$
I(y_i = \max\{t, u\}) I(|t - u| = 1) \tag{16.17}
$$

$$
\frac{\partial^2 \mathcal{L}}{\partial \kappa_t \partial \kappa_t} = \sum_{i=1}^{n} \left(\left[\frac{f'(\kappa_t - \xi_i)}{F(\kappa_t - \xi_i) - F(\kappa_{t-1} - \xi_i)} \right. \right.
$$
$$
\left. - \frac{f(\kappa_t - \xi_i) \{f(\kappa_t - \xi_i) - f(\kappa_{t-1} - \xi_i)\}}{\{F(\kappa_t - \xi_i) - F(\kappa_{t-1} - \xi_i)\}^2} \right] I(y_i = t)
$$
$$
- \left[\frac{f'(\kappa_t - \xi_i)}{F(\kappa_{t+1} - \xi_i) - F(\kappa_t - \xi_i)} \right.
$$
$$
\left. \left. - \frac{f(\kappa_t - \xi_i) \{f(\kappa_{t+1} - \xi_i) - f(\kappa_t - \xi_i)\}}{\{F(\kappa_{t+1} - \xi_i) - F(\kappa_t - \xi_i)\}^2} \right] I(y_i = t+1) \right) \tag{16.18}
$$

$$
\frac{\partial^2 \mathcal{L}}{\partial \kappa_t \partial \beta_u} = - \sum_{i=1}^{n} x_{ui} \mathcal{A}_i I(y_i = t) - \mathcal{B}_i I(y_i = t+1) \tag{16.19}
$$

where

$$
\mathcal{A}_i = \frac{f(\kappa_t - \xi_i) \{F(\kappa_t - \xi_i) - F(\kappa_{t-1} - \xi_i) + f(\kappa_t - \xi_i) - f(\kappa_{t-1} - \xi_i)\}}{\{F(\kappa_t - \xi_i) - F(\kappa_{t-1} - \xi_i)\}^2} \tag{16.20}
$$

$$
\mathcal{B}_i = \frac{f(\kappa_t - \xi_i) \{F(\kappa_{t+1} - \xi_i) - F(\kappa_t - \xi_i) + f(\kappa_{t+1} - \xi_i) - f(\kappa_t - \xi_i)\}}{\{F(\kappa_{t+1} - \xi_i) - F(\kappa_t - \xi_i)\}^2} \tag{16.21}
$$

In the derivation, we do not include a constant in the covariate list. Instead of absorbing the constant into the cutpoints, we may include the constant and then fit only $r - 2$ cutpoints, as we previously mentioned. We have presented the material to match the output of the Stata supporting commands.

The ordered outcome models include the proportional-odds assumption. This assumption dictates that the explanatory variables have the same effect on the odds of all levels of the response, thereby providing one coefficient for the table of parameter estimates with cutoff points defining ranges for the probability of being classified in a particular level of the response. The proportional-odds assumption is also called the parallel-lines assumption.

A natural approach for the interpretation of ordered-response models is through marginal effects. This is a measure of how a marginal change in a predictor changes the predicted probabilities

$$\frac{\partial P(y=j|x)}{\partial x_\ell} = \{f(\kappa_{j-1} - \xi) - f(\kappa_j - \xi)\}\,\beta_\ell \tag{16.22}$$

However, the interpretation of these marginal effects clearly depends on the particular values of the observation's covariates. Thus, if we instead focus on the relative marginal effect, the result is then given by β_ℓ/β_m, which does not depend on the particular observations; see Boes and Winkelmann (2006). The sign of the marginal effects can change only once as we proceed from the smallest to largest outcome categories. This limitation is the motivation for generalized ordered outcome models.

16.3 Ordered outcomes for specific links

Herein, we outline some of the details for the commonly used links in ordered outcome models. We describe specific functions available to Stata users for calculating necessary quantities in the log likelihood and derivatives. Interested users can use this information along with the Stata `ml` suite of commands to generate programs, though this is not necessary because Stata already provides commands to fit the models.

16.3.1 Ordered logit

Ordered logit uses the previous derivations, substituting

$$F(y) = \frac{\exp(y)}{1 + \exp(y)} \tag{16.23}$$

$$f(y) = \frac{\exp(y)}{\{1 + \exp(y)\}^2} = F(y)\{1 - F(y)\} \tag{16.24}$$

$$f'(y) = \frac{\exp(y) - \exp(2y)}{\{1 + \exp(y)\}^3} = F(y)\{1 - F(y)\}[F(y) - \{1 - F(y)\}] \tag{16.25}$$

$F(y)$ is the inverse of the logit function evaluated at y. Stata has the built-in function `invlogit()`, which calculates $F(y)$, so the calculations are straightforward for Stata users. The log likelihood is then

$$\mathcal{L} = \sum_{k=1}^{r}\sum_{i=1}^{n} \ln\{F(\kappa_k - \xi_i) - F(\kappa_{k-1} - \xi_i)\}\,I(y_i = k) \tag{16.26}$$

$$= \sum_{k=1}^{r}\sum_{i=1}^{n} \ln\{\texttt{invlogit}(\kappa_k - \xi_i) - \texttt{invlogit}(\kappa_{k-1} - \xi_i)\}\,I(y_i = k) \tag{16.27}$$

A foremost assumption of this model is that the coefficients do not vary, whereas the thresholds, or cutoffs, differ across the response values. This is commonly known as the parallel-lines assumption.

16.3.2   Ordered probit

Ordered probit uses the previous derivations, substituting

$$
\begin{aligned}
F(y) &= \Phi(y) & (16.28) \\
f(y) &= \phi(y) & (16.29) \\
f'(y) &= -y\phi(y) & (16.30)
\end{aligned}
$$

where $\phi(y)$ is the normal density function and $\Phi(y)$ is the cumulative normal distribution function.

$F(y)$ is the normal cumulative probability function evaluated at y, and $f(y)$ is the normal density function evaluated at y. Stata has the built-in functions `normal()` and `normalden()`, which calculate $F(\cdot)$ and $f(\cdot)$, respectively, so the calculations are straightforward for Stata users. The log likelihood is then

$$
\begin{aligned}
\mathcal{L} &= \sum_{k=1}^{r}\sum_{i=1}^{n}\ln\left\{F(\kappa_k - \xi_i) - F(\kappa_{k-1} - \xi_i)\right\}I(y_i = k) & (16.31) \\
&= \sum_{k=1}^{r}\sum_{i=1}^{n}\ln\left\{\texttt{normal}(\kappa_k - \xi_i) - \texttt{normal}(\kappa_{k-1} - \xi_i)\right\}I(y_i = k) & (16.32)
\end{aligned}
$$

16.3.3   Ordered clog-log

Ordered clog-log uses the previous derivations, substituting

$$
\begin{aligned}
F(y) &= 1 - \exp\{-\exp(y)\} & (16.33) \\
f(y) &= \{F(y) - 1\}\ln\{1 - F(y)\} & (16.34) \\
f'(y) &= f(y)\left[1 + \ln\{1 - F(y)\}\right] & (16.35)
\end{aligned}
$$

$F(y)$ is the inverse clog-log function evaluated at y. Stata has the built-in function `invcloglog()` to calculate $F(y)$, so the calculations are relatively straightforward. The log likelihood is then

$$
\begin{aligned}
\mathcal{L} &= \sum_{k=1}^{r}\sum_{i=1}^{n}\ln\left\{F(\kappa_k - \xi_i) - F(\kappa_{k-1} - \xi_i)\right\}I(y_i = k) & (16.36) \\
&= \sum_{k=1}^{r}\sum_{i=1}^{n}\ln\left\{\texttt{invcloglog}(\kappa_k - \xi_i) - \texttt{invcloglog}(\kappa_{k-1} - \xi_i)\right\}I(y_i = k) & (16.37)
\end{aligned}
$$

16.3.4   Ordered log-log

Ordered log-log uses the previous derivations, substituting

$$
\begin{aligned}
F(y) &= \exp\{-\exp(-y)\} & (16.38)\\
f(y) &= -F(y)\ln\{F(y)\} & (16.39)\\
f'(y) &= f(y)\left[1 + \ln\{F(y)\}\right] & (16.40)
\end{aligned}
$$

$F(y)$ is one minus the inverse log-log function evaluated at $-y$; see (16.33). Stata has the built-in function `invcloglog()` to calculate $F(y)$ so that the calculations are relatively straightforward. The log likelihood is then

$$
\begin{aligned}
\mathcal{L} &= \sum_{k=1}^{r}\sum_{i=1}^{n} \ln\left\{F(\kappa_k - \xi_i) - F(\kappa_{k-1} - \xi_i)\right\} I(y_i = k) & (16.41)\\
&= \sum_{k=1}^{r}\sum_{i=1}^{n} \ln\left\{ \texttt{invcloglog}(-\kappa_{k-1} + \xi_i) \right. \\
&\qquad\qquad \left. -\texttt{invcloglog}(-\kappa_k + \xi_i) \right\} I(y_i = k) & (16.42)
\end{aligned}
$$

16.3.5   Ordered cauchit

Ordered cauchit, or ordered inverse Cauchy, uses the previous derivations, substituting

$$
\begin{aligned}
F(y) &= 0.5 + \pi^{-1}\mathrm{atan}(-y) & (16.43)\\
f(y) &= -\frac{1}{\pi(1 + y^2)} & (16.44)\\
f'(y) &= f(y)2\pi y & (16.45)
\end{aligned}
$$

$F(y)$ is the cumulative distribution function for the Cauchy distribution. Although there are no built-in functions for this, the necessary geometric functions are available so that the calculations are straightforward.

Stata's official ordered-response models are limited to the ordered logit (`ologit`) and ordered probit (`oprobit`) procedures. However, Williams (2006a) extended these to include a more general procedure called `oglm`,[1] which allows the user to model all five previously listed ordered models, with the particular model selected as an option. Williams (2006b) is also responsible for developing a multilinked generalized ordered program called `gologit2`,[2] providing the user with a host of generalized ordered binomial modeling options. We give an overview of these types of models in the following section.

---

1. `oglm` can be installed by typing `ssc install oglm`.
2. `gologit2` can be installed by typing `ssc install gologit2`.

16.4   Generalized ordered outcome models

Here we investigate the model first introduced by Fu (1998). Although the community-contributed `gologit2` command Williams (2006b) uses the term *logit*, the command actually supports several link functions in the generalized ordered outcome models. Users will also benefit from the useful collection of programs commonly called SPost, which is described by Long and Freese (2014).

The latest (2006 or later) version of `gologit2` provides the five links that we described above for standard ordered binomial models, plus many extremely useful extensions, including the ability to model partial proportional odds and constrained generalized ordered binomial models. One can also use the `gologit2` command to fit standard ordered binomial models with the `pl` option. There are many other fitting enhancements that we will use to highlight examples for our discussion of generalized ordered binomial models.

The generalized ordered logit differs from the ordered logit model in that it relaxes the proportional-odds assumption. It allows that the explanatory variables may have a different effect on the odds that the outcome is above a cutpoint depending on how the outcomes are dichotomized. To accomplish this, the approach is similar to the unordered outcome multinomial logit model in that there will be $r-1$ estimated coefficient vectors that correspond to the effect of changing from one set of outcomes to a higher outcome not in the set.

The sets of coefficient vectors are defined for the cutpoints between the r outcomes. These cutpoints partition the outcomes into two groups. Therefore, the first coefficient vector corresponds to partitioning the outcomes into the sets $\{1\}$ and $\{2, \ldots, r\}$. The second coefficient vector corresponds to partitioning the outcomes into the sets $\{1, 2\}$ and $\{3, \ldots, r\}$. The $(r-1)$th coefficient vector corresponds to partitioning the outcomes into the sets $\{1, \ldots, r-1\}$ and $\{r\}$. To reflect this partitioning, we use the (nonstandard) subscript notation $\boldsymbol{\beta}_{\{1\},\{2,\ldots,r\}}$ to denote the coefficient vector that draws the partition between outcome $k=1$ and outcome $k=2$. In general, we will denote

$$\boldsymbol{\beta}_{\{1,\ldots,j\},\{j+1,\ldots,r\}} \tag{16.46}$$

for $j = 1, \ldots, r-1$ as the coefficient vector that draws the partition between outcomes j and $j+1$.

The sets of coefficient vectors correspond to a set of cumulative distribution functions

$$\Pr(y \le k) = F\left(-\mathbf{X}\boldsymbol{\beta}_{\{1,\ldots,k\},\{k+1,\ldots,r\}}\right) \tag{16.47}$$

for $k = 1, \ldots, r-1$. The distribution functions admit probabilities that are then defined as

$$\Pr(y = 1) \quad = \quad F\left(-\mathbf{X}\boldsymbol{\beta}_{\{1\},\{2,\ldots,r\}}\right) \tag{16.48}$$

$$\Pr(y = 2) \quad = \quad F\left(-\mathbf{X}\boldsymbol{\beta}_{\{1,2\},\{3,\ldots,r\}}\right) - F\left(-\mathbf{X}\boldsymbol{\beta}_{\{1\},\{2,\ldots,r\}}\right) \tag{16.49}$$

$$\vdots$$

$$\Pr(y = r) \quad = \quad 1 - F\left(-\mathbf{X}\boldsymbol{\beta}_{\{1,\ldots,r-1\},\{r\}}\right) \tag{16.50}$$

In general, we may write

$$\Pr(y = j) = F\left(-\mathbf{X}\boldsymbol{\beta}_{\{1,\ldots,j\},\{j+1,\ldots,r\}}\right) - F\left(-\mathbf{X}\boldsymbol{\beta}_{\{1,\ldots,j-1\},\{j,\ldots,r\}}\right) \tag{16.51}$$

for $j = 1, \ldots, r$ if we also define

$$F\left(-\mathbf{X}\boldsymbol{\beta}_{\{\},\{1,\ldots,r\}}\right) \quad = \quad 0 \tag{16.52}$$

$$F\left(-\mathbf{X}\boldsymbol{\beta}_{\{1,\ldots,r\},\{\}}\right) \quad = \quad 1 \tag{16.53}$$

The generalized ordered logit model assumes a logit function for $F(\cdot)$, but there is no reason that we could not use another link such as probit or clog-log. Fitting this model with other link functions such as the log link can be hard because the covariate values must restrict the above probabilities to the interval $[0, 1]$. Out-of-sample predictions can produce out-of-range probabilities.

The generalized ordered logit fits $r-1$ simultaneous logistic regression models, where the dependent variables for these models are defined by collapsing the outcome variable into new binary dependent variables defined by the partitions described above.

16.5   Example: Synthetic data

In the following example, we first create a synthetic dataset for illustration. We follow the same approach for generating the synthetic dataset as that appearing in Greene (2002). Our synthetic dataset is created by setting the random-number seed so that the analyses may be re-created.

```
. set obs 100
number of observations (_N) was 0, now 100
. set seed 12345
. generate double x1 = 3*runiform()+1
. generate double x2 = 2*runiform()-1
. generate double y = 1 + .5*x1 + 1.2*x2 + rnormal()
. generate int ys = cond(y<=2.5,1,cond(y<=3,2,cond(y<=4,3,4)))
```

Our dataset defines an outcome variable in the set $\{1, 2, 3, 4\}$ that depends on two covariates and a constant. We first fit an ordered logit model and obtain predicted probabilities and predicted classifications according to the maximum probability. Because there are $r = 4$ outcomes, there are four predicted probabilities associated with the outcomes.

```
. ologit ys x1 x2, nolog
Ordered logistic regression                       Number of obs   =        100
                                                  LR chi2(2)      =      15.68
                                                  Prob > chi2     =     0.0004
Log likelihood = -103.7709                        Pseudo R2       =     0.0703
```

ys	Coef.	Std. Err.	z	P>|z|	[95% Conf. Interval]	
x1	.2691789	.2276435	1.18	0.237	-.1769942	.715352
x2	1.411459	.3877474	3.64	0.000	.6514884	2.17143
/cut1	1.031253	.6092239			-.162804	2.22531
/cut2	1.556974	.6191686			.3434261	2.770522
/cut3	3.435817	.7165602			2.031385	4.840249

```
. estat ic
Akaike´s information criterion and Bayesian information criterion
```

Model	Obs	ll(null)	ll(model)	df	AIC	BIC
.	100	-111.613	-103.7709	5	217.5418	230.5677

```
                 Note: N=Obs used in calculating BIC; see [R] BIC note.
. predict double (olpr1 olpr2 olpr3 olpr4), pr
```

Our parameterization of this model does not contain a constant. Other presentations of this model do include a constant that is equal to the negative of the first cutpoint, κ_1, labeled /cut1 in the above output. The remaining cutpoints, κ_j^*, in the alternative (with constant) presentation are formed as

$$\kappa_j^* = \kappa_{j+1} - \kappa_1 \tag{16.54}$$

For comparison, we also fit ordered probit, clog-log, and log-log models with their respective predicted probabilities. These models are fit using either specific commands such as ologit or the general oglm program.

The oglm command is a generalized ordered binomial procedure supporting six different link functions including logit, probit, clog-log, log-log, cauchit, and log. We will not model the cauchit or log links (we did not show the log link since we believe it to have undesirable properties when modeling ordered binomial models). Comparison of results with the ordered logit will provide an insight into how the models differ.

```
. oprobit ys x1 x2, nolog
```

Ordered probit regression

				Number of obs	=	100
				LR chi2(2)	=	16.75
				Prob > chi2	=	0.0002
Log likelihood = -103.23556 | | | | Pseudo R2 | = | 0.0751 |

ys	Coef.	Std. Err.	z	P>|z|	[95% Conf. Interval]	
x1	.1577064	.1333975	1.18	0.237	-.103748	.4191608
x2	.8636432	.2226426	3.88	0.000	.4272718	1.300015
/cut1	.615321	.3575548			-.0854736	1.316116
/cut2	.9351996	.3606646			.2283101	1.642089
/cut3	2.017881	.403265			1.227497	2.808266

```
. estat ic
```
Akaike´s information criterion and Bayesian information criterion

Model	Obs	ll(null)	ll(model)	df	AIC	BIC
.	100	-111.613	-103.2356	5	216.4711	229.497

Note: N=Obs used in calculating BIC; see [R] BIC note.

```
. predict double (oppr1 oppr2 oppr3 oppr4), pr
```

We fit an ordered clog-log model by using the `oglm` command, which fits ordered outcome models through specifying a `link` option.

```
. oglm ys x1 x2, link(cloglog) nolog
```

Ordered Cloglog Regression

				Number of obs	=	100
				LR chi2(2)	=	15.13
				Prob > chi2	=	0.0005
Log likelihood = -104.04646 | | | | Pseudo R2 | = | 0.0678 |

ys	Coef.	Std. Err.	z	P>|z|	[95% Conf. Interval]	
x1	.14899	.1633625	0.91	0.362	-.1711946	.4691747
x2	1.079675	.2941718	3.67	0.000	.5031091	1.656241
/cut1	1.052725	.4501151	2.34	0.019	.1705159	1.934935
/cut2	1.452181	.4601092	3.16	0.002	.5503831	2.353978
/cut3	3.074762	.5595415	5.50	0.000	1.978081	4.171443

```
. estat ic
```
Akaike´s information criterion and Bayesian information criterion

Model	Obs	ll(null)	ll(model)	df	AIC	BIC
.	100	-111.613	-104.0465	5	218.0929	231.1188

Note: N=Obs used in calculating BIC; see [R] BIC note.

```
. predict double (ocpr1 ocpr2 ocpr3 ocpr4), pr
```

We fit an ordered log-log model by using the same `oglm` command, this time with the `link(loglog)` option. The binary log-log model is rarely used in research; however, it is a viable model that should be checked against others when deciding on a binary-response probability model. We have found many examples that used a probit or clog-log model to model data when using a log-log model would have been preferable. The authors simply failed to try the model. The same motivation for model checking holds for the ordered log-log model.

The clog-log and log-log fitted values are asymmetric, unlike those from logit and probit. Remember this when interpreting predicted probabilities.

```
. oglm ys x1 x2, link(loglog) nolog
Ordered Loglog Regression                          Number of obs    =         100
                                                   LR chi2(2)       =       16.90
                                                   Prob > chi2      =      0.0002
Log likelihood = -103.16072                        Pseudo R2        =      0.0757
```

ys	Coef.	Std. Err.	z	P>|z|	[95% Conf. Interval]	
x1	.1578883	.1275	1.24	0.216	-.0920071	.4077837
x2	.8125202	.2100811	3.87	0.000	.4007689	1.224271
/cut1	.2323641	.3366745	0.69	0.490	-.4275058	.8922339
/cut2	.5625211	.3336765	1.69	0.092	-.0914728	1.216515
/cut3	1.543967	.363646	4.25	0.000	.8312335	2.2567

```
. estat ic
```

Akaike's information criterion and Bayesian information criterion

Model	Obs	ll(null)	ll(model)	df	AIC	BIC
.	100	-111.613	-103.1607	5	216.3214	229.3473

Note: N=Obs used in calculating BIC; see [R] BIC note.
```
. predict double (ozpr1 ozpr2 ozpr3 ozpr4), pr
```

Finally, we may also fit a generalized ordered logit model to the data. We do this for illustration only, because we generated the data ourselves without violating the proportional-odds assumption.

```
. gologit2 ys x1 x2, nolog
Generalized Ordered Logit Estimates          Number of obs    =        100
                                             LR chi2(6)       =      19.03
                                             Prob > chi2      =     0.0041
Log likelihood = -102.10029                  Pseudo R2        =     0.0852
```

ys	Coef.	Std. Err.	z	P>|z|	[95% Conf. Interval]	
1						
x1	.199079	.2375022	0.84	0.402	-.2664167	.6645747
x2	1.374575	.3989607	3.45	0.001	.5926268	2.156524
_cons	-.857838	.6312853	-1.36	0.174	-2.095134	.3794584
2						
x1	.2660786	.2592639	1.03	0.305	-.2420693	.7742265
x2	1.211024	.4084366	2.97	0.003	.4105027	2.011545
_cons	-1.526718	.7042903	-2.17	0.030	-2.907102	-.1463348
3						
x1	.6614196	.4606793	1.44	0.151	-.2414952	1.564334
x2	2.306304	.8934817	2.58	0.010	.5551115	4.057495
_cons	-4.9348	1.436373	-3.44	0.001	-7.750039	-2.119562

```
. estat ic
Akaike's information criterion and Bayesian information criterion
```

Model	Obs	ll(null)	ll(model)	df	AIC	BIC
.	100	-111.613	-102.1003	9	222.2006	245.6471

```
                Note: N=Obs used in calculating BIC; see [R] BIC note.
```

We now want to test the proportional-odds assumption. If the proportional-odds assumption is valid, then the coefficient vectors should all be equal. We can use Stata's test command to investigate.

```
. test [1=2], notest
 ( 1)  [1]x1 - [2]x1 = 0
 ( 2)  [1]x2 - [2]x2 = 0
. test [1=3], accumulate
 ( 1)  [1]x1 - [2]x1 = 0
 ( 2)  [1]x2 - [2]x2 = 0
 ( 3)  [1]x1 - [3]x1 = 0
 ( 4)  [1]x2 - [3]x2 = 0
           chi2( 4) =     3.37
         Prob > chi2 =   0.4979
```

The χ^2 test fails to reject the proportional-odds assumption. Thus, we prefer the ordered logit (and ordered probit) model over the generalized ordered logit model. Because we synthesized the data ourselves, we know that the proportional-odds assumption is valid, and the outcome of this test reflects that.

Because we rejected the generalized ordered logit model, we now compare the two remaining ordered models. The coefficients from the ordered logit and ordered probit models are different, but the predicted probabilities, and thus the predicted classifications, are similar. Choosing between the two models is then arbitrary. The pseudo-R^2 listed in the output for the models is the McFadden likelihood-ratio index. See section 4.6.4.3 for an alternative measure.

We also tested each of the above models by using the AIC statistic. Stata does not automatically calculate or display this statistic with its maximum likelihood models. We use a postestimation AIC command created specifically (Hilbe 2011) for comparing maximum likelihood models. Here the ordered probit model has the lowest AIC statistic, indicating a better-fitted model. However, none of the four statistics appears to significantly differ from one another.

We list the predicted probabilities created by the **predict** commands for each of the logit, probit, clog-log, and log-log models. We summarize the coding given to the models as

olpr1	logit	**oppr1**	probit
ocpr1	clog-log	**ozpr1**	log-log

Level 1 predicted probabilities are given by

```
. list ys olpr1 oppr1 ocpr1 ozpr1 in 55/62
```

	ys	olpr1	oppr1	ocpr1	ozpr1
55.	1	.40858917	.41116357	.41284614	.42826653
56.	1	.58462448	.58224738	.59043994	.56969367
57.	1	.60352543	.60192716	.61548941	.58570397
58.	2	.61389029	.61396099	.63643378	.59414277
59.	2	.46270926	.46074214	.44789587	.47115241
60.	1	.82052206	.82377857	.80927668	.80996085
61.	3	.39696525	.39727969	.38058869	.42025953
62.	3	.78562844	.78811883	.78666663	.76748234

Level 2 predicted probabilities are given by

```
. list ys olpr2 oppr2 ocpr2 ozpr2 in 55/62
```

	ys	olpr2	oppr2	ocpr2	ozpr2
55.	1	.13031224	.1268084	.13963169	.11231228
56.	1	.11960247	.11883803	.11187345	.12090904
57.	1	.11676401	.11651407	.10667049	.12079617
58.	2	.1150823	.11495456	.10211671	.12064103
59.	2	.13026656	.12683405	.1356159	.11665408
60.	1	.06498122	.07052416	.05840821	.09078785
61.	3	.12990769	.12643701	.14255136	.1113438
62.	3	.07547614	.08047947	.06468385	.10110959

Level 3 predicted probabilities are given by

```
. list ys olpr3 oppr3 ocpr3 ozpr3 in 55/62
```

	ys	olpr3	oppr3	ocpr3	ozpr3
55.	1	.34549758	.34263125	.33700161	.33391241
56.	1	.23547999	.24523822	.23031143	.26569142
57.	1	.22370877	.2331919	.21560906	.25553231
58.	2	.21728229	.22578228	.20338196	.25004108
59.	2	.31211577	.31630639	.31561491	.31822759
60.	1	.0951264	.09585801	.10468919	.09714762
61.	3	.35249288	.34959076	.35681081	.3362381
62.	3	.11484686	.11758569	.11738276	.12695961

Level 4 predicted probabilities are given by

```
. list ys olpr4 oppr4 ocpr4 ozpr4 in 55/62
```

	ys	olpr4	oppr4	ocpr4	ozpr4
55.	1	.11560101	.11939678	.11052056	.12550879
56.	1	.06029306	.05367637	.06737517	.04370588
57.	1	.05600179	.04836687	.06223105	.03796755
58.	2	.05374512	.04530218	.05806754	.03517512
59.	2	.0949084	.09611742	.10087332	.09396592
60.	1	.01937032	.00983925	.02762592	.00210368
61.	3	.12063418	.12669254	.12004914	.13215857
62.	3	.02404856	.01381601	.03126677	.00444846

16.6   Example: Automobile data

In the following analyses, we investigate the automobile dataset that is included with the Stata software. Like Fu (1998), we also collapse the `rep78` variable from five outcomes to three to ensure having enough of each outcome.

Our outcome variable is collapsed and the data in memory are compressed by using the following commands:

```
. sysuse auto, clear
(1978 Automobile Data)
. replace rep78=3 if rep78<=3
(10 real changes made)
. drop if rep78==.
(5 observations deleted)
. label define replab 3 "poor-avg" 4 "good" 5 "best"
. label values rep78 replab
. tabulate rep78
```

Repair Record 1978	Freq.	Percent	Cum.
poor-avg	40	57.97	57.97
good	18	26.09	84.06
best	11	15.94	100.00
Total	69	100.00	

We also applied labels to the outcome variable to make the tabulation easier to read.

We wish to investigate a car's repair record (rep78) based on foreign, the foreign or domestic origin of the car; length, the length in inches of the car; mpg, the miles per gallon for the car; and displacement, the displacement of the engine in the car in cubic inches. First, we fit an ordered logit model.

```
. ologit rep78 foreign length displacement, nolog
```

Ordered logistic regression

			Number of obs	=	69
			LR chi2(3)	=	30.28
			Prob > chi2	=	0.0000
Log likelihood = -51.054656			Pseudo R2	=	0.2287

rep78	Coef.	Std. Err.	z	P>|z|	[95% Conf. Interval]	
foreign	2.831937	.8602215	3.29	0.001	1.145934	4.51794
length	.0333366	.0255179	1.31	0.191	-.0166776	.0833509
displacement	-.0094944	.0069388	-1.37	0.171	-.0230941	.0041054
/cut1	5.602613	4.022656			-2.281648	13.48687
/cut2	7.65387	4.087475			-.3574337	15.66517

```
. estat ic
```

Akaike's information criterion and Bayesian information criterion

Model	Obs	ll(null)	ll(model)	df	AIC	BIC
.	69	-66.19463	-51.05466	5	112.1093	123.2798

Note: N=Obs used in calculating BIC; see [R] BIC note.

Next, we fit an ordered probit model.

```
. oprobit rep78 foreign length displacement, nolog
Ordered probit regression                      Number of obs   =          69
                                               LR chi2(3)      =       30.26
                                               Prob > chi2     =      0.0000
Log likelihood = -51.064296                    Pseudo R2       =      0.2286
```

rep78	Coef.	Std. Err.	z	P>|z|	[95% Conf. Interval]	
foreign	1.492901	.4309192	3.46	0.001	.648315	2.337487
length	.0171283	.0144131	1.19	0.235	-.0111208	.0453773
displacement	-.006107	.0040351	-1.51	0.130	-.0140157	.0018017
/cut1	2.688937	2.167935			-1.560138	6.938012
/cut2	3.836788	2.18205			-.4399521	8.113528

```
. estat ic
Akaike's information criterion and Bayesian information criterion
```

Model	Obs	ll(null)	ll(model)	df	AIC	BIC
.	69	-66.19463	-51.0643	5	112.1286	123.2991

Note: N=Obs used in calculating BIC; see [R] BIC note.

Both models indicate that foreign cars have a higher probability of having better repair records. They also indicate that longer cars and cars with smaller displacement engines might have a higher probability of better repair records, but the results are not significant.

To assess the proportional-odds assumption, we fit a generalized ordered logit model.

```
. gologit2 rep78 foreign length displacement
Generalized Ordered Logit Estimates          Number of obs    =        69
                                             LR chi2(6)       =     40.65
                                             Prob > chi2      =    0.0000
Log likelihood =   -45.8704                  Pseudo R2        =    0.3070
```

rep78	Coef.	Std. Err.	z	P>|z|	[95% Conf. Interval]	
poor-avg						
foreign	3.659307	1.029914	3.55	0.000	1.640713	5.677902
length	.0372637	.0278299	1.34	0.181	-.0172819	.0918093
displacement	-.0068156	.0069198	-0.98	0.325	-.020378	.0067469
_cons	-7.093281	4.459775	-1.59	0.112	-15.83428	1.647717
good						
foreign	.1758222	1.162968	0.15	0.880	-2.103553	2.455197
length	.1879727	.0678996	2.77	0.006	.054892	.3210535
displacement	-.0946161	.0315209	-3.00	0.003	-.1563959	-.0328362
_cons	-21.75103	7.993136	-2.72	0.007	-37.41729	-6.084769

```
WARNING! 6 in-sample cases have an outcome with a predicted probability that
is less than 0. See the gologit2 help section on Warning Messages for more
information.
. estat ic
Akaike´s information criterion and Bayesian information criterion
```

Model	Obs	ll(null)	ll(model)	df	AIC	BIC
.	69	-66.19463	-45.8704	8	107.7408	125.6137

```
Note: N=Obs used in calculating BIC; see [R] BIC note.
```

Note the warning message. The output from the command reminds us of an oddity for generalized ordered models. The oddity is that a model may produce predicted probabilities outside $[0, 1]$. If the number of such predictions is a large proportion of the total number of observations used in the estimation, then we would investigate further to see if there were a better model.

Next, we test the proportional-odds assumption by obtaining a Wald test that the coefficient vectors are the same:

```
. test [poor-avg=good]
 ( 1)  [poor-avg]foreign - [good]foreign = 0
 ( 2)  [poor-avg]length - [good]length = 0
 ( 3)  [poor-avg]displacement - [good]displacement = 0
          chi2( 3) =    8.31
        Prob > chi2 =   0.0399
```

The results show that the data violate the proportional-odds assumption. Here we prefer the generalized ordered logit model, which accounts for the different effects of 1) moving from poor-avg to a higher classification and 2) moving from a lower classification to best.

Our results now indicate that the probability of having a better repair record if the car is foreign is important only when moving from `poor-avg` to a higher category. Also, the length of the car and engine displacement size have a significant effect only when moving from `poor-avg` to `best` or from `good` to `best`.

Williams (2006b) developed a generalized ordered binomial model with optional links of logit, probit, clog-log, log-log, and cauchit. The `gologit2` command is an extension (partial proportional odds) of the `oglm` command. The default for `gologit2` is the generalized logit model, but it can fit proportional odds or parallel lines `oglm` models using the `pl` option.

Here we illustrate the odds-ratio parameterization of the previous `gologit2` output.

```
. gologit2 rep78 foreign length displacement, link(logit) eform nolog
Generalized Ordered Logit Estimates           Number of obs   =          69
                                              LR chi2(6)      =       40.65
                                              Prob > chi2     =      0.0000
Log likelihood =    -45.8704                  Pseudo R2       =      0.3070
```

rep78	exp(b)	Std. Err.	z	P>|z|	[95% Conf. Interval]	
poor-avg						
foreign	38.83443	39.99612	3.55	0.000	5.158846	292.3353
length	1.037967	.0288865	1.34	0.181	.9828665	1.096156
displacement	.9932076	.0068728	-0.98	0.325	.9798282	1.00677
_cons	.0008307	.0037046	-1.59	0.112	1.33e-07	5.195105
good						
foreign	1.192226	1.38652	0.15	0.880	.1220222	11.64873
length	1.206801	.0819413	2.77	0.006	1.056426	1.378579
displacement	.9097222	.0286753	-3.00	0.003	.8552206	.967697
_cons	3.58e-10	2.86e-09	-2.72	0.007	5.62e-17	.0022773

```
WARNING! 6 in-sample cases have an outcome with a predicted probability that
is less than 0. See the gologit2 help section on Warning Messages for more
information.
```

Note the extremely high odds ratio for foreign cars in the `poor-avg` response level. Because of this, we advise trying other links. We next fit a generalized ordered probit model on the identical data used with `gologit2` above but will return to showing nonexponentiated parameter estimates.

```
. gologit2 rep78 foreign length displacement, link(probit) nolog
Generalized Ordered Probit Estimates           Number of obs   =        69
                                               LR chi2(6)      =     41.35
                                               Prob > chi2     =    0.0000
Log likelihood = -45.520115                    Pseudo R2       =    0.3123
```

rep78	Coef.	Std. Err.	z	P>|z|	[95% Conf. Interval]	
poor-avg						
foreign	2.166388	.5469985	3.96	0.000	1.094291	3.238486
length	.0212237	.0154616	1.37	0.170	-.0090805	.0515279
displacement	-.0039511	.0040961	-0.96	0.335	-.0119793	.004077
_cons	-4.087608	2.399035	-1.70	0.088	-8.789631	.6144147
good						
foreign	.1308595	.6611055	0.20	0.843	-1.164883	1.426602
length	.1096383	.0375006	2.92	0.003	.0361385	.1831382
displacement	-.0551583	.017555	-3.14	0.002	-.0895655	-.0207511
_cons	-12.72151	4.369989	-2.91	0.004	-21.28653	-4.156485

```
WARNING! 6 in-sample cases have an outcome with a predicted probability that
is less than 0. See the gologit2 help section on Warning Messages for more
information.
. estat ic
Akaike's information criterion and Bayesian information criterion
```

Model	Obs	ll(null)	ll(model)	df	AIC	BIC
.	69	-66.19463	-45.52011	8	107.0402	124.9131

Note: N=Obs used in calculating BIC; see [R] BIC note.

The AIC statistic indicates that the generalized ordered probit may be a better-fitted model than the generalized ordered logit model. However, the difference is neither appreciable nor significant. The coefficient of foreign cars for the `poor-avg` response level (exponentiated coefficient equal to 8.73) does not indicate the problems implied by the large coefficient (and odds ratio) in the generalized ordered logit model.

We show the output for the clog-log and log-log links to demonstrate the difference between all the models. Also, we list the exponentiated coefficient for foreign cars in the poor-avg response level for comparison.

```
. gologit2 rep78 foreign length displacement, link(cloglog) nolog
Generalized Ordered Cloglog Estimates         Number of obs    =         69
                                              LR chi2(6)       =      39.13
                                              Prob > chi2      =     0.0000
Log likelihood = -46.630058                   Pseudo R2        =     0.2956
```

rep78	Coef.	Std. Err.	z	P>|z|	[95% Conf. Interval]	
poor-avg						
foreign	2.334334	.6826142	3.42	0.001	.9964343	3.672233
length	.0259075	.0191936	1.35	0.177	-.0117112	.0635263
displacement	-.0047653	.0052544	-0.91	0.364	-.0150638	.0055332
_cons	-5.399207	3.135615	-1.72	0.085	-11.5449	.7464853
good						
foreign	.2292758	1.110919	0.21	0.836	-1.948085	2.406637
length	.1276791	.0478079	2.67	0.008	.0339772	.2213809
displacement	-.0656225	.0228494	-2.87	0.004	-.1104066	-.0208384
_cons	-15.28615	5.680261	-2.69	0.007	-26.41926	-4.153047

```
WARNING! 5 in-sample cases have an outcome with a predicted probability that
is less than 0. See the gologit2 help section on Warning Messages for more
information.
. estat ic
Akaike´s information criterion and Bayesian information criterion
```

Model	Obs	ll(null)	ll(model)	df	AIC	BIC
.	69	-66.19463	-46.63006	8	109.2601	127.133

Note: N=Obs used in calculating BIC; see [R] BIC note.

```
. display exp(2.334334) /* Exponentiated coefficient for Poor-Average foreign */
10.322583
```

```
. gologit2 rep78 foreign length displacement, link(loglog) nolog
Generalized Ordered Loglog Estimates              Number of obs   =        69
                                                  LR chi2(6)      =     42.80
                                                  Prob > chi2     =    0.0000
Log likelihood = -44.794841                       Pseudo R2       =    0.3233
```

rep78	Coef.	Std. Err.	z	P>|z|	[95% Conf. Interval]
poor-avg					
foreign	2.659394	.6925441	3.84	0.000	1.302033	4.016756
length	.0180377	.0143903	1.25	0.210	-.0101667	.0462421
displacement	-.0035022	.0039727	-0.88	0.378	-.0112887	.0042842
_cons	-3.217335	2.11529	-1.52	0.128	-7.363228	.9285575
good					
foreign	.0533791	.6130687	0.09	0.931	-1.148214	1.254972
length	.1187579	.0419275	2.83	0.005	.0365815	.2009343
displacement	-.0598442	.0201469	-2.97	0.003	-.0993314	-.020357
_cons	-13.18729	4.665233	-2.83	0.005	-22.33098	-4.043603

```
WARNING! 6 in-sample cases have an outcome with a predicted probability that
is less than 0. See the gologit2 help section on Warning Messages for more
information.
. estat ic
Akaike's information criterion and Bayesian information criterion
```

Model	Obs	ll(null)	ll(model)	df	AIC	BIC
.	69	-66.19463	-44.79484	8	105.5897	123.4625

Note: N=Obs used in calculating BIC; see [R] BIC note.

```
. display exp(2.659394) /* Exponentiated coefficient for Poor-Average foreign */
14.287628
```

The log-log linked model had the lowest value for the AIC statistic. Although this indicates a better-fitted model, there is little difference between values for all four models. When evaluating models, we would next look at the probabilities and consider which model is easiest to interpret.

16.7   Partial proportional-odds models

The generalized ordered logistic model is a viable alternative to the ordered logistic model when the assumption of proportional odds is untenable. The generalized model eliminates the restriction that the regression coefficients are the same for each outcome for all the variables. This may be a rather drastic solution when the proportional-odds assumption is violated by only one or a few of the regressors.

The gologit2 command allows partial proportional-odds models. Such a model is somewhere between the ordered logistic model and the generalized ordered logistic model. Of particular use in determining the type of model we should fit to the data is the test described in Brant (1990) and supported in the community-contributed brant command (Long and Freese 2014). The command provides both the global test of the proportional-odds assumption and a test of the proportional-odds assumption for each of the covariates. To install the brant and related commands, type net install spost13_ado, from(http://www.indiana.edu/~jslsoc/stata).

Using Stata's automobile dataset, we now investigate an ordered outcome model on categories of headroom size. We load the data and create the outcome variable of interest by typing

```
. sysuse auto, clear
(1978 Automobile Data)
. generate headsize = int(headroom+.5)
. replace headsize = 4 if headsize==5
(5 real changes made)
. tabulate headsize
```

headsize	Freq.	Percent	Cum.
2	17	22.97	22.97
3	27	36.49	59.46
4	30	40.54	100.00
Total	74	100.00	

A standard proportional-odds ordered logistic model may be fit using the ologit command. The estimates and the results of Brant's test of parallel lines is given in the following listing.

```
. ologit headsize foreign mpg price displacement, nolog
```

Ordered logistic regression

				Number of obs	=	74
				LR chi2(4)	=	23.03
				Prob > chi2	=	0.0001
Log likelihood = -67.799401				Pseudo R2	=	0.1452

headsize	Coef.	Std. Err.	z	P>|z|	[95% Conf. Interval]	
foreign	.9710014	.7389785	1.31	0.189	-.4773698	2.419373
mpg	-.0782365	.0551093	-1.42	0.156	-.1862487	.0297757
price	-.000117	.0001062	-1.10	0.271	-.0003252	.0000912
displacement	.0133369	.0051225	2.60	0.009	.003297	.0233768
/cut1	-1.052366	1.896215			-4.768879	2.664147
/cut2	.9785758	1.887007			-2.71989	4.677041

```
. brant
Brant test of parallel regression assumption
```

	chi2	p>chi2	df
All	12.61	0.013	4
foreign	6.96	0.008	1
mpg	3.80	0.051	1
price	0.27	0.605	1
displacement	2.05	0.152	1

```
A significant test statistic provides evidence that the parallel
regression assumption has been violated.
```

The test indicates that the parallel-lines assumption is violated. Use the `detail` option to request details of the test.

```
. brant, detail
Estimated coefficients from binary logits
```

Variable	y_gt_2	y_gt_3
foreign	2.129	-1.203
	2.28	-1.02
mpg	-0.010	-0.245
	-0.13	-2.32
price	-0.000	-0.000
	-0.05	-0.82
displacement	0.017	0.005
	2.23	0.82
_cons	-2.094	4.396
	-0.72	1.45

```
                               legend: b/t
Brant test of parallel regression assumption
```

	chi2	p>chi2	df
All	12.61	0.013	4
foreign	6.96	0.008	1
mpg	3.80	0.051	1
price	0.27	0.605	1
displacement	2.05	0.152	1

```
A significant test statistic provides evidence that the parallel
regression assumption has been violated.
```

The results of the test are shown along with the fitted coefficients for the individual logistic regression models. Because we have an indication that the ordered logistic regression model assumptions are violated, we can fit a generalized ordered logistic regression model by using either the `gologit` or `gologit2` command. This model is the least restrictive where all covariates are allowed to change for each outcome.

```
. gologit2 headsize foreign mpg price displacement
```

Generalized Ordered Logit Estimates

	Number of obs	=	74
	LR chi2(8)	=	39.26
	Prob > chi2	=	0.0000
Log likelihood = -59.684516 | Pseudo R2 | = | 0.2475 |

headsize	Coef.	Std. Err.	z	P>|z|	[95% Conf. Interval]	
2						
foreign	1.978872	.9039067	2.19	0.029	.2072475	3.750497
mpg	-.0291	.0717361	-0.41	0.685	-.1697001	.1115002
price	-.0000388	.0001806	-0.22	0.830	-.0003927	.0003151
displacement	.0140528	.0065791	2.14	0.033	.0011579	.0269476
_cons	-1.025763	2.46282	-0.42	0.677	-5.852801	3.801275
3						
foreign	-.9988688	1.092489	-0.91	0.361	-3.140108	1.14237
mpg	-.1918918	.091883	-2.09	0.037	-.3719792	-.0118044
price	-.0001025	.0001248	-0.82	0.411	-.0003471	.0001421
displacement	.0067022	.0056262	1.19	0.234	-.0043249	.0177294
_cons	2.948465	2.607146	1.13	0.258	-2.161447	8.058378

We can also fit a partial proportional-odds model where we allow the software to search for which of the covariates should be allowed to change across the outcomes. The `gologit2` command will perform such a search:

```
. gologit2 headsize foreign mpg price displacement, autofit lrforce
```

```
Testing parallel lines assumption using the .05 level of significance...
Step  1:  Constraints for parallel lines imposed for price (P Value = 0.7293)
Step  2:  Constraints for parallel lines imposed for mpg (P Value = 0.1144)
Step  3:  Constraints for parallel lines imposed for displacement
> (P Value = 0.4775)
Step  4:  Constraints for parallel lines are not imposed for
          foreign (P Value = 0.00331)
Wald test of parallel lines assumption for the final model:
 ( 1)  [2]price - [3]price = 0
 ( 2)  [2]mpg - [3]mpg = 0
 ( 3)  [2]displacement - [3]displacement = 0
            chi2( 3) =     2.93
          Prob > chi2 =    0.4019
An insignificant test statistic indicates that the final model
does not violate the proportional odds/ parallel lines assumption
If you re-estimate this exact same model with gologit2, instead
of autofit you can save time by using the parameter
pl(price mpg displacement)
```

```
Generalized Ordered Logit Estimates          Number of obs   =       74
                                             LR chi2(5)      =    35.23
                                             Prob > chi2     =   0.0000
Log likelihood = -61.700114                  Pseudo R2       =   0.2221
 ( 1)  [2]price - [3]price = 0
 ( 2)  [2]mpg - [3]mpg = 0
 ( 3)  [2]displacement - [3]displacement = 0
```

headsize	Coef.	Std. Err.	z	P>|z|	[95% Conf. Interval]	
2						
foreign	1.942757	.8554556	2.27	0.023	.2660953	3.61942
mpg	-.0936841	.0590846	-1.59	0.113	-.2094878	.0221196
price	-.0000759	.0001113	-0.68	0.496	-.000294	.0001423
displacement	.0096062	.005053	1.90	0.057	-.0002975	.01951
_cons	1.422395	1.946279	0.73	0.465	-2.392242	5.237032
3						
foreign	-.8960264	1.021318	-0.88	0.380	-2.897774	1.105721
mpg	-.0936841	.0590846	-1.59	0.113	-.2094878	.0221196
price	-.0000759	.0001113	-0.68	0.496	-.000294	.0001423
displacement	.0096062	.005053	1.90	0.057	-.0002975	.01951
_cons	.2172205	1.942194	0.11	0.911	-3.58941	4.023851

Although this shortcut is useful, we also could have applied the knowledge gained from the Brant test. Applying such knowledge requires that we either stipulate the variables to hold constant (using the `pl` option) or stipulate the variables that are allowed to change (using the `npl` option). Here we show that the final model could have been obtained through such stipulation.

```
. gologit2 headsize foreign mpg price displacement, npl(foreign)
Generalized Ordered Logit Estimates              Number of obs    =        74
                                                 LR chi2(5)       =     35.23
                                                 Prob > chi2      =    0.0000
Log likelihood = -61.700114                      Pseudo R2        =    0.2221
 ( 1)  [2]mpg - [3]mpg = 0
 ( 2)  [2]price - [3]price = 0
 ( 3)  [2]displacement - [3]displacement = 0
```

headsize	Coef.	Std. Err.	z	P>|z|	[95% Conf. Interval]	
2						
foreign	1.942757	.8554556	2.27	0.023	.2660953	3.61942
mpg	-.0936841	.0590846	-1.59	0.113	-.2094878	.0221196
price	-.0000759	.0001113	-0.68	0.496	-.000294	.0001423
displacement	.0096062	.005053	1.90	0.057	-.0002975	.01951
_cons	1.422395	1.946279	0.73	0.465	-2.392242	5.237032
3						
foreign	-.8960264	1.021318	-0.88	0.380	-2.897774	1.105721
mpg	-.0936841	.0590846	-1.59	0.113	-.2094878	.0221196
price	-.0000759	.0001113	-0.68	0.496	-.000294	.0001423
displacement	.0096062	.005053	1.90	0.057	-.0002975	.01951
_cons	.2172205	1.942194	0.11	0.911	-3.58941	4.023851

16.8 Continuation-ratio models

The ordered models we have been discussing assume that the response categories are all proportional to one another, that none are assumed to be proportional, or that some are proportional. Regardless of these choices, categories are all assumed to have some type of order but with no specification of degree of difference. But what if a person must pass through a lower level before achieving, or being classified in, a current level? For instance, consider the ascending ranks within the classification of General in the United States Army. One starts as a brigadier, or one-star, general, advances to a major general, lieutenant general, general, and finally general of the Army. There are five ordered categories in all. However, except for the brigadier general, before achieving any given rank, one must have been previously classified as a lower-ranked general. The models we have thus far discussed do not account for such a relationship between categories and therefore lose information intrinsic to an understanding of the response.

A continuation-ratio ordered binomial model is designed to accommodate the above-described response. The community-contributed command ocratio (Wolfe 1998) fits design ordered continuation-ratio models with logit, probit, and clog-log links. Fitted probabilities for each category or level of response, crucial when assessing the model fit, are obtained using the postestimation ocrpred command. We download these commands by typing net install sg86, from(http://www.stata.com/stb/stb44).

We will demonstrate the model with a simple example. The data are from a study relating educational level to religiosity, controlling for age, having children, and gender. For purposes of illustration, we will assume that the data are from individuals who begin by getting an associate's degree, possibly move on to getting a bachelor's degree, and then subsequently possibly move on to a master's degree or PhD. The three levels of education are tabulated as

```
. use http://www.stata-press.com/data/hh4/edreligion, clear

. tabulate educlevel
   educlevel |      Freq.     Percent        Cum.
-------------+-----------------------------------
          AA |        205       34.11       34.11
          BA |        204       33.94       68.05
      MA/PhD |        192       31.95      100.00
-------------+-----------------------------------
       Total |        601      100.00
```

The `age` predictor consists of 9 age levels ranging from 17.5 to 57 in intervals of 5 years. The remaining predictors are binary with the following counts, that is, the number of people in each level.

	Yes ($= 1$)	No ($= 0$)
religious	260	341
kids	430	171
male	286	315

We first model `educlevel` by using an ordered logistic regression model.

```
. ologit educlevel religious kids age male, nolog
Ordered logistic regression                     Number of obs   =        601
                                                LR chi2(4)      =     102.85
                                                Prob > chi2     =     0.0000
Log likelihood = -608.57623                     Pseudo R2       =     0.0779

-------------+----------------------------------------------------------------
   educlevel |      Coef.   Std. Err.      z    P>|z|     [95% Conf. Interval]
-------------+----------------------------------------------------------------
   religious |  -.377072   .1609611    -2.34   0.019    -.6925499   -.0615941
        kids |  -.3991817   .1897659    -2.10   0.035    -.7711161   -.0272473
         age |   .0288753   .0099135     2.91   0.004     .0094452    .0483054
        male |   1.452014   .1648728     8.81   0.000     1.128869    1.775159
-------------+----------------------------------------------------------------
       /cut1 |   .3890563   .2864187                     -.172314    .9504265
       /cut2 |   2.022015   .2994223                     1.435158    2.608872
------------------------------------------------------------------------------
```

```
. estat ic
Akaike's information criterion and Bayesian information criterion
```

Model	Obs	ll(null)	ll(model)	df	AIC	BIC
.	601	-660.0029	-608.5762	6	1229.152	1255.544

```
         Note: N=Obs used in calculating BIC; see [R] BIC note.
```

No other link produces a significantly better-fitting model. We assess the proportional-odds assumption of the model.

```
. brant, detail
Estimated coefficients from binary logits
```

Variable	y_gt_1	y_gt_2
religious	-0.476	-0.251
	-2.56	-1.27
kids	-0.658	0.015
	-2.87	0.06
age	0.023	0.031
	2.00	2.75
male	1.217	1.700
	6.39	8.49
_cons	0.091	-2.631
	0.27	-7.15

```
                        legend: b/t
Brant test of parallel regression assumption
```

	chi2	p>chi2	df
All	19.98	0.001	4
religious	1.24	0.266	1
kids	6.63	0.010	1
age	0.48	0.487	1
male	5.44	0.020	1

```
A significant test statistic provides evidence that the parallel
regression assumption has been violated.
```

Because the proportional-odds assumption of the model has been violated, we can run a partial proportional-odds model, holding those predictors constant across categories that are nonsignificant in the Brant table.

. gologit2 educlevel religious kids age male, autofit lrforce nolog

Testing parallel lines assumption using the .05 level of significance...
Step 1: Constraints for parallel lines imposed for religious
> (P Value = 0.3627)
Step 2: Constraints for parallel lines imposed for age (P Value = 0.2774)
Step 3: Constraints for parallel lines are not imposed for
          kids (P Value = 0.00300)
          male (P Value = 0.01609)
Wald test of parallel lines assumption for the final model:
 (1) [AA]religious - [BA]religious = 0
 (2) [AA]age - [BA]age = 0

          chi2(2) =   2.01
          Prob > chi2 =   0.3657

An insignificant test statistic indicates that the final model
does not violate the proportional odds/ parallel lines assumption

If you re-estimate this exact same model with gologit2, instead
of autofit you can save time by using the parameter

pl(religious age)

Generalized Ordered Logit Estimates   Number of obs   =   601
                                                    LR chi2(6)   =   121.01
                                                    Prob > chi2   =   0.0000
Log likelihood = -599.50008   Pseudo R2   =   0.0917
 (1) [AA]religious - [BA]religious = 0
 (2) [AA]age - [BA]age = 0

educlevel	Coef.	Std. Err.	z	P>|z|	[95% Conf. Interval]	
AA						
religious	-.3898915	.1615269	-2.41	0.016	-.7064784	-.0733046
kids	-.7513673	.2251507	-3.34	0.001	-1.192655	-.3100801
age	.0296937	.0099014	3.00	0.003	.0102874	.0491
male	1.204864	.1893592	6.36	0.000	.8337264	1.576001
_cons	-.0874313	.3009205	-0.29	0.771	-.6772246	.5023621
BA						
religious	-.3898915	.1615269	-2.41	0.016	-.7064784	-.0733046
kids	-.0362744	.2303568	-0.16	0.875	-.4877655	.4152166
age	.0296937	.0099014	3.00	0.003	.0102874	.0491
male	1.701747	.2004172	8.49	0.000	1.308937	2.094558
_cons	-2.489748	.3321644	-7.50	0.000	-3.140778	-1.838717

. estat ic

Akaike's information criterion and Bayesian information criterion

Model	Obs	ll(null)	ll(model)	df	AIC	BIC
.	601	-660.0029	-599.5001	8	1215	1250.189

Note: N=Obs used in calculating BIC; see [R] BIC note.

The model accounts for predictor odds that are both proportional and nonproportional across categories. The partial-proportional model results in a slightly lower AIC statistic, which is preferred because of its accommodation of the proportionality violations found in other models.

However, the model did not account for the continuation-ratio nature of the response. When the ordered outcomes are such that an outcome is a measure that can improve (or degrade) from one level to the next (without skipping levels), the continuation-ratio model is most appropriate. Examples include cancer stage or military rank.

Strictly speaking, one does not get an associate's degree before receiving a bachelor's degree; that is, a student could earn a bachelor's degree without earning an associate's degree. We are modeling these data under a continuation-ratio setting for which the outcome represents a level of amassed educational content and training. The amount of content and training required for each degree is assumed to be amassed continuously. There is no possibility to directly amass content and training for a bachelor's degree without first having amassed that amount of content and training represented by the associate's degree.

We examine the results of modeling the education level in a continuation-ratio model and compare the results with those of the partial-proportional model above. To calculate the AIC statistic following `ocratio`, we use our community-contributed `aic` command, which is downloadable by typing `net install aic, from(http://www.stata.com/users/jhardin)`.

```
. ocratio educlevel religious kids age male, link(logit)
Continuation-ratio logit Estimates           Number of obs =      997
                                             chi2(4)       =   109.20
                                             Prob > chi2   =   0.0000
Log Likelihood = -605.4009                   Pseudo R2     =   0.0827

    educlevel │     Coef.   Std. Err.      z    P>|z|    [95% Conf. Interval]
──────────────┼────────────────────────────────────────────────────────────
    religious │  -.320949    .143897    -2.23   0.026    -.602982   -.038916
         kids │ -.2768327   .1696096    -1.63   0.103   -.6092614   .0555959
          age │  .0276343   .0088566     3.12   0.002    .0102756   .0449929
         male │  1.311213   .1448946     9.05   0.000    1.027225   1.595201
──────────────┼────────────────────────────────────────────────────────────
        _cut1 │  .4384813   .2626932              (Ancillary parameters)
        _cut2 │  1.416228   .2820904

. aic
AIC Statistic =    1.222469
```

The number of physical observations in the dataset is 601 (as can be seen in the output from previous models). Here the number of observations listed is 997 and is a function of the number of physical observations and of the number of levels in the response variable.

The model was run using alternative links, but none were significantly better than the logit model. Notice how much smaller this adjusted AIC is than previous models.

It is smaller than the partial-proportional model and reflects what we know of the response—that a person with a higher level of education must have previously been classified with a lower level of education. The exception, of course, is the lowest level.

To facilitate interpretation, we next display model results as estimated odds ratios using the eform option.

```
. ocratio educlevel religious kids age male, link(logit) eform
Continuation-ratio logit Estimates        Number of obs =      997
                                          chi2(4)       =   109.20
                                          Prob > chi2   =   0.0000
    Log Likelihood = -605.4009            Pseudo R2     =   0.0827
```

educlevel	Odds ratio	Std. Err.	z	P>|z|	[95% Conf. Interval]	
religious	.7254602	.1043916	-2.23	0.026	.5471775	.9618315
kids	.7581813	.1285948	-1.63	0.103	.5437524	1.05717
age	1.02802	.0091048	3.12	0.002	1.010329	1.04602
male	3.710672	.5376563	9.05	0.000	2.793303	4.92932
_cut1	.4384813	.2626932		(Ancillary parameters)		
_cut2	1.416228	.2820904				

```
. aic
AIC Statistic =    1.222469
```

Being religious is associated with 27% less odds of transitioning to higher a education level—based on those with an associate's degree versus those with higher degrees (bachelor, master, or PhD) and based on those with a bachelor's degree versus those with a higher degree (master or PhD). Males had the higher education, and those who were older tended to have more education. The relationship between having children and educational level is not significant. Each of these relationships is consistent with findings from other studies.

These data were better modeled using a continuation-ratio model. Such a model reflects the nature of the response. Ordered categories are often modeled using standard ordered or even generalized ordered binomial models when a continuation-ratio model would be better. Care must be taken to assess the relationship between ordered categories.

16.9   Adjacent category model

Whereas the usual ordered outcome model seeks to model the probability of a specific outcome, the adjacent category model seeks to compare the likelihood of a response category j with the next higher category $j + 1$. The adjacent category logistic model can be described as

$$g_j(\mathbf{x}) \;=\; \log \left\{ \frac{P(Y = j + 1 | \mathbf{x})}{P(Y = j | \mathbf{x})} \right\} \tag{16.55}$$

$$\;=\; \alpha_j + \mathbf{x}\boldsymbol{\beta} \tag{16.56}$$

for $j = 1, \ldots, r - 1$. Note that no matter which two response categories, the coefficients of $\boldsymbol{\beta}$ are constant, but the intercepts α_j vary. Odds ratios $\exp(\beta_k)$ relate to the odds ratio comparing a difference of one in the associated covariate x_k for response 2 versus 1, response 3 versus 2, ..., response r versus $r - 1$.

Necessarily, this means that the odds ratio of category j versus 1 is equal to $(j - 1)$ times the odds ratio of response 2 versus 1 for response level $3, \ldots, r$. We point out, this also illuminates the constraints to be applied to the multinomial model to achieve this model; namely, that $\beta_{jk} = (j - 1)\beta_{2k}$ for $j = 3, \ldots, r$ and $k = 1 \ldots, p$.

Using the same religion data analyzed throughout this chapter, we fit the adjacent category model for education level as a function of whether the individual is religious, whether the individual has kids, the individual's age in years, and whether the individual is male. Here we illustrate the adjacent category ordered logistic model as a constrained multinomial logistic regression model.

```
. constraint 1 [3]religious = 2*[2]religious
. constraint 2 [3]kids     = 2*[2]kids
. constraint 3 [3]age      = 2*[2]age
. constraint 4 [3]male     = 2*[2]male
. mlogit educlevel religious kids age male, constraints(1,2,3,4) rrr nolog
```

```
Multinomial logistic regression            Number of obs    =       601
                                            Wald chi2(4)     =     86.15
Log likelihood = -609.17998                 Prob > chi2      =    0.0000
 ( 1)   - 2*[BA]religious + [MA_PhD]religious = 0
 ( 2)   - 2*[BA]kids + [MA_PhD]kids = 0
 ( 3)   - 2*[BA]age + [MA_PhD]age = 0
 ( 4)   - 2*[BA]male + [MA_PhD]male = 0
```

educlevel	RRR	Std. Err.	z	P>|z|	[95% Conf. Interval]	
AA	(base outcome)					
BA						
religious	.7717565	.087722	-2.28	0.023	.6176298	.964345
kids	.7855116	.1060533	-1.79	0.074	.6028794	1.023469
age	1.019279	.0069354	2.81	0.005	1.005776	1.032963
male	2.69545	.312877	8.54	0.000	2.146976	3.384039
_cons	.5088133	.1080197	-3.18	0.001	.3356222	.7713762
MA_PhD						
religious	.5956081	.1354	-2.28	0.023	.3814665	.9299612
kids	.6170285	.1666122	-1.79	0.074	.3634636	1.047489
age	1.03893	.0141382	2.81	0.005	1.011586	1.067013
male	7.265449	1.686688	8.54	0.000	4.609506	11.45172
_cons	.1849303	.0757468	-4.12	0.000	.0828627	.4127215

Note: _cons estimates baseline relative risk for each outcome.

In the fitted model, the coefficients are constrained to illustrate the association of the covariates with the likelihood of being in one category relative to all lower categories. This model indicates that religious individuals have 0.77 times the risk of having a higher educational level, males have 2.7 times the risk of females of having a higher educational level, and for each year older in age, the average person has 1.02 times the odds of higher educational level.

Part VI

Extensions to the GLM

17 Extending the likelihood

Contents

This chapter includes examples of fitting quasilikelihood models by using the accompanying software. Because quasilikelihood variance functions are not included in the software, the user must program appropriate functions to fit the model. See chapter 21 for information on the programs, and review the Stata documentation for programming assistance.

17.1 The quasilikelihood

We have previously presented the algorithms for fitting GLMs. We also mentioned that although our illustrations generally begin from a specified distribution, we need only specify the first two moments. Our original exponential-family density was specified as

$$f(y) = \exp\left\{\frac{y\theta - b(\theta)}{a(\phi)} + c(y, \phi)\right\} \tag{17.1}$$

with the log likelihood given by

$$\mathcal{L} = \sum_{i=1}^{n}\left\{\frac{y_i\theta_i - b(\theta_i)}{a(\phi)} + c(y_i, \phi)\right\} \tag{17.2}$$

We further showed that estimates could be obtained by solving the estimating equation given by

$$\frac{\partial \mathcal{L}}{\partial \mu} = \frac{\partial \mathcal{L}}{\partial \theta}\frac{\partial \theta}{\partial \mu} = \frac{y - b'(\theta)}{a(\phi)}\frac{1}{b''(\theta)} = \frac{y - \mu}{V(y)} = 0 \tag{17.3}$$

We assume that the quasilikelihood is defined by

$$Q(y; \mu) = \int_y^\mu \frac{y - \mu}{\phi V(\mu)} d\mu \tag{17.4}$$

If there exists a distribution (from the exponential family of distributions) that has $V(\mu)$ as its variance function, then the quasilikelihood reduces to a standard log likelihood. If not, then the resulting estimates are more properly called maximum quasilikelihood (MQL) estimates. A corresponding deviance called the quasideviance may be defined as

$$D = 2 \int_{\widehat{\mu}}^{y} \frac{y - \mu}{V(\mu)} d\mu \tag{17.5}$$

More extensions (not covered here) to the GLM approach include methods based on the quasilikelihood that allow for estimation of the variance function. Nelder and Pregibon (1987) describe a method for estimating the mean-variance relationship by using what they call the extended quasilikelihood. The extension adds a term to the quasilikelihood that allows comparison of quasideviances calculated from distributions with different variance functions. It does this by normalizing the quasilikelihood so that it has the properties of a log likelihood as a function of the scalar variance. This normalization is a saddle point approximation of the exponential family members derived, assuming that the distribution exists.

17.2   Example: Wedderburn's leaf blotch data

Here we present a quasilikelihood analysis of data from Wedderburn (1974). This analysis parallels the analysis presented in McCullagh and Nelder (1989). The data represent the proportion of leaf blotch on 10 varieties of barley grown at nine sites in 1965. The amount of leaf blotch is recorded as a percentage, and thus we can analyze the data by using binomial family variance with a logit link.

```
. use http://www.stata-press.com/data/hh4/accident
. glm y i.site i.variety, family(binomial) link(logit) vce(robust) nolog
note: y has noninteger values
```

Generalized linear models		No. of obs	=	90
Optimization : ML		Residual df	=	72
		Scale parameter =		1
Deviance = 6.125989621		(1/df) Deviance =		.0850832
Pearson = 6.391999102		(1/df) Pearson =		.0887778

Variance function: V(u) = u*(1-u/1)   [Binomial]
Link function   : g(u) = ln(u/(1-u))   [Logit]

| | | AIC | = | .8620201 |
| Log pseudolikelihood = -20.79090427 | | BIC | = | -317.8603 |

y	Coef.	Robust Std. Err.	z	P>|z|	[95% Conf. Interval]	
site						
2	1.639066	.3122966	5.25	0.000	1.026976	2.251156
3	3.326515	.3698515	8.99	0.000	2.601619	4.051411
4	3.58223	.4643503	7.71	0.000	2.67212	4.49234
5	3.583056	.229163	15.64	0.000	3.133904	4.032207
6	3.893253	.4159024	9.36	0.000	3.078099	4.708407
7	4.730018	.377958	12.51	0.000	3.989234	5.470802
8	5.52269	.2978881	18.54	0.000	4.93884	6.10654
9	6.794584	.3395895	20.01	0.000	6.129001	7.460167
variety						
2	.1500848	.2683793	0.56	0.576	-.3759289	.6760985
3	.6894659	.3256037	2.12	0.034	.0512944	1.327637
4	1.048159	.4023927	2.60	0.009	.2594842	1.836835
5	1.614706	.5646016	2.86	0.004	.5081071	2.721305
6	2.371158	.4884278	4.85	0.000	1.413857	3.328459
7	2.570528	.3787715	6.79	0.000	1.82815	3.312907
8	3.342041	.4315229	7.74	0.000	2.496271	4.18781
9	3.500049	.348527	10.04	0.000	2.816948	4.183149
10	4.253008	.3421847	12.43	0.000	3.582338	4.923677
_cons	-8.054648	.3424603	-23.52	0.000	-8.725858	-7.383438

Because the data are proportions instead of counts, there is no a priori theoretical reason that ϕ should be near one. Because of this, we use a quasilikelihood approach where we allow the Pearson deviance to scale the standard errors.

```
. glm y i.site i.variety, family(binomial) link(logit) scale(x2) nolog
note: y has noninteger values
```

Generalized linear models   No. of obs   =   90
Optimization   : ML   Residual df   =   72
                                                   Scale parameter =   1
Deviance   = 6.125989621   (1/df) Deviance = .0850832
Pearson   = 6.391999102   (1/df) Pearson = .0887778

Variance function: V(u) = u*(1-u/1)   [Binomial]
Link function   : g(u) = ln(u/(1-u))   [Logit]

                                                   AIC   = .8620201
Log likelihood = -20.79090427   BIC   = -317.8603

y	Coef.	OIM Std. Err.	z	P>|z|	[95% Conf. Interval]
site						
2	1.639066	1.443291	1.14	0.256	-1.189733	4.467865
3	3.326515	1.349218	2.47	0.014	.682096	5.970934
4	3.58223	1.344435	2.66	0.008	.9471863	6.217273
5	3.583056	1.344421	2.67	0.008	.9480384	6.218073
6	3.893253	1.340208	2.90	0.004	1.266494	6.520012
7	4.730018	1.334779	3.54	0.000	2.1139	7.346136
8	5.52269	1.334622	4.14	0.000	2.906879	8.138501
9	6.794584	1.340692	5.07	0.000	4.166877	9.422291
variety						
2	.1500848	.7236759	0.21	0.836	-1.268294	1.568464
3	.6894659	.6723591	1.03	0.305	-.6283337	2.007265
4	1.048159	.6494329	1.61	0.107	-.2247057	2.321024
5	1.614706	.6256624	2.58	0.010	.38843	2.840982
6	2.371158	.6090174	3.89	0.000	1.177506	3.56481
7	2.570528	.6064804	4.24	0.000	1.381849	3.759208
8	3.342041	.6015372	5.56	0.000	2.16305	4.521032
9	3.500049	.6013491	5.82	0.000	2.321426	4.678672
10	4.253008	.6042298	7.04	0.000	3.068739	5.437276
_cons	-8.054648	1.421953	-5.66	0.000	-10.84163	-5.267671

```
(Standard errors scaled using square root of Pearson X2-based dispersion.)
```

To investigate the fit, we graph the Pearson residuals versus the fitted values and the Pearson residuals versus the log of the variance function.

```
. predict double eta, eta
. predict double pearson, pearson
. predict double mu
(option mu assumed; predicted mean y)
. generate double logvar = log(mu*mu*(1-mu)*(1-mu))
. label var pearson "Pearson residuals"
. label var eta     "Linear predictor"
. label var logvar  "Log(variance)"
. twoway scatter pearson eta, yline(0)
. twoway scatter pearson logvar, yline(0)
```

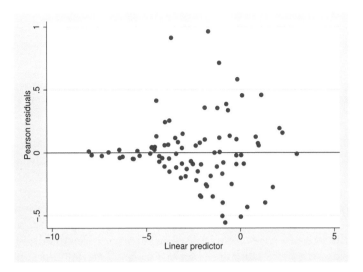

Figure 17.1. Pearson residuals versus linear predictor

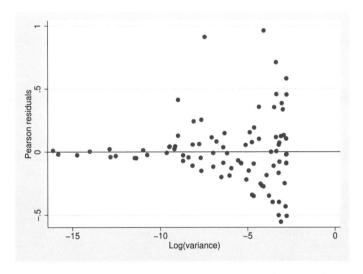

Figure 17.2. Pearson residuals versus log(variance)

Our diagnostic graphs in figures 17.1 and 17.2 indicate that there is a lack of fit for very small and very large values. Wedderburn (1974) suggests using a variance function with form

$$V(\mu) = \mu^2(1 - \mu)^2 \tag{17.6}$$

This is the square of the usual binomial variance function. Implementing this solution with the accompanying software uses the most sophisticated support mechanisms of the `glm` command. To successfully implement a solution, we must program the respective support. For this particular case, the quasideviance function cannot be defined because of the zeros in the data. This will ultimately mean that our solution will not support the `irls` option of the command. The quasilikelihood is then given as

$$Q(\mu; y) = (2y - 1) \ln \left(\frac{\mu}{1 - \mu} \right) - \frac{y}{\mu} - \frac{1 - y}{1 - \mu} \tag{17.7}$$

The easiest path to the solution is to build on the fact that the variance function of interest is the square of the usual binomial variance. We copy the `glim_v2.ado` file (which defines the support for the binomial variance function) to the file `binsq.ado` and edit the eight sections of the command. The eight sections correspond to various support functions for the iteratively reweighted least-squares and Newton–Raphson methods. Our responsibility will be to edit these sections to calculate appropriate quantities for the target variance function. In actuality, we need only edit the sections for the variance function calculation, the derivative of the variance, and the likelihood. The remaining sections either are removed (and thus not supported) or are left the same.

The support file `binsq.ado` is listed below.

Listing 17.1. Code for user-written `binsq` variance program

```
 1   program binsq
 2       args todo eta mu return touse
 3
 4       if 'todo' == -1 {
 5           if "$SGLM_m" != "1" {                    /* Title */
 6               global SGLM_vt "Quasi"
 7               global SGLM_vf "u^2*(1-u/$SGLM_m)^2"
 8           }
 9           else {
10               global SGLM_vt "Quasi"
11               global SGLM_vf "u^2*(1-u)^2"
12           }
13
14           local y      "$SGLM_y"
15           local m      "$SGLM_m"
16           local touse  "'eta'"
17
18           capture assert 'm'>0 if 'touse'      /* sic, > */
19           if _rc {
20               di as error '"'m' has nonpositive values"'
21               exit 499
22           }
23           capture assert 'm'==int('m') if 'touse'
24           if _rc {
25               di as error'"'m' has noninteger values"'
26               exit 499
27           }
28           capture assert 'y'>=0 if 'touse'
29           if _rc {
```

```
30              di as error '"dependent variable 'y'
31                  has negative values"'
32              exit 499
33          }
34          capture assert 'y'<='m' if 'touse'
35          if _rc {
36              noi di as error '"'y'>'m' in cases"'
37              exit 499
38          }
39
40          global SGLM_mu "glim_mu 0 $SGLM_m"
41          exit
42      }
43      if 'todo' == 0 {                  /* Initialize eta */
44          if "$SGLM_L" == "glim_l01" {          /* Identity */
45              generate double 'eta' = 'mu'/$SGLM_m
46          }
47          else if "$SGLM_L" == "glim_l02" {   /* Logit */
48              generate double 'eta' = ln('mu'/($SGLM_m-'mu'))
49          }
50          else if "$SGLM_L" == "glim_l03" {   /* Log */
51              generate double 'eta' = ln('mu'/$SGLM_m)
52          }
53          else if "$SGLM_L" == "glim_l05" {   /* Log compl */
54              generate double 'eta' = ln(1-'mu'/$SGLM_m)
55          }
56          else if "$SGLM_L" == "glim_l06" {   /* Log-log */
57              generate double 'eta' = -ln(-ln('mu'/$SGLM_m))
58          }
59          else if "$SGLM_L" == "glim_l07" {   /* Clog-log */
60              generate double 'eta' = ln(-ln(1-'mu'/$SGLM_m))
61          }
62          else if "$SGLM_L" == "glim_l08" {   /* Probit */
63              generate double 'eta' = invnormal('mu'/$SGLM_m)
64          }
65          else if "$SGLM_L" == "glim_l09" {   /* Reciprocal */
66              generate double 'eta' = $SGLM_m/'mu'
67          }
68          else if "$SGLM_L" == "glim_l10" {   /* Power(-2) */
69              generate double 'eta' = ($SGLM_m/'mu')^2
70          }
71          else if "$SGLM_L" == "glim_l11" {   /* Power(a) */
72              generate double 'eta' = ('mu'/$SGLM_m)^$SGLM_a
73          }
74          else if "$SGLM_L" == "glim_l12" {   /* OPower(a) */
75              generate double 'eta' = /*
76                  */ (('mu'/($SGLM_m-'mu'))^$SGLM_a-1) / $SGLM_a
77          }
78          else {
79              generate double 'eta' = $SGLM_m*($SGLM_y+.5)/($SGLM_m+1)
80          }
81          exit
82      }
83      if 'todo' == 1 {                  /* V(mu) */
84          generate double 'return' =  'mu'*'mu'*(1-'mu'/$SGLM_m) /*
85              */ *(1-'mu'/$SGLM_m)
```

```
86           exit
87      }
88      if 'todo' == 2 {                    /* (d V)/(d mu) */
89          generate double 'return' =  2*'mu' - /*
90              */ 6*'mu'*'mu'/$SGLM_m + /*
91              */ 4*'mu'*'mu'*'mu'/($SGLM_m*$SGLM_m)
92          exit
93      }
94      if 'todo' == 3 {                    /* deviance */
95          noi di as error "deviance calculation not supported"
96          generate double 'return' = 0
97          exit
98      }
99      if 'todo' == 4 {                    /* Anscombe */
100         noi di as error "Anscombe residuals not supported"
101         exit 198
102     }
103     if 'todo' == 5 {                    /* ln-likelihood */
104         local y "$SGLM_y"
105         local m "$SGLM_m"
106         if "'y'" == "" {
107             local y "'e(depvar)'"
108         }
109         if "'m'" == "" {
110             local m "'e(m)'"
111         }
112         generate double 'return' = (2*'y'-1)* /*
113             */ ln('mu'/('m'-'mu')) - 'y'/'mu' - /*
114             */ ('m'-'y')/('m'-'mu')
115         exit
116     }
117     if 'todo' == 6 {
118         noi di as error "adj. deviance residuals not supported"
119         exit 198
120     }
121     noi di as error "Unknown call to glim variance function"
122     error 198
123  end
```

We may then specify this variance function with the `glm` command and then obtain the same graphs as before to analyze the fit of the resulting model.

```
. glm y i.site i.variety, family(binsq) link(logit) nolog
deviance calculation not supported
```

Generalized linear models			No. of obs	=	90
Optimization	: ML		Residual df	=	72
			Scale parameter	=	.9885464
Deviance	=	0	(1/df) Deviance	=	0
Pearson	=	71.17534179	(1/df) Pearson	=	.9885464

```
Variance function: V(u) = u^2*(1-u)^2        [Quasi]
Link function    : g(u) = ln(u/(1-u))        [Logit]

                                             AIC        =  -.5527161
Log likelihood   =  42.87222528              BIC        =  -323.9863
```

y	Coef.	OIM Std. Err.	z	P>|z|	[95% Conf. Interval]	
site						
2	1.383119	.455533	3.04	0.002	.4902907	2.275947
3	3.860061	.4516254	8.55	0.000	2.974891	4.74523
4	3.557	.4577697	7.77	0.000	2.659788	4.454212
5	4.10786	.4495166	9.14	0.000	3.226824	4.988897
6	4.305356	.4611785	9.34	0.000	3.401463	5.209249
7	4.918099	.4451046	11.05	0.000	4.04571	5.790488
8	5.694892	.4538692	12.55	0.000	4.805325	6.584459
9	7.067632	.4512644	15.66	0.000	6.18317	7.952094
variety						
2	-.4673533	.4877992	-0.96	0.338	-1.423422	.4887157
3	.0788063	.4795614	0.16	0.869	-.8611168	1.018729
4	.9540754	.4765712	2.00	0.045	.0200131	1.888138
5	1.35263	.4661998	2.90	0.004	.438895	2.266365
6	1.328541	.5237571	2.54	0.011	.3019957	2.355086
7	2.340071	.4502318	5.20	0.000	1.457633	3.222509
8	3.262581	.4553686	7.16	0.000	2.370075	4.155087
9	3.135486	.4829722	6.49	0.000	2.188878	4.082094
10	3.887267	.4612816	8.43	0.000	2.983171	4.791362
_cons	-7.922378	.4418132	-17.93	0.000	-8.788316	-7.05644

The resulting graphs are obtained using the following commands:

```
. predict double eta, eta
. predict double pearson, pearson
. predict double mu
(option mu assumed; predicted mean y)
. generate double logvar = log(mu*mu*(1-mu)*(1-mu))
. label var eta     "Linear predictor"
. label var pearson "Pearson residuals"
. label var logvar  "Log(variance)"
. twoway scatter pearson eta, yline(0)
. twoway scatter pearson logvar, yline(0)
```

We can see in figures 17.3 and 17.4 that the effect of the extreme fitted values has been substantially reduced using this model.

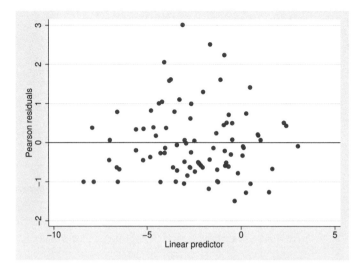

Figure 17.3. Pearson residuals versus linear predictor

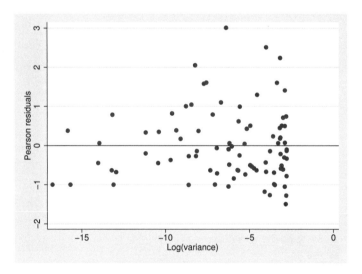

Figure 17.4. Pearson residuals versus log(variance)

17.3   Example: Tweedie family variance

The support file tweedie.ado is listed below. Here, we illustrate estimation of this user-written variance function to fit models that are equivalent of linear regression and Poisson regression. This illustration merely captures those particular instances where

the Tweedie variance is the same as other known variance functions to give the reader a
foundation for understanding Tweedie variance when it falls between or outside of these
known variances.

Listing 17.2. Code for user-written Tweedie variance program

```
 1  program tweedie
 2      version 15
 3      args todo eta mu return
 4
 5      if 'todo' == -1 {                    /* Title */
 6          global SGLM_vt "Tweedie"
 7          global SGLM_vf "phi*u^($SGLM_fa)"
 8          global SGLM_mu "glim_mu 0 ."
 9          exit
10      }
11      if 'todo' == 0 {
12          if $SGLM_p < 1e-7 {
13              generate double 'eta' = ln('mu')
14          }
15          else {
16              generate double 'eta' = 'mu'^$SGLM_p
17          }
18          exit
19      }
20      if 'todo' == 1 {                    /* V(mu) */
21          generate double 'return' =  'mu'^$SGLM_fa
22          noisily summarize 'return'
23          exit
24      }
25      if 'todo' == 2 {                    /* (d V)/(d mu) */
26          generate double 'return' =  ($SGLM_fa)*'mu'^($SGLM_fa-1)
27          noisily summarize 'return'
28          exit
29      }
30      if 'todo' == 3 {                    /* deviance */
31          local y "$SGLM_y"
32          if (abs($SGLM_fa-2) < 1e-7) {
33              noi di as err "Tweedie variance does not support p=2"
34              exit 198
35          }
36          if abs($SGLM_fa-1) < 1e-7 {
37              gen double 'return' = 2*('y'*ln('y'/'mu')-('y'-'mu'))
38          }
39          else {
40              local p = $SGLM_fa
41              gen double 'return' = 2*('y'^(2-'p')/((1-'p')*(2-'p')) ///
42              -('y'*'mu'^(1-'p'))/(1-'p')+'mu'^(2-'p')/(2-'p'))
43          }
44          exit
45      }
46      if 'todo' == 4 {                    /* Anscombe */
47          gen double 'return' = .
48          exit
49      }
50      if 'todo' == 5 {                    /* ln-likelihood */
```

```
51          noisily display as err "Tweedie family supported by IRLS only"
52          exit 198
53      }
54      if 'todo' == 6 {              /* Adjustment to deviance */
55          gen double 'return' = 0
56          exit
57      }
58
59      noisily display as err "Unknown call to glim variance function"
60      error 198
61  end
```

The Tweedie distribution with power 1 is equivalent to Poisson variance. Because the canonical link function for the Tweedie family is the power link, we specify power $= 0$, which is the same as the log link.

```
. use http://www.stata-press.com/data/hh4/medpar, clear
. glm los hmo white age, family(poisson) irls nolog
Generalized linear models                    No. of obs       =      1,495
Optimization      : MQL Fisher scoring       Residual df      =      1,491
                    (IRLS EIM)               Scale parameter  =          1
Deviance       =    8807.81667              (1/df) Deviance  =   5.907322
Pearson        =   11538.00986              (1/df) Pearson   =   7.738437

Variance function: V(u) = u                  [Poisson]
Link function    : g(u) = ln(u)              [Log]

                                             BIC              =  -2091.217
```

		EIM				
los	Coef.	Std. Err.	z	P>|z|	[95% Conf. Interval]	
hmo	-.1425641	.0237347	-6.01	0.000	-.1890833	-.0960449
white	-.1787383	.0278099	-6.43	0.000	-.2332447	-.124232
age	-.0113177	.0050028	-2.26	0.024	-.0211229	-.0015124
_cons	2.530371	.0334326	75.69	0.000	2.464845	2.595898

```
. glm los hmo white age, family(tweedie 1) link(power 0) irls nolog
Generalized linear models                    No. of obs       =      1,495
Optimization      : MQL Fisher scoring       Residual df      =      1,491
                    (IRLS EIM)               Scale parameter  =          1
Deviance       =    8807.81667              (1/df) Deviance  =   5.907322
Pearson        =   11538.00986              (1/df) Pearson   =   7.738437

Variance function: V(u) = phi*u^(1)          [Tweedie]
Link function    : g(u) = ln(u)              [Log]

                                             BIC              =  -2091.217
```

		EIM				
los	Coef.	Std. Err.	z	P>|z|	[95% Conf. Interval]	
hmo	-.1425641	.0660253	-2.16	0.031	-.2719713	-.0131569
white	-.1787383	.0773616	-2.31	0.021	-.3303643	-.0271123
age	-.0113177	.0139167	-0.81	0.416	-.038594	.0159586
_cons	2.530371	.0930029	27.21	0.000	2.348089	2.712654

The Tweedie distribution with power 0 is equivalent to Gaussian variance. We specify power = 1, which is the same a the Gaussian family canonical identity-link.

```
. glm los hmo white age, family(gauss) irls nolog
Generalized linear models                    No. of obs      =        1,495
Optimization       : MQL Fisher scoring      Residual df     =        1,491
                     (IRLS EIM)              Scale parameter =            1
Deviance        =     115626.416            (1/df) Deviance =     77.54957
Pearson         =     115626.416            (1/df) Pearson  =     77.54957

Variance function: V(u) = 1                  [Gaussian]
Link function    : g(u) = u                  [Identity]

                                             BIC             =     104727.4
```

los	Coef.	EIM Std. Err.	z	P>|z|	[95% Conf. Interval]
hmo	-1.330448	.6225106	-2.14	0.033	-2.550547 -.1103499
white	-1.933	.8313126	-2.33	0.020	-3.562343 -.3036574
age	-.1125751	.1387278	-0.81	0.417	-.3844765 .1593263
_cons	12.42505	.9828581	12.64	0.000	10.49868 14.35141

```
. glm los hmo white age, family(tweedie 0) link(power 1) irls nolog
Generalized linear models                    No. of obs      =        1,495
Optimization       : MQL Fisher scoring      Residual df     =        1,491
                     (IRLS EIM)              Scale parameter =            1
Deviance        =     115626.416            (1/df) Deviance =     77.54957
Pearson         =     115626.416            (1/df) Pearson  =     77.54957

Variance function: V(u) = phi*u^(0)          [Tweedie]
Link function    : g(u) = u^(1)              [Power]

                                             BIC             =     104727.4
```

los	Coef.	EIM Std. Err.	z	P>|z|	[95% Conf. Interval]
hmo	-1.330448	.6225106	-2.14	0.033	-2.550547 -.1103499
white	-1.933	.8313126	-2.33	0.020	-3.562343 -.3036574
age	-.1125751	.1387278	-0.81	0.417	-.3844765 .1593263
_cons	12.42505	.9828581	12.64	0.000	10.49868 14.35141

17.4   Generalized additive models

Hastie and Tibshirani (1986) discuss a generalization of GLMs called generalized additive models (GAMs). In a GAM approach, we attempt to overcome an inability of our covariates to adequately model the response surface. With this shortcoming, the fitted values deviate from our response surface in some systematic manner. The GAM modification is to replace the linear predictor with the additive model given by

$$\eta = \eta_0 + \sum_j f_j(x_j) \tag{17.8}$$

where the $f_j(x_j)$ are smooth functions that are estimated from the data. Our model is still additive in the covariates but is no longer linear, hence the change in the name of the analysis to GAM. The smooth functions are identifiable only up to a constant, which we denote η_0.

The examples included in the first edition of this book were generated with user-written software that is no longer available for modern operating systems. There is, as yet, no official program for fitting GAMs, but this could change quickly when the Stata community sets its mind on development. Periodically, use the `search` command, along with keeping your copy of Stata up to date, to learn of associated enhancements to the base software.

18 Clustered data

Contents

This chapter introduces several extensions to GLM that address clustered (panel) data. The topic of the chapter actually merits its own book, so our goal here is to relay introductory information to give an overview of these modern methods. We hope that by introducing the topics here under a notation consistent with our treatment of GLM, the conceptual relationships will be clear even if the implementation details are not.

18.1 Generalization from individual to clustered data

There are several methods to address clustered data. Clustered data occur when there is a natural classification to observations such that data may be organized according to generation or sampling from units. For example, we may collect data on loans where we have multiple observations from different banks. Addressing the dependence in the data on the bank itself would be natural, and there are several methods that we might use to take this dependence into account.

Each method for addressing the panel structure of the data has advantages and limitations, and recognizing the assumptions and inferences that are available benefits the researcher.

In a clustered dataset, we assume that we have $i = 1, \ldots, n$ panels (clusters), where each panel has $t = 1, \ldots, n_i$ correlated observations. Our exponential-family notation is amended to read

$$\exp \left\{ \frac{y_{it}\theta_{it} - b(\theta_{it})}{a(\phi)} - c(y_{it}, \phi) \right\} \tag{18.1}$$

We will focus on the Poisson model as an example. We will use a ship accident dataset from McCullagh and Nelder (1989, 205). This dataset contains a count of the number of reported damage incidents, `accident`; the aggregate months of service by ship type, `service`; the period of operation, `op_75_79`; the year of construction, `co_##_##`; and the ship type, `ship`. We use the number of months in the `service` variable as the exposure in our model. We use the indicator variable `op_75_79` to reflect definition by the starting and ending years of operation and the indicator variables `co_##_##` to reflect definition by the starting and ending years of construction. For example, the indicator variable `op_75_79` records whether the observation is for a ship in operation from 1975 to 1979, and the indicator variable `co_65_69` records whether the ship was in construction from 1965 to 1969.

Our focus is on the models and estimation methods, so we include output from the examples without the associated inference or usual diagnostic plots.

18.2   Pooled estimators

A simple approach to modeling panel data is simply to ignore the panel dependence that might be present. The result of this approach is called a *pooled* estimator because the data are simply pooled together without regard to which cluster the data naturally belong. The resulting estimated coefficient vector is consistent, but it is inefficient (Liang and Zeger 1986). As a result, the estimated (naïve) standard errors will not be a reliable measure for testing purposes. To address the standard errors, we may use a modified sandwich (see section 3.6.4) estimate of variance or another variance estimate that adjusts for the clusters that naturally occur in the data.

When using a pooled estimator with a modified variance estimate, remember that using the modified variance estimate is a declaration that the underlying likelihood is not correct. This is obvious because there is no alteration of the optimized estimating equation and because the modification of the results is limited to modifying the estimated variance of the coefficient vector.

The modified sandwich estimate of variance is robust to correlation within panels and is robust to any type of correlation. This variance estimate may result in smaller or larger standard errors than the naïve (Hessian) standard errors of the pooled estimator, depending on the type of correlation present. The calculation of the modified sandwich estimate of variance uses the sums of the residuals from each cluster. If the residuals are negatively correlated, the sums will be small, and the modified sandwich estimate of variance will produce smaller standard errors than the naïve estimator.

With this approach, our pooled estimator for a Poisson model results in a log likelihood given by

$$\mathcal{L} = \sum_{i=1}^{n} \sum_{t=1}^{n_i} \left[y_{it} \ln\{\exp(\mathbf{x}_{it}\boldsymbol{\beta})\} - \exp(\mathbf{x}_{it}\boldsymbol{\beta}) - \ln \Gamma(y_{it} + 1) \right] \qquad (18.2)$$

which we may solve using the usual GLM methods presented in this text. Suspecting that there may be correlation of repeated observations within ship types, we fit our model of interest by specifying the modified sandwich estimate of variance. Our estimator is consistent but is inefficient, and our standard errors are calculated to be robust to correlation within panels (ship types).

Applying the pooled estimator model to the ship accident data provides estimates given by

```
. webuse ships

. glm accident op_75_79 co_65_69 co_70_74 co_75_79, family(poisson) link(log)
> vce(cluster ship) exposure(service) eform nolog
Generalized linear models                       No. of obs        =         34
Optimization     : ML                           Residual df       =         30
                                                Scale parameter   =          1
Deviance         =  62.36534078                 (1/df) Deviance   =   2.078845
Pearson          =  82.73714004                 (1/df) Pearson    =   2.757905

Variance function: V(u) = u                     [Poisson]
Link function    : g(u) = ln(u)                 [Log]

                                                AIC               =   4.947995
Log pseudolikelihood = -80.11591605             BIC               =  -43.42547

                                 (Std. Err. adjusted for 5 clusters in ship)
```

accident	IRR	Robust Std. Err.	z	P>|z|	[95% Conf. Interval]	
op_75_79	1.47324	.1287036	4.44	0.000	1.2414	1.748377
co_65_69	2.125914	.2850531	5.62	0.000	1.634603	2.764897
co_70_74	2.860138	.6213563	4.84	0.000	1.868384	4.378325
co_75_79	2.021926	.4265285	3.34	0.001	1.337221	3.057227
_cons	.0009609	.0000277	-240.66	0.000	.000908	.0010168
ln(service)	1	(exposure)				

```
Note: _cons estimates baseline incidence rate.
```

In addition to applying the modified sandwich estimate of variance (the vce(cluster *clustvar*) option), we might apply the variable jackknife (the vce(jackknife) option) or grouped bootstrap (the vce(bootstrap) option) estimate of variance. The weighted sandwich estimates of variance are not available for grouped data in the glm command. The application of these variance estimates for our example is for illustration only. There are only five ships in this dataset! The asymptotic properties of these variance estimates depend on the number of groups being large.

18.3   Fixed effects

To address the panel structure in our data, we may include an effect for each panel in our estimating equation. We may assume that these effects are fixed effects or random effects. With regard to assuming fixed effects, we may use conditional or unconditional fixed-effects estimators. Unconditional fixed-effects estimators simply include an indicator variable for the panel in our estimation. Conditional fixed-effects estimators alter the likelihood to a conditional likelihood, which removes the fixed effects from the estimation.

18.3.1   Unconditional fixed-effects estimators

If there are a finite number of panels in a population and each panel is represented in our sample, we could use an unconditional fixed-effects estimator. If there are an infinite number of panels (or uncountable), then we would use a conditional fixed-effects estimator, because using an unconditional fixed-effects estimator would result in inconsistent estimates. Applying the unconditional estimator to our Poisson model results in a log likelihood given by

$$\mathcal{L} = \sum_{i=1}^{n} \sum_{t=1}^{n_i} \left[y_{it} \ln\{\exp(\mathbf{x}_{it}\boldsymbol{\beta} + \nu_i)\} - \exp(\mathbf{x}_{it}\boldsymbol{\beta}) - \ln\Gamma(y_{it} + 1) \right] \qquad (18.3)$$

Fitting this likelihood to our ship accident dataset provides the following results:

```
. glm accident op_75_79 co_65_69 co_70_74 co_75_79 i.ship, family(poisson)
> link(log) exposure(service) eform nolog
Generalized linear models                     No. of obs      =          34
Optimization     : ML                         Residual df     =          25
                                              Scale parameter =           1
Deviance         =  38.69505154               (1/df) Deviance =    1.547802
Pearson          =  42.27525312               (1/df) Pearson  =     1.69101

Variance function: V(u) = u                   [Poisson]
Link function    : g(u) = ln(u)               [Log]

                                              AIC             =    4.545928
Log likelihood   = -68.28077143               BIC             =   -49.46396
```

accident	IRR	OIM Std. Err.	z	P>|z|	[95% Conf. Interval]	
op_75_79	1.468831	.1737218	3.25	0.001	1.164926	1.852019
co_65_69	2.008002	.3004803	4.66	0.000	1.497577	2.692398
co_70_74	2.26693	.384865	4.82	0.000	1.625274	3.161912
co_75_79	1.573695	.3669393	1.94	0.052	.9964273	2.485397
ship						
2	.5808026	.1031447	-3.06	0.002	.4100754	.8226088
3	.502881	.1654716	-2.09	0.037	.2638638	.9584087
4	.926852	.2693234	-0.26	0.794	.5244081	1.638141
5	1.384833	.3266535	1.38	0.168	.8722007	2.198762
_cons	.0016518	.0003592	-29.46	0.000	.0010786	.0025295
ln(service)	1	(exposure)				

Note: _cons estimates baseline incidence rate.

18.3.2 Conditional fixed-effects estimators

A conditional fixed-effects estimator is formed by conditioning out the fixed effects from the estimation. This allows a much more efficient estimator at the cost of placing constraints on inference in the form of the conditioning imposed on the likelihood.

To derive the conditional fixed-effects Poisson model, we begin by specifying the associated probability of a specific outcome defined as

$$\Pr(Y_{it} = y_{it}) = \exp\{-\exp(\nu_i + \mathbf{x}_{it}\boldsymbol{\beta})\}\exp(\nu_i + \mathbf{x}_{it}\boldsymbol{\beta})^{y_{it}}/y_{it}! \qquad (18.4)$$

$$= \frac{1}{y_{it}!}\exp\{-\exp(\nu_i)\exp(\mathbf{x}_{it}\boldsymbol{\beta}) + \nu_i y_{it}\}\exp(\mathbf{x}_{it}\boldsymbol{\beta})^{y_{it}} \qquad (18.5)$$

Because we know that the observations are independent, we may write the joint probability for the observations within a panel as

$$\Pr(Y_{i1} = y_{i1}, \ldots, Y_{in_i} = y_{in_i}) =$$

$$\prod_{t=1}^{n_i} \frac{1}{y_{it}!} \exp\{-\exp(\nu_i)\exp(\mathbf{x}_{it}\boldsymbol{\beta}) + \nu_i y_{it}\} \exp(\mathbf{x}_{it}\boldsymbol{\beta})^{y_{it}} \tag{18.6}$$

$$= \prod_{t=1}^{n_i} \frac{\exp(\mathbf{x}_{it}\boldsymbol{\beta})^{y_{it}}}{y_{it}!} \exp\left\{-\exp(\nu_i)\sum_t \exp(\mathbf{x}_{it}\boldsymbol{\beta}) + \nu_i \sum_t y_{it}\right\} \tag{18.7}$$

We also know that the sum of n_i Poisson independent random variables, each with parameter λ, is distributed as Poisson with parameter $n_i\lambda$ so that we have

$$\Pr\left(\sum_t y_{it}\right) =$$

$$\frac{1}{(\sum_t y_{it})!} \exp\left\{-\exp(\nu_i)\sum_t \exp(\mathbf{x}_{it}\boldsymbol{\beta}) + \nu_i \sum_t y_{it}\right\}\left\{\sum_t \exp(\mathbf{x}_{it}\boldsymbol{\beta})\right\}^{\sum_t y_{it}} \tag{18.8}$$

Thus, the conditional log likelihood is conditioned on the sum of the outcomes in the panel (because that is the sufficient statistic for ν_i). The appropriate conditional probability is given by

$$\Pr\left(Y_{i1} = y_{i1}, \ldots, Y_{in_i} = y_{in_i}\Big|\nu_i, \boldsymbol{\beta}, \sum_t y_{it}\right) = \left(\sum_t y_{it}\right)! \prod_{t=1}^{n_i} \frac{\exp(\mathbf{x}_{it}\boldsymbol{\beta})^{y_{it}}}{y_{it}!\{\sum_k \exp(\mathbf{x}_{ik}\boldsymbol{\beta})\}^{y_{it}}} \tag{18.9}$$

which is free of the fixed effect ν_i. The conditional log likelihood (with offsets) is then derived as

$$\mathcal{L}_c = \ln\left[\prod_{i=1}^{n}\left(\sum_{t=1}^{n_i} y_{it}\right)! \prod_{t=1}^{n_i} \frac{\exp(\mathbf{x}_{it}\boldsymbol{\beta} + \text{offset}_{it})^{y_{it}}}{y_{it}!\{\sum_{\ell=1}^{n_\ell}\exp(\mathbf{x}_{i\ell}\boldsymbol{\beta} + \text{offset}_{i\ell})\}^{y_{it}}}\right] \tag{18.10}$$

$$= \ln\left\{\prod_{i=1}^{n} \frac{(\sum_t y_{it})!}{\prod_{t=1}^{n_i} y_{it}!} \prod_{t=1}^{n_i} p_{it}^{y_{it}}\right\} \tag{18.11}$$

$$= \sum_{i=1}^{n}\left[\ln\Gamma\left(\sum_{t=1}^{n_i} y_{it} + 1\right) - \sum_{t=1}^{n_i}\ln\Gamma(y_{it} + 1)\right.$$

$$\left. + \sum_{t=1}^{n_i}\left\{y_{it}(\mathbf{x}_{it}\boldsymbol{\beta} + \text{offset}_{it}) - y_{it}\ln\sum_{\ell=1}^{n_i}\exp(\mathbf{x}_{i\ell}\boldsymbol{\beta} + \text{offset}_{i\ell})\right\}\right] \tag{18.12}$$

$$p_{it} = e^{\mathbf{x}_{it}\boldsymbol{\beta} + \text{offset}_{it}}\Big/ \sum_\ell e^{\mathbf{x}_{i\ell}\boldsymbol{\beta} + \text{offset}_{i\ell}} \tag{18.13}$$

Applying this estimator to our ship accident dataset provides the following results:

```
. xtset ship
       panel variable:  ship (balanced)
. xtpoisson accident op_75_79 co_65_69 co_70_74 co_75_79, fe exposure(service)
> eform nolog
Conditional fixed-effects Poisson regression    Number of obs      =         34
Group variable: ship                            Number of groups   =          5

                                                Obs per group:
                                                             min =          6
                                                             avg =        6.8
                                                             max =          7

                                                Wald chi2(4)       =      48.44
Log likelihood  = -54.641859                    Prob > chi2        =     0.0000
```

accident	IRR	Std. Err.	z	P>|z|	[95% Conf. Interval]	
op_75_79	1.468831	.1737218	3.25	0.001	1.164926	1.852019
co_65_69	2.008002	.3004803	4.66	0.000	1.497577	2.692398
co_70_74	2.26693	.384865	4.82	0.000	1.625274	3.161912
co_75_79	1.573695	.3669393	1.94	0.052	.9964273	2.485397
ln(service)	1	(exposure)				

Earlier, we saw a special case of conditional fixed-effects logistic regression wherein only one member of each group had a positive choice. With this type of data, one can investigate using either the `mlogit` or the `clogit` command. Whereas the `clogit` model assumes that only the best choice is marked for a group, another model rank orders all the choices. Such a model is commonly referred to as an exploding logit model, a Plackett–Luce model, or a rank-ordered logit model. Stata can fit such models through the `rologit` command.

18.4   Random effects

A random-effects estimator parameterizes the random effects according to an assumed distribution for which the parameters of the distribution are estimated. These models are called subject-specific models, because the likelihood estimates the individual observations instead of the panels.

18.4.1   Maximum likelihood estimation

A maximum likelihood solution to GLMs with random effects is possible depending on the assumed distribution of the random effect. Stata has several random-effects estimators for individual models that fall within the GLM framework.

A standard approach assumes a normally distributed random effect. We can then build the likelihood on the basis of the joint distribution of the outcome and the random effect. With the normal distribution, we write out the likelihood, complete the square, and obtain a likelihood involving the product of functions of the form

$$\int_{-\infty}^{\infty} e^{-z^2} f(z) dz \tag{18.14}$$

This is numerically approximated using Gauss–Hermite quadrature. The accuracy of the approximation is affected by the number of points used in the quadrature calculation and the smoothness of the function $f(z)$. Stata provides the `quadchk` command for users to assess the effect of the number of points used in the quadrature. Increasing the number of points used in the quadrature attempts to overcome a poorly behaved function $f(z)$, but it is not a panacea. The number of observations in a panel affects the smoothness of $f(z)$, and the *Stata Base Reference Manual* recommends (from cited experience) that we should be suspicious of results where panel sizes are greater than 50.

The derivation of the approximate log likelihood $\mathcal{L}_a$ for a Gaussian random-effects model is given by

$$\mathcal{L} = \ln \prod_{i=1}^{n} \int_{-\infty}^{\infty} \frac{e^{-\nu_i^2/2\sigma_\nu^2}}{\sqrt{2\pi}\sigma_\nu} \left\{ \prod_{t=1}^{n_i} \mathcal{F}(\mathbf{x}_{it}\boldsymbol{\beta} + \nu_i) \right\} d\nu_i \tag{18.15}$$

$$\int_{-\infty}^{\infty} e^{-x^2} f(x) dx \approx \sum_{m=1}^{M} w_m^* f(x_m^*) \tag{18.16}$$

$$z = \frac{\nu}{\sqrt{2}\sigma_\nu} \tag{18.17}$$

$$\rho = \frac{\sigma_\nu^2}{\sigma_\nu^2 + \sigma_\epsilon^2} \tag{18.18}$$

$$\nu = z\sqrt{2\frac{\rho}{1-\rho}} \tag{18.19}$$

$$dz = d\nu / \sqrt{2\frac{\rho}{1-\rho}} \tag{18.20}$$

$$\mathcal{L} = \ln \prod_{i=1}^{n} \left\{ \int_{-\infty}^{\infty} e^{-z^2} \frac{1}{\sqrt{\pi}} \prod_{t=1}^{n_i} \mathcal{F}\left(\mathbf{x}_{it}\boldsymbol{\beta} + \sqrt{2\frac{\rho}{1-\rho}}\, z\right) dz \right\}^{w_i} \tag{18.21}$$

$$\approx \ln \prod_{i=1}^{n} \left\{ \sum_{m=1}^{M} w_m^* \frac{1}{\sqrt{\pi}} \prod_{t=1}^{n_i} \mathcal{F}\left(\mathbf{x}_{it}\boldsymbol{\beta} + \sqrt{2\frac{\rho}{1-\rho}}\, x_m^*\right) \right\}^{w_i} \tag{18.22}$$

$$\mathcal{L}_a = \sum_{i=1}^{n} w_i \ln \frac{1}{\sqrt{\pi}} \sum_{m=1}^{M} w_m^* \prod_{t=1}^{n_i} \mathcal{F}\left(\mathbf{x}_{it}\boldsymbol{\beta} + \sqrt{2\frac{\rho}{1-\rho}}\, x_m^*\right) \tag{18.23}$$

where (w_m^*, x_m^*) are the weights and abscissas of the Gauss–Hermite quadrature. We have not explicitly given the form of the function $\mathcal{F}(\cdot)$. This generalization facilitates deriving specific models. Stata has random-effects models for logit, probit, clog-log, Poisson, regression, tobit, and interval regression.

Specifically, for the Gaussian random-effects Poisson model, we may substitute

$$\mathcal{F}(z) = \exp\{-\exp(z)\} \exp(z)^{y_{it}} / y_{it}! \tag{18.24}$$

Applying this model specification to our ship accident dataset results in

```
. xtpoisson accident op_75_79 co_65_69 co_70_74 co_75_79, re normal
> exposure(service) eform nolog skip
Random-effects Poisson regression              Number of obs      =        34
Group variable: ship                           Number of groups   =         5
Random effects u_i ~ Gaussian                  Obs per group:
                                                            min =         6
                                                            avg =       6.8
                                                            max =         7
Integration method: mvaghermite                Integration pts.   =        12
                                               Wald chi2(4)       =     50.95
Log likelihood  = -74.780982                   Prob > chi2        =    0.0000
```

accident	IRR	Std. Err.	z	P>|z|	[95% Conf. Interval]	
op_75_79	1.466677	.1734403	3.24	0.001	1.163259	1.849236
co_65_69	2.032604	.3040933	4.74	0.000	1.516025	2.725205
co_70_74	2.357045	.3998397	5.05	0.000	1.690338	3.286717
co_75_79	1.646935	.3820235	2.15	0.031	1.045278	2.594905
_cons	.0013075	.0002775	-31.28	0.000	.0008625	.001982
ln(service)	1	(exposure)				
/lnsig2u	-2.351868	.8586262			-4.034745	-.6689918
sigma_u	.3085306	.1324562			.1330045	.7156988

```
Note: Estimates are transformed only in the first equation.
Note: _cons estimates baseline incidence rate (conditional on zero random
       effects).
LR test of sigma_u=0: chibar2(01) = 10.67          Prob >= chibar2 = 0.001
```

Other random-effects models may also be derived assuming a distribution other than Gaussian for the random effect. Alternative distributions are chosen so that the joint likelihood may be analytically solved. For example, assuming gamma random effects in a Poisson model allows us to derive a model that may be calculated without the need for numeric approximations. Stata's xtpoisson command will therefore fit either Gaussian random effects or gamma random effects.

In the usual Poisson model, we hypothesize that the mean of the outcome variable y is given by $\lambda_{it} = \exp(\mathbf{x}_{it}\boldsymbol{\beta})$. In the panel setting, we assume that each panel has a different mean that is given by $\exp(\mathbf{x}_{it}\boldsymbol{\beta} + \eta_i) = \lambda_{it}\nu_i$. We refer to the random effect as entering multiplicatively rather than additively as in random-effects linear regression.

Because the random effect $\nu_i = \exp(\eta_i)$ is positive, we select a gamma distribution and add the restriction that the mean of the random effects equals one so that we have only one more parameter, θ, to estimate.

$$f(\nu_i) = \frac{\theta^\theta}{\Gamma(\theta)}\nu_i^{\theta-1}\exp(-\theta\nu_i) \qquad (18.25)$$

Because the conditional mean of the outcome given the random effect is Poisson and the random effect is gamma(θ,θ), we have that the joint density function for a panel is given by

$$f(\nu_i, y_{i1}, y_{i2}, \ldots, y_{in_i}) = \frac{\theta^\theta}{\Gamma(\theta)}\nu_i^{\theta-1}\exp(-\theta\nu_i)\prod_{t=1}^{n_i}\exp(-\nu_i\lambda_{it})(\nu_i\lambda_{it})^{y_{it}}/y_{it}! \qquad (18.26)$$

and the joint density for all the panels is the product because the panels are independent.

The log likelihood for gamma-distributed random effects may then be derived by integrating over ν_i. The log likelihood is derived as

$$
\begin{aligned}
\mathcal{L} &= \ln\prod_{i=1}^{n}\left\{\int_0^\infty \frac{\theta^\theta}{\Gamma(\theta)}\nu_i^{\theta-1}\exp(-\theta\nu_i)\prod_{t=1}^{n_i}\frac{(\nu_i\lambda_{it})^{y_{it}}}{y_{it}!}\exp(-\nu_i\lambda_{it})d\nu_i\right\} & (18.27) \\
&= \sum_{i=1}^{n}\left\{\ln\Gamma\left(\theta + \sum_{t=1}^{n_i}y_{it}\right) - \ln\Gamma(\theta) - \sum_{t=1}^{n_i}\ln\Gamma(y_{it}+1) + \theta\ln u_i\right. \\
&\quad \left. + \left(\sum_{t=1}^{n_i}y_{it}\right)\ln(1-u_i) - \left(\sum_{t=1}^{n_i}y_{it}\right)\ln\left(\sum_{t=1}^{n_i}\lambda_{it}\right) + \sum_{t=1}^{n_i}y_{it}\xi_{it}\right\} & (18.28) \\
u_i &= \frac{\theta}{\theta + \sum_{t=1}^{n_i}\lambda_{it}} & (18.29) \\
\lambda_{it} &= \exp(\mathbf{x}_{it}\boldsymbol{\beta}) & (18.30)
\end{aligned}
$$

Applying this model specification to our ship accident dataset results in

```
. xtpoisson accident op_75_79 co_65_69 co_70_74 co_75_79, re exposure(service)
> eform nolog
```

Random-effects Poisson regression	Number of obs	=	34
Group variable: ship	Number of groups	=	5

Random effects u_i ~ Gamma

Obs per group:

min =		6
avg =		6.8
max =		7

	Wald chi2(4)	=	50.90
Log likelihood = -74.811217	Prob > chi2	=	0.0000

accident	IRR	Std. Err.	z	P>|z|	[95% Conf. Interval]	
op_75_79	1.466305	.1734005	3.24	0.001	1.162957	1.848777
co_65_69	2.032543	.304083	4.74	0.000	1.515982	2.72512
co_70_74	2.356853	.3999259	5.05	0.000	1.690033	3.286774
co_75_79	1.641913	.3811398	2.14	0.033	1.04174	2.58786
_cons	.0013724	.0002992	-30.24	0.000	.0008952	.002104
ln(service)	1	(exposure)				
/lnalpha	-2.368406	.8474597			-4.029397	-.7074155
alpha	.0936298	.0793475			.0177851	.4929165

Note: Estimates are transformed only in the first equation.
Note: _cons estimates baseline incidence rate (conditional on zero random
      effects).
LR test of alpha=0: chibar2(01) = 10.61   Prob >= chibar2 = 0.001

The above command is implemented in Stata in internal (fast) code. It has fast execution times for small to medium datasets and is specific to the Poisson model. Stata has other commands that are specific to other family–link combinations.

Stata users also have available a command for fitting general linear latent and mixed models: gllamm (Rabe-Hesketh, Pickles, and Taylor 2000). With this approach, we may specify an arbitrarily sophisticated model including GLM with random-effects models. The command will also allow for fitting the ordered and unordered outcomes not available as part of the glm command. The method of estimation includes Gaussian quadrature. gllamm is a powerful community-contributed command that one can obtain using Stata's Internet capabilities:

```
. ssc describe gllamm
```

The only downside to the command is that it is implemented using a derivative-free method. The resulting need for calculating accurate numeric derivatives results in slow execution times.

We can use this command to fit a Gaussian random-effects Poisson model for our ship accident dataset by using Gaussian quadrature for the random-effects estimation. The results are

```
. generate double exposure = ln(service)
(6 missing values generated)
. gllamm accident op_75_79 co_65_69 co_70_74 co_75_79, family(poisson)
> link(log) i(ship) offset(exposure) eform
Iteration 0:    log likelihood = -78.987823
Iteration 1:    log likelihood = -73.980982
Iteration 2:    log likelihood = -73.889371
Iteration 3:    log likelihood = -73.889346
Iteration 4:    log likelihood = -73.889346

number of level 1 units = 34
number of level 2 units = 5

Condition Number = 6.7997707

gllamm model

log likelihood = -73.889346
```

accident	exp(b)	Std. Err.	z	P>|z|	[95% Conf. Interval]	
op_75_79	1.469583	.1737318	3.26	0.001	1.165645	1.852772
co_65_69	2.025639	.302931	4.72	0.000	1.511007	2.715551
co_70_74	2.333301	.394834	5.01	0.000	1.674685	3.250936
co_75_79	1.638866	.3771262	2.15	0.032	1.043927	2.572865
_cons	.0012012	.0001684	-47.98	0.000	.0009127	.001581
exposure	1	(offset)				

```
Note: Estimates are transformed only in the first equation.

Variances and covariances of random effects
------------------------------------------------------------------------------
***level 2 (ship)
    var(1): .17662891 (.0937864)
------------------------------------------------------------------------------
```

Results from this derivative-free implementation will not match the Gaussian random effects xtpoisson command. Calculating numeric derivatives will result in a reliable comparison of approximately four digits between the two methods.

18.4.2 Gibbs sampling

Instead of analytically deriving or numerically approximating (via quadrature) the joint distribution of the model and random effect, we can also fit the model using Gibbs sampling. This approach takes a Bayesian view of the problem and focuses on the conditional distributions. It then uses Monte Carlo calculation techniques to estimate the conditional probabilities.

Zeger and Karim (1991) describe a general implementation for fitting GLMs with random effects. An advantage of their approach is that we are not limited to one random effect. This generalization costs computational efficiency; implementation will require much greater execution times. However, the Gibbs sampling approach will fit models that occasionally cause numeric problems for quadrature-implemented methods.

In the following, we assume that there is one random effect to facilitate comparison with the other models in this chapter. As usual, we assume that y_{it} follows an exponential-family form

$$f(y_{it}; \nu_i) = \exp\left\{ \frac{y_{it}\theta_{it} - b(\theta_{it})}{a(\phi)} - c(y_{it}, \phi) \right\} \tag{18.31}$$

and that our link function is given by

$$g(\mu_{it}) = \mathbf{x}_{it}\boldsymbol{\beta} + \nu_i \tag{18.32}$$

We further assume that the random effects ν_i are independent observations from some parametric distribution $\mathcal{F}$ with zero mean and variance σ_ν^2. Our objective is to derive the posterior distribution $f(\boldsymbol{\beta}, \sigma_\nu^2)$ given by

$$f(\boldsymbol{\beta}, \sigma_\nu^2) = \frac{\prod_{i=1}^n \int f(y_i; \nu_i, \boldsymbol{\beta}) g(\nu_i; \sigma_\nu^2) p(\boldsymbol{\beta}, \sigma_\nu^2) \, d\nu_i}{\int \prod_{i=1}^n \int f(y_i; \nu_i, \boldsymbol{\beta}) g(\boldsymbol{\beta}; \sigma_\nu^2) p(\boldsymbol{\beta}, \sigma_\nu^2) \, d\nu_i \, d\boldsymbol{\beta} \, d\sigma_\nu^2} \tag{18.33}$$

See Zeger and Karim (1991) for the implementation details of the Monte Carlo methods based on the above-defined conditional distributions. There are no facilities in Stata at this point for using this approach.

18.5 Mixed-effect models

Stata has a large collection of commands for fitting models that incorporate both fixed and random effects. In the previous section, we outlined specific approaches to single-level random effects, but random errors can be considered for more than one level. A concise overview of relevant commands for fitting such models is listed in table 18.1.

Table 18.1. Stata commands for mixed-effects modeling

Outcome	Commands
continuous	`mixed, sem, metobit, meintreg`
binary	`mecloglog, melogit, meprobit`
ordinal	`meologit, meoprobit`
count	`mepoisson, menbreg`
event time	`mestreg`
general	`meglm, gsem`

Clearly, the commands most relevant for the subject addressed in this book are covered by `meglm`; extensions are addressed by `meologit` and `meoprobit`. Although there is currently no specific command to fit mixed-effect multinomial models, Stata documentation illustrates how such a model can be fit using the `gsem` generalized structural equation model command. The rest of the commands listed in table 18.1 can be used for specific mixed-effect generalized linear models.

While it is beyond the scope of this book to completely cover structural equation models (SEMs), we provide a brief outline of the syntax of the `sem` and `gsem` commands in table 18.2 so that specification of equivalent models can be seen. In that way, we hope that the generalizations of SEMs can be handled without too much difficulty.

Table 18.2. Equivalent commands in Stata

Command	SEM command
`glm y x, family(gaussian) link(identity)`	`gsem (y <- x)`
`glm y x, family(gamma) link(log)`	`gsem (y <- x, gamma)`
`glm y x, family(binomial) link(logit)`	`gsem (y <- x, logit)`
`glm y x, family(binomial) link(probit)`	`gsem (y <- x, probit)`
`glm y x, family(poisson) link(log)`	`gsem (y <- x, poisson)`
`glm y x, family(nbinomial ml) link(log)`	`gsem (y <- x, nbreg)`
`glm y x, family(`*fam*`) link(`*lnk*`)`	`gsem (y <- x, family(`*fam*`) link(`*lnk*`))`
`ologit y x`	`gsem (y <- x, ologit)`
`oprobit y x`	`gsem (y <- x, oprobit)`
`mlogit y x, base(1)`	`gsem (b1.y <- x, mlogit)`

Each specification assumes suitably assigned values to the specified variables; in particular, the multinomial logistic regression example assumes that y has values 1, 2, or 3.

The addition of a single random effects in the SEM specification is to create a name for the latent random effect; that name cannot coincide with an extant variable. So, the following commands in table 18.3 are equivalent:

Table 18.3. Equivalent random-effects logistic regression commands in Stata

Command
`xtlogit y x, i(group) re`
`meglm y x || group:, family(bin) link(logit)`
`melogit y x || group:`
`gsem (y <- x M[group], logit)`

There is no variable in the dataset named M. This should make it obvious that a random-effects multinomial logistic regression model can be fit in Stata by typing

```
gsem (b1.y <- x M[group], mlogit)
```

There are more generalizations and specifications available to those interested in pursuing SEMs. The Stata documentation is your best resource along with a large collection of example models. Readers interested in learning more about generalized SEMs are encouraged to consult these sources of information.

To illustrate, we synthesize random-effects data:

```
. clear
. set seed 2394
. set obs 2000
number of observations (_N) was 0, now 2,000
. generate id  = floor((_n-1)/4) + 1
. generate rep = mod((_n-1),4) + 1
. sort id rep
. generate x1  = runiform()<.5
. generate x2  = runiform()<.5
. generate ui  = rgamma(1,1)
. quietly by id: generate lnui = ln(ui[1])
. generate xb1 = 1
. generate xb2 = exp(1 -.5*x1 -.5*x2 + lnui)
. generate xb3 = exp(1 + x1 - x2 + lnui)
. generate p1  = xb1/(xb1+xb2+xb3)
. generate p2  = xb2/(xb1+xb2+xb3)
. generate p3  = xb3/(xb1+xb2+xb3)
. generate u   = runiform()
. generate y   = 1 + (u>=p1) + (u>=p1+p2)
```

First, we illustrate a multinomial logit model. Because the model does not incorporate a random effect (present in the data), the standard errors are smaller than they should be given that the data are not all independent.

```
. mlogit y x1 x2, baseoutcome(1) nolog
```

Multinomial logistic regression   Number of obs   =   2,000
                                                 LR chi2(4)   =   198.65
                                                 Prob > chi2   =   0.0000
Log likelihood = -2054.2398   Pseudo R2   =   0.0461

y	Coef.	Std. Err.	z	P>|z|	[95% Conf. Interval]	
1	(base outcome)					
2						
x1	-.5021927	.1229752	-4.08	0.000	-.7432196	-.2611658
x2	-.2911434	.1203704	-2.42	0.016	-.527065	-.0552219
_cons	.0858046	.1011677	0.85	0.396	-.1124804	.2840896
3						
x1	.8549589	.1084426	7.88	0.000	.6424154	1.067502
x2	-.7706289	.1078603	-7.14	0.000	-.9820313	-.5592265
_cons	.1562477	.0962826	1.62	0.105	-.0324628	.3449582

```
. gsem (b1.y <- x1 x2 M[id], mlogit), nolog
```

Generalized structural equation model   Number of obs   =   2,000
Response   : y
Base outcome   : 1
Family   : multinomial
Link   : logit
Log likelihood = -1980.6805

 (1) [2.y]M[id] = 1

	Coef.	Std. Err.	z	P>|z|	[95% Conf. Interval]	
1.y	(base outcome)					
2.y						
x1	-.4377415	.1430275	-3.06	0.002	-.7180703	-.1574127
x2	-.4820666	.1428555	-3.37	0.001	-.7620583	-.2020749
M[id]	1	(constrained)				
_cons	.4186655	.1380258	3.03	0.002	.1481399	.6891912
3.y						
x1	.9233937	.1337851	6.90	0.000	.6611797	1.185608
x2	-.972387	.134831	-7.21	0.000	-1.236651	-.708123
M[id]	1.056892	.0872772	12.11	0.000	.8858314	1.227952
_cons	.4665141	.1384961	3.37	0.001	.1950668	.7379613
var(M[id])	1.833442	.3556713			1.25355	2.681592

18.6 GEEs

A summary of generalized estimating equations (GEEs) is presented in Hilbe and Hardin (2008), or Hardin and Hilbe (2013) present an entire text for interested readers.

Recall our derivation of the GLM estimating algorithm, in which we wrote

$$\boldsymbol{\beta}^{(r)} = \boldsymbol{\beta}^{(r-1)} - \left\{ \mathcal{L}''\left(\boldsymbol{\beta}^{(r-1)}\right)\right\}^{-1} \mathcal{L}'\left(\boldsymbol{\beta}^{(r-1)}\right) \tag{18.34}$$

for $r = 1, 2, \ldots$ and a reasonable starting value $\boldsymbol{\beta}^{(0)}$. We further have that

$$\mathcal{L}'(\boldsymbol{\beta}^{(r-1)}) = \mathbf{X}^T \mathbf{W}^{(r-1)} \mathbf{z}^{(r-1)} \tag{18.35}$$

$$\mathcal{L}''(\boldsymbol{\beta}^{(r-1)}) = \mathbf{X}^T \mathbf{W}^{(r-1)} \mathbf{X} \tag{18.36}$$

$$\mathbf{W}^{(r-1)} = D\left\{ \frac{1}{v(\mu)a(\phi)} \left(\frac{\partial\mu}{\partial\eta}\right)^2 \right\} \tag{18.37}$$

$$\mathbf{z}^{(r-1)} = (y - \mu)\left(\frac{\partial\eta}{\partial\mu}\right) + \left(\eta^{(r-1)} - \text{offset}\right) \tag{18.38}$$

where $D()$ denotes a diagonal matrix.

In that original derivation, we assumed that we had n independent observations. Here we investigate how we might estimate $\boldsymbol{\beta}$ if we have n panels (clusters), where each panel $i = 1, 2, \ldots, n$ has n_i correlated observations. We have n independent panels, but the observations within panels are correlated.

We may rewrite the updating equation given in (18.34) as

$$\boldsymbol{\beta}^{(r)} = \boldsymbol{\beta}^{(r-1)} - \left(\mathbf{X}^T D\left[\{v(\mu_i)a(\phi)\}^{-1} \left(\frac{\partial\mu}{\partial\eta}\right)_i^2 \right] \mathbf{X} \right)^{-1}$$

$$\times \ \mathbf{X}^T D\left[\{v(\mu_i)a(\phi)\}^{-1} \left(\frac{\partial\mu}{\partial\eta}\right)_i^2 \right]$$

$$\times \ \left\{ (y_i - \mu_i)\left(\frac{\partial\mu}{\partial\eta}\right)_i + \left(\eta_i^{(r-1)} - \text{offset}_i\right) \right\} \tag{18.39}$$

We now convert this equation for use with a panel dataset. Beginning with a pooled dataset, the observations within a panel are independent. Thus, we can write the variance as

$$V(\mathbf{y}_i) = D\left\{ v(\mu_{it})a(\phi)\right\} \tag{18.40}$$

$$= D\left\{ v(\mu_{it})a(\phi)\right\}^{1/2} \ \mathbf{I} \ D\left\{ v(\mu_{it})a(\phi)\right\}^{1/2} \tag{18.41}$$

without loss of generality. Making this substitution, our pooled GLM estimator may be found from the updating equation given by

$$\boldsymbol{\beta}^{(r)} = \boldsymbol{\beta}^{(r-1)}$$

$$- \left[\sum_{i=1}^{n} \mathbf{X}_i^T D \left(\frac{\partial \mu}{\partial \eta} \right)_{it} D \left\{ v(\mu_{it})a(\phi) \right\}^{-1/2} \mathbf{I} \, D \left\{ v(\mu_{it})a(\phi) \right\}^{-1/2} D \left(\frac{\partial \mu}{\partial \eta} \right)_{it} \mathbf{X}_i \right]^{-1}$$

$$\times \sum_{i=1}^{n} \left[\mathbf{X}_i^T D \left(\frac{\partial \mu}{\partial \eta} \right)_{it} D \left\{ v(\mu_{it})a(\phi) \right\}^{-1/2} \mathbf{I} \, D \left\{ v(\mu_{it})a(\phi) \right\}^{-1/2} D \left(\frac{\partial \mu}{\partial \eta} \right)_{it} \right.$$

$$\left. \times \left\{ D(y_{it} - \mu_{it}) \left(\frac{\partial \mu}{\partial \eta} \right)_{it} + \left(\boldsymbol{\eta}_i^{(r-1)} - \text{offset}_i \right) \right\} \right] \tag{18.42}$$

Liang and Zeger (1986) present an extension to GLMs called generalized estimating equations (GEEs). What they show for panel data is a construction of the covariance matrix for a panel of observations in terms of the usual variance function of the mean and an introduced "working" correlation matrix. The correlation matrix is then treated as a matrix of ancillary parameters, so standard errors are not calculated for the estimated correlations. The covariance matrix for the entire set of observations is then block diagonal with the individual blocks for $i = 1, \dots, n$ defined by

$$V(\mathbf{y}_i) = D \left\{ v(\mu_i)a(\phi) \right\}^{1/2} \mathbf{R} \, D \left\{ v(\mu_i)a(\phi) \right\}^{1/2} \tag{18.43}$$

where $\mathbf{R}$ is the within-panel correlation matrix (generalized from our identity matrix for a pooled GLM).

We can then write our GEE estimation algorithm, assuming that there are $i = 1, \dots, n$ panels, where each panel has $t = 1, \dots, n_i$ observations

$$\boldsymbol{\beta}^{(r)} = \boldsymbol{\beta}^{(r-1)}$$

$$- \left[\sum_{i=1}^{n} \mathbf{X}_i^T D \left(\frac{\partial \mu}{\partial \eta} \right)_i D \left\{ v(\boldsymbol{\mu}_i)a(\phi) \right\}^{-1/2} \mathbf{R} \, D \left\{ v(\boldsymbol{\mu}_i)a(\phi) \right\}^{-1/2} D \left(\frac{\partial \mu}{\partial \eta} \right)_i \mathbf{X}_i \right]^{-1}$$

$$\times \sum_{i=1}^{n} \left[\mathbf{X}_i^T D \left(\frac{\partial \mu}{\partial \eta} \right)_i D \left\{ v(\boldsymbol{\mu}_i)a(\phi) \right\}^{-1/2} \mathbf{R} \, D \left\{ v(\boldsymbol{\mu}_i)a(\phi) \right\}^{-1/2} D \left(\frac{\partial \mu}{\partial \eta} \right)_i \right.$$

$$\left. \times \left\{ D(\mathbf{y}_i - \boldsymbol{\mu}_i) \left(\frac{\partial \mu}{\partial \eta} \right)_i + \left(\boldsymbol{\eta}_i^{(r-1)} - \text{offset}_i \right) \right\} \right] \tag{18.44}$$

The estimation proceeds alternating between updating $\boldsymbol{\beta}$ and updating an estimate of the correlation matrix $\mathbf{R}$ (which is usually parameterized to have some user-specified structure). The form of the GEE shows that the focus is on the marginal distribution and that the estimator sums the panels after accounting for within-panel correlation. As such, this is called a population-averaged estimator. It is also easy to see that when the correlation is assumed to be independent, we have an ordinary pooled GLM estimator.

Zeger, Liang, and Albert (1988) and other sources provide further theoretical insight into this model. One can find good comparisons of panel-data estimators in Neuhaus, Kalbfleisch, and Hauck (1991) or Neuhaus (1992). Users can find complete coverage of GEE and its extensions in Hardin and Hilbe (2013).

Recall from chapter 3 the general form of the modified sandwich estimate of variance. There we presented arguments for modifying the usual sandwich estimate of variance that would allow for independent panels (clusters) that had some unspecified form of intrapanel correlation. That is exactly the case here. Therefore, we may specify a modified sandwich estimate of variance by using

$$
\widehat{B}_{\mathrm{MS}} =
$$
$$
\left(\sum_{i=1}^{n} a(\phi)^2 \mathbf{X}_i \mathcal{S}\mathcal{S}^T \mathbf{X}_i^T \right) \tag{18.45}
$$
$$
\mathcal{S} =
$$
$$
D\left(\frac{\partial \mu}{\partial \eta}\right) D\left\{v(\boldsymbol{\mu}_i)a(\phi)\right\}^{-1/2} \mathbf{R}^{-1} D\left\{v(\boldsymbol{\mu}_i)a(\phi)\right\}^{-1/2} D\left(\mathbf{y}_i - \boldsymbol{\mu}_i\right) \tag{18.46}
$$

as the middle of the sandwich and allowing that V_H^{-1} is the first term of the product in (18.44). This is V_{EH}^{-1} and is the usual choice in software (including Stata) because it is the variance matrix implemented in the modified iteratively reweighted least-squares optimization algorithm.

$$
\widehat{V}_{\mathrm{MS}} = \widehat{V}_H^{-1} \widehat{B}_{\mathrm{MS}} \widehat{V}_H^{-1} \tag{18.47}
$$

We assume a specific form of correlation in estimating the coefficient vector where we also estimate that specified correlation structure. This allows us to obtain a more efficient estimator of $\boldsymbol{\beta}$ if the true correlation follows the structure of our specified model. The modified sandwich estimate of variance provides standard errors that are robust to possible misspecification of this correlation structure. We therefore recommend that the modified sandwich estimate of variance always be used with this model. In Stata, this means that you should always specify the vce(robust) option when using the xtgee command or any command that calls xtgee, as we illustrate below.

Applying this model specification to our ship accident dataset results in

```
. webuse ships, clear
. xtpoisson accident op_75_79 co_65_69 co_70_74 co_75_79, pa exposure(service)
> vce(robust) eform nolog
GEE population-averaged model              Number of obs      =         34
Group variable:                     ship  Number of groups   =          5
Link:                                log  Obs per group:
Family:                          Poisson                 min =          6
Correlation:                 exchangeable                avg =        6.8
                                                         max =          7
                                           Wald chi2(4)      =     252.94
Scale parameter:                       1   Prob > chi2       =     0.0000
                                (Std. Err. adjusted for clustering on ship)
```

accident	IRR	Robust Std. Err.	z	P>|z|	[95% Conf. Interval]	
op_75_79	1.483299	.1197901	4.88	0.000	1.266153	1.737685
co_65_69	2.038477	.1809524	8.02	0.000	1.712955	2.425859
co_70_74	2.643467	.4093947	6.28	0.000	1.951407	3.580962
co_75_79	1.876656	.33075	3.57	0.000	1.328511	2.650966
_cons	.0010255	.0000721	-97.90	0.000	.0008935	.001177
ln(service)	1	(exposure)				

```
Note: _cons estimates baseline incidence rate (conditional on zero random
      effects).
```

18.7   Other models

Since the first edition of this book, Stata has continued development of clustered-data commands. It has also developed a powerful matrix programming language that can be used in direct matrix manipulations, as well as through indirect definition of matrices from data; see the documentation on `mata`. If we limit discussion to linear models, another approach to models is through the specification of various levels of correlation. Such specification amounts to a mixed model in which various fixed and random effects are posited for a specific model. Stata's `mixed` command will fit such models.

Through a National Institutes of Health–funded effort with academia, Stata developed extensions to GLMs that allow specification of measurement error. Though not strictly for clustered data, this software includes various approaches to model estimation including regression calibration and simulation extrapolation. The software developed from this effort includes commands `qvf`, `rcal`, and `simex`. Full description of the software and theory behind the estimators was published in the *Stata Journal*; see Hardin and Carroll (2003) for details. The associated collection of software can be installed by typing `net install st0049_1, from(http://www.stata-journal.com/software/sj11-3)`. This collection of software basically replicates the `glm` command, introducing various support mechanisms for consideration of measurement error in one or more predictor variables. Stata users may also use the `gllamm` command, which is fully described with other related Stata commands in Rabe-Hesketh and Skrondal (2012).

To motivate interest in these extensions to GLMs, we present a slightly expanded presentation of an example in the previously cited paper. Here we will synthesize data to illustrate the issues faced when covariates are measured with error.

We will generate data for a linear model. That is, the data will be for an identity link from the Gaussian family. We begin with

```
. drop _all
. set seed 1239
. set more off
. set obs 500
number of observations (_N) was 0, now 500
. generate x1 = runiform()*10
. generate x2 = runiform()*5
. generate x3 = runiform()
. generate x4 = runiform()
. generate x5 = runiform()
. generate err = rnormal()
. generate y =   1*x1 + 2*x2 + 3*x3 + 4*x4 + 5 + err
. generate a1 = x3 + 0.25*rnormal()
. generate a2 = x3 + 0.25*rnormal()
. generate b1 = x4 + 0.25*rnormal()
. generate b2 = x4 + 0.25*rnormal()
```

Thus, we have generated data for a linear regression model. We now assume that instead of actually obtaining the x3 and x4 variables, we could obtain only two error-prone proxy measures for each of these covariates. Instead of observing x3, we instead obtained two observations, a1 and a2. Likewise, instead of observing x4, we have the two error-prone proxy measures b1 and b2. To be clear, had we actually obtained the x3 and x4 covariates, we would simply fit

```
. glm y x1 x2 x3 x4, noheader
Iteration 0:   log likelihood = -718.55956
```

		OIM				
y	Coef.	Std. Err.	z	P>|z|	[95% Conf.	Interval]
x1	1.005124	.015735	63.88	0.000	.9742843	1.035964
x2	2.068741	.031883	64.89	0.000	2.006251	2.13123
x3	3.284629	.157679	20.83	0.000	2.975584	3.593675
x4	4.015878	.1620628	24.78	0.000	3.698241	4.333515
_cons	4.670417	.1648023	28.34	0.000	4.34741	4.993424

Note how the estimated coefficients are near the values used in synthesizing the data, just as we would expect. However, faced with the data at hand, we may not know exactly how to proceed. One might consider simply using one of the two proxy measures for each of the "unobserved" covariates.

```
. glm y x1 x2 a1 b1, noheader
Iteration 0:   log likelihood = -872.39889
```

y	Coef.	OIM Std. Err.	z	P>|z|	[95% Conf. Interval]	
x1	.9726765	.0213475	45.56	0.000	.930836	1.014517
x2	2.088238	.0434078	48.11	0.000	2.003161	2.173316
a1	2.076503	.1644456	12.63	0.000	1.754196	2.398811
b1	2.269671	.1676637	13.54	0.000	1.941057	2.598286
_cons	6.248809	.1982844	31.51	0.000	5.860179	6.637439

```
. glm y x1 x2 a2 b2, noheader
Iteration 0:   log likelihood = -889.30557
```

y	Coef.	OIM Std. Err.	z	P>|z|	[95% Conf. Interval]	
x1	.9800165	.0220888	44.37	0.000	.9367233	1.02331
x2	2.095727	.0448924	46.68	0.000	2.007739	2.183715
a2	1.86148	.1710242	10.88	0.000	1.526279	2.196681
b2	2.249416	.1708848	13.16	0.000	1.914488	2.584344
_cons	6.304609	.2073387	30.41	0.000	5.898233	6.710986

However, the results do not agree with what we know about the data. This is a well-known phenomenon where regression coefficients are attenuated toward the null; estimated coefficients for error-prone proxy covariates tend toward zero.

One approach to this particular problem involves having replicate (error-prone) measures. Because we obtained replicates of the error-prone measures, we have a way of estimating the error in those proxy variables. The simulation–extrapolation method estimates that error and then runs a series of experiments. In each experiment, the approach generates even more measurement error, adds it to the error-prone covariates, and then fits the model. After running these experiments, the approach gains knowledge about how the fitted coefficients are related to the measurement error. This is the simulation part of the algorithm. The extrapolation part is in forecasting what the coefficients would be for no measurement error, thus extrapolating the relationship built up through simulation. The results of the approach along with a plot to illustrate the extrapolation help to illustrate the technique.

```
. simex (y=x1 x2) (w3: a1 a2) (w4: b1 b2), brep(299) seed(12345)
Simulation extrapolation                       No. of obs        =        500
                                               Bootstraps reps   =        299
Residual df    =        495                    Wald F(4,495)     =    1252.89
                                               Prob > F          =     0.0000
Variance Function: V(u) = 1                       [Gaussian]
Link Function    : g(u) = u                       [Identity]

                           Bootstrap
           y       Coef.   Std. Err.       t    P>|t|     [95% Conf. Interval]

          x1    .987933   .0213545    46.26    0.000     .9459763     1.02989
          x2   2.107543   .0389649    54.09    0.000     2.030986      2.1841
          w3   3.172564    .232039    13.67    0.000     2.716661    3.628466
          w4    3.65793   .2258524    16.20    0.000     3.214183    4.101678
       _cons    4.86325   .2340398    20.78    0.000     4.403416    5.323084

. simexplot w3
```

The graph generated by the `simexplot` command is given in figure 18.1. Bootstrap standard errors are calculated in this example; see the documentation of the software for other options. Note how the estimated coefficients are much closer to the values used to generate the original data even though the estimation was performed without using the x3 and x4 covariates.

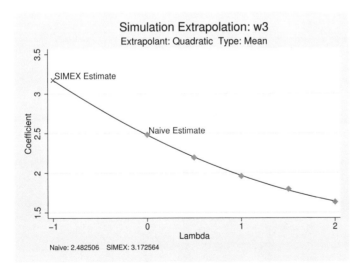

Figure 18.1. Simulation–extrapolation results

Finally, another great source for developments in longitudinal and panel-data analysis is the Stata documentation. Specifically, the *Longitudinal-Data/Panel-Data Reference Manual* includes statistical derivations, technical descriptions, documentation, and example usage for many different models.

19 Bivariate and multivariate models

Contents

19.1   Bivariate and multivariate models for binary outcomes

Stata has several built-in commands for only a couple of bivariate models; see the documentation for the `biprobit` and `mprobit` commands for information on bivariate and multivariate probit models, respectively. In addition, the `mvreg` and `sureg` commands fit multivariate and seemingly unrelated linear regression models.

In the following sections, we introduce a series of bivariate count data models. The log likelihood of each regression models is calculated as the sum of the log of the probabilities of the observed outcomes. In subsequent sections, we discuss bivariate models resulting from specific bivariate distribution functions, but before that, we discuss calculation of bivariate probabilities when we have only the marginal distributions of the two outcomes. This remarkable simplification even allows consideration of bivariate outcomes with different marginal distributions. These initial bivariate models utilize copula functions to calculate joint probabilities for models in which we specify the marginal distributions.

19.2   Copula functions

Assume that we have q possibly dependent continuous random uniform$(0, 1)$ variables $U_1, U_2, \ldots, U_q$. Also, assume that a multivariate cumulative distribution function for these q variates may be written as

$$\Pr\left(U_1 \le u_1, \ldots, U_q \le u_q\right) = C(u_1, \ldots, u_q) \tag{19.1}$$

where $C(\cdot, \cdots, \cdot)$ is the copula, representation of the cumulative distribution function (CDF), and u_j is a particular realization of U_j for $j = 1, \ldots, q$. For more information, note that the theory of copula functions was introduced in Sklar (1959) and, discussed again, later in Sklar (1973).

For the q marginal CDFs $F_1(\cdot)$, ..., $F_q(\cdot)$ and arbitrary values at which to evaluate the marginals $(x_1, \ldots, x_q)$, we can write

$$\begin{aligned} C\{F_1(x_1), \ldots, F_q(x_q)\} &= \left\{F_1^{-1}(U_1) \le x_1, \ldots, F_q^{-1}(U_q) \le x_q\right\} \tag{19.2} \\ &= F(X_1, \ldots, X_q) \tag{19.3} \end{aligned}$$

Thus, the basic idea is that we can estimate a joint CDF of random variates using specified marginal distributions of the individual variates using a special function—the copula function. Astute readers will note that the marginals (when discrete) may not have inverses, but the theory extends to discrete distributions using differencing.

We have that $F(\cdot, \cdots, \cdot)$ is the joint CDF for the q variables $X_1, \ldots, X_q$. The joint CDF is computed with a copula function construction of a specific set of marginals. Necessarily, a regression model built from this specification requires that the marginal distributions are correct.

Sklar's theorem states that for any multivariate CDF, there exists a copula function such that the multivariate CDF can be represented as a function of its marginals through this copula. If the multivariate CDF is continuous, then the copula representation is unique (and not unique for discrete random variables).

19.3 Using copula functions to calculate bivariate probabilities

For the particular case of $C(u_1, u_2)$, a two-dimensional probability distribution function defined on the unit square with univariate marginals on $[0, 1]$, Sklar's theorem states that there exists a copula function C such that

$$F(x_1, x_2) = C\{F_1(x_1), F_2(x_2)\} \tag{19.4}$$

where $F(x_1, x_2) = \Pr(X_1 \le x_1, X_1 \le x_1)$ is a bivariate distribution function of random variables X_1 and X_2. Also, $F_1(x_1)$ and $F_2(x_2)$ denote the marginal distribution functions. If F is the distribution of a discrete random variable (as is the case in which we are interested), then the copula function

$$C(u_1, u_2) = F\{F_1^{-1}(u_1), F_2^{-1}(u_2)\} \tag{19.5}$$

may be parameterized to include measures of dependence between the marginal distributions. Table 19.1 provides a list of bivariate copula functions.

Table 19.1. Bivariate copula functions

Copula type	Function $C(u_1, u_2)$	Dependence
Product	$u_1 u_2$	NA
Frank	$-\theta^{-1} \log[\{\varsigma - (1 - e^{-\theta u_1})(1 - e^{-\theta u_2})\}/\varsigma]$	$\theta \in \Re \backslash \{0\}$
Normal	$\Phi_2 \{\Phi^{-1}(u_1), \Phi^{-1}(u_2); \theta\}$	$-1 \leq \theta \leq +1$
Kimeldorf and Sampson	$(u_1^{-\theta} + u_2^{-\theta} - 1)^{-1/\theta}$	$\theta \in [-1, \infty)\backslash\{0\}$

Note: $\varsigma = 1 - \exp(-\theta)$

Like all multivariate distribution functions, bivariate copulas must obey Fréchet–Hoeffding lower and upper bounds so that joint probabilities satisfy

$$\max \{F_1(x_1) + F_2(x_2) - 1, 0\} \leq F(x_1, x_2) \leq \min \{F_1(x_1), F_2(x_2)\} \qquad (19.6)$$

Because of roundoff error and negative results, these bounds should be imposed on calculation.

With the cumulative distribution function defined via the copula function, the probability mass function $c_{12}(\cdot)$ can be defined by taking finite differences

$$c_{12} \{F_1(x_1), F_2(x_2); \theta\} = \qquad (19.7)$$

$$C \{F_1(x_1), F_2(x_2); \theta\} - C \{F_1(x_1 - 1), F_2(x_2); \theta\}$$
$$-C \{F_1(x_1), F_2(x_2 - 1); \theta\} + C \{F_1(x_1 - 1), F_2(x_2 - 1); \theta\} \quad (19.8)$$

A regression model may then be defined in terms of $\log(c_{12}[\cdot])$ for which we need only use a copula function and specific marginal CDFs.

19.4   Synthetic datasets

In this section, we briefly discuss the algorithm behind the code used to generate datasets for estimation of bivariate count-data models. The desire is to generate a dataset from a population with known parameters. If we could invert the joint probability functions of each of the supported models, we could use that inverse function to produce variates from random uniform quantiles. This is not possible. Instead, we utilize rejection sampling. Interested readers should consult Robert and Casella (2004) for detailed information, because we give only a brief introduction in the following.

A rejection sampling approach can be illustrated in a unidimensional sense by imagining a simple distribution function given by $P(X = 1) = 0.1$, $P(X = 2) = 0.2$, $P(X = 3) = 0.3$, and $P(X = 4) = 0.4$. We know that the support of the distribution is the set of outcomes $\{1, 2, 3, 4\}$, and we know the probability of each of the outcomes. The technique works using the following steps:

1. Generate a random discrete outcome, `x=runiformint(1,4)`.

2. Generate a random continuous uniform, `u=runiform()`.

3. Accept `x` into the random sample if $P(X = \texttt{x}) < \texttt{u}$.

4. Repeat the first three steps until the desired sample size is achieved.

Each outcome is considered for the sample equally but will only be accepted into the sample at the rate dictated by its probability. This simple idea is easily extended to multiple dimensions. The speed with which a random sample can be constructed is driven by how often candidate outcomes are rejected from the sample. The probability of rejecting a candidate in our simple example is

$$\frac{1}{4}\left\{(1 - 0.10) + (1 - 0.20) + (1 - 0.30) + (1 - 0.40)\right\} = 0.75$$

This implies that, on average, we would need to generate 400 candidate values to build a 100-observation sample.

To increase the efficiency, we note that the largest probability of our distribution function is 0.40. Thus, in step 2, we could generate a random uniform between 0 and 0.40 instead of between 0 and 1. Equivalently, we could alter step 3 to, "Accept `x` into the random sample if $P(X = \texttt{x}) < \texttt{u/m}$ where $\texttt{m} = \max_{i=1,2,3,4} P(X = i)$." With this change, the probability of rejecting a candidate in our simple example is cut in half:

$$\frac{1}{4}\left\{(1 - 0.25) + (1 - 0.50) + (1 - 0.75) + 0\right\} = 0.375 \tag{19.9}$$

This implies that, on average, we would need to generate 160 candidate values to build a 100-observation sample.

Of course, in cases like this one where we have access to the inverse of the CDF (the quantile function), we could just use that inverse function in an approach requiring generation of 100 values to build the 100-observations sample.

The last simplification we need for our application is that the range of our outcomes is infinite, which makes step 1 in our algorithm untenable. To get around this issue, we generate a large number of variates from the marginal distribution. We then use the maximum of the large sample as the maximum possible value that we will consider for our random bivariate pairs; we do this for each of the marginal distributions.

In this way, we created a general rejection sampling program called `rejectsample` that takes as arguments

1. A marginal function to generate random variates (for each of the two variates)

2. A function to generate the joint probability for two observations

The program is not intended to be bulletproof, but it is good enough to generate sample datasets to verify the `ml` programs provided here. The `rejectsample` command

and the helper programs listed in table 19.2 can be installed by typing `net install rejectsample, from(http://www.stata.com/users/jhardin)`. We did not create help files for these commands. In most cases, the ancillary parameters are directly comparable, but look at the provided examples to see whether they might be reparameterized.

Based on the example specification,

```
rejectsample y1 y2, mu1(mean1) mu2(mean2) lambda(12)
    marg1(marg_pmf_program_name) marg2(marg_pmf_program_name)
    jointpmf(pmf_program_name)
```

the basic use of the program involves the user specifying two new variable names (`y1` and `y2`) into which the bivariate outcomes will be stored. The marginal conditional means are specified as existing variable names in the `mu1()` and `mu2()` options. The names of the programs that should be called to generate the random variants following the marginal pmf for each of the bivariate outcomes are specified in the `marg1()` and `marg2()` options. We have provided programs `margpoisson` for generating Poisson random variates and `margnbinom` for generating negative binomial variates. These programs are called and passed the marginal means; for example, the program specified in `marg1()` is passed the conditional mean specified in `mu1`. Finally, the rejection sampling needs access to a joint pmf; the joint pmfs that we have supplied for the five examples in this chapter are listed at the bottom of table 19.2. Finally, the user can also specify a value for λ (model-specific) and can specify a value for the negative binomial α parameter. Note that what gets printed in the examples is the logarithm of the negative binomial parameter.

Table 19.2. Programs for generating bivariate outcomes with `rejectsample`

Program name	Purpose
`margpoisson`	generates random Poisson variates
`margnbinom`	generates random negative binomial variates
`jointfcoppp`	calculates $P(Y_1 = y_1, Y_2 = y_2)$ using the Frank copula function for marginal Poisson distributions
`jointfcopnn`	calculates $P(Y_1 = y_1, Y_2 = y_2)$ using the Frank copula function for marginal negative binomial distributions
`jointfamp`	calculates $P(Y_1 = y_1, Y_2 = y_2)$ using Famoye's bivariate Poisson distribution function
`jointfamn`	calculates $P(Y_1 = y_1, Y_2 = y_2)$ using Famoye's bivariate negative binomial distribution function
`jointmo`	calculates $P(Y_1 = y_1, Y_2 = y_2)$ using the Marshall and Olkin bivariate negative binomial distribution function

Included with `rejectsample` are helper programs listed in table 19.2. This collection of helper programs is not complete. Interested readers can extend the functionality of the `rejectsample` program by writing programs to calculate joint probabilities for other copula functions and for other marginal distributions. We use one of the generated datasets in section 20.9.2.9 in an example of Bayesian GLMs.

19.5   Examples of bivariate count models using copula functions

Using the ideas and programs outlined in section 19.4, we generate a suitable program for the estimation of the log likelihood. The log likelihood for the model is based on the bivariate probabilities of specific outcomes. For this first example, we assume that each marginal distribution is Poisson. The log-likelihood function is written using the linear form (see the reference manual and Gould, Pitblado, and Poi [2010]). We ensure that the Fréchet–Hoeffding bounds are enforced in the calculation of the joint probabilities, and then define the observation-level contributions to the overall model log likelihood.

After defining the program, we fit individual Poisson models for the two dependent variables and save the estimated coefficient vectors to use as starting values for the bivariate model. Finally, we use Stata's `ml` subcommands to specify and then fit the model of interest.

```
. program bipoissll
  1.          version 15
  2.          args todo b lnf
  3.          quietly {
  4.                  local y1 "$ML_y1"
  5.                  local y2 "$ML_y2"
  6.                  tempvar  xb1 xb2
  7.                  tempname theta eta
  8.                  mleval `xb1´   = `b´, eq(1)
  9.                  mleval `xb2´   = `b´, eq(2)
 10.                  mleval `theta´ = `b´, eq(3) scalar
 11.

     .          // Bound theta for exp() function
     .          scalar `theta´ = max(min(`theta´,18),-18)
 12.                  scalar `eta´   = 1 - exp(-`theta´)
 13.

     .          // SECTION: Calculate marginal probabilities
     .          // (marginals are Poisson)
     .          tempvar f10 f11 f20 f21
 14.                  generate double `f10´ = poisson(exp(`xb1´), `y1´)
 15.                  generate double `f11´ = cond(`y1´==0, 0,
  >                      poisson(exp(`xb1´), `y1´-1))
 16.                  generate double `f20´ = poisson(exp(`xb2´), `y2´)
 17.                  generate double `f21´ = cond(`y2´==0, 0,
  >                      poisson(exp(`xb2´), `y2´-1))
 18.

     .          // Calculate the Frank Copula function
     .          tempvar ttt t1 t2 t3 t4
```

```
 19.              generate double `t1´ = - (1/`theta´) *
 >                    log((`eta´ - (1-exp(-`theta´*`f10´)) *
 >                    (1-exp(-`theta´*`f20´)))/`eta´)
 20.              generate double `t2´ = - (1/`theta´) *
 >                    log((`eta´ - (1-exp(-`theta´*`f11´)) *
 >                    (1-exp(-`theta´*`f20´)))/`eta´)
 21.              generate double `t3´ = - (1/`theta´) *
 >                    log((`eta´ - (1-exp(-`theta´*`f10´)) *
 >                    (1-exp(-`theta´*`f21´)))/`eta´)
 22.              generate double `t4´ = - (1/`theta´) *
 >                    log((`eta´ - (1-exp(-`theta´*`f11´)) *
 >                    (1-exp(-`theta´*`f21´)))/`eta´)
 23.
  .               // Set up for reasonable bounds of probabilities
  .               tempname maxv minv
 24.                scalar `minv´ = 1e-6
 25.                scalar `maxv´ = 1.0-`minv´
 26.
  .               // Impose Frechet-Hoeffding bounds
  .               replace `t1´ = min(max(`t1´, max(
 >                    `f10´+`f20´-1,`minv´)),
 >                    min(`f10´, `f20´,`maxv´))
 27.              replace `t2´ = min(max(`t2´, max(
 >                    `f11´+`f20´-1,`minv´)),
 >                    min(`f11´, `f20´,`maxv´))
 28.              replace `t3´ = min(max(`t3´, max(
 >                    `f10´+`f21´-1,`minv´)),
 >                    min(`f10´, `f21´,`maxv´))
 29.              replace `t4´ = min(max(`t4´, max(
 >                    `f11´+`f21´-1,`minv´)),
 >                    min(`f11´, `f21´,`maxv´))
 30.
  .               // Use difference calculation to get joint pmf
  .               generate double `ttt´ = `t1´-`t2´-`t3´+`t4´
 31.
  .               // Calculate log likelihood
  .               replace `lnf´ = log(`ttt´)
 32.        }
 33. end

. drop _all

. set seed 29348

. set obs 1000
number of observations (_N) was 0, now 1,000

. gen byte x11 = rbinomial(1,.5)

. gen byte x12 = rbinomial(1,.5)

. gen byte x21 = rbinomial(1,.5)

. gen byte x22 = rbinomial(1,.5)

. gen double mu1 = exp(1-x11+x12)

. gen double mu2 = exp(1+x21-x22)

. rejectsample y1 y2, mu1(mu1) mu2(mu2) lambda(12) marg1(margpoisson)
> marg2(margpoisson) jointpmf(jointfcoppp)
```

```
. * Run models to get starting values for each outcome
. poisson y1 x11 x12, nolog
```

Poisson regression   Number of obs   =   1,000
                                                      LR chi2(2)   =   1395.46
                                                      Prob > chi2   =   0.0000
Log likelihood = -1891.0498   Pseudo R2   =   0.2695

y1	Coef.	Std. Err.	z	P>|z|	[95% Conf. Interval]	
x11	-.9504469	.037357	-25.44	0.000	-1.023665	-.8772285
x12	.9616862	.0376796	25.52	0.000	.8878356	1.035537
_cons	1.032362	.0334892	30.83	0.000	.9667246	1.098

```
. matrix b1 = e(b)
. poisson y2 x21 x22, nolog
```

Poisson regression   Number of obs   =   1,000
                                                      LR chi2(2)   =   1650.42
                                                      Prob > chi2   =   0.0000
Log likelihood = -1874.3466   Pseudo R2   =   0.3057

y2	Coef.	Std. Err.	z	P>|z|	[95% Conf. Interval]	
x21	.9942201	.0376978	26.37	0.000	.9203338	1.068106
x22	-1.054908	.0383926	-27.48	0.000	-1.130156	-.9796594
_cons	1.046974	.0336729	31.09	0.000	.9809759	1.112971

```
. matrix b2 = e(b)

.
. * Define (near zero = independent) starting value for theta
. matrix tt = (0.1)

.
. * Paste together the starting value vector
. matrix bb = b1, b2, tt

.
. * Use ml to fit a bivariate model where
. * each marginal distribution is Poisson
. ml model lf0 bipoissll (y1:y1 = x11 x12) (y2:y2 = x21 x22) (theta:)
. ml init bb, copy
```

```
. * Turn search off because we have starting values
. ml maximize, search(off) nolog title(Bivariate Poisson-Poisson regression)
numerical derivatives are approximate
flat or discontinuous region encountered
```

Bivariate Poisson-Poisson regression	Number of obs	=	1,000
	Wald chi2(2)	=	3948.74
Log likelihood = -3175.3411	Prob > chi2	=	0.0000

	Coef.	Std. Err.	z	P>|z|	[95% Conf. Interval]
y1					
x11	-.9656972	.0207379	-46.57	0.000	-1.006343 -.9250517
x12	.9828914	.0209891	46.83	0.000	.9417535 1.024029
_cons	1.011313	.023111	43.76	0.000	.966016 1.056609
y2					
x21	1.017444	.0211147	48.19	0.000	.9760599 1.058828
x22	-1.011529	.0215178	-47.01	0.000	-1.053703 -.9693551
_cons	1.000696	.0231028	43.32	0.000	.9554154 1.045976
theta					
_cons	11.27714	.474611	23.76	0.000	10.34692 12.20736

Note how the coefficients for each outcome and the dependency parameter estimate are all close to the specified parameter values of the `rejectsample` data synthesis program. The independent Poisson regression models also estimate the means well, but they estimate larger standard errors.

Making fairly simple changes to the bivariate Poisson log likelihood in the previous example, we create a bivariate negative binomial log likelihood. In this example, we fit a bivariate model for the correlated negative binomial outcomes. This time, we skip estimation of the individual models and appeal to Stata's `ml search` subcommand to search for starting values. Afterward, we again fit the model of interest.

```
. program binbinomll
  1.            version 15
  2.            args todo b lnf
  3.            quietly {
  4.                    local y1 "$ML_y1"
  5.                    local y2 "$ML_y2"
  6.                    tempvar  xb1 xb2
  7.                    tempname theta eta lnalpha1 lnalpha2
  8.                    mleval `xb1´      = `b´, eq(1)
  9.                    mleval `lnalpha1´ = `b´, eq(2) scalar
 10.                    mleval `xb2´      = `b´, eq(3)
 11.                    mleval `lnalpha2´ = `b´, eq(4) scalar
 12.                    mleval `theta´    = `b´, eq(5) scalar
 13.
   .                    // Bound theta for exp() function
   .                    scalar `theta´ = max(min(`theta´,18),-18)
 14.                    scalar `eta´   = 1 - exp(-`theta´)
 15.
   .                    tempname m1 p1 m2 p2
 16.                    scalar `m1´    = exp(-`lnalpha1´)
```

```
 17.             generate double `p1´ = 1/(1+exp(`xb1´+`lnalpha1´))
 18.
   .          scalar `m2´      = exp(-`lnalpha2´)
 19.             generate double `p2´ = 1/(1+exp(`xb2´+`lnalpha2´))
 20.
   .          // Calculate marginal probabilities
   .          tempvar f10 f11 f20 f21
 21.             generate double `f10´ = nbinomial(`m1´, `y1´, `p1´)
 22.             generate double `f11´ = cond(`y1´==0, 0,
  >                 nbinomial(`m1´, `y1´-1, `p1´))
 23.             generate double `f20´ = nbinomial(`m2´, `y2´, `p2´)
 24.             generate double `f21´ = cond(`y2´==0, 0,
  >                 nbinomial(`m2´, `y2´-1, `p2´))
 25.
   .          // Calculate the Frank Copula function
   .          tempvar ttt t1 t2 t3 t4
 26.             generate double `t1´ = - (1/`theta´) * log((`eta´ -
  >                 (1-exp(-`theta´*`f10´)) *
  >                 (1-exp(-`theta´*`f20´)))/`eta´)
 27.             generate double `t2´ = - (1/`theta´) * log((`eta´ -
  >                 (1-exp(-`theta´*`f11´)) *
  >                 (1-exp(-`theta´*`f20´)))/`eta´)
 28.             generate double `t3´ = - (1/`theta´) * log((`eta´ -
  >                 (1-exp(-`theta´*`f10´)) *
  >                 (1-exp(-`theta´*`f21´)))/`eta´)
 29.             generate double `t4´ = - (1/`theta´) * log((`eta´ -
  >                 (1-exp(-`theta´*`f11´)) *
  >                 (1-exp(-`theta´*`f21´)))/`eta´)
 30.
   .          // Set up for reasonable bounds of probabilities
   .          tempname maxv minv
 31.             scalar `minv´ = 1e-6
 32.             scalar `maxv´ = 1.0-`minv´
 33.
   .          // Impose Frechet-Hoeffding bounds
   .          replace `t1´ = min(max(`t1´, max(
  >                 `f10´+`f20´-1,`minv´)),
  >                 min(`f10´, `f20´,`maxv´))
 34.             replace `t2´ = min(max(`t2´, max(
  >                 `f11´+`f20´-1,`minv´)),
  >                 min(`f11´, `f20´,`maxv´))
 35.             replace `t3´ = min(max(`t3´, max(
  >                 `f10´+`f21´-1,`minv´)),
  >                 min(`f10´, `f21´,`maxv´))
 36.             replace `t4´ = min(max(`t4´, max(
  >                 `f11´+`f21´-1,`minv´)),
  >                 min(`f11´, `f21´,`maxv´))
 37.
   .          // Use difference calculation to get joint pmf
   .          generate double `ttt´ = `t1´-`t2´-`t3´+`t4´
 38.
   .          // Calculate log likelihood
   .          replace `lnf´ = log(`ttt´)
 39.      }
 40. end
```

```
. drop _all

. set seed 89238

. set obs 1000
number of observations (_N) was 0, now 1,000

. gen byte x11 = rbinomial(1,.5)

. gen byte x12 = rbinomial(1,.5)

. gen byte x21 = rbinomial(1,.5)

. gen byte x22 = rbinomial(1,.5)

. gen double mu1 = exp(1-x11+x12)

. gen double mu2 = exp(1+x21-x22)

. rejectsample y1 y2, mu1(mu1) mu2(mu2) lambda(2) ap1(0.5) ap2(4.0) marg1(margn
> binom) marg2(margnbinom) jointpmf(jointfcopnn)

.

. * Use ml to fit a bivariate model where
. * each marginal distribution is negative binomial
. ml model lf0 binbinomll (y1:y1 = x11 x12) (lnalpha1:)
> (y2:y2 = x21 x22) (lnalpha2:) (theta:)

. * Turn search on instead of using starting values
. ml maximize, search(on) ltol(1e-2) tol(1e-2) nolog
> title(Bivariate Nbinom-Nbinom regression)
initial:       log likelihood =       -<inf>  (could not be evaluated)
feasible:      log likelihood = -4734.983
rescale:       log likelihood = -4734.983
rescale eq:    log likelihood = -4534.9776
```

```
Bivariate Nbinom-Nbinom regression                Number of obs    =       1,000
                                                  Wald chi2(2)     =      602.69
Log likelihood = -4172.929                        Prob > chi2      =      0.0000
```

	Coef.	Std. Err.	z	P>|z|	[95% Conf.	Interval]
y1						
x11	-1.033412	.058349	-17.71	0.000	-1.147774	-.9190499
x12	1.061596	.0594911	17.84	0.000	.9449959	1.178197
_cons	1.006201	.0529178	19.01	0.000	.9024842	1.109918
lnalpha1						
_cons	-.6177576	.0749822	-8.24	0.000	-.76472	-.4707951
y2						
x21	.9271686	.1280989	7.24	0.000	.6760994	1.178238
x22	-1.337176	.1282771	-10.42	0.000	-1.588595	-1.085758
_cons	1.309538	.1159159	11.30	0.000	1.082347	1.536729
lnalpha2						
_cons	1.340178	.0602965	22.23	0.000	1.221999	1.458357
theta						
_cons	2.232059	.2314664	9.64	0.000	1.778393	2.685724

We emphasize that one could easily combine the two illustrated log-likelihood functions to create a bivariate count-data model (for use with Stata's `ml` program) in which one marginal is Poisson and the other is negative binomial. We could also write code for marginals following the negative binomial-P, generalized Poisson, Waring, or other distributions.

19.6   The Famoye bivariate Poisson regression model

The Famoye bivariate Poisson regression model allows negative, zero, or positive correlation.

Famoye (2010) presents a bivariate Poisson distribution given by

$$\Pr(y_1, y_2) \quad = \quad \mu_1^{y_1} \mu_2^{y_2} e^{-\mu_1 - \mu_2} \left\{ 1 + \lambda(e^{-y_1} - e^{-d\mu_1})(e^{-y_2} - e^{-d\mu_2}) \right\} / y_1! y_2! \quad (19.10)$$

where $d = 1 - \exp(-1)$. In this presentation, the two outcomes are independent when $\lambda = 0$. The sign of λ indicates the sign of the correlation between the outcomes. The correlations can be calculated per observation as

$$\rho_i = \lambda d^2 \sqrt{\mu_{1i} \mu_{2i}} \exp\{-d(\mu_{1i} + \mu_{2i})\} \quad (19.11)$$

The average of ρ_i can then be used to summarize the correlation of the outcome variables. However, note that the variation in the correlation per covariate pattern will vary widely based on the scalar value of λ, so as a summary this works only in a constant-only association. The log likelihood is then given in terms of the sum of the logs of the observation-specific probabilities.

The code for the log likelihood is given by typing

```
. program famoyepbll
  1.          version 15
  2.          args todo b lnf
  3.          quietly {
  4.                  local y1 "$ML_y1"
  5.                  local y2 "$ML_y2"
  6.                  tempname lambda
  7.                  tempvar  xb1 xb2
  8.                  mleval `xb1'    = `b', eq(1)
  9.                  mleval `xb2'    = `b', eq(2)
 10.                  mleval `lambda' = `b', eq(3) scalar
 11.
  .                  tempvar mu1 mu2
 12.                  generate double `mu1' = exp(`xb1')
 13.                  generate double `mu2' = exp(`xb2')
 14.
  .                  tempname dd
 15.                  scalar `dd'     = 1 - exp(-1)
```

```
 16.
   .                  replace `lnf´ = `y1´*`xb1´+`y2´*`xb2´-`mu1´-`mu2´-
   >                  lngamma(`y1´+1) - lngamma(`y2´+1) +
   >                  log(1+`lambda´*(exp(-`y1´)-exp(-`dd´*`mu1´)) *
   >                  (exp(-`y2´)-exp(-`dd´*`mu2´)))
 17.          }
 18. end
```

After defining the log-likelihood program, we fit each individual Poisson model for the two dependent variables. We save the estimated coefficients to use as starting values for the bivariate model, then we use Stata's `ml` commands to fit the defined model.

```
. drop _all

. set seed 74238

. set obs 1000
number of observations (_N) was 0, now 1,000

. gen byte x11 = rbinomial(1,.5)

. gen byte x12 = rbinomial(1,.5)

. gen byte x21 = rbinomial(1,.5)

. gen byte x22 = rbinomial(1,.5)

. gen double mu1 = exp(1-x11+x12)

. gen double mu2 = exp(1+x21-x22)

. rejectsample y1 y2, mu1(mu1) mu2(mu2) lambda(20) marg1(margpoisson)
> marg2(margpoisson) jointpmf(jointfamp)

. poisson y1 x11 x12, nolog
```

```
Poisson regression                           Number of obs   =      1,000
                                             LR chi2(2)      =    1504.58
                                             Prob > chi2     =     0.0000
Log likelihood = -1920.3491                  Pseudo R2       =     0.2815
```

y1	Coef.	Std. Err.	z	P>|z|	[95% Conf. Interval]	
x11	-.9466079	.0379384	-24.95	0.000	-1.020966	-.87225
x12	1.001019	.0379899	26.35	0.000	.9265603	1.075478
_cons	.9967071	.0341807	29.16	0.000	.9297142	1.0637

```
. matrix a1 = e(b)

. poisson y2 x21 x22, nolog
```

```
Poisson regression                           Number of obs   =      1,000
                                             LR chi2(2)      =    1397.86
                                             Prob > chi2     =     0.0000
Log likelihood = -1873.5306                  Pseudo R2       =     0.2717
```

y2	Coef.	Std. Err.	z	P>|z|	[95% Conf. Interval]	
x21	.9143322	.0378699	24.14	0.000	.8401085	.9885559
x22	-.9842342	.0377572	-26.07	0.000	-1.058237	-.9102316
_cons	1.06533	.0337839	31.53	0.000	.9991152	1.131546

```
. matrix a2 = e(b)

. matrix tt = (0.0)
```

```
. matrix aa = a1, a2, tt
. ml model lf0 famoyepbll (y1:y1 = x11 x12) (y2:y2 = x21 x22) (lambda:)
. ml init aa, copy
. ml maximize, search(off) title(Famoye Biv Poisson regression) nolog
```

Famoye Biv Poisson regression			Number of obs	=	1,000
			Wald chi2(2)	=	2408.71
Log likelihood = -3369.1273			Prob > chi2	=	0.0000

	Coef.	Std. Err.	z	P>|z|	[95% Conf.	Interval]
y1						
x11	-1.039468	.0228664	-45.46	0.000	-1.084285	-.9946506
x12	.9693075	.0211947	45.73	0.000	.9277666	1.010848
_cons	1.023958	.0112279	91.20	0.000	1.001952	1.045965
y2						
x21	.9322954	.0219269	42.52	0.000	.8893195	.9752714
x22	-1.019038	.0226755	-44.94	0.000	-1.063481	-.9745949
_cons	1.059124	.015588	67.94	0.000	1.028573	1.089676
lambda						
_cons	30.30875	.5055295	59.95	0.000	29.31793	31.29957

19.7   The Marshall–Olkin bivariate negative binomial regression model

Marshall and Olkin (1985) present a "shared frailty" model for which the dispersion parameters of the marginal negative binomial distributions are set to be equal; there is no additional parameter directly measuring correlation between the two outcomes. The joint probability mass function is given by

$$
f(y_1, y_2 | \lambda_1, \lambda_2, \alpha) = \frac{\Gamma(y_1 + y_2 + \alpha)}{y_1! y_2! \Gamma(\alpha)} \left(\frac{\lambda_1}{\lambda_1 + \lambda_2 + 1} \right)^{y_1}
$$
$$
\left(\frac{\lambda_2}{\lambda_1 + \lambda_2 + 1} \right)^{y_2} \left(\frac{1}{\lambda_1 + \lambda_2 + 1} \right)^{\alpha} \quad (19.12)
$$

where λ_1 and λ_2 are the conditional marginal means of the two negative binomial outcomes and α is the (common) overdispersion parameter.

This approach defines a specific distribution for which estimates can easily be computed, the marginals are negative binomial, the correlation between the outcomes is positive, and the heterogeneity in each marginal is assumed to be equal. The correlation between the outcomes for this model is given by

$$
\text{Corr}(y_1, y_2) = \frac{\lambda_1 \lambda_2}{\sqrt{(\lambda_1^2 + \alpha \lambda_1)(\lambda_2^2 + \alpha \lambda_2)}} \quad (19.13)
$$

The code for the log likelihood is given below. Note that we reparameterized the model so that it matches the marginal output of Stata's `nbreg` command. As such, the exponential of the recovered `lnalpha` coefficient should match the inverse of the `lambda` argument in the `resample` command. It should also closely match the `lnalpha` parameter of the marginal models estimated using `nbreg`.

```
. program mobinbll
  1.          version 15
  2.          args todo b lnf
  3.          quietly {
  4.                  local y1 "$ML_y1"
  5.                  local y2 "$ML_y2"
  6.                  tempvar  xb1 xb2
  7.                  tempname lnalpha
  8.                  mleval `xb1´    = `b´, eq(1)
  9.                  mleval `xb2´    = `b´, eq(2)
 10.                  mleval `lnalpha´ = `b´, eq(3) scalar
 11.
  .                  tempvar mu1 mu2 alpha den
 12.                  generate double `mu1´   = exp(`xb1´+`lnalpha´)
 13.                  generate double `mu2´   = exp(`xb2´+`lnalpha´)
 14.                  generate double `alpha´ = exp(-`lnalpha´)
 15.                  generate double `den´   = `mu1´+`mu2´+1
 16.
  .                  tempvar p1 p2 p3
 17.                  generate double `p1´ = `mu1´/`den´
 18.                  generate double `p2´ = `mu2´/`den´
 19.                  generate double `p3´ = 1/`den´
 20.
  .                  replace `lnf´ = `y1´*log(`p1´) +
  >                          `y2´*log(`p2´) + `alpha´*log(`p3´) +
  >                          lngamma(`y1´+`y2´+`alpha´) -
  >                          lngamma(`y1´+1) - lngamma(`y2´+1) -
  >                          lngamma(`alpha´)
 21.          }
 22. end
```

To fit the model, we first fit the individual models to obtain starting values. Once obtained, we set the dependency parameter to a small value and then fit the model of interest using Stata's `ml` subcommands.

```
. drop _all
. set seed 82222
. set obs 1000
number of observations (_N) was 0, now 1,000
. gen byte x11 = rbinomial(1,.5)
. gen byte x12 = rbinomial(1,.5)
. gen byte x21 = rbinomial(1,.5)
. gen byte x22 = rbinomial(1,.5)
. gen double mu1 = exp(1-x11+x12)
. gen double mu2 = exp(1+x21-x22)
. rejectsample y1 y2, mu1(mu1) mu2(mu2) lambda(2.0) marg1(margnbinom)
> marg2(margnbinom) jointpmf(jointmo)
```

```
. poisson y1 x11 x12, nolog
Poisson regression                            Number of obs      =      1,000
                                              LR chi2(2)         =    1627.38
                                              Prob > chi2        =     0.0000
Log likelihood = -2535.989                    Pseudo R2          =     0.2429
```

y1	Coef.	Std. Err.	z	P>|z|	[95% Conf. Interval]	
x11	-.9424179	.0387707	-24.31	0.000	-1.018407	-.8664288
x12	1.095815	.0389023	28.17	0.000	1.019568	1.172062
_cons	.9069491	.0352457	25.73	0.000	.8378689	.9760294

```
. matrix a1 = e(b)
. poisson y2 x21 x22, nolog
Poisson regression                            Number of obs      =      1,000
                                              LR chi2(2)         =    1477.63
                                              Prob > chi2        =     0.0000
Log likelihood = -2493.7389                   Pseudo R2          =     0.2286
```

y2	Coef.	Std. Err.	z	P>|z|	[95% Conf. Interval]	
x21	1.003541	.038204	26.27	0.000	.928662	1.078419
x22	-.9996012	.0379422	-26.35	0.000	-1.073967	-.9252359
_cons	1.005871	.0338517	29.71	0.000	.9395231	1.072219

```
. matrix a2 = e(b)
. matrix tt = (0.0)
. matrix aa = a1, a2, tt
. ml model lf0 mobinbll (y1:y1 = x11 x12) (y2:y2 = x21 x22) (lnalpha:)
. ml init aa, copy
. ml maximize, search(off) title(Marshall-Olkin Biv Negbin regression) nolog
Marshall-Olkin Biv Negbin regression          Number of obs      =      1,000
                                              Wald chi2(2)       =     830.36
Log likelihood = -4136.294                    Prob > chi2        =     0.0000
```

	Coef.	Std. Err.	z	P>|z|	[95% Conf. Interval]	
y1						
x11	-.9993851	.0498404	-20.05	0.000	-1.09707	-.9016997
x12	1.020162	.0496475	20.55	0.000	.9228547	1.117469
_cons	.9782036	.0472584	20.70	0.000	.8855788	1.070828
y2						
x21	.9984861	.0491725	20.31	0.000	.9021098	1.094862
x22	-1.014135	.0489985	-20.70	0.000	-1.11017	-.9180997
_cons	1.017582	.0456162	22.31	0.000	.928176	1.106988
lnalpha						
_cons	-.6658851	.0617097	-10.79	0.000	-.786834	-.5449362

19.8   The Famoye bivariate negative binomial regression model

The specification of a bivariate negative binomial regression model is more general in Famoye (2010) than in Marshall and Olkin (1985). The Famoye model allows negative, zero, or positive correlation. In addition, this more general model allows the two negative binomial dispersion parameters to differ.

The joint probabilities for a bivariate negative binomial distribution given by

$$
\Pr(y_1, y_2) \;=\; \left\{ \prod_{k=1}^{2} \binom{m_k^{-1} + y_k - 1}{y_k} \left(\frac{\mu_k}{m_k^{-1} + \mu_k} \right)^{y_k} \left(\frac{m_k^{-1}}{m_k^{-1} + \mu_k} \right)^{m_k^{-1}} \right\}
$$
$$
\times \left\{ 1 + \lambda \left(e^{-y_1} - c_1 \right) \left(e^{-y_2} - c_2 \right) \right\} \tag{19.14}
$$

where $c_k = \{(1 - \theta_k)/(1 - \theta_k e^{-1})\}^{m_k^{-1}}$ and $\theta_k = \mu_k/(m_k^{-1} + \mu_k)$. The two outcomes are independent when $\lambda = 0$. Also, note that the sign of λ indicates the sign of the correlation between the outcomes and that as $m_k \to 0$, the negative binomial marginal model reduces to that of Poisson marginal model. The correlations can be calculated per observation as

$$
\rho_i = \frac{\lambda d^2 \sqrt{\mu_{1i}\mu_{2i}(1 + m_1\mu_{1i})(1 + m_2\mu_{2i})}}{(1 + dm_1\mu_{1i})^{1+1/m_1} \; (1 + dm_2\mu_{2i})^{1+1/m_2}} \tag{19.15}
$$

where $d = 1 - \exp(-1)$. Subsequently, as was the case for the bivariate Poisson specification, the average of ρ_i can be used to summarize the correlation of the outcome variables.

The code for the log likelihood is given by typing

```
. program fbinbll
  1.            version 15
  2.            args todo b lnf
  3.            quietly {
  4.                    local y1 "$ML_y1"
  5.                    local y2 "$ML_y2"
  6.                    tempvar  xb1 xb2
  7.                    tempname lnalpha1 lnalpha2 lambda
  8.                    mleval `xb1´       = `b´, eq(1)
  9.                    mleval `lnalpha1´ = `b´, eq(2) scalar
 10.                    mleval `xb2´       = `b´, eq(3)
 11.                    mleval `lnalpha2´ = `b´, eq(4) scalar
 12.                    mleval `lambda´   = `b´, eq(5) scalar
 13.

                       tempname alpha1 alpha2
 14.                       scalar `alpha1´ = exp(-`lnalpha1´)
 15.                       scalar `alpha2´ = exp(-`lnalpha2´)
 16.

                       tempname dd m1 m2
 17.                       scalar `dd´ = 1 - exp(-1)
 18.                       scalar `m1´ = 1/`alpha1´
```

```
19.                     scalar `m2´ = 1/`alpha2´
20.
 .              tempvar mu1 mu2 c1 c2
21.                 generate double `mu1´ = exp(`xb1´)
22.                 generate double `mu2´ = exp(`xb2´)
23.                 generate double `c1´  = (1+`dd´*`mu1´*`m1´)^(-1/`m1´)
24.                 generate double `c2´  = (1+`dd´*`mu2´*`m2´)^(-1/`m2´)
25.
 .              tempname j1 j2
26.                 summarize `y1´
27.                 local j1 = r(max)-1
28.                 summarize `y2´
29.                 local j2 = r(max)-1
30.
 .              replace `lnf´ = log(1+`lambda´*(exp(-`y1´)-`c1´) *
>                         (exp(-`y2´)-`c2´))
31.                 forvalues t=1/2 {
32.                         forvalues j=0/`j`t´´ {
33.                                 replace `lnf´ = `lnf´ + log(
>                                     1/`m`t´´ + `j´) if `y`t´´ > `j´
34.                         }
35.                         replace `lnf´ = `lnf´ + `y`t´´*log(`mu`t´´) -
>                             (1/`m`t´´)*log(`m`t´´) -
>                             (`y`t´´+1/`m`t´´) *
>                             log(`mu`t´´+1/`m`t´´) - lngamma(`y`t´´+1)
36.                 }
37.         }
38. end
```

To fit the model, we first fit the individual models to obtain starting values. Once obtained, we set the dependency parameter to a small value and then fit the model of interest using Stata's `ml` subcommands.

```
. drop _all
. set seed 13002
. set obs 1000
number of observations (_N) was 0, now 1,000
. gen byte x11 = rbinomial(1,.5)
. gen byte x12 = rbinomial(1,.5)
. gen byte x21 = rbinomial(1,.5)
. gen byte x22 = rbinomial(1,.5)
. gen double mu1 = exp(1-x11+x12)
. gen double mu2 = exp(1+x21-x22)
```

```
. rejectsample y1 y2, mu1(mu1) mu2(mu2) ap1(0.5) ap2(1.5) lambda(2.0)
> marg1(margnbinom) marg2(margnbinom) jointpmf(jointfamn)
. nbreg y1 x11 x12, nolog
```

| Negative binomial regression | | | | Number of obs | = | 1,000 |

Dispersion = mean				LR chi2(2)	=	194.00
				Prob > chi2	=	0.0000
Log likelihood = -1993.6757				Pseudo R2	=	0.0464

y1	Coef.	Std. Err.	z	P>|z|	[95% Conf. Interval]	
x11	-1.050332	.1002168	-10.48	0.000	-1.246754	-.8539109
x12	1.045634	.1002386	10.43	0.000	.8491698	1.242098
_cons	.8618841	.086264	9.99	0.000	.6928099	1.030958
/lnalpha	.6852004	.0661933			.5554639	.814937
alpha	1.984169	.1313388			1.742749	2.259033

```
LR test of alpha=0: chibar2(01) = 2657.01          Prob >= chibar2 = 0.000
. matrix b1 = e(b)
. nbreg y2 x21 x22, nolog
```

| Negative binomial regression | | | | Number of obs | = | 1,000 |

Dispersion = mean				LR chi2(2)	=	350.36
				Prob > chi2	=	0.0000
Log likelihood = -2235.4975				Pseudo R2	=	0.0727

y2	Coef.	Std. Err.	z	P>|z|	[95% Conf. Interval]	
x21	.9631178	.066112	14.57	0.000	.8335406	1.092695
x22	-.9488326	.0658073	-14.42	0.000	-1.077812	-.8198527
_cons	1.025018	.0571896	17.92	0.000	.9129284	1.137108
/lnalpha	-.389404	.0723777			-.5312616	-.2475464
alpha	.6774605	.049033			.5878629	.780714

```
LR test of alpha=0: chibar2(01) = 1122.64          Prob >= chibar2 = 0.000
. matrix b2 = e(b)
. matrix tt = (0.5)
. matrix bb = b1, b2, tt
. ml model lf0 fbinbll (y1:y1 = x11 x12) (lnalpha1:) (y2:y2 = x21 x22)
> (lnalpha2:) (lambda:)
. ml init bb, copy
```

```
. ml maximize, search(off) title(Famoye Biv Negbin regression) nolog
Famoye Biv Negbin regression                        Number of obs    =      1,000
                                                    Wald chi2(2)     =     209.64
Log likelihood = -4183.5469                         Prob > chi2      =     0.0000
```

	Coef.	Std. Err.	z	P>|z|	[95% Conf.	Interval]
y1						
x11	-1.0302	.0986123	-10.45	0.000	-1.223477	-.8369237
x12	1.012751	.0988905	10.24	0.000	.8189291	1.206573
_cons	.8675672	.0857364	10.12	0.000	.699527	1.035607
lnalpha1						
_cons	.7010901	.0659447	10.63	0.000	.5718409	.8303392
y2						
x21	.9726579	.0644803	15.08	0.000	.8462789	1.099037
x22	-.9659647	.0642543	-15.03	0.000	-1.091901	-.8400286
_cons	1.024277	.0561179	18.25	0.000	.9142876	1.134266
lnalpha2						
_cons	-.3937908	.0719746	-5.47	0.000	-.5348583	-.2527232
lambda						
_cons	1.904298	.1408202	13.52	0.000	1.628295	2.1803

Specific bivariate models have been developed under different assumption sets. For the specific case of a bivariate negative binomial regression model, consider the following:

- Famoye's bivariate Poisson and negative binomial
- Marshall–Olkin shared frailty negative binomial
- Bivariate negative binomial via a copula function

The copula function approach is superior because it allows the marginals to be any distribution (not just negative binomial). The Marshall–Olkin function requires the same dispersion parameter across the two outcomes and requires the correlation between the outcomes to be nonnegative.

Xu and Hardin (2016) present the `bivcnto` command to fit the models highlighted in this chapter. Their support for bivariate models fit with copula functions includes marginal outcomes for the Poisson, negative binomial, and generalized Poisson models. The command also includes support for the three copula functions listed in table 19.1, as well as Famoye's bivariate Poisson, Famoye's bivariate negative binomial, and the Marshall–Olkin shared frailty negative binomial functions.

The `bivcnto` command is a convenient alternative to the cumbersome approach illustrated throughout this chapter. In fact, we could have presented all the examples here using `bivcnto`. However, by illustrating estimation of these models using Stata's `ml` commands, we hope the reader is inspired to extend the collection of supported models. We clearly illustrate how each marginal distribution is specified in each `ml` command so that it should be clear how to incorporate Poisson-inverse Gaussian, negative binomial(P), Famoye's generalized negative binomial, Waring's generalized negative binomial, or other distributions; see chapter 14 for details.

20 Bayesian GLMs

Contents

20.1   Brief overview of Bayesian methodology

Bayesian modeling has been gaining steady popularity since about 1990 and more rapidly since the middle of the first decade of the 21st century. Bayesian modeling is popular in many areas of research. For example, most new statistical analysis of astrophysical data entails Bayesian modeling. Ecologists are also basing more and more of their research on Bayesian methods.

The increasing popularity of Bayesian methods is in part due to personal computers having sufficient speed and memory to run, in a timely manner, the many iterations required for most Bayesian models. Additionally, user-friendly software has been developed for implementing the required simulation algorithms.

Bayesian analysis approaches the estimation of model parameters in a different manner than has been discussed so far in this book. Many books and articles have recently been published that provide analysts with a good sense of the difference between traditional frequentist-based and Bayesian approaches to modeling. Stata's *Bayesian Analysis Reference Manual* also provides an excellent overview of the differences in approach and interpretation. In standard maximum likelihood estimation (MLE), estimated regression coefficients are assumed to be fixed (that is, nonrandom) parameters. These unknown parameters are associated with an underlying probability density function describing the model data. Because the coefficient estimators are random variables, regression models estimate their variance matrix. By combining estimated coefficients with the associated estimated variance matrix, confidence intervals for individual parameters may be constructed. The GLM models discussed earlier in this book are based on this approach to statistical modeling.

The Bayesian approach takes the viewpoint that regression coefficients are associated with the conditional distributions of parameters (which are not fixed). Bayesian analysis estimates the distribution of each parameter. Specifically, Bayesians specify a prior distribution of the parameter and then use information from the current dataset (the likelihood) to update knowledge about that parameter in the form of a posterior distribution. Prior distributions are specified based on existing knowledge about the parameter, for example, estimates from a previous researcher.

The foundation of Bayesian analysis comes from Bayes's rule. The definition of the conditional probability of event A given B is written as

$$P(A|B) = \frac{P(A, B)}{P(B)} \tag{20.1}$$

which says that the conditional probability of event A given the status of event B is the ratio of the joint probability of events A and B, $P(A, B)$, divided by the marginal probability of event B, $P(B)$. Bayes's rule states that the conditional probability of B given A can be derived as

$$
\begin{aligned}
P(B|A) &= \frac{P(A, B)}{P(A)} & (20.2)\\[2mm]
&= \frac{P(A, B)}{P(B)} \frac{P(B)}{P(A)} & (20.3)\\[2mm]
&= P(A|B)\frac{P(B)}{P(A)} & (20.4)\\[2mm]
&= \frac{P(A|B)P(B)}{P(A)} & (20.5)\\[2mm]
&\propto P(A|B)P(B) & (20.6)
\end{aligned}
$$

The importance of (20.6) is that the proportionality holds even if the probabilities, $P(A)$ and $P(B)$, are multiplied by a constant. Moving from writing equations in terms of events A and B, to observed data, $\mathbf{y}$, and parameters, θ.

$$p(\boldsymbol{\theta}|\mathbf{y}) \;=\; \frac{p(\mathbf{y}|\boldsymbol{\theta})p(\boldsymbol{\theta})}{p(\mathbf{y})} \tag{20.7}$$

$$=\; \frac{p(\mathbf{y}|\boldsymbol{\theta})\pi(\boldsymbol{\theta})}{m(\mathbf{y})} \tag{20.8}$$

$$=\; \frac{L(\boldsymbol{\theta}|\mathbf{y})\pi(\boldsymbol{\theta})}{m(\mathbf{y})} \tag{20.9}$$

Bayesian analysis describes this result as

$$\text{Posterior distribution} \quad \propto \quad \text{Likelihood} \times \text{Prior distribution} \tag{20.10}$$

$$p(\boldsymbol{\theta}|\mathbf{y}) \quad \propto \quad L(\boldsymbol{\theta}|\mathbf{y})\pi(\boldsymbol{\theta}) \tag{20.11}$$

That is, the distribution of the parameters given the data is proportional to the product of the distribution of the data given the parameter and the prior distribution of the parameter. Bayesians interpret this equation as "the information about the parameters is proportional to the prior information about the parameters combined with the current information from the data given the parameter".

The full specification of the posterior distribution is given by

$$p(\boldsymbol{\theta}|\mathbf{y}) = \frac{p(\mathbf{y}|\boldsymbol{\theta})\pi(\boldsymbol{\theta})}{\int p(\mathbf{y}|\boldsymbol{\theta}')\pi(\boldsymbol{\theta}')d\boldsymbol{\theta}'} \tag{20.12}$$

Various measures of center for the posterior distribution can be used as point estimates, including the posterior mean,

$$E(\boldsymbol{\theta}|\mathbf{y}) = \int \boldsymbol{\theta} p(\boldsymbol{\theta}|\mathbf{y})d\boldsymbol{\theta} \tag{20.13}$$

the posterior median ($q_{0.50}$),

$$P\{\boldsymbol{\theta} \leq q_{0.50}(\boldsymbol{\theta})\} = 0.50 \tag{20.14}$$

and the posterior mode, which is the numerator of (20.12).

The mean of the posterior distribution of a parameter is generally used as the measure of centrality for the purposes of estimation. We do this when the posterior distribution of a parameter is (approximately) normally distributed. When the distribution is skewed, either the median or the mode can be used.

Rather than constructing confidence intervals for regression parameters as in frequentist modeling, Bayesians construct interval estimates of regression coefficients based on the estimated posterior distribution of that parameter. These intervals are referred to as credible intervals. Two different approaches are commonly used in constructing credible intervals. For example, the quantile-based 95% interval is defined by ($q_{0.025}, q_{0.975}$). Alternatively, given that posterior density functions are not necessarily symmetric, an infinite number of regions could be constructed to contain 95% of the density; a few examples include $(q_{0.01}, q_{0.96})$, $(q_{0.01525}, q_{0.98475})$, and $(q_{0.04}, q_{0.99})$. The shortest such interval has the highest density and is called the highest posterior density (HPD) interval. This interval is unique for unimodal distributions but not necessarily otherwise. Frequentists determine the utility of a parameter in the model based on whether 0 is or is not included within the lower and upper confidence intervals. This is not relevant in Bayesian modeling. In Bayesian analysis, credible intervals indicate the spread of the marginal posterior distributions for the corresponding parameters. The 95% credible interval can be interpreted as the region in which the parameter falls 95% of the time. This cannot be said for maximum likelihood confidence intervals, even though some analysts wish it could be the case.

20.1.1   Specification and estimation

The basic specification of a Bayesian regression model consists of three parts: 1) specification of the dependent and independent variables, 2) specification of the likelihood $p(\mathbf{y}|\boldsymbol{\theta})$, and 3) specification of the prior distributions of the regression parameters $\pi(\boldsymbol{\theta})$, including priors for each predictor, intercept, scale, variance, and dispersion parameter. The prior distribution of the parameters, $\pi(\theta)$, provides a method of incorporating existing knowledge about a parameter into the model.

Care must be taken in assigning priors to model predictors. There is substantial literature on the subject, but it extends beyond the scope of our discussion. If there is little existing information about a parameter, priors that reflect this lack of information can be specified. Noninformative priors are probability distributions that spread the probability so wide that the prior has effectively no influence on the posterior distribution. Noninformative priors are also called diffuse, vague, or reference priors. Some statisticians give each of these terms a slightly different import or meaning, though none of them significantly contribute useful additional information to the likelihood.

One type of noninformative prior is known as a flat prior. A flat prior is a uniform distribution where all values in the range of the distribution are equally likely. When the prior distribution integrates to infinity, the assumed prior distribution is referred to as improper. In some instances, improper priors result in the posterior not being a viable probability distribution. This is particularly the case when the parameter of interest is continuous and the prior provides values over an unbounded domain. The *Stata Bayesian Analysis Reference Manual* has a nice discussion on the dangers related to flat or uniform priors. At times, such priors are perfectly legitimate; at other times, they are not.

Some analysts prefer to use what sometimes is called a diffuse prior, for example, a wide normal distribution such as normal(0,1000). Normal priors of this form are considered to be noninformative priors by most analysts. The key here is that the prior has a wide variance, spreading out any information throughout a wide range. Note that the parameters specified in a prior are called hyperparameters by Bayesians. Here the hyperparameters for the normal prior are mean = 0 and variance = 1000.

We expect that a Bayesian model having noninformative priors on all of its parameters will produce posterior estimates that are similar to those produced by standard frequentist-based MLE. For instance, the coefficients, standard errors, and confidence intervals calculated by logistic regression will be similar to the mean, standard deviation, and credible intervals from the posterior distribution of each parameter in the model.

Another noteworthy type of prior distribution is conjugate priors. Much of the difficulty associated with Bayesian analysis is centered on the evaluation of posterior distributions (to extract modes, quantiles, etc.). When a prior distribution admits an analytic solution for which the posterior distribution is in the same family, the two distributions are called conjugate distributions, and the prior distribution is called the conjugate prior for the likelihood function. For example, the binomial and beta distributions are conjugate in that the binomial function is given as

$$\frac{\Gamma(n+1)}{\Gamma(y+1)\Gamma(n-y+1)}\theta^y(1-\theta)^{n-1} \tag{20.15}$$

and the beta probability density function as

$$\frac{\Gamma(a+b)}{\Gamma(a)\Gamma(b)}\theta^{a-1}(1-\theta)^{b-1} \tag{20.16}$$

Multiplying (20.15) by (20.16) results in another beta distribution

$$\frac{\Gamma(n+a+b)}{\Gamma(y+a)\Gamma(n-y+b)}\theta^{y+a-1}(1-\theta)^{n-y+b-1} \tag{20.17}$$

The resulting beta parameters are given by $a' = a + y$ and $b' = b + n - y$. It is from this distribution that the mean and variance statistics for the posterior distribution can be derived. The posterior mean or expectation is $a'/(a' + b')$, and the variance is $(a' + b')/(a' + b')^2(a' + b' + 1)$. The beta distribution is a conjugate prior for the mean parameters of the binomial. Unfortunately, such algebraic conveniences are not all that common.

In cases where analytic solutions are not available, it becomes necessary to numerically evaluate integral expressions (for example, marginal distributions and posterior moments). A common approach is to use Markov chain Monte Carlo (MCMC) sampling methods to produce random samples from the posterior distribution. Each draw is subject to random variation, so repeated draws are made and the mean is used as the estimate of the statistic of interest. The Metropolis–Hastings (MH) algorithm, which Stata's Bayesian commands (`bayes` and `bayesmh`) implement, is quite general and adequately addresses many of the models in which we have an interest. Readers interested in the details of the specific techniques used for approximation (Markov chain Monte Carlo methods) should consult the Stata documentation and Tanner and Wong (1987).

Unlike MLE, there are no clear rules for determining when an MCMC chain has converged to a stable distribution. Because early sampling is likely to be far from the true posterior mean value, many initial iterations are run but do not figure in the calculation of the posterior mean. These initial iterations are referred to as burn-in iterations. The time it takes for the algorithm to converge to stable posterior distributions depends in part on the mixing properties of the Markov chain. However, diagnostic tools exist for assessing whether MCMC chains have converged. Evidence of an improper posterior distribution will be seen in nonconvergence of the MCMC algorithm. Thus, it is very important to check diagnostics, especially when specifying flat priors. Convergence diagnostics will be discussed further throughout the chapter but receive special attention in section 20.2.2.

Once models have been fit, analysts of Bayesian models apply a generalization of the Akaike information criterion (AIC) statistic called the deviance information criterion (DIC) statistic, which is defined as

$$\text{DIC} = D(\overline{\boldsymbol{\theta}}) + 2p_D \tag{20.18}$$

where $D(\overline{\boldsymbol{\theta}})$ is the deviance of the model evaluated at the mean of $\boldsymbol{\theta}$ and $p_D = E\{D(\boldsymbol{\theta})\} - D(\overline{\boldsymbol{\theta}})$. Note that $D(\overline{\boldsymbol{\theta}})$ measures the distance of the model from the data (the amount of lack of fit). Thus, the smaller the value, the better the fit. In that way, the DIC statistic can be used to select between comparable models, like we have used the AIC and Bayesian information criterion (BIC) statistics in other contexts.

20.1.2 Bayesian analysis in Stata

Stata users will be pleased to discover that Stata's Bayesian modeling facility is much easier to use than other software, particularly with the release of Stata 15. The built-in power of modeling capability and diagnostics is comparable, if not superior, with other software.

Bayesian modeling capabilities first became available for researchers using Stata statistical software in April 2015 with the release of Stata 14. The `bayesmh` command was designed to be easy to use but flexible enough to fit a wide variety of models. Stata 15 was released in mid-2017 and provides Stata users with an even easier interface for fitting Bayesian models. The new `bayes` command is used as a prefix in front of an otherwise maximum likelihood model. Stata 15 continues to support the `bayesmh` command to be used for modeling, but users will likely use `bayesmh` only for models not directly supported by the `bayes` prefix. In this chapter, we demonstrate how both the `bayes` prefix and `bayesmh` command can be used for the standard GLM family models and for several extended GLMs. Toward the end of the chapter, we will discuss extensions that can be estimated only using the `bayesmh` command.

Nearly all official maximum likelihood models offered by Stata allow the use of the `bayes` prefix, which turns the model from maximum likelihood to Bayesian. That is, the `bayes` prefix uses the log likelihood that is embedded in each of the respective MLE models as the log likelihood of the Bayesian parameters. Tables 20.1 and 20.2 provide a list of official Stata maximum likelihood models that can be used by the `bayes` command. This information can also be found in the online help (`help bayesian estimation`) and PDF documentation.

Table 20.1. Built-in support for the `bayes` prefix

Linear regression models

`regress`	Linear regression
`hetregress`	Heteroskedastic linear regression
`tobit`	Tobit regression
`intreg`	Interval regression
`truncreg`	Truncated regression
`mvreg`	Multivariate regression

Binary response regression models

`logistic`	Logistic regression, reporting odds ratios
`logit`	Logistic regression, reporting coefficients
`probit`	Probit regression
`cloglog`	Complementary log-log regression
`hetprobit`	Heteroskedastic probit regression
`binreg`	GLM for the binomial family
`biprobit`	Bivariate probit regression

Ordinal response regression models

`ologit`	Ordered logistic regression
`oprobit`	Ordered probit regression
`zioprobit`	Zero-inflated ordered probit regression

Categorical response regression models

`mlogit`	Multinomial (polytomous) logistic regression
`mprobit`	Multinomial probit regression
`clogit`	Conditional logistic regression

Table 20.2. Additional built-in support for the `bayes` prefix

Count response regression models
`poisson`	Poisson regression
`nbreg`	Negative binomial regression
`gnbreg`	Generalized negative binomial regression
`tpoisson`	Truncated Poisson regression
`tnbreg`	Truncated negative binomial regression
`zip`	Zero-inflated Poisson regression
`zinb`	Zero-inflated negative binomial regression

Generalized linear models
`glm`	Generalized linear models

Fractional response regression models
`fracreg`	Fractional response regression
`betareg`	Beta regression

Survival regression models
`streg`	Parametric survival models

Sample-selection regression models
`heckman`	Heckman selection model
`heckprobit`	Probit regression with sample selection
`heckoprobit`	Ordered probit model with sample selection

Multilevel regression models
`mixed`	Multilevel linear regression
`metobit`	Multilevel tobit regression
`meintreg`	Multilevel interval regression
`melogit`	Multilevel logistic regression
`meprobit`	Multilevel probit regression
`mecloglog`	Multilevel complementary log-log regression
`meologit`	Multilevel ordered logistic regression
`meoprobit`	Multilevel ordered probit regression
`mepoisson`	Multilevel Poisson regression
`menbreg`	Multilevel negative binomial regression
`meglm`	Multilevel generalized linear model
`mestreg`	Multilevel parametric survival regression

The developers of the `bayes` prefix have made every effort to make the task of using Bayesian modeling on relatively simple models as seamless as possible. The overall syntax of the `bayes` prefix is

`bayes` $\left[\,,\ bayes\_options\right]$: $mle\_command$ $\left[\,,\ mle\_command\_options\right]$

Priors and other estimation options that are appropriate for Bayesian modeling follow the `bayes` command name, and a colon is placed at the end of the final option to the `bayes` prefix, signaling the end of the `bayes` specifications. The maximum likelihood command that is to be converted into a Bayesian model is placed to the right of the colon with its options.

The `bayesmh` command provides many Bayesian modeling capabilities. The command allows Bayesian estimation using both built-in and user-defined log-likelihood functions as well as for both built-in and user-defined priors. With respect to GLMs, built-in log-likelihood evaluators are provided for common models, including `normal`, `logit`, `probit`, grouped `logit`, and `poisson` models. In Stata 15, the t distribution was added. A full listing of the built-in `bayesmh` log-likelihood evaluators is displayed in table 20.3.

Table 20.3. Built-in log likelihoods $L(\boldsymbol{\theta}|y) = p(\boldsymbol{\theta}|y)$

likelihood()	GLM	Common regression model name
normal(*var*)	Gaussian identity	linear
t(*sigma2*, *df*)		
lognormal(*var*)	Gaussian log	lognormal
exponential		exponential
mvnormal(*Sigma*)		multivariate linear
probit	binomial probit	probit
logit	binomial logit	logistic
binlogit(*n*)	binomial logit	binomial
oprobit		ordered probit
ologit		ordered logit
poisson	Poisson log	Poisson

The `bayesmh` command allows user specification of additional functions. This feature is useful when the model log likelihood is not one for which there is a built-in option, for example, in developing gamma and inverse Gaussian models. The `llf()` likelihood specification allows direct specification of a log-likelihood expression, and the `llevaluator()` option allows specification of a program name to be called for the evaluation of the log likelihood. Table 20.4 lists the log-likelihood evaluators that were developed and are illustrated in this chapter. Some of these examples implement models that are available using built-in specifications and are included only for pedagogical purposes; others implement models that currently require the use of more custom evaluators in at least some versions of Stata.

Table 20.4. Illustrated (user-specified) log likelihoods

Command	Description
blogitll	logit-link Bernoulli log likelihood
bclogll	clog-log-link Bernoulli log likelihood
bpoill	log-link Poisson log likelihood
bnbll	log-link negative binomial log likelihood
bzinbll	log-link zero-inflated negative binomial log likelihood
bgammall	reciprocal-link gamma log likelihood
bivgll	power(-2)-link inverse Gaussian log likelihood
bpoi0ll	log-link zero-truncated Poisson log likelihood
bfambpll	log-link bivariate Poisson log likelihood

All Bayesian models require priors for the parameters. Recall that care must be taken in assigning priors to model predictors. Prior distributions are entered into the `bayes` and `bayesmh` commands as probability distributions with specified values. They may or may not affect the centrality measure of the parameter posterior distributions in the model. The `bayesmh` command requires the user to specify priors for each parameter. By default, the `bayes` prefix provides so-called noninformative priors for the Bayesian model. The `bayes` prefix also allows for the user to learn which defaults are associated with a particular model by using the `dryrun` option. This option does not actually estimate the posteriors but displays the model structure and default priors and their values on the screen.

Both `bayes` and `bayesmh` allow the user to easily specify common priors; see table 20.5 for list of built-in prior distributions. Prior distributions not included in the built-in collection may be programmed and then specified by using the `density()` or `logdensity()` option. Stata 15 offers Cauchy, t, and Laplace priors, but Stata 14 does not. Below, we show examples of how to specify priors by hand.

Table 20.5. Built-in prior distributions $\pi(\boldsymbol{\theta})$

Univariate continuous	Multivariate continuous	Discrete
normal(*mu*, *var*)	mvnormal(d, *mean*, *Sigma*)	bernoulli(*p*)
lognormal(*mu*, *var*)	zellnersg(d, *g*, *mean*, *Sigma*)	index(*p1*,...,*pk*)
uniform(*a*, *b*)	wishart(d, *df*, *V*)	poisson(*mu*)
gamma(*alpha*, *beta*)	iwishart(d, *df*, *V*)	
igamma(*alpha*, *beta*)	jeffreys(d)	
t(*mu*, *sigma2*, *df*)		
cauchy(*loc*, *beta*)		
laplace(*mu*, *beta*)		
exponential(*beta*)		
beta(*a*, *b*)		
chi2(*df*)		
jeffreys		

With the `bayes` prefix, priors can be specified using either the `prior()` option or the specialized prior options. `normalprior(#)` specifies the standard deviation of the default normal or Gaussian priors. `normalprior(100)` is the default, which is equivalent to `normal(0,10000)`. Thus, the variance is specified as 10,000. The option `igammaprior(α β)` is also new, with a default of $\alpha = 0.01$, $\beta = 0.01$. Inserting a missing value, ., for either hyperparameter (parameter of the prior distribution) is equivalent to specifying 0.01.

Both `bayes` and `bayesmh` allow the user to control the manner in which the Markov chain is constructed using the `block` option. That is, rather than updating all parameters together, the user can specify a sequence of blocks—sets of correlated parameters. Blocking parameters can be helpful when the efficiency of the chain is low. The MH algorithm then updates the parameters in the order of the specified blocks. The idea is to specify the parameters as nearly independent sets (blocks) of correlated parameters. Of course, the increase in efficiency is paid for by a decrease in speed of the algorithm. Thus, a poor specification of blocks (one which fails to adequately lower required sample size through increased efficiency) can end up being slower to converge. An excellent discussion of blocking and effective sample-size statistics can be found in Gamerman and Lopes (2006)

Convergence diagnostics play a key role in Bayesian modeling because the MCMC algorithm does not have one clear convergence criterion. Standard diagnostics in most Bayesian modeling programs provide information regarding the shape of the posterior, the evenness of the sampling process, evidence of autocorrelation, and so forth. In Stata, this information can be easily displayed in graphical form using the `bayesgraph diagnostics` command. This quartet of plots should be assessed for each of the parameters in a Bayesian model. We will see how this works with models from a variety of distributions, but the graphs are discussed in more detail when Bayesian logistic regression models are discussed in section 20.2.2.

This chapter provides the analyst with a good starting point in using Bayesian modeling with any of the members of the GLM family of distributions. We provide example specifications of Bayesian GLMs and illustrate how to develop log-likelihood evaluators for specification of log likelihoods that are outside of the built-in collection. We start with one of the simplest distributions—the Bernoulli. We follow that with modeling multiple members of the Bernoulli and binomial families. Subsequently, we turn to the Poisson and negative binomial count models. These are all single parameter distributions. The continuous family members—gamma and inverse Gaussian—have a scale parameter to be estimated. Because they are more complex, we address them later in the chapter. We also give examples of four Bayesian extensions to GLMs, the ordered logistic model, zero-truncated Poisson, zero-inflated Poisson, and a bivariate Poisson model.

Most model examples are illustrated with flat priors merely for convenience and illustration. We also provide several of the models with priors, most of which are associated with models having built-in prior distributions.

The `bayes` and `bayesmh` commands have a host of very useful options, many of which we do not illustrate for the models described in this chapter. Review the current documentation for additional features not described in this chapter. The *Bayesian Analysis Reference Manual* provides information related to the use of commands for Bayesian estimation.

20.2 Bayesian logistic regression

A variable having values only of 1 and 0 can be modeled by a Bernoulli distribution, which is a subset of the binomial distribution. As detailed in part III, with link functions the binomial distribution defines logistic regression, as well as probit, log-log, and clog-log regressions. Refer to the chapter 9 on logistic regression for details of the model.

The model will be (equivalently) fit using the Stata 15 `bayes` command, as well as with the `bayesmh` built-in logit function. With the `bayes` command, the user need only prefix the standard `logit`, `logistic`, or `glm` commands with `bayes, <options>:`. In Stata 14, this function is specified with the `bayesmh` command through its `likelihood()` option. The `bayesmh` command knows that a Bayesian logistic model is being requested when the terms `logit` or `logistic` are used as arguments for the likelihood function.

For an example of constructing a Bayesian logistic model, we use `rwm1984.dta`. The data are from the 1984 German health reform data, and this example is also discussed in Hilbe (2014). The data used in the model include the response variable, `outwork`, an indicator of whether the subject was unemployed at least 50% of the year 1984. Predictors include `female`, an indicator of whether the subject is female; `kids`, an indicator of whether the subject has children; `cdocvis`, a centered variable of the number of visits to the doctor in 1984; and `cage`, a centered variable of the age in years of the subject.

20.2.1 Bayesian logistic regression—noninformative priors

We first model the German health data using a Bayesian logistic model with noninformative priors on all parameters. We will primarily use the flat priors and normal priors with mean 0 and variance 1,000 (normal(0, 1000)) as noninformative priors. Recall that noninformative priors can lead to improper posterior distributions; evidence of this can be seen in nonconvergence of the MCMC algorithm.

Because we are using noninformative priors on all of its parameters, we expect that the posterior estimates from our Bayesian model will be similar to those produced by standard frequentist-based MLE. We will compare the coefficients, standard errors, and confidence intervals calculated by logistic regression with the mean, standard deviation, and credible intervals from the posterior distribution of each parameter in the model.

We first model the data using a standard logistic model. Note that both continuous predictors are centered. It is generally important to either center or standardize continuous predictors when they are taking the role as predictors in Bayesian models.

```
. use http://www.stata-press.com/data/hh4/rwm1984
(German health data for 1984; Hardin & Hilbe, GLM and Extensions, 4th ed)
. quietly summarize docvis, meanonly
. generate cdoc = docvis - r(mean)
. quietly summarize age, meanonly
. generate cage = age - r(mean)
. glm outwork female kids cdoc cage, family(binomial) link(logit) nolog
```

Generalized linear models			No. of obs	=	3,874
Optimization	: ML		Residual df	=	3,869
			Scale parameter	=	1
Deviance	=	3918.21514	(1/df) Deviance	=	1.01272
Pearson	=	4205.000001	(1/df) Pearson	=	1.086844
Variance function: V(u) = u*(1-u)			[Bernoulli]		
Link function : g(u) = ln(u/(1-u))			[Logit]		
			AIC	=	1.013995
Log likelihood	=	-1959.10757	BIC	=	-28047.63

outwork	Coef.	OIM Std. Err.	z	P>|z|	[95% Conf. Interval]	
female	2.256804	.0827624	27.27	0.000	2.094593	2.419016
kids	.3579762	.0899639	3.98	0.000	.1816503	.5343022
cdoc	.0244323	.0062628	3.90	0.000	.0121574	.0367071
cage	.0543791	.0041591	13.07	0.000	.0462274	.0625308
_cons	-2.010276	.0810901	-24.79	0.000	-2.16921	-1.851342

Next, we will use the `bayes` command to fit a Bayesian logistic regression model. The `bayes` command can be used with default priors to immediately run a Bayesian logistic model. Be sure to drop `nolog` from the options used in the `glm` command, because the `bayes` prefix will return an error if it is specified. A seed has been given so that the reader may reproduce the results exactly as observed. Finally, we can let the command provide default starting values for estimation, or we may specify them ourselves.

```
. bayes, rseed(2017): glm outwork female kids cdoc cage, family(binomial)
> link(logit)

Burn-in ...
Simulation ...

Model summary
```

Likelihood: outwork ~ glm(xb_outwork)	
Prior: {outwork:female kids cdoc cage _cons} ~ normal(0,10000)	(1)

```
(1) Parameters are elements of the linear form xb_outwork.
```

Bayesian generalized linear models	MCMC iterations	=	12,500
Random-walk Metropolis-Hastings sampling	Burn-in	=	2,500
	MCMC sample size	=	10,000
Family : Bernoulli	Number of obs	=	3,874
Link : logit	Scale parameter	=	1
	Acceptance rate	=	.2363
	Efficiency: min	=	.02665
	avg	=	.05244
Log marginal likelihood = -2001.0895	max	=	.101

outwork	Mean	Std. Dev.	MCSE	Median	Equal-tailed [95% Cred. Interval]	
female	2.260566	.0803699	.002529	2.25819	2.103102	2.421029
kids	.3562706	.0893508	.004221	.3583318	.1783027	.5281715
cdoc	.024483	.0061758	.000329	.0244374	.0130205	.0369231
cage	.0545722	.0040688	.000174	.0547603	.0464829	.0622289
_cons	-2.01027	.082349	.005044	-2.00927	-2.172552	-1.852391

```
Note: Default priors are used for model parameters.
```

By default, `bayes` sets the priors for all parameters to normal(0, 10000). We will discuss priors with the `bayesmh` command because they must be included in the model setup. The options are the same for use with `bayes`, except that the priors are listed before the colon as options.

The `bayesmh` command is set up so that the response and predictors follow the command name as in most other Stata regression commands. Following the list of predictors, we type the likelihood function with `logit` as its argument. Then the priors are declared. Recall that every parameter has a prior, although we are giving them all noninformative priors for this example.

Priors are declared by specifying the outcome variable name within parentheses, followed by a comma, and then the distribution and parameters. If no predictor names are specified, then the software assumes that all parameters not specified elsewhere have the indicated prior. Hence, `prior({outwork:}, normal(0,10000))` tells the command to give normal prior with a mean of 0, and a variance of 10,000 applies to all parameters in the equation for `outwork`.

A Bayesian logistic regression with a built-in evaluator can be specified as

```
. bayesmh outwork female kids cdoc cage, likelihood(logit)
> prior({outwork:}, normal(0,10000)) rseed(2017)

Burn-in ...
Simulation ...

Model summary
```

```
Likelihood:
  outwork ~ logit(xb_outwork)
Prior:
  {outwork:female kids cdoc cage _cons} ~ normal(0,10000)                (1)
```

(1) Parameters are elements of the linear form xb_outwork.

```
Bayesian logistic regression                    MCMC iterations   =    12,500
Random-walk Metropolis-Hastings sampling        Burn-in           =     2,500
                                                MCMC sample size  =    10,000
                                                Number of obs     =     3,874
                                                Acceptance rate   =     .2363
                                                Efficiency:  min  =    .02665
                                                             avg  =    .05244
Log marginal likelihood = -2001.0895                         max  =      .101
```

outwork	Mean	Std. Dev.	MCSE	Median	Equal-tailed [95% Cred. Interval]	
female	2.260566	.0803699	.002529	2.25819	2.103102	2.421029
kids	.3562706	.0893508	.004221	.3583318	.1783027	.5281715
cdoc	.024483	.0061758	.000329	.0244374	.0130205	.0369231
cage	.0545722	.0040688	.000174	.0547603	.0464829	.0622289
_cons	-2.01027	.082349	.005044	-2.00927	-2.172552	-1.852391

Looking at the output from both **bayes** and **bayesmh**, notice the near identity of the standard frequentist-based logistic regression and the **bayes** model with noninformative priors for all four predictors and intercept.

Again, and this is important: Bayesian results are given in terms of the mean and median (or mode), standard deviation, and credible intervals of the posterior distributions of each parameter. The mean is popular unless the posterior distribution is not symmetric; in this case, the median should be preferred. Recall that credible intervals indicate the spread of the marginal posterior distributions for the corresponding parameters. Bayesian analysts do not judge relevance of a parameter by whether 0 is or is not included within the lower and upper credible intervals.

The DIC statistic below is calculated as 3928.039. Because the DIC statistic is a comparative statistic, its value is not meaningful unless another model is being compared. Note also that the marginal log-likelihood value is displayed as −2001.09.

```
. bayesstats ic
Bayesian information criteria
```

	DIC	log(ML)	log(BF)
active	3928.038	-2001.09	.

```
Note: Marginal likelihood (ML) is computed
      using Laplace-Metropolis approximation.
```

20.2.2   Diagnostic plots

Because the posterior distribution sample is generated by the MCMC algorithm, values sampled from the posterior distribution can be examined graphically for signs of nonconvergence. Stata illustrates four plots to visualize the behavior of the MCMC process. All four plots can be obtained using the command `bayesgraph diagnostics`. Typing the command without arguments will display graphs of all four predictors and the intercept. For space concerns, we have chosen to display the diagnostic graphs of only the two continuous predictors. We use the postestimation command `bayesgraph diagnostics` to indicate which parameters we wish to test. The first is the centered `docvis` variable (`cdoc`).

```
. bayesgraph diagnostics {outwork: cdoc}
```

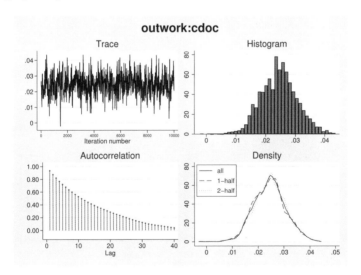

Figure 20.1. Centered number of doctor visits

A trace plot graphs the parameter estimate at each iteration. If the specified model does not lead to an improper posterior distribution, the sample mean should converge to a steady state and vary at a specific value; there should be no long-term signal obvious in the trace plot.

Below, the trace plot is a plot showing the autocorrelation of draws at successive lags. Ideally, the correlation between draws will decline rapidly as the number of lags increases.

At the upper right of the graph is a histogram of the posterior density. This histogram should have a shape that is consistent with the specified prior. A unimodal distribution for which the mode is defined and the tails of the distribution are not too large indicates that a reliable calculation of the highest density credible interval can be made for the parameter. Directly below the histogram is an overlay of three kernel density estimates, the entire sample, the first half of the sample, and the last half of the sample. Discrepancies between the distributions for the halves suggest convergence problems.

Looking at the diagnostic plots for cdoc, a researcher would think these plots satisfactory for most studies. The trace in the upper left appears to display a fairly random iteration log. In this case, the histogram and density curve are fairly bell-shaped, though this is not a necessary condition. The estimate is the mean value of the distribution. The graph of centered age (displayed below) is not as good as for doctor visits. If we were to run this model for a real research project, we could add more iterations to the sampling, and thin the selection of values by perhaps 3 or 5 to reduce autocorrelation.

```
. bayesgraph diagnostics {outwork: cage}
```

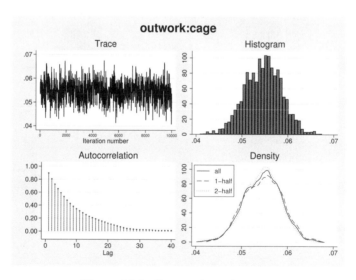

Figure 20.2. Centered age in years

Another important diagnostic command that we advise using after all Bayesian models is `bayesstats ess`, which provides effective sample sizes, correlation times, and efficiencies for each specified parameter in the model. Note that the minimum, maximum, and mean efficiency statistics are also displayed in the model header. Information on model efficiency can be used to evaluate the effectiveness of sampling. In Bayesian modeling, the posterior mean of a model parameter is determined by the mean of the MCMC sampling of the posterior distribution of the parameter. The standard error of an estimated sample mean is proportional to $1/\sqrt{N}$, where N is the size of a random sample of independent observations or cases. However, the problem with such sampling is that positive autocorrelation is usually inherent in the data. That is, sample observations are not independent, and the formula for determining sample mean standard errors must be amended as a result. The Bayesian standard error for the estimated posterior mean is proportional to $1/\sqrt{\text{ESS}}$, where ESS is the effective sample size of the posterior mean sampled data.

```
. bayesstats ess
Efficiency summaries     MCMC sample size =     10,000
```

outwork	ESS	Corr. time	Efficiency
female	1009.74	9.90	0.1010
kids	448.19	22.31	0.0448
cdoc	352.37	28.38	0.0352
cage	545.17	18.34	0.0545
_cons	266.54	37.52	0.0267

Note that the ESS values are low. Ideally, they should be fairly close to the MCMC sample size T; in the current case, this is $T = 10000$. The efficiency is determined by multiplying the ESS value by $1/T$ or $1/10000$. The ESS value for `female` is 1009.74; the efficiency is therefore 0.101. An average efficiency value of 0.01 is the minimum efficiency for an acceptable model, unless the analyst is running Gibbs sampling. Values above 0.1 are regarded as highly efficient for the random-walk Metropolis algorithm. The correlation time is the inverse of the efficiency and may be thought of as the estimated lag after which autocorrelation is low. Lower values are better. For `female`, it is 9.90. The correlation time for `kids` is 22.31, which is quite high, and even higher values are obtained for `cdoc` and the intercept.

Finally, we may be only interested in assessing the ESS statistics for a subset of parameters from a model with many more. We could have used `bayesstats` to check just `cdoc`; in which case, the command would be the following:

```
. bayesstats ess {outwork:cdoc}
```

These statistics provide a more quantitative evaluation of model adequacy than simply inspecting the diagnostic plots above, so we advise using this postestimation command. However, if the MCMC sampling algorithm fails to sample the entire range of the distribution, then it is possible for ESS values to be high, when the MCMC chain has not converged. Bi- or multimodal distributions can lead to this result. This information can generally be taken from inspection of the diagnostic plots given from `bayesgraph`. A low efficiency value does not invalidate a model, though it could be regarded as a signal that longer runs are needed.

If there is evidence of autocorrelation among the samples, thinning can be used to reduce autocorrelation. Thinning is the process of subsetting the generated sample by keeping only every kth generated sample. The default is to use every value generated in the iteration procedure as a sampling candidate. Thinning has the sampling algorithm select, for example, every 5th observation to be sampled. Thinning helps prevent auto-correlation in the MCMC procedure. Note that the default number of sampling iterations we have been using for our example is 10,000.

For pedagogical purposes, we will not be overly concerned in this chapter with whether our developed models have high or even acceptable effective sample-size statistics. Rather, our goal is to demonstrate how to use the `bayes` and `bayesmh` commands to develop GLM-based Bayesian models, including illustrating the use of many of the modeling options available to the user. Many times, a model with low ESS statistics can be improved by thinning, or amending previously specified thinning values, by blocking, increasing the burnin period, and so forth. Interestingly, it appears that amending a model in this way may result in substantially higher ESS values, but with little to negligible change in posterior means and credible interval values. Nevertheless, models, or parameters, with very low ESS values should be evaluated as to their worth. What also may be the case when the average model efficiency is very low is that the model may not be appropriate for the data. As an analogy, consider a maximum-likelihood Poisson model with every predictor having a p-value of 0.000, but with a high value for the dispersion statistic. The model appears on the surface to be well-fit, but the evidence of overdispersion reveals that the model is not appropriate for the data. Depending on the cause of the overdispersion, another model may be much more suitable for the data. Keep this fact in mind when confronted with a model having low ESS parameter values.

20.2.3 Bayesian logistic regression—informative priors

In this example, we fit the same model but with Cauchy priors on both `cdoc` and `cage`. Ordinarily, we would want an informative prior such as the Cauchy prior to be based on solid outside information. In Stata 15, the Cauchy distribution is built-in;[1] for this example, however, we use the following general specification with arbitrary values so that we can demonstrate how a distribution that is not explicitly provided for by `bayesmh` can be used in the model:

```
prior({outwork:cage}, logdensity(ln(11)-ln(11^2+({outwork_cage})^2-ln(_pi))))
```

---

1. To specify a Cauchy distribution in Stata 15, type the following with the `bayes` prefix or the `bayesmh` command: `prior({outwork:cage}, cauchy(0,11))`.

To refer to parameter {outwork:cage} in the substitutable expression within the logdensity() option, we use {outwork_cage}; that is, an underscore (_) is used rather than a colon (:). Specifying a colon would indicate the linear combination of the variable cage and its coefficient.

The Cauchy prior is specified using location parameter a and scale parameter b:

$$\log\{b\} - \log\{b^2 + y - a^2\} - \log(\pi) \tag{20.19}$$

For the variables cdoc and cage, we set $a = 0$. For cdoc, we set the scale parameter $b = 6$, and for cage, we let $b = 11$. We specify the variables female and kids with normal(0,1000) priors. Note that this is the default prior for these coefficients when bayes is used.

A Bayesian logistic regression with informative priors can be specified with bayes as the following:

```
. bayes, prior({outwork:female kids _cons}, normal(0, 1000))
> prior({outwork:cdoc},
> logdensity(ln(6)-ln(6^2+({outwork_cdoc})^2)-ln(_pi)))
> prior({outwork:cage},
> logdensity(ln(11)-ln(11^2+{outwork_cage}^2)-ln(_pi)))
> rseed(35841): glm outwork female kids cdoc cage, family(binomial)

Burn-in ...
Simulation ...

Model summary
_____

Likelihood:
  outwork ~ glm(xb_outwork)

Priors:
  {outwork:female kids _cons} ~ normal(0,1000)                        (1)
                {outwork:cdoc} ~ logdensity(<expr1>)                  (1)
                {outwork:cage} ~ logdensity(<expr2>)                  (1)

Expressions:
  expr1 : ln(6)-ln(6^2+({outwork:cdoc})^2)-ln(_pi)
  expr2 : ln(11)-ln(11^2+{outwork:cage}^2)-ln(_pi)
_____

(1) Parameters are elements of the linear form xb_outwork.
```

```
Bayesian generalized linear models            MCMC iterations    =      12,500
Random-walk Metropolis-Hastings sampling      Burn-in            =       2,500
                                              MCMC sample size   =      10,000
Family : Bernoulli                            Number of obs      =       3,874
Link   : logit                                Scale parameter    =           1
                                              Acceptance rate    =       .2057
                                              Efficiency:  min   =      .03496
                                                           avg   =      .04486
Log marginal likelihood = -1993.0855                       max   =      .05717
```

| | | | | | Equal-tailed | |
outwork	Mean	Std. Dev.	MCSE	Median	[95% Cred. Interval]	
female	2.260141	.0842407	.004506	2.254865	2.097708	2.431913
kids	.3540421	.0906731	.003792	.3560368	.1811122	.5379961
cdoc	.0248758	.0063243	.000318	.0247956	.0130655	.0382379
cage	.0544032	.0039354	.00017	.0543025	.0469003	.0626092
_cons	-2.012612	.0799112	.004051	-2.011676	-2.171733	-1.862155

The same Bayesian logistic regression with informative priors can be specified with bayesmh as the following:

```
. bayesmh outwork female kids cdoc cage, likelihood(logit)
> prior({outwork:female kids _cons}, normal(0, 1000))
> prior({outwork:cdoc},
> logdensity(ln(6)-ln(6^2+({outwork_cdoc})^2)-ln(_pi)))
> prior({outwork:cage},
> logdensity(ln(11)-ln(11^2+{outwork_cage}^2)-ln(_pi)))
> rseed(35841)

Burn-in ...
Simulation ...

Model summary
```
```
Likelihood:
  outwork ~ logit(xb_outwork)

Priors:
  {outwork:female kids _cons} ~ normal(0,1000)                             (1)
            {outwork:cdoc} ~ logdensity(<expr1>)                           (1)
            {outwork:cage} ~ logdensity(<expr2>)                           (1)

Expressions:
  expr1 : ln(6)-ln(6^2+({outwork:cdoc})^2)-ln(_pi)
  expr2 : ln(11)-ln(11^2+{outwork:cage}^2)-ln(_pi)
```

(1) Parameters are elements of the linear form xb_outwork.

```
Bayesian logistic regression                  MCMC iterations   =      12,500
Random-walk Metropolis-Hastings sampling      Burn-in           =       2,500
                                              MCMC sample size  =      10,000
                                              Number of obs     =       3,874
                                              Acceptance rate   =       .1844
                                              Efficiency:  min  =       .03402
                                                           avg  =       .05568
Log marginal likelihood = -1993.0309                       max  =       .07315
```

					Equal-tailed	
outwork	Mean	Std. Dev.	MCSE	Median	[95% Cred.	Interval]
female	2.258607	.0839686	.004552	2.261023	2.0928	2.423481
kids	.3565683	.0899923	.003727	.3605532	.1768716	.5279933
cdoc	.024811	.0064852	.000282	.0244694	.0122816	.0378834
cage	.0547071	.0039411	.000146	.0547988	.0469502	.06197
_cons	-2.010109	.0794933	.003246	-2.009283	-2.171436	-1.854177

```
. bayesstats ic
Bayesian information criteria
```

	DIC	log(ML)	log(BF)
active	3928.181	-1993.031	.

Note: Marginal likelihood (ML) is computed
      using Laplace-Metropolis approximation.

The fit of the model with informative priors is statistically identical to the previous noninformative prior models. The DIC statistics are nearly the same. The values of the mean of the posteriors vary somewhat between models with noninformative priors and this model with informative priors. Table 20.6 displays the values for comparison.

Table 20.6. Results from `bayesmh` with informative and noninformative priors

	noninformative	informative
female	2.60	2.259
kids	0.356	0.356
cdoc	0.024	0.025
cage	0.055	0.055
_cons	−2.01	−2.01

Stata has postestimation commands that allow the user to perform hypothesis tests based on the posterior probabilities of models being compared. The `bayestest model` command displays the marginal likelihoods and prior and posterior probabilities for the models being comparatively tested. We do not discuss this test here, but interested readers should check the online Stata reference manual for `bayestest` and `bayestest model`.

Finally, we review diagnostic graphs of the two continuous predictors. We are particularly interested in whether the graph of `cage` is better distributed than before. If
so, the Cauchy prior is likely the cause.

```
. bayesgraph diagnostics {outwork: cdoc}
```

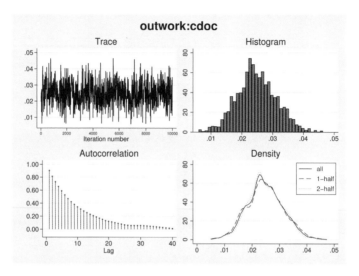

Figure 20.3. Centered number of doctor visits

```
. bayesgraph diagnostics {outwork: cage}
```

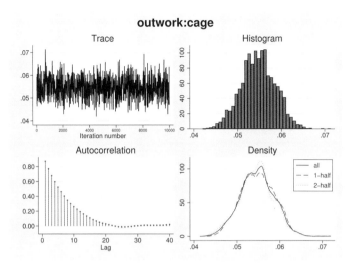

Figure 20.4. Centered age in years

This graph looks much better than before. The earlier graph for centered age was nearly bimodal. Note also that the credible interval of `cage` is more narrow (because the prior is slightly more informative) with this model than for the model with noninformative priors. The model using the Cauchy prior definitely seems to have a better fit for centered age than the model having a flat prior.

20.3   Bayesian probit regression

Probit regression differs from logistic regression by having a different link function; see table A.2. As a function of the conditional mean, the logistic regression model associates the linear predictor to be given by $\ln\{\mu/(1-\mu)\}$, the probit model uses the inverse CDF of the standard normal distribution $\Phi^{-1}(\mu)$. The predicted values of probit regression are generally close to those of logistic regression.

Stata's `bayes` and `bayesmh` commands both have a built-in probit function that is used in the same way as the `logit` function. To illustrate, we model the same data as we did with the Bernoulli logistic regression above. The code for executing `bayes` using the GLM binomial probit family as well as for the maximum-likelihood `probit` command is provided below. The code for creating variables and the `bayesmh` command and output are displayed afterward.

```
bayes, rseed(23840): ///
    glm outwork female kids cdoc cage, family(binomial) link(probit)

bayes, rseed(23840): ///
    probit outwork female kids cdoc cage
```

The Bayesian probit model with built-in evaluator using bayesmh is as follows:

```
. use http://www.stata-press.com/data/hh4/rwm1984, clear
(German health data for 1984; Hardin & Hilbe, GLM and Extensions, 4th ed)
. quietly summarize docvis, meanonly
. generate cdoc = docvis - r(mean)
. quietly summarize age, meanonly
. generate cage = age - r(mean)
. bayesmh outwork female kids cdoc cage, likelihood(probit)
> prior({outwork:}, flat) rseed(23840)

Burn-in ...
Simulation ...

Model summary
```

Likelihood:	
outwork ~ probit(xb_outwork)	
Prior:	
{outwork:female kids cdoc cage _cons} ~ 1 (flat)	(1)

(1) Parameters are elements of the linear form xb_outwork.

Bayesian probit regression		MCMC iterations	=	12,500
Random-walk Metropolis-Hastings sampling		Burn-in	=	2,500
		MCMC sample size	=	10,000
		Number of obs	=	3,874
		Acceptance rate	=	.2611
		Efficiency: min	=	.03696
		avg	=	.05405
Log marginal likelihood = -1983.1754		max	=	.08442

outwork	Mean	Std. Dev.	MCSE	Median	Equal-tailed [95% Cred. Interval]	
female	1.332289	.0448123	.001542	1.332267	1.24314	1.41771
kids	.1402861	.0496876	.002057	.1418047	.0387234	.2362415
cdoc	.0137776	.0034098	.000173	.0137531	.0072373	.0207638
cage	.0290757	.0022217	.000116	.0291653	.0247221	.0332143
_cons	-1.14347	.0415692	.001829	-1.144272	-1.223891	-1.063053

```
. bayesstats ic
Bayesian information criteria
```

	DIC	log(ML)	log(BF)
active	3940.95	-1983.175	.

Note: Marginal likelihood (ML) is computed
      using Laplace-Metropolis approximation.

The DIC statistic for the Bayesian logistic model on the same data was 3928.038, compared with the probit model DIC of 3940.95. The 13-point difference in the models shows a significantly better fit for the logistic model than for the probit. Of the two models, we prefer the logistic model.

We now compare the above Bayesian probit model with a traditional frequentist-based model. The results should be similar. Bayesian results using noninformative priors are closer to the MLE model results.

```
. glm outwork female kids cdoc cage, family(binomial) link(probit) noheader
> nolog
```

		OIM				
outwork	Coef.	Std. Err.	z	P>|z|	[95% Conf.	Interval]
female	1.332467	.0466329	28.57	0.000	1.241068	1.423866
kids	.1396832	.0513311	2.72	0.007	.039076	.2402904
cdoc	.0141333	.0035524	3.98	0.000	.0071706	.0210959
cage	.0290342	.0022821	12.72	0.000	.0245614	.0335069
_cons	-1.142121	.042402	-26.94	0.000	-1.225228	-1.059015

20.4 Bayesian complementary log-log regression

The complementary log-log (clog-log) model is an asymmetric link function used for binary response models. Starting in Stata 15, the `bayes` prefix can be used with the `cloglog` command to run a Bayesian clog-log regression model. In Stata 14, there is no built-in clog-log log-likelihood evaluator associated with the `bayesmh` command. In section 20.9.2.2, we illustrate a clog-log log-likelihood evaluator to fit Bayesian clog-log models using `bayesmh`.

Here we show an example of how to fit this model using the `bayes` prefix. For this command, we will specify a flat prior on each parameter.

To fit the model using the `cloglog` command, type

```
bayes, prior(outwork:, flat) rseed(34872): ///
    cloglog outwork female kids cdoc cage
```

Alternatively, the `glm` command can be used

```
. use http://www.stata-press.com/data/hh4/rwm1984, clear
(German health data for 1984; Hardin & Hilbe, GLM and Extensions, 4th ed)
. quietly summarize docvis, meanonly
. generate cdoc = docvis - r(mean)
. quietly summarize age, meanonly
. generate cage = age - r(mean)
```

```
. bayes, prior({outwork:}, flat) rseed(34872):
> glm outwork female kids cdoc cage, family(binomial) link(cloglog)

Burn-in ...
Simulation ...
Model summary
```

```
Likelihood:
  outwork ~ glm(xb_outwork)
Prior:
  {outwork:female kids cdoc cage _cons} ~ 1 (flat)                              (1)
```

```
(1) Parameters are elements of the linear form xb_outwork.
```

Bayesian generalized linear models		MCMC iterations	=	12,500
Random-walk Metropolis-Hastings sampling		Burn-in	=	2,500
		MCMC sample size	=	10,000
Family : Bernoulli		Number of obs	=	3,874
Link : complementary log-log		Scale parameter	=	1
		Acceptance rate	=	.2473
		Efficiency: min	=	.01261
		avg	=	.0361
Log marginal likelihood = -1970.7461		max	=	.07037

outwork	Mean	Std. Dev.	MCSE	Median	Equal-tailed [95% Cred. Interval]	
female	1.784155	.0704918	.006278	1.781017	1.64424	1.927227
kids	.3517064	.0653117	.002462	.350771	.2226274	.4791738
cdoc	.0115392	.0035466	.000194	.0116997	.0044004	.018193
cage	.0428226	.0031672	.000149	.0429182	.0365953	.0490006
_cons	-2.062277	.068843	.005005	-2.062176	-2.199849	-1.929954

20.5 Bayesian binomial logistic regression

Both Stata's bayes and bayesmh commands have built-in methods of fitting a Bayesian binomial logistic regression model. When working with bayesmh, the log-likelihood evaluator binlogit(n) can be used. The bayes prefix can be used to fit the model using glm. bayesmh does not have built-in log-likelihood evaluators for binomial probit or clog-log regression. However, the glm command with the binomial n family, where n is the binomial denominator, can be modeled using bayes. The GLM approach is an excellent way to model this class of grouped data because it is simple to alternatively use the probit, cloglog, or loglog links.

Recall that each observation in a grouped model has a response variable consisting of two components: a denominator specifying the number of observations with the same covariate pattern, and a numerator depicting the number of cases in each observation for which the response variable has a value of 1. The log-likelihood function is given as

$$\mathcal{L}(\beta; y, m) = \sum_{i=1}^{n} \left[y_i(x_i\beta) - m_i \ln\{1 + \exp(x_i\beta)\} + \ln\binom{m_i}{y_i} \right] \tag{20.20}$$

We use the 1912 Titanic shipping disaster data in grouped format for an example. The variable `cases` tells us the number of observations having an identical pattern of covariate values. The variable `survive` tells us how many people in each covariate pattern survived. With binary variables `age` and `sex` and a three level categorical predictor, `class`, as predictors, the model may be set using the code below. A flat prior is specified for each predictor.

Stata has a built-in evaluator for the binomial log likelihood called `binlogit()`. The option comes with an argument, identifying the variable in the data that defines group sizes (number of trials or the binomial denominators). Analysts will note that whether data are stored individually or grouped, the results of an analysis will be the same (except for sampling error).

We display the `bayesmh` output after loading the Titanic data. The `bayes` code to model the grouped data can be given as follows. Note that we use the same seed number, resulting in identical output:

```
bayes, prior(survive:, flat) rseed(39422): ///
     glm survive i.age i.sex i.class, family(bin cases)

bayes, prior(survive:, flat) rseed(39422): ///
     binreg survive i.age i.sex i.class, n(cases)
```

The Titanic data and **bayesmh** command for modeling the data is given as

```
. use http://www.stata-press.com/data/hh4/titanicgrp, clear
(Generalized Linear Models and Extensions, 4 ed - Titanic passengers)
. bayesmh survive i.age i.sex i.class, likelihood(binlogit(cases))
> prior({survive:}, flat) rseed(39422)

Burn-in ...
Simulation ...

Model summary
```

Likelihood:	
survive ~ binlogit(xb_survive,cases)	
Prior:	
{survive:1.age 1.sex i.class _cons} ~ 1 (flat)	(1)

```
(1) Parameters are elements of the linear form xb_survive.
```

Bayesian binomial regression	MCMC iterations =	12,500
Random-walk Metropolis-Hastings sampling	Burn-in =	2,500
	MCMC sample size =	10,000
	Number of obs =	12
	Acceptance rate =	.2255
	Efficiency: min =	.0321
	avg =	.04151
Log marginal likelihood = -79.00623	max =	.053

survive	Mean	Std. Dev.	MCSE	Median	Equal-tailed [95% Cred. Interval]	
age						
adults	-1.060595	.2364765	.01192	-1.047095	-1.544944	-.6057152
sex						
man	-2.382315	.1462014	.006351	-2.383875	-2.672569	-2.095639
class						
2nd class	-1.019393	.1903787	.008684	-1.024982	-1.377257	-.6434665
3rd class	-1.782275	.1766445	.009436	-1.772677	-2.153385	-1.437651
_cons	3.083371	.2991788	.016699	3.070692	2.514807	3.697764

We compare the above Bayesian logistic binomial model with a standard logistic binomial regression. We expect that the mean values for the posterior distribution of each predictor are nearly the same as the parameter estimates in the standard model. For space purposes, the header statistics will not be displayed.

```
. glm survive i.age i.sex i.class, family(binomial cases) noheader nolog
```

survive	Coef.	OIM Std. Err.	z	P>|z|	[95% Conf. Interval]	
age adults	-1.055608	.2426561	-4.35	0.000	-1.531205	-.5800106
sex man	-2.369465	.1452512	-16.31	0.000	-2.654152	-2.084778
class 2nd class 3rd class	-1.010558 -1.766372	.1949348 .1707174	-5.18 -10.35	0.000 0.000	-1.392623 -2.100971	-.6284927 -1.431772
_cons	3.061882	.2980054	10.27	0.000	2.477802	3.645962

As anticipated, the values from the model estimated using maximum likelihood are similar to those from a Bayesian model with noninformative priors.

20.6   Bayesian Poisson regression

Poisson regression, as discussed earlier in the text, is the basic model applied to count data. The Poisson regression model assumes that the conditional mean and conditional variance of the outcome are equal. When the value of the conditional variance exceeds the conditional mean, the data are said to be overdispersed; when the variance is less than the mean, the data are underdispersed. A full discussion of Poisson regression is found in chapter 12.

The Poisson log likelihood is given by

$$\mathcal{L}(\beta; y) = \sum_{i=1}^{n} \{y_i \ln(x_i\beta) - \exp(x_i\beta) - \ln\Gamma(y_i + 1)\} \tag{20.21}$$

In this section, we model the same dataset, `rwm1984.dta`, as was considered for logistic regression. We will use the number of doctor visits (`docvis`) as the response. The predictors are an indicator of having been out of work, `outwork`, and age in years that has been standardized, `sage`. Recall that we suggest that continuous predictor variables be centered or standardized to help affect better mixing during sampling. Interpretation needs should hinge on whether you center or standardize.

20.6.1 Bayesian Poisson regression with noninformative priors

Stata's bayes prefix can be used with poisson or glm with the Poisson family option. To fit the model described above using bayes with default priors, we can type

```
bayes, rseed(239048) : glm docvis outwork sage, family(poisson)
```

or

```
bayes, rseed(239048) : poisson docvis outwork sage
```

When the bayes prefix is used, the offset() and exposure() options of poisson and glm can be used as usual. If the Bayesian posterior parameters are to be interpreted as incidence-rate ratios, we can use the irr option with the poisson or eform option with glm to effect this result. See the chapter on Poisson regression earlier in the text for additional discussion.

Stata's bayesmh command provides a built-in evaluator for Bayesian Poisson regression. Below, we specify a model that is equivalent to the one specified in the bayes command above.

```
. use http://www.stata-press.com/data/hh4/rwm1984, clear
(German health data for 1984; Hardin & Hilbe, GLM and Extensions, 4th ed)
. egen sage = std(age)
. bayesmh docvis outwork sage, likelihood(poisson)
> prior({docvis:}, normal(0,10000)) rseed(239048)

Burn-in ...
Simulation ...
Model summary

Likelihood:
  docvis ~ poissonreg(xb_docvis)
Prior:
  {docvis:outwork sage _cons} ~ normal(0,10000)                          (1)

(1) Parameters are elements of the linear form xb_docvis.
Bayesian Poisson regression                    MCMC iterations  =     12,500
Random-walk Metropolis-Hastings sampling       Burn-in          =      2,500
                                               MCMC sample size =     10,000
                                               Number of obs    =      3,874
                                               Acceptance rate  =      .1782
                                               Efficiency:  min =     .06108
                                                            avg =     .07252
Log marginal likelihood = -15663.547                        max =      .0827
```

docvis	Mean	Std. Dev.	MCSE	Median	Equal-tailed [95% Cred. Interval]	
outwork	.4069786	.019332	.000672	.4071802	.3692588	.4429259
sage	.2481176	.0095319	.000351	.2478607	.229565	.2677239
_cons	.938687	.0129775	.000525	.9385642	.9137222	.9634572

The suboptions `offset()` and `exposure()` may be used within the `likelihood()` option to specify offset or exposure variables. After `bayesmh`, the `bayesstats summary` command can be used to obtain incidence-rate ratios.

```
. bayesstats summary (outwork:exp({docvis:outwork})) (sage:exp({docvis:sage}))
Posterior summary statistics                        MCMC sample size =    10,000
      outwork : exp({docvis:outwork})
         sage : exp({docvis:sage})
```

	Mean	Std. Dev.	MCSE	Median	Equal-tailed [95% Cred. Interval]	
outwork	1.502553	.0290171	.00101	1.502575	1.446662	1.557257
sage	1.281669	.0122243	.00045	1.281281	1.258053	1.306986

We now compare the Bayesian model with a standard maximum likelihood model.

```
. glm docvis outwork sage, family(poisson) nolog
Generalized linear models                    No. of obs       =       3,874
Optimization       : ML                      Residual df      =       3,871
                                             Scale parameter =           1
Deviance           =   24190.35807           (1/df) Deviance =    6.249124
Pearson            =   43909.61516           (1/df) Pearson  =    11.34322

Variance function: V(u) = u                  [Poisson]
Link function    : g(u) = ln(u)              [Log]
                                             AIC             =    8.074027
Log likelihood     = -15636.39048            BIC             =    -7792.01
```

docvis	Coef.	OIM Std. Err.	z	P>|z|	[95% Conf. Interval]	
outwork	.4079314	.0188447	21.65	0.000	.3709965	.4448663
sage	.2482285	.0094161	26.36	0.000	.2297733	.2666838
_cons	.9380965	.0127571	73.54	0.000	.913093	.9630999

The values of the posterior means, standard deviations, and credible intervals are close to the values of the MLE coefficients, standard errors, and confidence intervals. The Poisson dispersion statistic of 11.34322 indicates that the data are highly overdispersed. As such, we evaluate the data using a negative binomial in section 20.7.

20.6.2   Bayesian Poisson with informative priors

As an example of a Poisson model with informative priors, we use the same data as in section 20.6.1, but with the added categorical predictor, `edlevel`, included in the model. The `edlevel` categorical variable has four values indicating level of education.

```
. use http://www.stata-press.com/data/hh4/rwm1984, clear
(German health data for 1984; Hardin & Hilbe, GLM and Extensions, 4th ed)

. tabulate edlevel
```

Level of education	Freq.	Percent	Cum.
Not HS grad	3,152	81.36	81.36
HS grad	203	5.24	86.60
Coll/Univ	289	7.46	94.06
Grad School	230	5.94	100.00
Total	3,874	100.00	

We specify a beta prior for `outwork` and a normal(0,1) prior for the standardized age variable, `sage`. The beta prior for `outwork` will constrain the posterior of the parameter to be within the range (0,1). We specify a beta(3,30) prior for `docvis:outwork`, characterized as having a mean of $3/(3 + 30) \approx 0.09$. The choice of informative priors for `outwork` and `sage` was arbitrary but allows us to demonstrate the use of priors. Preferably, the analyst should have specific prior information before using priors. Flat priors are given to the indicators of the `edlevel` factor and the intercept. Recall that flat priors give equal probability to the entire range of possible values. In some instances, this results in the posterior not being a viable probability distribution.

Levels of education have been blocked, so that they are sampled together, but separately from the other parameters, by MCMC.

```
. egen sage = std(age)

. bayesmh docvis outwork sage i.edlevel, likelihood(poisson)
> prior({docvis:_cons i.edlevel}, flat)
> prior({docvis:sage}, normal(0,1))
> prior({docvis:outwork}, beta(3,30)) rseed(14)
> block({docvis:i.edlevel})

Burn-in ...
Simulation ...

Model summary
```

```
Likelihood:
  docvis ~ poissonreg(xb_docvis)
Priors:
  {docvis:_cons i.edlevel} ~ 1 (flat)                                    (1)
            {docvis:sage} ~ normal(0,1)                                  (1)
         {docvis:outwork} ~ beta(3,30)                                   (1)
```

```
(1) Parameters are elements of the linear form xb_docvis.
```

```
Bayesian Poisson regression              MCMC iterations   =    12,500
Random-walk Metropolis-Hastings sampling  Burn-in          =     2,500
                                          MCMC sample size =    10,000
                                          Number of obs    =     3,874
                                          Acceptance rate  =      .227
                                          Efficiency:  min =    .06762
                                                       avg =    .07685
Log marginal likelihood = -15627.938                   max =    .09095
```

| | | | | | Equal-tailed | |
docvis	Mean	Std. Dev.	MCSE	Median	[95% Cred. Interval]	
outwork	.3791305	.0183346	.00067	.3796882	.3439628	.4148068
sage	.2414087	.0091153	.000336	.2412038	.2240879	.2596208
edlevel						
HS grad	-.0620811	.0423815	.00163	-.0600315	-.1458024	.0225617
Coll/Univ	-.2215646	.0392457	.001413	-.2207948	-.2967829	-.1422705
Grad School	-.2876148	.0472449	.001567	-.2871876	-.3812494	-.1948024
_cons	.9834335	.0131284	.000473	.9835215	.9576303	1.009859

The bayes formulation of the above model may be given as

```
bayes, prior({docvis:_cons i.edlevel}, flat)        ///
    prior({docvis:sage}, normal(0,1))               ///
    prior({docvis:outwork}, beta(3,30)) rseed(14)   ///
    block({docvis:i.edlevel}):                      ///
    glm docvis outwork sage i.edlevel, family(poisson)
```

Traditional Poisson regression output is given by

```
. glm docvis outwork sage i.edlevel, family(poisson) noheader nolog
```

| | | OIM | | | | |
docvis	Coef.	Std. Err.	z	P>|z|	[95% Conf. Interval]	
outwork	.3936652	.0189679	20.75	0.000	.3564888	.4308416
sage	.2397415	.0095103	25.21	0.000	.2211016	.2583814
edlevel						
HS grad	-.0575221	.0422577	-1.36	0.173	-.1403457	.0253015
Coll/Univ	-.2209144	.0395723	-5.58	0.000	-.2984747	-.143354
Grad School	-.2827698	.0479484	-5.90	0.000	-.376747	-.1887926
_cons	.9763507	.013581	71.89	0.000	.9497325	1.002969

Notice that although we used strong priors on outwork and sage, beta(3,30) and normal(0,1), the estimates are not that different from those with a flat prior. We might expect that they would be considerably different. The reason is that the model is based on a large sample, 3,874 observations, and the likelihood (the data) overwhelms the prior. To have the priors influence the estimates more, we need to subsample the data. We sample 10% for a new model.

```
. use http://www.stata-press.com/data/hh4/rwm1984, clear
(German health data for 1984; Hardin & Hilbe, GLM and Extensions, 4th ed)

. set seed 54969

. sample 10
(3,487 observations deleted)

. egen sage = std(age)

. bayesmh docvis outwork sage i.edlevel, likelihood(poisson)
> prior({docvis:_cons i.edlevel}, flat)
> prior({docvis:sage}, normal(0,1))
> prior({docvis:outwork}, beta(3,30)) rseed(14)
> block({docvis:i.edlevel})

Burn-in ...
Simulation ...

Model summary
```

```
Likelihood:
  docvis ~ poissonreg(xb_docvis)
Priors:
  {docvis:_cons i.edlevel} ~ 1 (flat)                                    (1)
          {docvis:sage} ~ normal(0,1)                                    (1)
       {docvis:outwork} ~ beta(3,30)                                     (1)
```

(1) Parameters are elements of the linear form xb_docvis.

```
Bayesian Poisson regression                     MCMC iterations  =     12,500
Random-walk Metropolis-Hastings sampling        Burn-in          =      2,500
                                                MCMC sample size =     10,000
                                                Number of obs    =        387
                                                Acceptance rate  =      .2651
                                                Efficiency:  min =     .07224
                                                             avg =      .0865
Log marginal likelihood = -1590.5841                         max =       .106
```

docvis	Mean	Std. Dev.	MCSE	Median	Equal-tailed [95% Cred. Interval]	
outwork	.131385	.0440809	.00164	.1303241	.0495689	.2233928
sage	.1424261	.0286004	.000879	.1421966	.0859119	.1990601
edlevel						
HS grad	-.15167	.1240352	.004394	-.1462209	-.4073397	.0764131
Coll/Univ	-.5185033	.1393277	.004457	-.5165809	-.7977753	-.2590323
Grad School	-.5906947	.1970404	.0069	-.5808253	-.9881177	-.2247898
_cons	1.166479	.0355216	.001241	1.166426	1.098543	1.234206

The reduced sample size did affect the posterior estimates, as we expected it would.

We display the effective sample-size statistics for this more complex model with informative priors on the binary `outwork` and continuous standardized age, `sage`.

```
. bayesstats ess
Efficiency summaries      MCMC sample size =     10,000

        docvis │        ESS   Corr. time   Efficiency
───────────────┼──────────────────────────────────────
       outwork │     722.38        13.84       0.0722
          sage │    1059.62         9.44       0.1060
               │
       edlevel │
       HS grad │     796.74        12.55       0.0797
      Coll/Univ │     977.21        10.23       0.0977
    Grad School │     815.54        12.26       0.0816
               │
         _cons │     818.76        12.21       0.0819
───────────────┴──────────────────────────────────────
```

The efficiency of `outwork` is 0.0722 and `sage` is 0.1060. These two parameters displayed much lower efficiency values when noninformative priors were used compared with those produced with these priors. This provides evidence on the value of using them when modeling these data. The efficiency is also a bit better, but not much. Blocking the levels of `edlevel` also increased efficiency. Moreover, the efficiency values of all parameters are good, with `sage` being considered quite good.

We should always inspect the graphs of the posterior parameters to determine whether convergence has occurred, whether the shape of the posterior is as expected based on the prior and likelihood, and whether there is evidence of autocorrelation. The commands below requests graphs for the `outwork` and `sage` parameters. The traces appear fine, as does the check on autocorrelation.

```
. bayesgraph diagnostics {docvis: outwork}
```

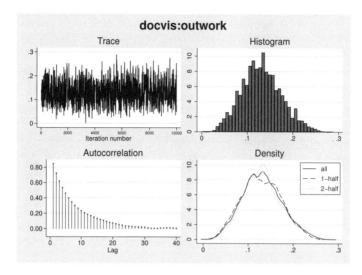

Figure 20.5. Out of work

```
. bayesgraph diagnostics {docvis: sage}
```

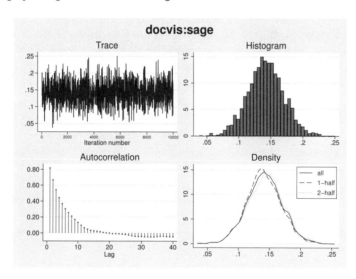

Figure 20.6. Standardized age in years

20.7 Bayesian negative binomial likelihood

We next turn to Bayesian negative binomial regression. An analyst will want to use this model if there is evidence of overdispersion relative to the Poisson or excessive autocorrelation in the data. Most count data are overdispersed, so the negative binomial is an important Bayesian model to be considered for this class of data.

Starting in Stata 15, this model can be fit using the `bayes` prefix with the `nbreg` command. When used with the `bayes` prefix, `nbreg` accepts common options such as `dispersion()`, `exposure()`, and `irr`. For users of Stata 14, `bayesmh` does not have a built-in negative binomial likelihood evaluator. In section 20.9.2.4, we show how to write an evaluator to fit a negative binomial model.

For an example, we will continue to use the 1984 German health data, `rwm1984.dta`. The number of visits made by a patient to a physician during 1984 are given in the response variable, `docvis`. The binary variable, `outwork`, indicates if the patient was unemployed in 1984, and `sage` is the standardized `age` variable. Again, it is recommended to always center or standardize continuous predictors in a Bayesian model. It assists with convergence.

```
. bayes, rseed(4385): nbreg docvis outwork sage

Burn-in ...
Simulation ...

Model summary
────────────────────────────────────────────────────────────────
Likelihood:
  docvis ~ nbreg(xb_docvis,{lnalpha})

Priors:
  {docvis:outwork sage _cons} ~ normal(0,10000)                    (1)
                   {lnalpha} ~ normal(0,10000)
────────────────────────────────────────────────────────────────
(1) Parameters are elements of the linear form xb_docvis.
```

```
Bayesian negative binomial regression        MCMC iterations  =    12,500
Random-walk Metropolis-Hastings sampling      Burn-in          =     2,500
                                              MCMC sample size =    10,000
                                              Number of obs    =       387
                                              Acceptance rate  =     .3025
Dispersion           =       mean            Efficiency:  min =     .07558
                                                          avg =      .1174
Log marginal likelihood = -879.80402                      max =      .2215
```

	Mean	Std. Dev.	MCSE	Median	Equal-tailed [95% Cred. Interval]	
docvis						
outwork	.1724304	.1744478	.005876	.1655078	−.1508635	.5168519
sage	.1717035	.0902068	.003281	.170503	−.0008067	.3381311
_cons	1.098552	.1144462	.003935	1.107003	.8764144	1.322934
lnalpha	.862146	.0960139	.00204	.8629605	.6727028	1.055582

Note: Default priors are used for model parameters.

The output above displays `lnalpha` but can obtain summary information on the posterior distribution of alpha using the `bayesstats summary` command.

```
. bayesstats summary (alpha: exp({lnalpha}))

Posterior summary statistics                    MCMC sample size =    10,000
        alpha : exp({lnalpha})
```

	Mean	Std. Dev.	MCSE	Median	Equal-tailed [95% Cred. Interval]	
alpha	2.379186	.2292247	.004872	2.370167	1.959526	2.873647

20.7.1   Zero-inflated negative binomial logit

A zero-inflated negative binomial logit model is a well-used method of modeling count data having an excess of zero counts. Hurdle models are also used on such data, but zero-inflated models have been preferred by many statisticians. See section 14.3 for an overview of the logic underlying zero-inflated regression and the zero-inflated negative binomial logit models in particular.

`bayes` can fit the zero-inflated model by simply calling the `zinb` command following the colon. Users of Stata 14 can fit a zero-inflated negative binomial logit model using `bayesmh` and specifying the likelihood. Section 20.9.1.2 describes how to fit this model using the `llf()` option, and section 20.9.2.5 demonstrates how to fit the model using an evaluator program with the `llevaluator()` option.

It is worth noting that as of this writing the defaults of the `bayes` code for the `zinb` are not correct for the inflated component. It is generally preferred to give flat priors to the inflated terms as a default, unless there is good reason to specify otherwise. Doing so results in posteriors that are close in value to the coefficients in the inflated component of the model. In the example below, we will specify flat priors for the parameters in the inflation portion of the model.

We will continue to use the 1984 German health data, `rwm1984.dta`. The dependent variable will be `docvis`. The independent variables include `outwork` and the binary variables `female` and `married`. `sage` and `outwork` are included in the inflation equation. Below, we fit this model using `bayes` and use `bayesstats summary` to show `alpha` rather than `lnalpha`.

```
. bayes, rseed(1944) prior({inflate:}, flat):
> zinb docvis outwork female married, inflate(outwork sage)

Burn-in ...
Simulation ...

Model summary
```

```
Likelihood:
  docvis ~ zinb(xb_docvis,xb_inflate,{lnalpha})

Priors:
  {docvis:outwork female married _cons} ~ normal(0,10000)          (1)
         {inflate:outwork sage _cons} ~ 1 (flat)                   (2)
                        {lnalpha} ~ normal(0,10000)
```

(1) Parameters are elements of the linear form xb_docvis.
(2) Parameters are elements of the linear form xb_inflate.

Bayesian zero-inflated negative binomial model	MCMC iterations	=	12,500
Random-walk Metropolis-Hastings sampling	Burn-in	=	2,500
	MCMC sample size	=	10,000
Inflation model: logit	Number of obs	=	387
	Acceptance rate	=	.3013
	Efficiency: min	=	.02089
	avg	=	.06856
Log marginal likelihood = -861.42742	max	=	.2115

	Mean	Std. Dev.	MCSE	Median	\[95% Cred.	Equal-tailed Interval]
docvis						
outwork	-.0304493	.1917976	.007877	-.0276291	-.4181968	.3242994
female	.6320867	.1900685	.007532	.6303794	.2538312	1.015773
married	-.1026869	.2068477	.008911	-.0950759	-.5383433	.2780424
_cons	.9132528	.2011271	.008644	.9084735	.5400902	1.31221
inflate						
outwork	41.10651	237.4449	16.4296	42.99047	-452.7466	480.1014
sage	-10.85812	73.66614	4.26349	-9.785416	-162.1918	132.4274
_cons	-365.9602	156.361	6.64926	-369.3628	-634.4178	-64.95135
lnalpha	.8295301	.0980104	.002131	.8307152	.6403786	1.019691

Note: Default priors are used for some model parameters.

```
. bayesstats summary (alpha: exp({lnalpha}))
```

| Posterior summary statistics | | | | | MCMC sample size = | 10,000 |

alpha : exp({lnalpha})

	Mean	Std. Dev.	MCSE	Median	\[95% Cred.	Equal-tailed Interval]
alpha	2.303281	.2263683	.004912	2.29496	1.897199	2.772338

We will not say more about the model here, but we refer you to our previous discussion of the model in section 14.3. We invite you to check the frequentist-based ZINB model displayed in that section. It is remarkably similar to the Bayesian output above, especially given the difference in how the models are fit.

20.8   Bayesian normal regression

Examples of Stata's Bayesian normal model are provided throughout the *Stata Bayesian Analysis Reference Manual*; also see chapter 5. We need only to point out some of the major differences that exist in modeling a normal model compared with the other GLMs we discuss in this chapter.

When setting up the command for a Bayesian normal regression, it is important to remember that the normal model has two parameters to be estimated: the mean and the variance (σ^2). The `bayes` and `bayesmh` commands automatically parameterize the mean in calculating the linear predictor. Recall that the normal model has a canonical identity link, identifying the mean and linear predictor. Therefore, by calculating the linear predictor, the mean parameter is determined. On the other hand, the variance parameter must be estimated.

Using the `bayes` command, the user can specify either `regress` or `glm` with the default identity link as the Bayesian model. When using `glm`, the variance parameter is not estimated. The variance parameter is estimated when `regress` is used. An example of its use is given below.

This example also appears in the *Stata Bayesian Analysis Reference Manual*; `change` is a continuous response, `group` is a binary indicator, and `age` is continuous.

```
. webuse oxygen, clear
(Oxygen Uptake Data)
. bayes, rseed(123): regress change group age

Burn-in ...
Simulation ...
Model summary
```

```
Likelihood:
  change ~ regress(xb_change,{sigma2})
Priors:
  {change:group age _cons} ~ normal(0,10000)                          (1)
                  {sigma2} ~ igamma(.01,.01)
```

```
(1) Parameters are elements of the linear form xb_change.
```

```
Bayesian linear regression                    MCMC iterations    =      12,500
Random-walk Metropolis-Hastings sampling      Burn-in            =       2,500
                                              MCMC sample size   =      10,000
                                              Number of obs      =          12
                                              Acceptance rate    =       .2911
                                              Efficiency:  min   =      .02282
                                                           avg   =      .07095
Log marginal likelihood = -45.604378                       max   =      .09179
```

	Mean	Std. Dev.	MCSE	Median	Equal-tailed [95% Cred. Interval]	
change						
group	5.585808	2.08133	.074657	5.533031	1.45311	9.88814
age	1.877839	.3274079	.010826	1.872394	1.26436	2.548515
_cons	-46.3091	7.621248	.251548	-46.04428	-61.76206	-31.60638
sigma2	10.31546	7.816072	.517453	8.25175	3.558916	31.07108

Note: Default priors are used for model parameters.

When using `bayesmh`, the normal likelihood function includes a specification of the distribution and a designation for the variance, `likelihood(normal({var}))`. The remainder of the command shown below deals with assigning priors to any predictors in the model and assigning a prior to the variance parameter. In the example below, we assign the Jeffreys noninformative prior to the variance. At times, an `igamma(.01, .01)`, which is the default for the `bayes` command, is also used for the prior on the variance. As discussed elsewhere, the specification of priors is similar when the `bayes` command is used. To fit the Bayesian normal regression, we use

```
. webuse oxygen, clear
(Oxygen Uptake Data)

. bayesmh change group age, likelihood(normal({var}))
> prior({change:}, flat) prior({var}, jeffreys) rseed(123)

Burn-in ...
Simulation ...

Model summary
```
```
Likelihood:
  change ~ normal(xb_change,{var})

Priors:
  {change:group age _cons} ~ 1 (flat)                                    (1)
                   {var} ~ jeffreys
```
```
(1) Parameters are elements of the linear form xb_change.
```

```
Bayesian normal regression                              MCMC iterations     =      12,500
Random-walk Metropolis-Hastings sampling                Burn-in             =       2,500
                                                        MCMC sample size =         10,000
                                                        Number of obs       =          12
                                                        Acceptance rate     =       .2921
                                                        Efficiency:  min =          .01147
                                                                     avg =          .05454
Log marginal likelihood = -24.444381                                 max =          .08632
```

	Mean	Std. Dev.	MCSE	Median	Equal-tailed [95% Cred. Interval]	
change						
group	5.459675	2.157552	.0965	5.36734	1.365305	10.08589
age	1.884567	.3349281	.0114	1.890073	1.204147	2.557773
_cons	-46.46854	7.798519	.293925	-46.45018	-62.34541	-30.90944
var	10.44927	6.588704	.615169	8.654826	3.625464	30.55348

The same data modeled using a standard normal regression is given by

```
. glm change group age, family(gaussian) nolog
Generalized linear models                      No. of obs          =          12
Optimization      : ML                         Residual df         =           9
                                               Scale parameter =        7.820974
Deviance       =  70.38876803                  (1/df) Deviance =        7.820974
Pearson        =  70.38876803                  (1/df) Pearson  =        7.820974

Variance function: V(u) = 1                    [Gaussian]
Link function    : g(u) = u                    [Identity]

                                               AIC                 =    5.107004
Log likelihood   =  -27.64202473               BIC                 =    48.02461
```

	Coef.	OIM Std. Err.	z	P>|z|	[95% Conf. Interval]	
change						
group	5.442621	1.796453	3.03	0.002	1.921638	8.963603
age	1.885892	.295335	6.39	0.000	1.307046	2.464738
_cons	-46.4565	6.936531	-6.70	0.000	-60.05185	-32.86115

The variance, σ^2, is provided in GLM as e(phi) (in Stata's glm command) or—it is the scale parameter in the header when the header is shown. The variance is shown as the mean squared residual in OLS normal regression. GLMs fit the mean or location parameter. The normal or Gaussian variance is estimated in full maximum-likelihood regression and in Bayesian normal regression, as we observe.

For pedagogical purposes, we amend the above model by issuing a strong informative prior for the variance. Instead of using the Jeffreys prior on the variance, we now specify a standard normal prior, normal(0,1).

```
. bayesmh change group age, likelihood(normal({var}))
> prior({change:}, flat) prior({var}, normal(0, 1)) rseed(123)

Burn-in ...
Simulation ...
Model summary
```

```
Likelihood:
  change ~ normal(xb_change,{var})
Priors:
  {change:group age _cons} ~ 1 (flat)                                       (1)
                   {var} ~ normal(0,1)
```

(1) Parameters are elements of the linear form xb_change.

Bayesian normal regression	MCMC iterations	=	12,500
Random-walk Metropolis-Hastings sampling	Burn-in	=	2,500
	MCMC sample size	=	10,000
	Number of obs	=	12
	Acceptance rate	=	.2406
	Efficiency: min	=	.03799
	avg	=	.06589
Log marginal likelihood = -33.964412	max	=	.08524

						Equal-tailed
	Mean	Std. Dev.	MCSE	Median	[95% Cred. Interval]	
change						
group	5.447064	1.106004	.056741	5.442598	3.344929	7.640677
age	1.886231	.1849877	.006336	1.881924	1.527525	2.253723
_cons	-46.47589	4.344278	.155996	-46.47509	-55.12026	-38.0229
var	2.929283	.5288183	.02111	2.897402	2.006403	4.056847

Much more may be discussed regarding the Bayesian normal model, and we refer you to the rather comprehensive overview of this model and its variations in the *Stata Bayesian Analysis Reference Manual*.

20.9   Writing a custom likelihood

For models that are not supported by either the `bayes` prefix or the `bayesmh` command's built-in functions for the `likelihood()` option, you may write your own likelihood. The `bayesmh` command offers two options for specifying a custom likelihood, `likelihood(llf())` and `llevaluator()`. The `llf()` suboption allows the user to write the likelihood as part of the `bayesmh` command using substitutable expressions. For more complex likelihoods, the user can write a separate evaluator program, which is then called in the `llevaluator()` option.

When using either `likelihood(llf())` or `llevaluator()`, the user may specify priors and use other `bayesmh` options in the usual way. The standard `bayesmh` postestimation options, including convergence diagnostics and graphs are also available.

With the introduction of the `bayes` command in Stata 15, the number of models that can be easily fit using Bayesian estimation increased substantially. Whenever possible, we suggest users access the list of built-in models for `bayes` to determine if `bayes` can be used with your modeling task. For those using Stata 14, there are many models that require writing an evaluator, including the negative binomial, gamma, and zero-inflated negative binomial models; and mixed or hierarchical model.

When writing our own log likelihoods, we can test whether the code is correct by assessing 1) whether the sampling that constructs a posterior distribution converges, and 2) whether the coefficients and standard errors of the MLE model are nearly the same as the Bayesian model with noninformative priors. If one of these two criteria do not hold, we have made a mistake in programming the Bayesian model.

20.9.1   Using the llf() option

The `llf()` suboption is used to specify the observation-level log likelihood. Writing the likelihood in the `llf()` suboption using a substitutable expression requires less programming than writing an evaluator program. Note that the `llf()` suboption is itself embedded in a greater `likelihood()` option. For detailed information on writing substitutable expressions for use with `bayesmh`, see the *Bayesian Analysis Reference Manual*.

20.9.1.1   Bayesian logistic regression using llf()

We will start by using the familiar example of the logistic regression model to demonstrate the use of the `llf()` suboption. We pursue this derivation merely so that we can illustrate the construction of a user-specified log likelihood (see section 20.9.2.1) compared with the built-in log likelihood in the `bayesmh` command.

The inverse link function (see table A.2) converts the linear predictor to the fitted or predicted value, μ.

The general Bernoulli regression log likelihood is given by

$$\mathcal{L}(\mu; y) = \sum_{i=1}^{n} [y_i \ln\{\mu_i/(1 - \mu_i)\} + \ln(1 - \mu_i)] \qquad (20.22)$$

where μ_i is the mean; that is, it is $P(y_i) = 1$.

The logistic regression log likelihood can also be written as

$$\mathcal{L}(\boldsymbol{\beta}; \mathbf{y}) = \sum_{i=1}^{n} \ln[F\{y_i(x_i\boldsymbol{\beta}) + (1 - y_i)(-x_i\boldsymbol{\beta})\}] \qquad (20.23)$$

where $F(x) = e^x/(1 + e^x)$.

In substitutable expressions {lc:x} indicates a linear combination of coefficients and the variables listed after the colon plus an intercept. For example, {lc:x1 x2} indicates {lc:x1}*x1 + {lc:x2}*x2 + {lc:_cons}. lc is an arbitrary equation name. Using the cond() function, we can specify the value of μ_i dependent on the value of y logs of the probabilities for the two possible outcomes. The resulting individual-level log likelihood for the logit model can be written as

```
ln(invlogit(cond(y, {y: x},-{y: x})))
```

In the bayesmh command below, we construct a conditional log-likelihood function to be placed in the llf() suboption. As in section 20.2.1, we will model the variable binary outwork using the variables female, kids, cdoc, and cage in rwm1984.dta.

```
. use http://www.stata-press.com/data/hh4/rwm1984, clear
(German health data for 1984; Hardin & Hilbe, GLM and Extensions, 4th ed)
. quietly summarize docvis, meanonly
. generate cdoc = docvis - r(mean)
. quietly summarize age, meanonly
. generate cage = age - r(mean)
. bayesmh outwork, likelihood(llf(ln(invlogit(cond(outwork,
> {outwork: female kids cdoc cage}, -{outwork: female kids cdoc cage})))))
> prior({outwork:}, flat) rseed(2017)

Burn-in ...
Simulation ...
Model summary

Likelihood:
  outwork ~ logdensity(<expr1>)
Prior:
  {outwork:_cons female kids cdoc cage} ~ 1 (flat)
Expression:
  expr1 : ln(invlogit(cond(outwork,{outwork:female kids cdoc cage},-{outwork:f
          emale kids cdoc cage})))
```

Bayesian regression				MCMC iterations	=	12,500
Random-walk Metropolis-Hastings sampling				Burn-in	=	2,500
				MCMC sample size	=	10,000
				Number of obs	=	3,874
				Acceptance rate	=	.1964
				Efficiency: min	=	.0376
				avg	=	.0505
Log marginal likelihood = -1973.4817				max	=	.05827

					Equal-tailed	
outwork	Mean	Std. Dev.	MCSE	Median	[95% Cred. Interval]	
female	2.256131	.0808571	.00385	2.253758	2.102875	2.423835
kids	.354585	.0874051	.003621	.3546013	.1865255	.5305977
cdoc	.025052	.0061	.000255	.0248072	.0133201	.0375476
cage	.0546308	.0040554	.000172	.0547251	.0468982	.0625383
_cons	-2.007932	.0797606	.004114	-2.004352	-2.169057	-1.856861

Note: Adaptation tolerance is not met.

20.9.1.2 Bayesian zero-inflated negative binomial logit regression using llf()

As a second example, we fit a Bayesian zero-inflated negative binomial model using the same data shown in section 20.7.1. Recall that this model used t model count data with excess zero counts. In Stata 15, this model can be fit with the `bayes` prefix, as shown in 20.7.1, but in Stata 14, the likelihood must be specified by the user as shown here or in section 20.9.2.5.

As before, we will use the 1984 German health data, `rwm1984.dta`. The number of visits made by a patient to a physician during 1984 are given in the response variable, `docvis`. The binary variable, `outwork`, indicates if the patient was unemployed in 1984 and is used in both parts of the model. The variables `female` and `married` are used to predict the count portion of the model only. The standardized `age` variable, `sage`, is used only in the inflation part of the model. Again, it is recommended to always center or standardize continuous predictors in a Bayesian model. It assists with convergence.

With `bayesmh`, the code for a negative binomial logit evaluator is separated into two components: a count component and a binary component. The binary component consists of a logistic regression on zero counts; the count component is a full negative binomial regression. The mixture aspect of the model relates to the fact that zero counts are estimated using both components.

To write the log likelihood for the ZINB model, we need to add the intercept ourselves to the code. We create a variable called `cons` and give it the value of 1. Note that `cons` is a parameter to be sampled for each component of the ZINB model, as are parameters for each component predictor and the negative binomial dispersion parameter, `alpha`. Also be aware that using `llf()`, we model `alpha` directly and not `lnalpha`. The logic of the code should be clear to you. The log-likelihood code is set up as done for the `bzinbll` evaluator. The priors are the same as well.

```
. use http://www.stata-press.com/data/hh4/rwm1984, clear
(German health data for 1984; Hardin & Hilbe, GLM and Extensions, 4th ed)
. egen sage = std(age)
. gen cons = 1
. bayesmh docvis, noconstant
> likelihood(llf(cond(docvis==0, ln(invlogit({xb:}) +
> (1-invlogit({xb:}))/(1+{alpha}*exp({xc:}))^(1/{alpha})),
> log(1-invlogit({xb:})) + lngamma(docvis+1/{alpha}) -
> lngamma(docvis+1) - lngamma(1/{alpha})+1/{alpha} *
> ln(1/(1+{alpha}*exp({xc:}))) + docvis *
> ln(1-(1/(1+{alpha}*exp({xc:})))))))
> xbdefine(xb: outwork sage cons) xbdefine(xc: outwork female married cons)
> prior({xb:} {xc:}, flat)
> prior({alpha}, normal(0,100)) init({alpha} 1) dots
> mcmcs(20000) burnin(12000) thinning(3) rseed(1944)
note: discarding every 2 sample observations; using observations 1,4,7,...
```

```
Burn-in 12000 aaaaaaaaa1000aaaaaa...2000.........3000.........4000.........5000
> .........6000.........7000.........8000.........9000.........10000.........
> 11000.........12000 done
Simulation 59998 .........1000.........2000.........3000.........4000.........
> 5000.........6000.........7000.........8000.........9000.........10000.......
> ..11000.........12000.........13000.........14000.........15000.........16000.
> .........17000.........18000.........19000.........20000.........21000........
> .22000.........23000.........24000.........25000.........26000.........27000.
> .........28000.........29000.........30000.........31000.........32000.........
> .33000.........34000.........35000.........36000.........37000.........38000.
> .........39000.........40000.........41000.........42000.........43000........
> .44000.........45000.........46000.........47000.........48000.........49000.
> .........50000.........51000.........52000.........53000.........54000........
> .55000.........56000.........57000.........58000.........59000......... done
Model summary
```

Likelihood:
  docvis ~ logdensity(<expr1>)

Priors:
              {xb:outwork sage cons} ~ 1 (flat)   (1)
     {xc:outwork female married cons} ~ 1 (flat)   (2)
                        {alpha} ~ normal(0,100)

Expression:
  expr1 : cond(docvis==0,ln(invlogit(xb_xb) + (1-invlogit(xb_xb))/(1+{alpha}*e
          xp(xb_xc))^(1/{alpha})),log(1-invlogit(xb_xb)) + lngamma(docvis+1/{a
          lpha}) - lngamma(docvis+1) - lngamma(1/{alpha}) + 1/{alpha} * ln(1/(
          1+{alpha}*exp(xb_xc))) + docvis * ln(1-(1/(1+{alpha}*exp(xb_xc)))))

(1) Parameters are elements of the linear form xb_xb.
(2) Parameters are elements of the linear form xb_xc.

Bayesian regression   MCMC iterations   =   71,998
Random-walk Metropolis-Hastings sampling   Burn-in   =   12,000
                                             MCMC sample size =   20,000
                                             Number of obs   =   3,874
                                             Acceptance rate   =   .2678
                                             Efficiency: min =   .02408
                                                          avg =   .07517
Log marginal likelihood = -8347.2893   max =   .1369

	Mean	Std. Dev.	MCSE	Median	Equal-tailed [95% Cred. Interval]	
xb						
outwork	-.499655	.2689997	.012258	-.4919422	-1.064593	.0181693
sage	-.7081379	.1211428	.003523	-.7007736	-.9683917	-.4914465
cons	-1.878398	.2538691	.008855	-1.853743	-2.438672	-1.448478
xc						
outwork	.3563382	.0612067	.002442	.3555399	.2346967	.4774785
female	.2440498	.0554049	.001137	.2439115	.1346589	.3527864
married	-.0959022	.0637205	.001305	-.0954737	-.2235348	.0264651
cons	1.087265	.0735395	.001956	1.087095	.9460228	1.232791
alpha	1.778875	.1018871	.001947	1.77709	1.586264	1.98428

We will not say more about the model here, but we refer you to our previous discussion of the model in section 20.7.1.

20.9.2 Using the llevaluator() option

The `llevaluator()` option of the `bayesmh` command may be used when extending basic GLMs that are not built-in for `bayes` or `bayesmh`. Using the `llevaluator()` option requires writing a separate evaluator program and then calling that program using `bayesmh`. These programs are more general than the model-specific likelihoods used with the `llf()` suboption. This can be useful if you frequently fit the same type of model using Bayesian estimation.

20.9.2.1 Logistic regression model using llevaluator()

As in the section on the `llf()` suboption, we begin with the logit model because it represents a familiar likelihood. See section 20.9.1.1 for a review of the log likelihood for the logit model. Below, we named the logistic log-likelihood evaluator `blogitll`. Once defined, the `llevaluator()` function calls our program.

Using the `cond()` function, we specify logs of the probabilities for the two possible outcomes. The name of the response variable is stored in a global macro `MH_y` allowing our code access to that variable.

Listing 20.1. `blogitll` evaluator for logistic regression

```
program define blogitll
    version 15
    args lnf xb
    quietly{
        tempvar lnfj
        generate 'lnfj' = ln(invlogit( ///
            cond($MH_y,'xb',-'xb'))) if $MH_touse
        summarize 'lnfj' if $MH_touse, meanonly
        if r(N) < $MH_n {
            scalar 'lnf' = .
            exit
        }
        scalar 'lnf' = r(sum)
    }
end
```

The program `blogitll` accepts two arguments. The first argument is the name of a scalar into which the log likelihood is to be stored. The second is the name of a temporary variable in which the parameter specification values are stored. The program declares an additional temporary variable, `lnfj`, used to hold the observation-level log-likelihood values. Once calculated, the sum of the log likelihoods is stored.

Now we model the data using the `blogitll` program that we just wrote. As in previous examples of logit models, we will model the variable binary `outwork` using the

variables `female`, `kids`, `cdoc`, and `cage` from `rwm1984.dta`. This time, we specify the
`llevaluator()` option instead of the `likelihood()` option. Aside from that, these two
specifications are equivalent because we merely programmed the same model provided
by the built-in function.

```
. bayesmh outwork female kids cdoc cage, llevaluator(blogitll)
> prior({outwork:}, flat) rseed(32842)

Burn-in ...
Simulation ...
Model summary
```

```
Likelihood:
  outwork ~ blogitll(xb_outwork)
Prior:
  {outwork:female kids cdoc cage _cons} ~ 1 (flat)                            (1)
```

(1) Parameters are elements of the linear form xb_outwork.

```
Bayesian regression                        MCMC iterations    =     12,500
Random-walk Metropolis-Hastings sampling   Burn-in            =      2,500
                                           MCMC sample size   =     10,000
                                           Number of obs      =      3,874
                                           Acceptance rate    =      .1608
                                           Efficiency:   min  =      .01581
                                                         avg  =      .03899
Log marginal likelihood =  -1973.256                     max  =      .05945
```

					Equal-tailed	
outwork	Mean	Std. Dev.	MCSE	Median	[95% Cred.	Interval]
female	2.2632	.0783207	.003699	2.266164	2.106232	2.422944
kids	.3510605	.0953729	.00482	.3545452	.1703534	.5364152
cdoc	.0251792	.0062226	.000329	.0252219	.0131271	.0379541
cage	.0546341	.0045157	.000359	.0546352	.0460428	.0636336
_cons	-2.012066	.0820148	.003364	-2.01375	-2.16764	-1.851949

Note: Adaptation tolerance is not met.

The posterior means, or Bayesian parameter estimates, are close to the same as
we calculated for the standard logistic regression model and for the `bayes` model with
default priors and for the `bayesmh` logistic model with the built-in logit function. Below,
we display the DIC and log marginal-likelihood value for this model.

```
. bayesstats ic
Bayesian information criteria
```

	DIC	log(ML)	log(BF)
active	3928.76	-1973.256	.

Note: Marginal likelihood (ML) is computed
      using Laplace-Metropolis approximation.

Other diagnostic graphics and statistics will be similar to those used with the code using `bayes` and the `bayesmh` built-in `likelihood(logit)` function shown in section 20.2.

20.9.2.2 Bayesian clog-log regression with llevaluator()

A Bayesian clog-log regression model can be fit in Stata 15 or later using the `cloglog` command with the `bayes` prefix. In Stata 14, the `bayesmh` command can be used to specify this model, but the user must specify the likelihood. Below, we illustrate a clog-log log-likelihood evaluator to fit Bayesian clog-log models, which again matches the other two Bernoulli log likelihoods (logistic and probit) differing only in the link function relating the conditional mean to the linear predictor.

Listing 20.2. `bclogll` evaluator for complementary log-log regression

```
program define bclogll
    version 15
    args lnf xb
    quietly{
        tempvar lnfj
        generate 'lnfj' = cond($MH_y,///
            ln(1-exp(-exp('xb'))),-exp('xb')) if $MH_touse
        summarize 'lnfj' if $MH_touse, meanonly
        if r(N) < $MH_n {
            scalar 'lnf' = .
            exit
        }
        scalar 'lnf' = r(sum)
    }
end
```

```
. use http://www.stata-press.com/data/hh4/rwm1984, clear
(German health data for 1984; Hardin & Hilbe, GLM and Extensions, 4th ed)

. quietly summarize docvis, meanonly

. generate cdoc = docvis - r(mean)

. quietly summarize age, meanonly

. generate cage = age - r(mean)

. bayesmh outwork female kids cdoc cage, llevaluator(bclogll)
> prior({outwork:}, flat) rseed(34872)

Burn-in ...
Simulation ...
Model summary
```

```
Likelihood:
  outwork ~ bclogll(xb_outwork)
Prior:
  {outwork:female kids cdoc cage _cons} ~ 1 (flat)                          (1)
```

```
(1) Parameters are elements of the linear form xb_outwork.
```

```
Bayesian regression                             MCMC iterations   =      12,500
Random-walk Metropolis-Hastings sampling        Burn-in           =       2,500
                                                MCMC sample size  =      10,000
                                                Number of obs     =       3,874
                                                Acceptance rate   =       .1345
                                                Efficiency:  min  =      .03541
                                                             avg  =      .04148
Log marginal likelihood = -1970.7764                         max  =      .04776
```

outwork	Mean	Std. Dev.	MCSE	Median	Equal-tailed [95% Cred. Interval]	
female	1.786421	.0657532	.003009	1.785069	1.652619	1.925696
kids	.354318	.0646741	.003085	.35219	.2274263	.4820587
cdoc	.0114179	.003623	.000193	.0115025	.003437	.0177788
cage	.0429665	.0031766	.000156	.0429173	.0368986	.0491681
_cons	-2.064778	.0682101	.003471	-2.061188	-2.1936	-1.935438

The output can be compared with the frequentist-based model below.

```
. glm outwork female kids cdoc cage, family(binomial) link(cloglog) noheader
> nolog
```

outwork	Coef.	OIM Std. Err.	z	P>|z|	[95% Conf. Interval]	
female	1.783627	.06723	26.53	0.000	1.651859	1.915395
kids	.3532082	.0652455	5.41	0.000	.2253295	.481087
cdoc	.0117499	.0035781	3.28	0.001	.0047369	.0187629
cage	.0429462	.0031238	13.75	0.000	.0368236	.0490687
_cons	-2.061404	.0695593	-29.64	0.000	-2.197738	-1.92507

20.9.2.3   Bayesian Poisson regression with llevaluator()

Below, we demonstrate using the `bayesmh` log-likelihood evaluator to fit a Bayesian poisson regression model. This model can easily be fit using the `bayes` prefix or `bayesmh` with built-in likelihoods, as shown in section 20.6. We can create our own evaluator to create an identity-linked Poisson to help, for example, an analyst who desires to make changes.

Recall that the Poisson log likelihood is given by

$$\mathcal{L}(\beta; y) = \sum_{i=1}^{n} \{ y_i \ln(x_i \beta) - \exp(x_i \beta) - \ln \Gamma(y_i + 1) \} \qquad (20.24)$$

We implement this likelihood in the evaluator program `bpoill`.

Listing 20.3. `bpoill` evaluator for Poisson regression

```
program define bpoill
    args lnf xb
    quietly {
        tempvar lnfj
        generate 'lnfj' = -exp('xb') + $MH_y * ('xb') - ///
            lngamma($MH_y + 1) if $MH_touse
        summarize 'lnfj' if $MH_touse, meanonly
        if r(N) < $MH_n {
            scalar 'lnf' = .
            exit
        }
        scalar 'lnf' = r(sum)
    }
end
```

Note the similarities of this evaluator to the previously illustrated programs.

```
. use http://www.stata-press.com/data/hh4/rwm1984, clear
(German health data for 1984; Hardin & Hilbe, GLM and Extensions, 4th ed)
. egen sage = std(age)
. bayesmh docvis outwork sage, llevaluator(bpoill)
> prior({docvis:}, flat) rseed(14)

Burn-in ...
Simulation ...

Model summary
```

```
Likelihood:
  docvis ~ bpoill(xb_docvis)
Prior:
  {docvis:outwork sage _cons} ~ 1 (flat)                                   (1)
```

(1) Parameters are elements of the linear form xb_docvis.

```
Bayesian regression                        MCMC iterations   =   12,500
Random-walk Metropolis-Hastings sampling   Burn-in           =    2,500
                                           MCMC sample size  =   10,000
                                           Number of obs     =    3,874
                                           Acceptance rate   =    .1857
                                           Efficiency:  min  =   .06327
                                                        avg  =   .07207
Log marginal likelihood = -15647.075                    max  =   .07831
```

docvis	Mean	Std. Dev.	MCSE	Median	Equal-tailed [95% Cred. Interval]	
outwork	.4066746	.0190465	.000681	.4069928	.3693646	.4443897
sage	.2485969	.0091639	.000335	.2487928	.2302338	.2659337
_cons	.9387944	.0126867	.000504	.9385616	.9138956	.9639873

Again, the results are nearly identical.

20.9.2.4   Bayesian negative binomial regression using llevaluator()

The bayesmh command does not have a built-in negative binomial likelihood evaluator. It does have an nbinomialp() function that we may use, but the evaluator must be programmed. Users of Stata 14 are required to program the Bayesian negative binomial with their own likelihood, but with Stata 15 one may use bayes with nbreg or the glm command with the negative binomial family. This is much easier, but for pedagogical purposes, we describe how to program a negative binomial likelihood evaluator for bayesmh.

Below is a negative binomial likelihood evaluator. It can be used for any negative binomial model with bayesmh (though this implementation lacks support for an offset).

Listing 20.4. bnbll evaluator for negative binomial regression

```
program define bnbll
    version 15
    args lnf xb lnalpha
    quietly{
        tempvar lnfj
        generate double 'lnfj' = ln(nbinomialp(1/exp('lnalpha'),   ///
            $MH_y, 1/(1+exp('lnalpha'+'xb')))) if $MH_touse
        summarize 'lnfj' if $MH_touse, meanonly
        if r(N) < $MH_n {
            scalar 'lnf' = .
            exit
        }
        scalar 'lnf' = r(sum)
    }
end
```

Note the code above that maximizes for the log of alpha, which is inserted as the third argument to the program. In this way, the value of alpha is guaranteed to be nonnegative. This provides a direct relationship between the amount of correlation in the data and the value of the dispersion parameter as given in the variance, $\mu + \alpha\mu^2$.

For illustration, we specify flat priors on all parameters, except the dispersion parameter for which we specify a uniform prior from 0.001 to 5. Note that giving the negative binomial dispersion parameter an igamma(0.001, 0.001) prior equally spreads the variability wide enough to minimize any additional information to the posterior of the dispersion. The results are nearly identical to that when specifying the uniform(0.001, 5) prior.

```
. use http://www.stata-press.com/data/hh4/rwm1984, clear
(German health data for 1984; Hardin & Hilbe, GLM and Extensions, 4th ed)

. egen sage = std(age)

. bayesmh docvis outwork sage, llevaluator(bnbll, parameters({lnalpha}))
>   prior({docvis:}, flat)
>   prior({lnalpha}, uniform(0.001,5)) rseed(98342) init({lnalpha} 1)

Burn-in ...
Simulation ...
Model summary
```

```
Likelihood:
  docvis ~ bnbll(xb_docvis,{lnalpha})

Priors:
  {docvis:outwork sage _cons} ~ 1 (flat)                                    (1)
               {lnalpha} ~ uniform(0.001,5)
```

(1) Parameters are elements of the linear form xb_docvis.

```
Bayesian regression                            MCMC iterations   =      12,500
Random-walk Metropolis-Hastings sampling       Burn-in           =       2,500
                                               MCMC sample size  =      10,000
                                               Number of obs     =       3,874
                                               Acceptance rate   =       .2239
                                               Efficiency:  min  =      .02969
                                                            avg  =      .06825
Log marginal likelihood = -8344.3906                         max  =      .09555
```

	Mean	Std. Dev.	MCSE	Median	Equal-tailed [95% Cred. Interval]	
docvis						
outwork	.4159908	.0552918	.003209	.4146727	.3135698	.5284722
sage	.2491313	.0270081	.001065	.2498117	.1978718	.3007356
_cons	.9345319	.0341995	.001184	.9345088	.8668928	.9986375
lnalpha	.83284	.0307083	.000993	.8327162	.7749053	.8946471

Because of the transformation in the log-likelihood evaluator, the value of the log transform of alpha lnalpha is displayed. Thus, we must exponentiate this parameter to get alpha.

```
. bayesstats summary {docvis: outwork sage _cons}
> (lnalpha: {lnalpha}) (alpha: exp({lnalpha}))
Posterior summary statistics                      MCMC sample size =     10,000
        lnalpha : {lnalpha}
          alpha : exp({lnalpha})
```

	Mean	Std. Dev.	MCSE	Median	Equal-tailed [95% Cred. Interval]	
docvis						
outwork	.4159908	.0552918	.003209	.4146727	.3135698	.5284722
sage	.2491313	.0270081	.001065	.2498117	.1978718	.3007356
_cons	.9345319	.0341995	.001184	.9345088	.8668928	.9986375
lnalpha	.83284	.0307083	.000993	.8327162	.7749053	.8946471
alpha	2.300926	.0707452	.002287	2.299556	2.170387	2.446472

We compare the above Bayesian model with a standard MLE model of the data.

```
. nbreg docvis outwork sage, nolog
Negative binomial regression                      Number of obs    =      3,874
                                                  LR chi2(2)       =     184.89
Dispersion      = mean                            Prob > chi2      =     0.0000
Log likelihood = -8332.7623                       Pseudo R2        =     0.0110
```

docvis	Coef.	Std. Err.	z	P>|z|	[95% Conf. Interval]	
outwork	.4146624	.0553158	7.50	0.000	.3062455	.5230793
sage	.2489207	.0266265	9.35	0.000	.1967337	.3011077
_cons	.9347726	.0334834	27.92	0.000	.8691464	1.000399
/lnalpha	.8316161	.0308983			.7710565	.8921756
alpha	2.297028	.0709743			2.162049	2.440433

```
LR test of alpha=0: chibar2(01) = 1.5e+04           Prob >= chibar2 = 0.000
```

Offsets (or exposures) with the negative binomial cannot be set up in the likelihood as we showed for the Bayesian Poisson model. Instead the MHextravars global macro must be used to pass in the name of the variable containing the offset values. Both of these revisions are reflected in the nbll evaluator.

20.9.2.5 Zero-inflated negative binomial logit using llevaluator()

In this section, we will show how to write an evaluator program to fit a Bayesian zero-inflated negative binomial model for use with the llevaluator() suboption. Recall that zero-inflated negative binomial logit models are used to model data that contain both overdispersion and excess zeros. In Stata 15, this model can be fit with the bayes prefix, as shown in 20.7.1, but in Stata 14, the likelihood must be specified by the user as shown here or in section 20.9.1.2.

Listing 20.5. `bzinbll` evaluator for log-link zero-inflated negative binomial

```
program define bzinbll
        version 14.1
        args lnf xbnb xbl lnalpha
        tempvar lnfj
        tempname a
        scalar 'a' = exp('lnalpha')
        generate 'lnfj' =  cond($MH_y1, log(1-invlogit('xbl')) ///
                + lngamma(1/'a' + $MH_y1 )                      ///
                - lngamma($MH_y1 +1) - lngamma(1/'a')           ///
                + (1/'a') * log(1/(1+'a'*exp('xbnb')))          ///
                + $MH_y1 * log(1-(1/(1+'a'*exp('xbnb')))),      ///
                    log(invlogit('xbl') + (1-invlogit('xbl'))   ///
                / ((1+'a'*exp('xbnb'))^(1/'a'))))              ///
                if $MH_touse
        summarize 'lnfj', meanonly
        if r(N) < $MH_n {
                scalar 'lnf' = .
                exit
        }
        scalar 'lnf' = r(sum)
end
```

As in previous examples for this model, we will use the 1984 German health data, `rwm1984.dta`. The response variable, `docvis`, is the number of visits made by a patient to a physician during 1984. The binary variable, `outwork`, indicates if the patient was unemployed in 1984. Additional independent variables are `female` and `married`, which are both binary, and `sage`, which is the standardized `age` variable.

We also create a second copy of the response variable called `docvis2`. This variable is not used but is needed to satisfy the expectations of the equation arguments to the `bayesmh` command. As with all zero-inflated models, the likelihood of a zero outcome is modeled in the binary component.

The `outwork` variable is common to both components, and `female` and `married` are adjusters of the count component, whereas `sage` is a predictor of the binary `docvis2`. Below, we illustrate calling the evaluator for a specific dataset. Note that we changed the default burnin and sampling iterations and blocked the parameters in both components to be sampled at a unit. We also provided a random-number seed.

```
. use http://www.stata-press.com/data/hh4/rwm1984, clear
(German health data for 1984; Hardin & Hilbe, GLM and Extensions, 4th ed)

. egen sage = std(age)

. gen docvis2 = docvis

. bayesmh (docvis outwork female married) (inflate:docvis2 outwork sage),
> llevaluator(bzinbll, param({lnalpha})) prior({docvis:}, flat)
> prior({inflate:}, flat)  prior({lnalpha},normal(0,1000))
> block({docvis:}) block({inflate:}) burnin(5000) mcmcsize(8000)
> rseed(1944) title(Zero-inflated negative binomial)

Burn-in ...
Simulation ...

Model summary
```

```
Likelihood:
  docvis docvis2 ~ bzinbll(xb_docvis,xb_inflate,{lnalpha})

Priors:
  {docvis:outwork female married _cons} ~ 1 (flat)                          (1)
        {inflate:outwork sage _cons} ~ 1 (flat)                            (2)
                        {lnalpha} ~ normal(0,1000)
```

```
(1) Parameters are elements of the linear form xb_docvis.
(2) Parameters are elements of the linear form xb_inflate.
```

Zero-inflated negative binomial	MCMC iterations	=	13,000
Random-walk Metropolis-Hastings sampling	Burn-in	=	5,000
	MCMC sample size	=	8,000
	Number of obs	=	3,874
	Acceptance rate	=	.2873
	Efficiency: min	=	.02119
	avg	=	.03687
Log marginal likelihood = -8349.242	max	=	.05329

| | | | | | Equal-tailed | |
	Mean	Std. Dev.	MCSE	Median	[95% Cred. Interval]	
docvis						
outwork	.3586153	.0586133	.003062	.3572574	.2463498	.4760687
female	.2406522	.0524622	.002541	.2396473	.1417235	.3433338
married	-.1018777	.0618536	.003253	-.1038167	-.2203712	.0220215
_cons	1.09476	.0718187	.003994	1.095147	.951983	1.240938
inflate						
outwork	-.5096863	.2557408	.015314	-.5124026	-1.041994	-.0388206
sage	-.7061611	.1305698	.008195	-.6955899	-.977448	-.4699048
_cons	-1.859052	.2510148	.018731	-1.838916	-2.426483	-1.413603
lnalpha	.5691297	.0574863	.004415	.5689939	.4546503	.6822358

To display the estimate of the negative binomial dispersion parameter, we type the following commands:

```
. * Display alpha dispersion parameter
. matrix b = e(mean)
. display exp(b[1,8])
1.7667287
```

We can assess whether the code is likely correct if the Bayesian results are close to the maximum likelihood results on the same data. Of course, as mentioned before, this only the case when all Bayesian model parameters have diffuse priors. The `zinb` command can be used to fit maximum-likelihood zero-inflated negative binomial models.

```
. zinb docvis outwork female married, inflate(outwork sage) nolog
```

Zero-inflated negative binomial regression				Number of obs	=	3,874

				Nonzero obs	=	2,263
				Zero obs	=	1,611

Inflation model = logit				LR chi2(3)	=	80.34
Log likelihood = -8330.906				Prob > chi2	=	0.0000

docvis	Coef.	Std. Err.	z	P>|z|	[95% Conf.	Interval]
docvis						
outwork	.3503719	.0598199	5.86	0.000	.2331271	.4676166
female	.2404267	.0556911	4.32	0.000	.131274	.3495793
married	-.0954744	.0629868	-1.52	0.130	-.2189263	.0279775
_cons	1.101278	.0723939	15.21	0.000	.9593886	1.243167
inflate						
outwork	-.4910056	.2342102	-2.10	0.036	-.9500491	-.0319621
sage	-.6741582	.1129135	-5.97	0.000	-.8954645	-.4528518
_cons	-1.751627	.2242313	-7.81	0.000	-2.191112	-1.312142
/lnalpha	.5476475	.0573996	9.54	0.000	.4351464	.6601487
alpha	1.72918	.0992543			1.545189	1.93508

The Bayesian model means are close in value to the maximum-likelihood coefficient estimates, indicating good agreement between the models.

20.9.2.6 Bayesian gamma regression using llevaluator()

The canonical or natural form of the gamma log likelihood can be given as

$$\mathcal{L}(\beta, \phi; y) = \sum_{i=1}^{n} \left\{ \frac{1}{\phi} \ln\left(\frac{y_i}{\phi\mu_i}\right) - \left(\frac{y_i}{\phi\mu_i}\right) - \ln(y_i) - \ln\Gamma\left(\frac{1}{\phi}\right) \right\} \qquad (20.25)$$

Statisticians also use the log-linked form of the gamma model to deal with positive only continuous variables as an alternative to lognormal regression or log-transformed normal regression. However, we focus on the canonical form using the inverse link function, $1/\mu$. Notice that the evaluator below is similar to the others developed in this chapter. A difference, though, is that we bring in the scale parameter ϕ as an argument (like for the negative binomial). Otherwise, except for the differences in defining the log likelihoods, the coding is the same.

Note that the `bayesmh` command should be used with the gamma models. Use of the `bayes` prefix with `meglm` specifying the gamma family produces similar posterior parameters to the `glm` command coefficients, but the estimated scale (which you must exponentiate after being displayed) is similar to the deviance dispersion, which is a point estimate of the gamma scale. However, the code below gives a preferred scale posterior parameter value and is closer to a two-parameter maximum-likelihood estimate of the scale.

Listing 20.6. `bgammall` evaluator for gamma regression

```
program define bgammall
    version 15
    args lnf xb phi
    quietly {
        tempvar lnfj
        generate double 'lnfj' = (1/'phi')                          ///
            * log($MH_y/('phi'/'xb')) - ($MH_y/('phi'/'xb')) ///
            - log($MH_y) - lngamma(1/'phi') if $MH_touse
        summarize 'lnfj' if $MH_touse, meanonly
        if r(N) < $MH_n {
            scalar 'lnf' = .
            exit
        }
        scalar 'lnf' = r(sum)
    }
end
```

We again use `rwm1984.dta` with `docvis` as the response. Recall that the gamma and inverse Gaussian distributions do not support values less than or equal to 0. Therefore, for these examples, we drop all observations for which `docvis` is less than or equal to 0.

Bayesian gamma regression via the `llevaluator()` option produces the following output:

```
. use http://www.stata-press.com/data/hh4/rwm1984, clear
(German health data for 1984; Hardin & Hilbe, GLM and Extensions, 4th ed)

. egen sage = std(age)

. bayesmh docvis outwork sage if docvis>0,
> llevaluator(bgammall, parameters({phi})) prior({docvis:} {phi}, flat)
> rseed(123) block({phi}) init({docvis:_cons} 1 {phi} 1)

Burn-in ...
Simulation ...

Model summary
--------------------------------------------------------------------------
Likelihood:
  docvis ~ bgammall(xb_docvis,{phi})
Priors:
  {docvis:outwork sage _cons} ~ 1 (flat)                                (1)
                    {phi} ~ 1 (flat)
--------------------------------------------------------------------------
(1) Parameters are elements of the linear form xb_docvis.
```

```
Bayesian regression                                 MCMC iterations   =      12,500
Random-walk Metropolis-Hastings sampling            Burn-in           =       2,500
                                                    MCMC sample size  =      10,000
                                                    Number of obs     =       2,263
                                                    Acceptance rate   =       .2875
                                                    Efficiency:  min  =        .061
                                                                 avg  =       .1119
Log marginal likelihood =  -6028.105                         max  =       .2337
```

					Equal-tailed	
	Mean	Std. Dev.	MCSE	Median	[95% Cred.	Interval]
docvis						
outwork	-.0371086	.0074373	.000255	-.0373391	-.0522331	-.0224599
sage	-.0231736	.0035864	.000145	-.0231479	-.0302754	-.0159192
_cons	.2102746	.005152	.000198	.2103342	.200547	.2209667
phi	.7973221	.0213522	.000442	.7968925	.7567865	.8400569

We compare the Bayesian model above with a GLM gamma family:

```
. glm docvis outwork sage if docvis>0, family(gamma) nolog
Generalized linear models                           No. of obs        =       2,263
Optimization     : ML                               Residual df       =       2,260
                                                    Scale parameter   =    1.786454
Deviance         =     2028.15267                   (1/df) Deviance   =    .8974127
Pearson          =     4037.385167                  (1/df) Pearson    =    1.786454
Variance function: V(u) = u^2                       [Gamma]
Link function    : g(u) = 1/u                       [Reciprocal]
                                                    AIC               =    5.345961
Log likelihood   = -6045.955045                     BIC               =    -15429.1
```

		OIM				
docvis	Coef.	Std. Err.	z	P>|z|	[95% Conf.	Interval]
outwork	-.0372324	.0109244	-3.41	0.001	-.0586437	-.015821
sage	-.0232894	.0054283	-4.29	0.000	-.0339287	-.01265
_cons	.2103181	.0077956	26.98	0.000	.195039	.2255973

Like the other models we have discussed, the mean, standard deviation, and credible intervals of the posterior distributions in a Bayesian model are nearly identical to the coefficients, standard errors, and confidence intervals of the corresponding GLM.

20.9.2.7 Bayesian inverse Gaussian regression using llevaluator()

The inverse Gaussian distribution and inverse Gaussian regression are seldom used by analysts for research, although it has been used with success in transportation. However, simulation studies Hilbe (2016) have demonstrated that the distribution is best for modeling data in which the greatest density of values are in the very low numbers.

A Bayesian inverse Gaussian model can be fit using `bayes` with the `glm` command specifying the inverse Gaussian family. However, this method will not provide an estimate of the deviance dispersion statistic.

The log likelihood of the inverse Gaussian distribution, with the canonical inverse quadratic link $(1/\mu^2)$, is

$$\mathcal{L}(\mu, \sigma^2; y) = -1/2\{(y_i - \mu)^2/(\sigma^2\mu^2 y_i)\} - 1/2\ln(2\pi y_i^3\sigma^2) \qquad (20.26)$$

The inverse Gaussian evaluator equates $\eta_i = x_i\beta = 1/\mu^2$ so that we can substitute $1/\sqrt{\eta_i} = \mu$ and note that

Listing 20.7. `bivgll` evaluator for inverse Gaussian regression

```
program define bivgll
    version 15
    args lnf xb sig2
    quietly {
        tempvar lnfj
        generate double 'lnfj' = -0.5*($MH_y-1/sqrt('xb'))^2 / ///
            ('sig2'/'xb'*$MH_y) ///
            -0.5*log(_pi*$MH_y^3*'sig2') if $MH_touse
        summarize 'lnfj' if $MH_touse, meanonly
        if r(N) < $MH_n {
            scalar 'lnf' = .
            exit
        }
        scalar 'lnf' = r(sum)
    }
end
```

On the final line of the command below, we use the `init` option to supply starting values for model parameters. By default, the regression coefficients are initialized with 0, but then 'xb' would be a zero vector and the log likelihood would be missing because of the expression 'mu'=1/sqrt('xb'). The default value {sig2} = 0 is also invalid. That is why we initialize {docvis:_cons} with 1 (thus 'xb' becomes 1) and {sig2} with 1 using the `initial()` option. These choices ensure a valid initial state.

Bayesian inverse-Gaussian regression with support via `llevaluator()` is given by

```
. use http://www.stata-press.com/data/hh4/rwm1984, clear
(German health data for 1984; Hardin & Hilbe, GLM and Extensions, 4th ed)

. egen sage = std(age)

. bayesmh docvis outwork sage if docvis>0,
> llevaluator(bivgll, parameters({sig2}))
> prior({docvis:} {sig2}, flat) block({sig2})
> initial({docvis:_cons} 1 {sig2} 1) rseed(87324)

Burn-in ...
Simulation ...

Model summary
```

```
Likelihood:
  docvis ~ bivgll(xb_docvis,{sig2})
Priors:
  {docvis:outwork sage _cons} ~ 1 (flat)                                    (1)
                   {sig2} ~ 1 (flat)
```

```
(1) Parameters are elements of the linear form xb_docvis.
Bayesian regression                             MCMC iterations   =     12,500
Random-walk Metropolis-Hastings sampling        Burn-in           =      2,500
                                                MCMC sample size  =     10,000
                                                Number of obs     =      2,263
                                                Acceptance rate   =       .284
                                                Efficiency:  min  =     .06507
                                                             avg  =     .09758
Log marginal likelihood = -5695.8121                         max  =      .1824
```

					Equal-tailed	
	Mean	Std. Dev.	MCSE	Median	[95% Cred. Interval]	
docvis						
outwork	-.0130461	.003347	.00013	-.0129838	-.0200555	-.0069008
sage	-.0084625	.0016643	.00006	-.0084922	-.0117749	-.0051982
_cons	.0444448	.0024786	.000097	.0443713	.0397042	.0495807
sig2	.2257628	.0067558	.000158	.2257437	.2126759	.2390027

The GLM provides similar parameter estimates. Note also that the parameter σ^2 is point estimated by the GLM deviance dispersion statistic. The deviance dispersion below is 0.225, which is nearly identical to σ^2.

```
. glm docvis outwork sage if docvis>0, family(igaussian) nolog
Generalized linear models                          No. of obs      =      2,263
Optimization     : ML                              Residual df     =      2,260
                                                   Scale parameter =    .3417566
Deviance         =   508.8288567                   (1/df) Deviance =    .2251455
Pearson          =   772.3698525                   (1/df) Pearson  =    .3417566

Variance function: V(u) = u^3                      [Inverse Gaussian]
Link function     : g(u) = 1/(u^2)                 [Power(-2)]

                                                   AIC             =    5.736006
Log likelihood   = -6487.290395                    BIC             =   -16948.42
```

docvis	Coef.	OIM Std. Err.	z	P>|z|	[95% Conf. Interval]	
outwork	-.0133198	.0041779	-3.19	0.001	-.0215084	-.0051313
sage	-.008429	.0020533	-4.11	0.000	-.0124535	-.0044046
_cons	.0444786	.0031198	14.26	0.000	.0383639	.0505932

20.9.2.8 Bayesian zero-truncated Poisson using llevaluator()

We discussed this model earlier in the book; see section 14.2. Briefly, if there is no
possibility of a count variable having any zero counts, we recognize that the sum of
probabilities for all nonzero outcomes in a Poisson distribution is equal to one minus
the probability of a zero outcome. Thus, to create a valid probability function for
outcomes strictly greater than zero, we can divide Poisson probabilities by $1 - \exp(-\mu)$;
one minus the probability of a zero outcome.

Starting in Stata 15 however, bayes has tpoisson and tnbreg as optional models,
and the command can be used directly without the need to write our own log likelihood.
Also new in Stata 15, the tpoisson and tnbreg commands offer left-, right-, and
interval-truncation across the full range of observations in the data. Use of the zero-
truncated models is easy because you are not required to specify a lower boundary. The
default is zero-truncation.

To fit this model using bayesmh, we will need to supply a log-likelihood evaluator to
the llevaluator() function to fit the model.

$$\mathcal{L}(\beta; y) = \sum_{i=1}^{n} \left(y_i \ln(x_i\beta) - \exp(x_i\beta) - \ln \Gamma(y_i + 1) - \ln[1 - \exp\{-\exp(x_i\beta)\}] \right) \quad (20.27)$$

Listing 20.8. `bpoi0ll` evaluator for zero-truncated Poisson regression

```
program  bpoi0ll
    version 15
    args lnf xb
    quietly {
        tempvar lnfj
        generate `lnfj' =  -exp(`xb') +  $MH_y*`xb' - ///
            lngamma($MH_y + 1) - log(1-exp(-exp(`xb'))) if $MH_touse
        summarize `lnfj' if $MH_touse, meanonly
        if r(N) < $MH_n {
            scalar `lnf' = .
            exit
        }
        scalar `lnf' = r(sum)
    }
end
```

We must make certain that no zero values are given to the model.

```
. use http://www.stata-press.com/data/hh4/rwm1984, clear
(German health data for 1984; Hardin & Hilbe, GLM and Extensions, 4th ed)

. quietly summarize age, meanonly

. generate cage = age - r(mean)

. bayesmh docvis cage if docvis>0, llevaluator(bpoi0ll)
> prior({docvis: }, flat) rseed(28734)

Burn-in ...
Simulation ...

Model summary
```

```
Likelihood:
  docvis ~ bpoi0ll(xb_docvis)
Prior:
  {docvis:cage _cons} ~ 1 (flat)                                      (1)
```

```
(1) Parameters are elements of the linear form xb_docvis.
```

Bayesian regression				MCMC iterations	=	12,500
Random-walk Metropolis-Hastings sampling				Burn-in	=	2,500
				MCMC sample size =		10,000
				Number of obs	=	2,263
				Acceptance rate	=	.2621
				Efficiency: min =		.1144
				avg =		.1266
Log marginal likelihood = -9689.0231				max =		.1387

docvis	Mean	Std. Dev.	MCSE	Median	Equal-tailed [95% Cred. Interval]	
cage	.0145077	.0008095	.000024	.0145173	.0129474	.0161428
_cons	1.646316	.0098261	.000264	1.646088	1.627578	1.665429

Note: Adaptation tolerance is not met.

Comparing the results with Stata's `tpoisson` command, the results are nearly identical.

```
. tpoisson docvis cage if docvis>0, nolog
Truncated Poisson regression                      Number of obs    =      2,263
Limits:  lower =            0                      LR chi2(1)       =     320.44
         upper =         +inf                      Prob > chi2      =     0.0000
Log likelihood = -9679.0673                       Pseudo R2        =     0.0163
```

docvis	Coef.	Std. Err.	z	P>|z|	[95% Conf. Interval]
cage	.014493	.0008188	17.70	0.000	.0128882	.0160979
_cons	1.646516	.0096356	170.88	0.000	1.62763	1.665401

On your own, compare these results with the output of the following command:

```
bayes, rseed(28734): tpoisson docvis cage if docvis>0
```

20.9.2.9   Bayesian bivariate Poisson using llevaluator()

In chapter 19, we discussed a particular parameterization of a bivariate Poisson regression model. Here, we investigate how to specify and fit the same model in a Bayesian context.

A community-contributed command can be used to run a maximum-likelihood bivariate Poisson model, but it cannot be used with `bayes`. This discussion will therefore relate solely to the `bayesmh` command.

The log-likelihood evaluator program is a copy of the program specified in section 19.6. The differences are that the program has to calculate the log likelihood rather than filling in a variable with the observation-level log-likelihood values.

```
. capture program drop bfambpll
. program bfambpll
  1.         version 15
  2.         args lnf xb1 xb2 lambda
  3.         quietly {
  4.                 local y1 "$MH_y1"
  5.                 local y2 "$MH_y2"
  6.
  .                 tempname dd
  7.                 tempvar ll mu1 mu2
  8.                 generate double `mu1´ = exp(`xb1´) if $MH_touse
  9.                 generate double `mu2´ = exp(`xb2´) if $MH_touse
 10.                 scalar `dd´       = 1 - exp(-1)
 11.                 generate double `ll´   = `y1´*`xb1´+`y2´*`xb2´-`mu1´
  >                     -`mu2´- lngamma(`y1´+1) - lngamma(`y2´+1)
  >                     + log(1+`lambda´*(exp(-`y1´)-exp(-`dd´*`mu1´))
  >                     * (exp(-`y2´)-exp(-`dd´*`mu2´))) if $MH_touse
 12.
```

```
    .                       summarize `ll´ if $MH_touse, meanonly
  13.                           if r(N) < $MH_n {
  14.                                   scalar `lnf´ = .
  15.                                   exit
  16.                           }
  17.                           scalar `lnf´ = r(sum)
  18.                   }
  19. end
```

The results using noninformative priors are close to the previous results in section 19.6.

```
. use http://www.stata-press.com/data/hh4/corrcounts, clear

. bayesmh (ycp1 x11 x12) (ycp2 x21 x22),
> llevaluator(bfambpll, parameters({lambda})) prior({ycp1:} {ycp2:}, flat)
> prior({lambda}, normal(0,100)) thinning(10) rseed(3423)
note: discarding every 9 sample observations; using observations 1,11,21,...

Burn-in ...
Simulation ...

Model summary
─────────────────────────────────────────────────────────────────────────
Likelihood:
  ycp1 ycp2 ~ bfambpll(xb_ycp1,xb_ycp2,{lambda})
Priors:
  {ycp1:x11 x12 _cons} ~ 1 (flat)                                       (1)
  {ycp2:x21 x22 _cons} ~ 1 (flat)                                       (2)
            {lambda} ~ normal(0,100)
─────────────────────────────────────────────────────────────────────────
(1) Parameters are elements of the linear form xb_ycp1.
(2) Parameters are elements of the linear form xb_ycp2.
```

```
Bayesian regression                       MCMC iterations    =    102,491
Random-walk Metropolis-Hastings sampling  Burn-in            =      2,500
                                          MCMC sample size   =     10,000
                                          Number of obs      =      1,000
                                          Acceptance rate    =      .2126
                                          Efficiency:  min   =    .001104
                                                       avg   =      .0773
Log marginal likelihood = -5356.7042                   max   =      .2287
```

					Equal-tailed	
	Mean	Std. Dev.	MCSE	Median	[95% Cred.	Interval]
ycp1						
x11	-.4569859	.0247791	.002401	-.4568606	-.5062166	-.4085484
x12	.4748329	.0252318	.002341	.4753806	.4245793	.5235598
_cons	1.856029	.0216019	.001636	1.856293	1.814003	1.89796
ycp2						
x21	-.4233668	.0247649	.000898	-.4234495	-.4717716	-.374532
x22	.5309597	.0251011	.000525	.5309727	.4813177	.5797526
_cons	1.808225	.0220941	.0005	1.808238	1.764184	1.850831
lambda	29.16357	11.13604	3.35187	29.73252	4.466429	43.94822

Note: There is a high autocorrelation after 500 lags.

In a more realistic example, we investigate a bivariate Poisson model for doctor and hospital visits from the 1984 German health study.

```
. use http://www.stata-press.com/data/hh4/rwm1984, clear
(German health data for 1984; Hardin & Hilbe, GLM and Extensions, 4th ed)
. egen sage = std(age)
. bayesmh (docvis outwork sage) (hospvis outwork sage),
> llevaluator(bfambpll, parameters({lambda}))
> prior({docvis:} {hospvis:}, flat)
> prior({lambda}, normal(0,1000)) mcmcsize(50000) burnin(30000)
> rseed(23498)

Burn-in ...
Simulation ...
Model summary
```

Likelihood:
  docvis hospvis ~ bfambpll(xb_docvis,xb_hospvis,{lambda})
Priors:
  {docvis:outwork sage _cons} ~ 1 (flat)   (1)
  {hospvis:outwork sage _cons} ~ 1 (flat)   (2)
                  {lambda} ~ normal(0,1000)

(1) Parameters are elements of the linear form xb_docvis.
(2) Parameters are elements of the linear form xb_hospvis.

Bayesian regression   MCMC iterations   =   80,000
Random-walk Metropolis-Hastings sampling   Burn-in   =   30,000
                                        MCMC sample size   =   50,000
                                        Number of obs   =   3,874
                                        Acceptance rate   =   .1314
                                        Efficiency: min   =   .001141
                                                     avg   =   .01063
Log marginal likelihood = -17279.678   max   =   .0505

					Equal-tailed	
	Mean	Std. Dev.	MCSE	Median	[95% Cred. Interval]
docvis						
outwork	.4048455	.018837	.001107	.4053719	.367406	.4403119
sage	.2466942	.009456	.00048	.2468324	.2279034	.2652483
_cons	.9414583	.0130264	.000872	.9413856	.9158086	.9672146
hospvis						
outwork	.2998573	.0915961	.012126	.2921151	.1517537	.5239606
sage	.0995143	.0407582	.003805	.0989628	.0181501	.1806651
_cons	-1.996836	.0598409	.005424	-1.998479	-2.108879	-1.876464
lambda	1.19526	.0349578	.000696	1.200429	1.111462	1.248381

Note: There is a high autocorrelation after 500 lags.

The relevant statistics for the model are given by

```
. bayesstats ess
Efficiency summaries     MCMC sample size =     50,000
```

	ESS	Corr. time	Efficiency
docvis			
outwork	289.45	172.74	0.0058
sage	388.64	128.65	0.0078
_cons	223.33	223.89	0.0045
hospvis			
outwork	57.06	876.33	0.0011
sage	114.76	435.69	0.0023
_cons	121.70	410.86	0.0024
lambda	2524.75	19.80	0.0505

To compare these results with the MLE model, we generate the bivariate Poisson model using the code first introduced in section 19.6.

```
. use http://www.stata-press.com/data/hh4/rwm1984, clear
(German health data for 1984; Hardin & Hilbe, GLM and Extensions, 4th ed)
. egen sage = std(age)
. quietly poisson docvis outwork sage
. matrix a1 = e(b)
. quietly poisson hospvis outwork sage
. matrix a2 = e(b)
. matrix tt = (0)
. matrix aa = a1, a2, tt
. ml model lf0 famoyepbll (docvis:docvis=outwork sage)
> (hospvis:hospvis=outwork sage) (lambda:)
. ml init aa, copy
```

```
. ml maximize, search(off) title(Famoye Biv Poisson regression) nolog
Famoye Biv Poisson regression                    Number of obs    =      3,874
                                                 Wald chi2(2)     =    1607.73
Log likelihood = -17256.021                      Prob > chi2      =     0.0000
```

	Coef.	Std. Err.	z	P>|z|	[95% Conf.	Interval]
docvis						
outwork	.407155	.0187586	21.70	0.000	.3703888	.4439213
sage	.2464215	.0093789	26.27	0.000	.2280393	.2648038
_cons	.9405373	.0127247	73.91	0.000	.9155974	.9654773
hospvis						
outwork	.3101129	.0922154	3.36	0.001	.1293739	.4908518
sage	.0903086	.0452243	2.00	0.046	.0016706	.1789466
_cons	-1.997887	.0620891	-32.18	0.000	-2.119579	-1.876194
lambda						
_cons	1.21546	.031114	39.06	0.000	1.154478	1.276442

In both cases, the covariates are significantly associated with the incidence rates of the both doctor visits and hospital visits. There is also evidence that the two outcomes are correlated.

It should be noted, however, that the parameters of the Bayesian model have extremely low efficiency values. As discussed earlier in the chapter, this informs the user that the Bayesian model may not be reliable or suited for the data being modeled. The low average model efficiency may indicate that the data may be overdispersed; in which case, a Bayesian bivariate negative binomial model may be preferred for modeling these data. In any case, this Bayesian bivariate Poisson model with diffuse priors matches up with the maximum likelihood model as we would hope. The task for a researcher desiring to use a Bayesian bivariate count model on these data is to try various sampling strategies on the bivariate Poisson with the aim of increasing efficiency. Overdispersion should also be examined, and a bivariate negative binomial model should be tried as an alternative.

Part VII

Stata Software

21 Programs for Stata

The knowledge presented in the text would not be useful without the practical application of available supporting software. Readers are not required to use the following software, but other software will not address all the issues described here.

This chapter describes commands for the Stata software program. The entire chapter is devoted to Stata software and makes many references to Stata documentation (manuals, help files, the website, and NetCourses). We follow the notational and typographic conventions of the Stata reference manuals; [U] refers to the *Stata User's Guide* and [R] refers to the *Stata Base Reference Manual*.

In addition to this documentation on the `glm` command in Stata, we also used other user-written programs. Interested readers may locate, download, and install such programs by using Stata's `search` facilities.

21.1   The glm command

21.1.1   Syntax

glm *depvar* [*indepvars*] [*if*] [*in*] [*weight*] [, *options*]

options	Description
Model	
<u>fam</u>ily(*familyname*)	distribution of *depvar*; default is family(gaussian)
<u>link</u>(*linkname*)	link function; default is canonical link for family() specified
Model 2	
<u>noconst</u>ant	suppress constant term
<u>expo</u>sure(*varname*)	include ln(*varname*) in model with coefficient constrained to 1
<u>off</u>set(*varname*)	include *varname* in model with coefficient constrained to 1
<u>constra</u>ints(*constraints*)	apply specified linear constraints
<u>col</u>linear	keep collinear variables
asis	retain perfect predictor variables
mu(*varname*)	use *varname* as the initial estimate for the mean of *depvar*
<u>init</u>(*varname*)	synonym for mu(*varname*)
SE/Robust	
vce(*vcetype*)	*vcetype* may be oim, <u>r</u>obust, <u>cl</u>uster *clustvar*, eim, opg, <u>boot</u>strap, <u>jack</u>knife, hac *kernel*, jackknife1, or <u>unbi</u>ased
<u>vf</u>actor(*#*)	multiply variance matrix by scalar *#*
disp(*#*)	quasilikelihood multiplier
<u>sca</u>le(x2 | dev | *#*)	set the scale parameter
Reporting	
<u>level</u>(*#*)	set confidence level; default is level(95)
<u>ef</u>orm	report exponentiated coefficients
nocnsreport	do not display constraints
display_options	control columns and column formats, row spacing, line width, display of omitted variables and base and empty cells, and factor-variable labeling
Maximization	
ml	use maximum likelihood optimization; the default
irls	use iterated, reweighted least-squares optimization of the deviance
maximize_options	control the maximization process; seldom used
fisher(*#*)	use the Fisher scoring Hessian or expected information matrix (EIM)
search	search for good starting values

noheader	suppress header table from above coefficient table
notable	suppress coefficient table
nodisplay	suppress the output; iteration log is still displayed
coeflegend	display legend instead of statistics

familyname	Description
gaussian	Gaussian (normal)
igaussian	inverse Gaussian
binomial $\left[\,varname_N \mid \#_N\,\right]$	Bernoulli/binomial
poisson	Poisson
nbinomial $\left[\,\#_k \mid \texttt{ml}\,\right]$	negative binomial
gamma	gamma

linkname	Description
identity	identity
log	log
logit	logit
probit	probit
cloglog	complementary log-log
power $\#$	power
opower $\#$	odds power
nbinomial	negative binomial
loglog	log-log
logc	log-complement

indepvars may contain factor variables; see [U] **11.4.3 Factor variables**.

depvar and *indepvars* may contain time-series operators; see [U] **11.4.4 Time-series varlists**.

bayes, bootstrap, by, fmm, fp, jackknife, mfp, mi estimate, nestreg, rolling, statsby, stepwise, and svy are allowed; see [U] **11.1.10 Prefix commands**.

vce(bootstrap), vce(jackknife), and vce(jackknife1) are not allowed with the mi estimate prefix.

Weights are not allowed with the bootstrap prefix.

aweights are not allowed with the jackknife prefix.

vce(), vfactor(), disp(), scale(), irls, fisher(), noheader, notable, nodisplay, and weights are not allowed with the svy prefix.

fweights, aweights, iweights, and pweights are allowed; see [U] **11.1.6 weight**.

noheader, notable, nodisplay, and coeflegend do not appear in the dialog box.

See [U] **20 Estimation and postestimation commands** for additional capabilities of estimation commands.

21.1.2   Description

glm fits generalized linear models. It can fit models by using either IRLS (maximum quasilikelihood) or Newton–Raphson (maximum likelihood) optimization, which is the default.

See [U] **26 Overview of Stata estimation commands** for a description of all of Stata's estimation commands, several of which fit models that can also be fit using glm.

21.1.3   Options

⌐‾‾‾‾| Model |‾‾

family(*familyname*) specifies the distribution of *depvar*; family(gaussian) is the default.

link(*linkname*) specifies the link function; the default is the canonical link for the family() specified (except for family(nbinomial)).

⌐‾‾‾‾| Model 2 |‾‾‾

noconstant, exposure(*varname*), offset(*varname*), constraints(*constraints*), collinear; see [R] **estimation options**. constraints(*constraints*) and collinear are not allowed with irls.

asis forces retention of perfect predictor variables and their associated, perfectly predicted observations and may produce instabilities in maximization; see [R] **probit**. This option is only allowed with option family(binomial) with a denominator of 1.

mu(*varname*) specifies *varname* as the initial estimate for the mean of *depvar*. This option can be useful with models that experience convergence difficulties, such as family(binomial) models with power or odds-power links. init(*varname*) is a synonym.

⌐‾‾‾‾| SE/Robust |‾‾‾

vce(*vcetype*) specifies the type of standard error reported, which includes types that are derived from asymptotic theory, that are robust to some kinds of misspecification, that allow for intragroup correlation, and that use bootstrap or jackknife methods; see [R] *vce_option*.

In addition to the standard *vcetypes*, glm allows the following alternatives:

vce(eim) specifies that the EIM estimate of variance be used.

vce(jackknife1) specifies that the one-step jackknife estimate of variance be used.

vce(hac *kernel* [#]) specifies that a heteroskedasticity- and autocorrelation-consistent (HAC) variance estimate be used. HAC refers to the general form for combining weighted matrices to form the variance estimate. There are three kernels built into glm. *kernel* is a user-written program or one of

<u>nw</u>est |<u>gal</u>lant |<u>an</u>derson

specifies the number of lags. If # is not specified, $N - 2$ is assumed. If you wish to specify vce(hac ...), you must tsset your data before calling glm.

vce(unbiased) specifies that the unbiased sandwich estimate of variance be used.

vfactor(#) specifies a scalar by which to multiply the resulting variance matrix. This option allows you to match output with other packages, which may apply degrees of freedom or other small-sample corrections to estimates of variance.

disp(#) multiplies the variance of *depvar* by # and divides the deviance by #. The resulting distributions are members of the quasilikelihood family.

scale(x2|dev|#) overrides the default scale parameter. This option is allowed only with Hessian (information matrix) variance estimates.

By default, scale(1) is assumed for the discrete distributions (binomial, Poisson, and negative binomial), and scale(x2) is assumed for the continuous distributions (Gaussian, gamma, and inverse Gaussian).

scale(x2) specifies that the scale parameter be set to the Pearson chi-squared (or generalized chi-squared) statistic divided by the residual degrees of freedom, which is recommended by McCullagh and Nelder (1989) as a good general choice for continuous distributions.

scale(dev) sets the scale parameter to the deviance divided by the residual degrees of freedom. This option provides an alternative to scale(x2) for continuous distributions and overdispersed or underdispersed discrete distributions.

scale(#) sets the scale parameter to #. For example, using the scale(1) option in family(gamma) models results in exponential-errors regression. Additional use of link(log) rather than the default link(power -1) for family(gamma) essentially reproduces Stata's streg, dist(exp) nohr command (see [ST] **streg**) if all the observations are uncensored.

Table 21.1. Resulting standard errors

Options	Estimated variance matrix	Reference
nothing	EIM Hessian	Equation 3.31
irls	OIM Hessian	Equation 3.51[a]
vce(opg)	OPG	Section 3.6.2
vce(robust)	Sandwich	Section 3.6.3
irls vce(robust)	Sandwich	Section 3.6.3
vce(unbiased)	Unbiased sandwich	Section 3.6.5
irls vce(unbiased)	Unbiased sandwich	Section 3.6.5
vce(cluster *clustvar*)	Modified sandwich	Section 3.6.4
irls vce(cluster *clustvar*)	Modified sandwich	Section 3.6.4
vce(cluster *clustvar*) vce(unbiased)	Mod. unbiased sandwich	Section 3.6.6
irls vce(cluster *clustvar*) vce(unbiased)	Mod. unbiased sandwich	Section 3.6.6
vce(hac nwest)	Weighted sandwich	Section 3.6.7
vce(jackknife)	Usual jackknife	Section 3.6.8.1
vce(jackknife1)	One-step jackknife	Section 3.6.8.2
vce(jackknife, cluster(*clustvar*))	Variable jackknife	Section 3.6.8.4
vce(bootstrap)	Usual bootstrap	Section 3.6.9.1
vce(bootstrap, cluster(*clustvar*))	Grouped bootstrap	Section 3.6.9.2

[a] If the canonical link is specified, then the EIM is the same as the OIM. Output from the program will still label the variance matrix as EIM.

[b] Since the program assumes that the sandwich estimate of variance is calculated using the EIM instead of the OIM (even when the two are equal), the output will be labeled "Semi-Robust" instead of "Robust".

___| Reporting |_____

level(*#*); see [R] **estimation options**.

eform displays the exponentiated coefficients and corresponding standard errors and confidence intervals. For family(binomial) link(logit) (that is, logistic regression), exponentiation results in odds ratios; for family(poisson) link(log) (that is, Poisson regression), exponentiated coefficients are rate ratios.

nocnsreport; see [R] **estimation options**.

display_options: noci, nopvalues, noomitted, vsquish, noemptycells, baselevels, allbaselevels, nofvlabel, fvwrap(*#*), fvwrapon(*style*), cformat(%*fmt*), pformat(%*fmt*), sformat(%*fmt*), and nolstretch; see [R] **estimation options**.

___

| Maximization |

ml requests that optimization be carried out using Stata's ml commands and is the default.

irls requests iterated, reweighted least-squares (IRLS) optimization of the deviance instead of Newton–Raphson optimization of the log likelihood. If the irls option is not specified, the optimization is carried out using Stata's ml commands, in which case all options of ml maximize are also available.

maximize_options: difficult, technique(*algorithm_spec*), iterate(#), [no]log, trace, gradient, showstep, hessian, showtolerance, tolerance(#), ltolerance(#), nrtolerance(#), nonrtolerance, from(*init_specs*); see [R] **maximize**. These options are seldom used.

> Setting the optimization type to technique(bhhh) resets the default *vcetype* to vce(opg).

fisher(#) specifies the number of Newton–Raphson steps that should use the Fisher scoring Hessian or EIM before switching to the observed information matrix (OIM). This option is useful only for Newton–Raphson optimization (and not when using irls).

search specifies that the command search for good starting values. This option is useful only for Newton–Raphson optimization (and not when using irls).

The following options are available with glm but are not shown in the dialog box:

noheader suppresses the header information from the output. The coefficient table is still displayed.

notable suppresses the table of coefficients from the output. The header information is still displayed.

nodisplay suppresses the output. The iteration log is still displayed.

coeflegend; see [R] **estimation options**.

21.2   The predict command after glm

21.2.1   Syntax

predict [*type*] *newvar* [*if*] [*in*] [, *statistic options*]

statistic	Description
Main	
<u>mu</u>	expected value of y; the default
xb	linear prediction $\eta = \mathbf{x}\widehat{\boldsymbol{\beta}}$
<u>eta</u>	synonym of xb
stdp	standard error of the linear prediction
<u>ans</u>combe	Anscombe (1953) residuals
<u>c</u>ooksd	Cook's distance
<u>d</u>eviance	deviance residuals
<u>h</u>at	diagonals of the "hat" matrix
<u>l</u>ikelihood	a weighted average of standardized deviance and standardized Pearson residuals
<u>p</u>earson	Pearson residuals
<u>r</u>esponse	differences between the observed and fitted outcomes
<u>sc</u>ore	first derivative of the log likelihood with respect to $\mathbf{x}_j\boldsymbol{\beta}$
<u>w</u>orking	working residuals

options	Description
Options	
<u>nooff</u>set	modify calculations to ignore offset variable
adjusted	adjust deviance residual to speed up convergence
<u>standardiz</u>ed	multiply residual by the factor $(1 - h)^{-1/2}$
<u>studentiz</u>ed	multiply residual by one over the square root of the estimated scale parameter
<u>modi</u>fied	modify denominator of residual to be a reasonable estimate of the variance of *depvar*

These statistics are available both in and out of sample; type predict ... if e(sample) ...
    if wanted only for the estimation sample.

mu, xb, stdp, and score are the only statistics allowed with svy estimation results.

21.2.2   Options

     ⌐ Main ⌐

mu, the default, specifies that predict calculate the expected value of y, equal to
    $g^{-1}(\mathbf{x}\widehat{\boldsymbol{\beta}})$ $[ng^{-1}(\mathbf{x}\widehat{\boldsymbol{\beta}})$ for the binomial family].

xb calculates the linear prediction $\eta = \mathbf{x}\widehat{\boldsymbol{\beta}}$.

eta is a synonym for xb.

stdp calculates the standard error of the linear prediction.

anscombe calculates the Anscombe (1953) residuals to produce residuals that closely
follow a normal distribution.

cooksd calculates Cook's distance, which measures the aggregate change in the estimated coefficients when each observation is left out of the estimation.

deviance calculates the deviance residuals. Deviance residuals are recommended by
McCullagh and Nelder (1989) and by others as having the best properties for examining the goodness of fit of a GLM. They are approximately normally distributed
if the model is correct. They may be plotted against the fitted values or against a
covariate to inspect the model's fit. Also see the **pearson** option below.

hat calculates the diagonals of the "hat" matrix as an analog to simple linear regression.

likelihood calculates a weighted average of standardized deviance and standardized
Pearson residuals.

pearson calculates the Pearson residuals. Pearson residuals often have markedly skewed
distributions for nonnormal family distributions. Also see the **deviance** option
above.

response calculates the differences between the observed and fitted outcomes.

score calculates the equation-level score, $\partial \ln L / \partial(\mathbf{x}_j \boldsymbol{\beta})$.

working calculates the working residuals, which are response residuals weighted according to the derivative of the link function.

___

| Options |

nooffset is relevant only if you specified **offset**(*varname*) for **glm**. It modifies the
calculations made by **predict** so that they ignore the offset variable; the linear
prediction is treated as $\mathbf{x}_j \mathbf{b}$ rather than as $\mathbf{x}_j \mathbf{b} + \text{offset}_j$.

adjusted adjusts the deviance residual to speed up the convergence to the limiting
normal distribution. The adjustment deals with adding to the deviance residual
a higher-order term that depends on the variance function family. This option is
allowed only when **deviance** is specified.

standardized requests that the residual be multiplied by the factor $(1 - h)^{-1/2}$, where
h is the diagonal of the hat matrix. This operation is done to account for the
correlation between *depvar* and its predicted value.

studentized requests that the residual be multiplied by one over the square root of the
estimated scale parameter.

`modified` requests that the denominator of the residual be modified to be a reasonable estimate of the variance of *depvar*. The base residual is multiplied by the factor $(k/w)^{-1/2}$, where k is either one or the user-specified dispersion parameter and w is the specified weight (or one if left unspecified).

Table 21.2. Statistics for predict

Statistic	Description	Reference
mu	inverse link; the default	$g^{-1}(\mathbf{X}\widehat{\boldsymbol{\beta}})$, see table A.2
xb	linear prediction	$\widehat{\eta} = \mathbf{X}\widehat{\beta}$
eta	linear prediction (synonym for xb)	$\widehat{\eta} = \mathbf{X}\widehat{\beta}$
stdp	standard error	standard error of $\widehat{\eta}$
anscombe	Anscombe residual	See section 4.4.5 and table A.10
cooksd	Cook's distance	See section 4.2.1
deviance	deviance residual	See section 4.4.6 and table A.11
hat	hat diagonal	See section 4.2
likelihood	likelihood residual	See section 4.4.8
pearson	Pearson residual	See section 4.4.3
response	response residual	See section 4.4.1
score	score residual	See section 4.4.9
working	working residual	See section 4.4.2

21.3   User-written programs

This section is for advanced users who need to extend the functionality of the included software. You should be comfortable with programming in Stata to at least the level taught in Stata NetCourse 151, Introduction to Stata Programming; see https://www.stata.com/netcourse/programming-intro-nc151/ for information.

The functionality of the `glm` program may be extended in key areas. Users may write their own programs for the variance function, for the link function, and for the Newey–West kernel weights.

The following sections present the format of the user-written programs. Once written, these programs can be used in the `glm` program by specifying the program name in the appropriate option.

21.3.1   Global macros available for user-written programs

User-written programs will need access to information about the data and to the model specification to perform the necessary computations. In the following table, we present a list of global macros along with the information that they carry. User-written programs may use these macros to access needed information.

21.3.2   User-written variance functions

Users may write their own programs for variance functions. To take advantage of all the functionality of the `glm` command and for the predictions available postestimation, the program for variance functions should be able to calculate various quantities as well as set macros for output.

The following listing gives the outline of the program:

Listing 21.1. Skeleton code for a user-written variance program

```
1  program  program-name
2      args todo eta mu return
3
4      if 'todo' == -1 {
5          Macros are set for the output
6
7          Where arguments are defined as
8          eta   = indicator variable for whether each obs is in-sample
9          mu    = <ignored>
10         return = <ignored>
11
12         global SGLM_vt  "title assigned here"
13         global SGLM_vf  "subtitle showing function assigned here"
14
15         error checking code for range of y
16         limit checks based on 'eta' sample definition
17
18         global SGLM_mu  "program to enforce boundary conditions on μ"
19
20         exit
21     }
22     if 'todo' == 0 {
23         η is defined
24
25         Where arguments are defined as
26         eta   = variable name to define
27         mu    = μ
28         return = <ignored>
29
30         generate double variable named 'eta'
31
32         exit
33     }
34     if 'todo' == 1 {
35         V(μ) is defined
36
37         Where arguments are defined as
38         eta   = η
39         mu    = μ
```

```
40            return = variable name to define
41
42            generate double variable named 'return'
43
44       exit
45  }
46  if 'todo' == 2 {
47       ∂V(μ)/∂μ is defined
48
49       Where arguments are defined as
50       eta   = η
51       mu    = μ
52       return = variable name to define
53
54       generate double variable named 'return'
55
56       exit
57  }
58  if 'todo' == 3 {
59       deviance is defined
60
61       Where arguments are defined as
62       eta   = η
63       mu    = μ
64       return = variable name to define
65
66       generate double variable named 'return'
67       exit
68  }
69  if 'todo' == 4 {
70       Anscombe residual is defined
71
72       Where arguments are defined as
73       eta   = η
74       mu    = μ
75       return = variable name to define
76
77       generate double variable named 'return'
78       or issue an error if no support for Anscombe residuals
79
80       exit
81  }
82  if 'todo' == 5 {
83       Ln-likelihood L is defined
84
85       Where arguments are defined as
86       eta   = η
87       mu    = μ
88       return = variable name to define
```

```
89
90              generate double variable named 'return'
91              or issue an error if there is no true likelihood
92
93          exit
94      }
95      if 'todo' == 6 {
96          ρ₃(θ)/6 for adjusted deviance residuals is defined
97
98              Where arguments are defined as
99              eta    = η
100             mu     = μ
101             return = variable name to define
102
103             generate double variable named 'return'
104             or issue an error if no support for adjusted deviance residuals
105
106         exit
107     }
108     noisily display as error "Unknown glm variance function"
109         error 198
110 end
```

21.3.3 User-written programs for link functions

Users may write their own programs for link functions. To take advantage of all the functionality of the `glm` command and for the predictions available postestimation, the program for link functions should be able to calculate various quantities and set macros for output.

The following listing gives the outline of the program:

Listing 21.2. Skeleton code for a user-written link program

```
1  program program-name
2      args todo eta mu return
3
4      if 'todo' == -1 {
5          Macros are set for the output
6
7              Where arguments are defined as
8              eta    = <ignored>
9              mu     = <ignored>
10             return = <ignored>
11
12         global SGLM_lt  "title assigned here"
13         global SGLM_lf  "subtitle showing function assigned here"
14
15         exit
```

```
16        }
17        if 'todo' == 0 {
18            η is defined from g(μ)
19
20            Where arguments are defined as
21            eta   = variable name to define
22            mu    = μ
23            return = <ignored>
24
25            generate double variable named 'eta'
26
27            exit
28        }
29        if 'todo' == 1 {
30            μ is defined from g⁻¹(η)
31
32            Where arguments are defined as
33            eta   = η
34            mu    = variable name to define
35            return = <ignored>
36
37            generate double variable named 'mu'
38
39            exit
40        }
41        if 'todo' == 2 {
42            ∂μ/∂η is defined
43
44            Where arguments are defined as
45            eta   = η
46            mu    = μ
47            return = variable name to define
48
49            generate double variable named 'return'
50
51            exit
52        }
53        if 'todo' == 3 {
54            ∂²μ/∂η² is defined
55
56            Where arguments are defined as
57            eta   = η
58            mu    = μ
59            return = variable name to define
60
61            generate double variable named 'return'
62
63            exit
64        }
```

```
65        noisily display as error "Unknown glm link function"
66        exit 198
67    end
```

21.3.4   User-written programs for Newey–West weights

Users may write their own kernel evaluators for the Newey–West variance estimate. The program to be developed is an r-class program that will be passed four arguments. The program will return information via one scalar and two macros.

The following listing gives the outline of the program:

Listing 21.3. Skeleton code for user-written Newey–West kernel weights program

```
1    program program-name, rclass
2        args G j
3            Where arguments are defined as
4            G   = maximum lag
5            j   = current lag
6
7        your code goes here
8
9        return scalar wt    = computed weight value assigned here
10       return local setype  "Newey-West"
11       return local sewtype "name of kernel assigned here"
12   end
```

After defining the program, you can use the vce(hac nwest $\left[\,\#\,\right]$) option to specify that program for calculating the weights.

As an example, recall from section 3.6.7 the definition of the Tukey–Hanning kernel. Using the definitions of the maximum lag G and weights given in table A.8, we may develop the following do-file for Stata:

Listing 21.4. Example code for user-written Tukey–Hanning weights kernel

```
1    capture program drop thanning
2    program thanning, rclass
3        args G j
4
5        tempname z
6        scalar `z' = `j'/(`G'+1.0)
7
8        return scalar wt = (1.0 + cos(_pi*`z'))/2
9        return local setype  "Newey-West"
10       return local sewtype "Tukey-Hanning"
11   end
```

We can use this code in the `glm` command with, for example, the option `vce(hac thanning 1)`. This variance estimator depends on ordered data, which can be specified using the `tsset` command.

21.4   Remarks

21.4.1   Equivalent commands

Some `family()` and `link()` combinations with the `glm` command result in a model that is already fit by Stata:

Table 21.3. Equivalent Stata commands

family()	link()	Options	Equivalent Stata command
gaussian	identity	*nothing* | irls | irls vce(oim)	regress
gaussian	identity	t(*var*) vce(hac nwest #) vfactor($\#_v$)	newey, t(*var*) lag(#)
binomial	cloglog	*nothing* | irls vce(oim)	cloglog
binomial	probit	*nothing* | irls vce(oim)	probit
binomial	logit	*nothing* | irls | irls vce(oim)	logit
poisson	log	*nothing* | irls | irls vce(oim)	poisson
nbinomial	log	*nothing* | irls vce(oim)	nbreg
gamma	log	scale(1)	streg, dist(exp) nohr

In general, if a Stata command already exists for a family and link combination, then the results will match for the coefficients (will be close to numeric differences in the optimization) and for the standard errors (will be close to numeric differences in the optimization) if `irls` is not specified or if both `irls` and `oim` are specified. And, if the link is the canonical link for the specified family, the standard errors will match if just `irls` is specified because the EIM is equal to the OIM in this case. See section 3.6 for details.

21.4.2   Special comments on family(Gaussian) models

Although `glm` can be used to fit linear regression (`family(Gaussian) link(identity)` models) and, in fact, does so by default, it is better to use the `regress` command because it is quicker and because many postestimation commands are available to explore the adequacy of the fit.

21.4.3   Special comments on family(binomial) models

The binomial distribution can be specified as 1) `family(binomial)`,
2) `family(binomial #)`, or 3) `family(binomial varname)`.

In case 2, `#` is the value of the binomial denominator N, the number of trials. Specifying `family(binomial 1)` is the same as specifying `family(binomial)`; both mean that y has the Bernoulli distribution with values 0 and 1 only.

In case 3, *varname* is a variable containing the binomial denominator, thus allowing the number of trials to vary across observations.

For `family(binomial) link(logit)` models, we recommend also using the command `logistic`. Both produce the same answers, and each command provides useful postestimation commands (not found in both commands).

Log-log regression may be performed by subtracting the response from one and modeling as a complementary log-log model. Proportional log-log data may be transformed by transposing the proportion of successes and failures.

21.4.4   Special comments on family(nbinomial) models

The negative binomial distribution can be specified as 1) `family(nbinomial)`, or the ancillary parameter may be specified in 2) `family(nbinomial #)`.

`family(nbinomial)` is equivalent to `family(nbinomial 1)`. `#`, often called k, enters the variance and deviance functions; typical values range between 0.01 and 2.

`family(nbinomial) link(log)` models, also known as negative binomial regression, are used for data with an overdispersed Poisson distribution. Although `glm` can be used to fit such models, Stata's maximum-likelihood `nbreg` command is probably preferable. Under the `glm` approach, we must search for value of k that results in the deviance-based dispersion being 1. `nbreg`, on the other hand, finds the maximum likelihood estimate of k and reports its confidence interval.

The default ancillary parameter value of 1 is the same as geometric regression. A negative binomial distribution with a scale of one is a geometric distribution. Conversely, a negative binomial distribution parameterized as we have shown in the text with a scale of zero is a Poisson distribution.

21.4.5   Special comment on family(gamma) link(log) models

`glm` with `family(gamma)` and `link(log)` is identical to exponential regression. However, censoring is not available with this method. Censored exponential regression may be modeled using `glm` with `family(poisson)`. The log of the original response is entered into a Poisson model as an offset, whereas the new response is the censor variable. The result of such modeling is identical to the log relative hazard parameterization of the `streg` command in Stata.

Maximum-likelihood exponential regression estimation via `glm` with the `scale(1)` option will report a log likelihood that differs from the log likelihood reported by the `streg` command. The difference is documented in the manual and amounts to the `streg` command recentering the log likelihood to allow a comparison with Weibull regression log likelihoods.

22 Data synthesis

Contents

This chapter presents information on data synthesis and simulation. We illustrate how to generate data best suited for particular models. Our presentation begins by reviewing the generation of random values with population characteristics and the generation of data with sample characteristics. We present several examples illustrating the ideas as well as illustrating how one can use simulation to pursue analyses. This chapter does not introduce topics on generalized linear models (GLMs), but it provides tools and techniques that allow interested researchers to generate data suitable for illustrating and investigating GLMs.

22.1 Generating correlated data

Researchers often desire to generate a sample dataset from a multivariate normal distribution characterized by a given correlation matrix.

The solution is to rotate a set of variables using the Cholesky decomposition of the specified correlation matrix. That is, let's assume that we start with independent variables (that are uncorrelated) in a matrix $\mathbf{X}_o$ where without loss of generality each variable has unit variance. We want to transform the data matrix such that $\mathbf{X} = \mathbf{A}\mathbf{X}_o$ so that the columns of the matrix $\mathbf{X}$ have the specified correlations. That is, we should have

$$\text{Corr}(\mathbf{X}) \quad = \quad \mathbf{A}_1^T \text{Corr}(\mathbf{X}_o) \mathbf{A}_1 \qquad\qquad (22.1)$$

$$= \quad \mathbf{A}_1^T \mathbf{A}_1 \qquad\qquad (22.2)$$

which implies that $\mathbf{A}_1$ is the Cholesky decomposition of the desired correlation.

Thus, if you have a given correlation matrix $\mathbf{P}$, you find the Cholesky decomposition (matrix square root) and use it to multiply the data matrix.

```
. matrix P = (1,.2\.2,1)
. matrix A1 = cholesky(P)
. matrix list A1
A1[2,2]
           c1        c2
r1          1         0
r2         .2  .9797959
. set obs 40
number of observations (_N) was 0, now 40
. set seed 13215
. generate xo1 = rnormal()
. generate xo2 = rnormal()
```

Armed with the Cholesky decomposition of the desired correlation matrix, we can carry out the matrix multiplication of the original data to obtain the desired data.

```
. generate x1 = xo1
. generate x2 = A1[2,1]*xo1 + A1[2,2]*xo2
. correlate x1 x2
(obs=40)
                 x1        x2

        x1   1.0000
        x2   0.2606    1.0000
```

Note that the correlation of the data does not exactly match the desired correlation matrix $\mathbf{P}$. The reason is that we did not start with data that had sample correlation equal to zero. Thus, we end up with a sample of data from a population that is characterized by $\mathbf{P}$.

What if we want a sample with the stated correlation $\mathbf{P}$?

To do this, we must transform the original data to have a sample correlation of zero and then apply the transformation matrix. That is, we will have to transform the data twice. After tedious matrix algebra, the formula is given by

$$\mathbf{X} \;=\; (\mathbf{I} - \mathbf{J}\mathbf{J}^{\mathrm{T}} n)\mathbf{X}_o \mathbf{A}_2 \mathbf{A}_1^{\mathrm{T}} + \mathbf{J}\mathbf{B}^{\mathrm{T}} \tag{22.3}$$

where $\mathbf{A}_2$ is the Cholesky decomposition of the inverse of the sample covariance matrix, $\mathbf{A}_1$ is the Cholesky decomposition of the desired correlation–covariance matrix, $\mathbf{J}$ is a vector of ones, and $\mathbf{B}$ is a vector of desired means.

Disregarding components that center the mean at zero, generating a pair of variables with a unit sample variance and specified sample correlation can be done:

```
. correlate xo1 xo2, covariance
(obs=40)

             |      xo1      xo2
-------------+------------------
         xo1 |  1.25874
         xo2 |  .037611  .947423

. matrix A2 = cholesky(invsym(r(C)))

. generate z1 = xo1*A2[1,1]+xo2*A2[2,1]

. generate z2 = (xo1*A2[1,1]+xo2*A2[2,1])*A1[2,1] +
> (xo1*A2[1,2]+xo2*A2[2,2])*A1[2,2]

. correlate z1 z2
(obs=40)

             |       z1       z2
-------------+------------------
          z1 |   1.0000
          z2 |   0.2000   1.0000
```

Fortunately, we do not have to remember all of this algebra because Stata provides two commands for generating correlated data. The `drawnorm` command draws a random sample from a population with specified correlation (and mean) structure; the `corr2data` command generates a sample with the specified correlation (and mean) structure.

Both commands include the options `corr(`*name*`)`, which allows you to specify the name of a correlation matrix, and `cov(`*name*`)`, which allows you to specify the name of a covariance matrix.

We can use the `corr2data` command to create variables named `err1`, `err2`, …, `err`*t*. A variable named `panel` can be created that is equal to the observation number. If the data are reshaped from the wide format to the long format, the `panel` variable plays the role of the ID. The three variables we create below have the same correlation:

```
. drop _all

. set obs 20
number of observations (_N) was 0, now 20

. matrix exch = J(3,3,.4) + .6*I(3)

. corr2data err1 err2 err3, corr(exch)
```

```
. correlate err*
(obs=20)
                |   err1      err2      err3
     -----------+---------------------------------
          err1  |  1.0000
          err2  |  0.4000    1.0000
          err3  |  0.4000    0.4000    1.0000
```

The data are set up as panel data so that we can then use these variables to build a generalized estimating equation model with specified correlation.

```
. generate panel = _n

. reshape long err, i(panel) j(t)
(note: j = 1 2 3)

Data                                wide   ->   long
-----------------------------------------------------------------
Number of obs.                        20   ->       60
Number of variables                    4   ->        3
j variable (3 values)                      ->   t
xij variables:
                          err1 err2 err3   ->   err
-----------------------------------------------------------------

. quietly xtset panel

. xtgee err

Iteration 1: tolerance = 2.655e-17

GEE population-averaged model            Number of obs      =         60
Group variable:                  panel   Number of groups   =         20
Link:                         identity   Obs per group:
Family:                       Gaussian                 min =          3
Correlation:              exchangeable                 avg =        3.0
                                                       max =          3
                                         Wald chi2(0)       =          .
Scale parameter:                   .95   Prob > chi2        =          .

------------------------------------------------------------------------------
         err |      Coef.   Std. Err.      z    P>|z|     [95% Conf. Interval]
-------------+----------------------------------------------------------------
       _cons |  -7.99e-10   .1688194    -0.00   1.000     -.33088      .33088
------------------------------------------------------------------------------

. estat wcorrelation

Estimated within-panel correlation matrix R:

          |       c1        c2        c3
    ------+------------------------------
       r1 |        1
       r2 |       .4         1
       r3 |       .4        .4         1
```

The benefit of the xtgee command is that the errors may be generated separately from the linear predictor for studies of regression because the link function is the identity.

It is more complicated to generate correlated binary data. Interested readers should consult Pan (2001) and the references therein. In Pan's manuscript (and in the works of the cited authors), reference is made to assessing the ability of a statistic for model selection. In all cases, the linear predictor component of the model is given by

$$0.25 - 0.25a_{it} - 0.25b_{it} \tag{22.4}$$

which is composed of a binary predictor $a_i \sim$ Uniform(0.50) and a time-dependent covariate $b_{it} = t - 1$.

We developed `drawbinmodel`,[1] which can be used to generate correlated binary outcomes for this specified model. This command is similar to the `drawnorm` command in that the generated data are a sample from a population with the specified correlation properties. This command should be used to generate large numbers of panels where each panel is of modest size (less than 5).

The algorithm generates random values from a population for each of the covariate patterns in the data. This is the reason that all of these authors have used the same regression model in these manuscripts. Interested readers are welcome to expand the command to other models, though we point out the task is daunting. As such, even though the command generates data for only two covariates, one can investigate goodness of fit of models as well as estimation under different correlation structures.

```
. drop _all
. set seed 2394
. drawbinmodel 50000 3, corr(exch 0.25)
Creating datasets for 8 covariate patterns
----+----1----+----2----+----3----+----4----+----5
....+...
. xtset id
      panel variable:  id (balanced)
. xtgee y a b, family(binomial) link(logit) nolog
GEE population-averaged model              Number of obs      =     150,000
Group variable:                      id    Number of groups   =      50,000
Link:                             logit    Obs per group:
Family:                        binomial                        min =          3
Correlation:               exchangeable                        avg =        3.0
                                                               max =          3
                                           Wald chi2(2)        =    2728.79
Scale parameter:                      1    Prob > chi2         =     0.0000

-------------+----------------------------------------------------------------
           y |     Coef.   Std. Err.      z    P>|z|     [95% Conf. Interval]
-------------+----------------------------------------------------------------
           a | -.2575717   .0098791   -26.07   0.000    -.2769344   -.238209
           b | -.2524308   .0055351   -45.61   0.000    -.2632793   -.2415823
       _cons |  .2704477   .0097596    27.71   0.000     .2513193    .2895762
-------------+----------------------------------------------------------------
```

---

1. The `drawbinmodel` command can be installed by typing `net install drawbinmodel, from(http://www.stata.com/users/jhardin)`.

```
. estat wcorrelation
Estimated within-id correlation matrix R:
                  c1           c2           c3
      r1 |         1
      r2 |   .2515456           1
      r3 |   .2515456     .2515456           1
```

22.2   Generating data from a specified population

Data may be generated from a population for a specified GLM in simple steps:

- Predictors $\mathbf{X}$ are generated.
- The linear predictor η is calculated for a given set of coefficients $\boldsymbol{\beta}$.
- The means are calculated using the inverse of a specified link function $\mu = g^{-1}(\eta)$.
- Using the mean μ as the location parameter of a specified distribution (implied by a specific variance function), random-number generators can then produce outcomes Y.

The subsequent subsections illustrate generating data for specific models; see chapter 14 for more examples.

To highlight similarities in approaches for generating data with specific population properties and for generating data with specific sample properties, we use binary covariates in our examples to limit the number of unique covariate patterns. We do so because it is the number of unique covariate patterns that dictates the number of specific distributions (that is, the number of specific parameter values of a given distribution).

22.2.1   Generating data for linear regression

Using Stata's `rnormal()` function, we can generate data from the normal distribution with a specified mean and variance. Here we illustrate data generation for linear regression. We simulate data $y = 1.0 + 1.0x_1 + 1.0x_2$, where x_1 and x_2 are binary variables. The design variables are design blocking variables defining a balanced experiment. In this first synthesis, we generate a large dataset of 10,000 observations.

```
. drop _all
. set seed 2393
. set obs 10000
number of observations (_N) was 0, now 10,000
. generate x1  = mod((_n-1),4)==1 | mod((_n-1),4)==3
. generate x2  = mod((_n-1),4)==2 | mod((_n-1),4)==3
. generate eta = 1.00+1.00*x1+1.00*x2 // Linear predictor with specified beta
. generate mu  = eta                  // mu as the inverse link function of eta
```

Because the variances are common for each specific covariate pattern, the data for the population are easily generated:

```
. generate yp = rnormal(mu)
```

However, to generate data that will have the exact values for the sample is more difficult. We must center the data to the exact specified mean and rescale so that each subsample has the same standard deviation.

```
. generate ys = rnormal(mu)
. forvalues x1=0/1 {
  2.     forvalues x2=0/1 {
  3.         quietly summarize ys if x1==`x1´ & x2==`x2´
  4.         quietly replace ys = (ys-r(mean))/r(sd)+eta if x1==`x1´ & x2==`x2´
  5.     }
  6. }
```

We can see the differences in the two approaches using the `table` command. That command is used here to generate a table that summarizes the mean and standard deviation of the outcomes with the population properties and then to generate a similar table for the outcomes with the sample properties.

```
. table x1 x2, contents(mean yp sd yp)
```

x1	x2 0	1
0	1.009566 1.01345	1.957496 1.024266
1	1.996042 1.008208	2.982609 .9728627

```
. table x1 x2, contents(mean ys sd ys)
```

x1	x2 0	1
0	1 1	2 1
1	2 1	3 1

Finally, we regress each outcome on x1 and x2.

```
. glm yp x1 x2, noheader
Iteration 0:   log likelihood = -14236.544
```

yp	Coef.	OIM Std. Err.	z	P>|z|	[95% Conf. Interval]	
x1	1.005795	.0200976	50.05	0.000	.966404	1.045185
x2	.9672486	.0200976	48.13	0.000	.9278581	1.006639
_cons	.9999065	.017405	57.45	0.000	.9657933	1.03402

```
. glm ys x1 x2, noheader
Iteration 0:   log likelihood = -14187.385
```

ys	Coef.	OIM Std. Err.	z	P>|z|	[95% Conf. Interval]	
x1	1	.019999	50.00	0.000	.9608027	1.039197
x2	1	.019999	50.00	0.000	.9608027	1.039197
_cons	1	.0173196	57.74	0.000	.9660541	1.033946

For comparison, we could have changed the number of simulated observations to 50 in the commands at the beginning of section 22.2.1.

```
. glm yp x1 x2, noheader
Iteration 0:   log likelihood = -66.984973
```

yp	Coef.	OIM Std. Err.	z	P>|z|	[95% Conf. Interval]	
x1	.5893436	.2695057	2.19	0.029	.0611221	1.117565
x2	1.356639	.2697216	5.03	0.000	.827994	1.885283
_cons	1.081411	.2303871	4.69	0.000	.6298606	1.532961

```
. glm ys x1 x2, noheader
Iteration 0:   log likelihood = -68.862386
```

ys	Coef.	OIM Std. Err.	z	P>|z|	[95% Conf. Interval]	
x1	1	.2798176	3.57	0.000	.4515676	1.548432
x2	1	.2800417	3.57	0.000	.4511284	1.548872
_cons	1	.2392022	4.18	0.000	.5311724	1.468828

22.2.2 Generating data for logistic regression

Using Stata's runiform() function, we generate data from the Bernoulli distribution with a specified mean. Here we illustrate data generation for logistic regression.

We simulate data $g(y) = 1.0 + 1.0x_1$, where x_1 and x_2 are binary variables and $g(\cdot)$ is the inverse logit function. The design variables are design blocking variables defining a balanced experiment. In this synthesis, we generate a large dataset of 10,000 observations.

```
. drop _all
. set seed 8721
. set obs 10000
number of observations (_N) was 0, now 10,000
. generate x1  = mod((_n-1),2)==1
. generate eta = 1.00+1.00*x1  // Linear predictor with specified beta
. generate mu  = invlogit(eta) // mu as the inverse link function of eta
```

As in the previous subsection, we generate outcomes reflecting population properties and then other outcomes reflecting sample properties. For continuous outcomes, it is difficult to generate data with specific sample properties. For discrete data, it can be impossible, especially in small samples.

```
. generate yp = runiform() < mu
```

The limitation in generating specific parameter values lies in the nature of the underlying distributions. Given the discreteness of the outcomes, one cannot generate any specific coefficient. The coefficients are related to the tabulation of outcomes and those tabulations are integers. Here is an example that shows results for a smaller sample.

```
. generate ys   = .
(10,000 missing values generated)
. generate flag = 0
. forvalues x1=0/1 {
  2.       quietly {
  3.             replace flag = -(x1==`x1´)
  4.             count if flag==-1
  5.             local N = r(N)
  6.             sort flag
  7.             replace ys = _n < round(mu*`N´,1) in 1/`N´
  8.       }
  9. }
```

The fitted models are then given by

```
. glm yp x1, family(binomial) link(logit) noheader nolog
```

yp	Coef.	OIM Std. Err.	z	P>|z|	[95% Conf. Interval]
x1	1.041695	.0532294	19.57	0.000	.9373671 1.146023
_cons	.920775	.0313351	29.38	0.000	.8593594 .9821907

```
. glm ys x1, family(binomial) link(logit) noheader nolog
```

		OIM				
ys	Coef.	Std. Err.	z	P>|z|	[95% Conf.	Interval]
x1	.9994391	.0540257	18.50	0.000	.8935507	1.105327
_cons	.9986852	.0318844	31.32	0.000	.936193	1.061177

With a finite sample size, the coefficients reflect the association given in the following table:

```
. table ys x1, contents(freq)
```

	x1	
ys	0	1
0	1,346	597
1	3,654	4,403

From the table, we can see that the coefficient for male (x1) for outcome ys is given by $\ln\{(4403/597)/(3654/1346)\} = 0.9994391$ as was shown in the glm results above.

For comparison, we could have changed the number of simulated observations to 50 at the beginning of section 22.2.2. With this change, our glm results would be as follows:

```
. glm yp x1, family(binomial) link(logit) noheader nolog
```

		OIM				
yp	Coef.	Std. Err.	z	P>|z|	[95% Conf.	Interval]
x1	.6325226	.6586528	0.96	0.337	-.6584132	1.923458
_cons	.7537718	.4287465	1.76	0.079	-.0865558	1.594099

```
. glm ys x1, family(binomial) link(logit) noheader nolog
```

		OIM				
ys	Coef.	Std. Err.	z	P>|z|	[95% Conf.	Interval]
x1	.9044563	.6938606	1.30	0.192	-.4554856	2.264398
_cons	.7537718	.4287465	1.76	0.079	-.0865558	1.594099

Note that generated data for a sample of size 50 produced a sample coefficient that was not close to one. Although there are tabulations for a sample size of 50 that produce a coefficient that is closer to one, the individual means for those tabulations are much farther away from the specifications of x1=0 with mean invlogit(1)=0.7311, and mean invlogit(2)=0.8808 when x1=1.

22.2.3   Generating data for probit regression

As will be covered extensively in chapter 9 and chapter 10, there are other types of binary regression models. These other models are defined by alternate link functions discussed in chapter 3.

```
. drop _all
. set seed 23224
. set obs 50000
number of observations (_N) was 0, now 50,000
. generate x1  = mod((_n-1),2)==1
. generate eta = 1.00-1.50*x1  // Linear predictor with specified beta
. generate mu  = normal(eta)   // mu as the inverse link function of eta
. generate yp  = runiform() < mu
. glm yp x1, family(binomial) link(probit) noheader nolog
```

yp	Coef.	OIM Std. Err.	z	P>|z|	[95% Conf. Interval]	
x1	-1.491003	.0126449	-117.91	0.000	-1.515786	-1.466219
_cons	1.001718	.0095558	104.83	0.000	.9829886	1.020447

Here we simulate data for a binary regression model in which the association of the outcome is through the probit link. We illustrate a program to use with the `simulate` command that will compute the number of times the AIC statistic indicates a preference for the probit over the logistic model. In this simulation, we consider a single continuous covariate because the AIC statistics would be the same for a model with a single binary covariate.

```
. capture program drop Mysim
. program Mysim, rclass
  1.       drop _all
  2.       set obs 500
  3.       generate x1 = rnormal()
  4.       generate eta= 1.00-1.50*x1
  5.       generate mu = normal(eta)
  6.       generate yp = runiform() < mu
  7.       glm yp x1, family(binomial) link(logit) iterate(10)
  8.       return scalar logit = e(aic)
  9.       glm yp x1, family(binomial) link(probit) iterate(10)
 10.       return scalar probit = e(aic)
 11. end
. set seed 23224
. simulate probit=r(probit) logit=r(logit), reps(100): Mysim
      command:  Mysim
       probit:  r(probit)
        logit:  r(logit)
Simulations (100)
 ───────┼─── 1 ───┼─── 2 ───┼─── 3 ───┼─── 4 ───┼─── 5
.................................................    50
.................................................   100
```

```
. count if probit < logit
  74

. display "Probit is preferred " r(N) " times out of 100"
Probit is preferred 74 times out of 100
```

22.2.4 Generating data for complimentary log-log regression

An alternative binary model (to logistic and probit) is the complementary log-log regression model.

```
. drop _all
. set seed 82344
. set obs 50000
number of observations (_N) was 0, now 50,000
. generate x1  = mod((_n-1),2)==1
. generate eta = 0.50-0.50*x1        // Linear predictor with specified beta
. generate mu  = 1-exp(-exp(eta)) // mu as the inverse link function of eta
. generate yp  = runiform() < mu
. glm yp x1, family(binomial) link(cloglog) noheader nolog
```

		OIM				
yp	Coef.	Std. Err.	z	P>|z|	[95% Conf.	Interval]
x1	-.5000226	.0114229	-43.77	0.000	-.5224112	-.4776341
_cons	.5021957	.0078622	63.87	0.000	.4867861	.5176054

As in the other cases, we can simulate data for which the specified characteristics are as close as possible for the particular sample. We emphasize that it is not always possible to simulate discrete data with exact coefficients.

```
. generate ys = .
(50,000 missing values generated)
. generate flag = 0
. forvalues x1=0/1 {
  2.      quietly {
  3.          replace flag = -(x1==`x1´)
  4.          count if flag==-1
  5.          local N = r(N)
  6.          sort flag
  7.          replace ys = _n < round(mu*`N´,1) in 1/`N´
  8.      }
  9. }
. glm ys x1, family(binomial) link(cloglog) noheader nolog
```

		OIM				
ys	Coef.	Std. Err.	z	P>|z|	[95% Conf.	Interval]
x1	-.5000334	.0114255	-43.76	0.000	-.5224271	-.4776398
_cons	.4999232	.0078618	63.59	0.000	.4845143	.5153321

22.2.5 Generating data for Gaussian variance and log link

In chapter 5, we discuss two similar models. In one model, we investigate associations of covariates for an outcome with Gaussian variance and a log link. In a similar model, we investigate the log-transform of an outcome with Gaussian variance and an identity link. Initially, we generate data for the former specification.

```
. drop _all
. set seed 2393
. set obs 10000
number of observations (_N) was 0, now 10,000
. generate x1  = mod((_n-1),4)==1 | mod((_n-1),4)==3
. generate x2  = mod((_n-1),4)==2 | mod((_n-1),4)==3
. generate eta = 1.00+1.00*x1+0.00*x2 // Linear predictor with specified beta
. generate mu  = exp(eta)              // mu as the inverse link function of eta
. generate y   = rnormal(mu, 2.0)
. generate ly  = log(y)
(466 missing values generated)
```

Then we fit the two competing models.

```
. glm y  x1 x2, family(gaussian) link(log) noheader nolog
```

y	Coef.	OIM Std. Err.	z	P>|z|	[95% Conf. Interval]	
x1	1.00923	.0112657	89.58	0.000	.9871494	1.03131
x2	-.0077667	.0072497	-1.07	0.284	-.0219759	.0064425
_cons	.9917446	.0111969	88.57	0.000	.969799	1.01369

```
. glm ly x1 x2, family(gaussian) link(identity) noheader nolog
```

ly	Coef.	OIM Std. Err.	z	P>|z|	[95% Conf. Interval]	
x1	1.077381	.0128698	83.71	0.000	1.052156	1.102605
x2	-.0268599	.0128544	-2.09	0.037	-.052054	-.0016657
_cons	.8913033	.0113052	78.84	0.000	.8691455	.9134611

22.2.6 Generating underdispersed count data

In chapter 14, we discuss alternatives to Poisson models for count data. One such model discussed is the generalized Poisson. This alternative model is suitable for underdispersed, equidispersed, and overdispersed count data. Here we simulate underdispersed data and illustrate estimation of such data using the generalized Poisson model in comparison with the Poisson model.

The challenge is that there is no built-in function for generating random variates from the generalized Poisson distribution. As such, we created a program that returns a variate from the distribution for a random quantile. This approach is very slow. However, we are not planning to use the program other than to generate a few sample datasets, so we are willing to live with the slow execution speed.

First, we create the quantile function. We do not have the ability to write a function, so we have to write a program to work with scalars and return a scalar result.

```
. program GPquantile, rclass
  1.      args u mu k
  2.      quietly {
  3.          local mu = exp(`mu´)
  4.          local maxval = 10000
  5.          local x = 0
  6.          local p = exp(ln((1-`k´)*`mu´) + (`x´-1)*ln((1-`k´)*`mu´+`k´*`x´)
  >              - (1-`k´)*`mu´ - `k´*`x´ - lngamma(`x´+1))
  7.          while `p´ <`u´ & `x´ < `maxval´ {
  8.              local x = `x´+1
  9.              local p = `p´+ exp(ln((1-`k´)*`mu´) + (`x´-1)*ln((1-`k´)*`mu´+
  >                  `k´*`x´) - (1-`k´)*`mu´ - `k´*`x´ - lngamma(`x´+1))
  10.         }
  11.         return scalar x = `x´
  12.     }
  13. end
```

Then we write a program to loop over the observations and call the quantile function to generate each random variate with the associated arguments.

```
. program rgpoisson
  1.      args ry xb delta
  2.      quietly {
  3.          tempvar u
  4.          generate `u´ = runiform()
  5.          capture drop `ry´
  6.          generate `ry´ = .
  7.          describe, short
  8.          local N = r(N)
  9.          forvalues i=1/`N´ {
  10.             GPquantile (`u´[`i´]) (`xb´[`i´]) (`delta´)
  11.             replace `ry´ = r(x) in `i´
  12.         }
  13.     }
  14. end
```

To create a dataset, we set the observations and define a linear predictor. Once those are specified, we generate the random generalized Poisson variates.

```
. drop _all
. set seed 29843
. set obs 40
number of observations (_N) was 0, now 40
. generate x1 = runiform() < .5
. generate x2 = runiform() < .5
. generate eta = 1 + 0.25*x1 - 0.25*x2
. rgpoisson y eta (-0.3)
```

Finally, we demonstrate the utility of the generalized Poisson model by comparing it with a Poisson model when the data are underdispersed.

```
. gpoisson y x1 x2, nolog
Generalized Poisson regression                Number of obs   =         40
                                              LR chi2(2)      =       4.98
Dispersion       = -.4105379                  Prob > chi2     =     0.0827
Log likelihood = -61.27216                    Pseudo R2       =     0.0391
```

y	Coef.	Std. Err.	z	P>|z|	[95% Conf. Interval]	
x1	-.0891833	.1332121	-0.67	0.503	-.3502742	.1719076
x2	-.3118402	.1450438	-2.15	0.032	-.5961207	-.0275596
_cons	1.134418	.1083369	10.47	0.000	.9220818	1.346755
/atanhdelta	-.436258	.1715404			-.772471	-.100045
delta	-.4105379	.1426287			-.648364	-.0997125

```
Likelihood-ratio test of delta=0:  chi2(1) =     8.53      Prob>=chi2 = 0.0017
. glm y x1 x2, family(poisson) nolog
Generalized linear models                     No. of obs      =         40
Optimization     : ML                         Residual df     =         37
                                              Scale parameter =          1
Deviance         =   22.03660183              (1/df) Deviance =   .5955838
Pearson          =   19.14777597              (1/df) Pearson  =   .5175075
Variance function: V(u) = u                   [Poisson]
Link function    : g(u) = ln(u)               [Log]
                                              AIC             =   3.426911
Log likelihood   =  -65.5382133               BIC             =  -114.4519
```

		OIM				
y	Coef.	Std. Err.	z	P>|z|	[95% Conf. Interval]	
x1	-.0748567	.1954077	-0.38	0.702	-.4578487	.3081353
x2	-.3978742	.2153718	-1.85	0.065	-.8199951	.0242466
_cons	1.153085	.1560562	7.39	0.000	.8472204	1.45895

22.3 Simulation

This section discusses simulation and illustrates its use. Herein, we provide examples of using simulation to deduce distributions, of computing power, and for running experiments to illustrate properties of statistics and models.

22.3.1   Heteroskedasticity in linear regression

In the following example, we are interested in the effect of the violation of the constant variance (homoskedasticity) assumption on the test statistic of a regression coefficient. We simulate data for a regression model with known (population) coefficients. Repeatedly synthesizing data from the population, we collect the test statistics evaluating one of the regression coefficients.

In the first simulation, we synthesize data that do not violate the assumption of constant variance.

```
. set seed 28373

. capture program drop Mysim

. program Mysim, rclass
  1.      syntax [, CONStant]
  2.      quietly {
  3.          drop _all
  4.          set obs 30
  5.          if "`constant'" == "" {
  6.              local scale = 1
  7.          }
  8.          else {
  9.              tempvar scale
 10.              generate `scale' = 2*runiform()
 11.          }
 12.          generate x1  = runiform()
 13.          generate x2  = runiform()
 14.          generate eta = 0.00+1.00*x1+2.00*x2
 15.          generate mu  = eta
 16.          generate y = rnormal(mu)*sqrt(`scale')
 17.          glm y x1 x2, family(gaussian) link(identity)
 18.          return scalar z = (_b[x1]-1)/_se[x1]
 19.      }
 20. end
```

Then we define a dataset and execute the simulation. Once the simulated test statistics are compiled, a command is issued to illustrate the results.

```
. capture drop _all

. set obs 30
number of observations (_N) was 0, now 30

. foreach var in x1 x2 eta mu y {
  2.     quietly generate `var' = .
  3. }

. simulate z = r(z), reps(1000) nodots: Mysim
      command:  Mysim
            z:  r(z)

. label var z "Test statistic"
```

```
. histogram z, kdensity plot(function stdnorm = normalden(x,0,1),
> range(-3 3) lpattern(dash)) title(Distribution of test statistic)
> subtitle(Homoskedastic Variance)
(bin=29, start=-4.3689919, width=.26780664)
```

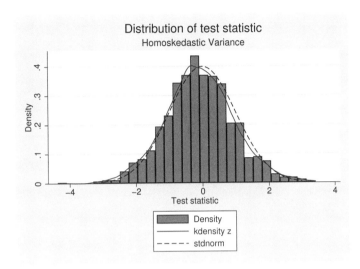

One can see that the histogram of the statistic and a kernel density estimate of the density of the statistic is close to the theoretical Gaussian density.

The question of interest is whether the sample density will match the theoretical density of the same statistic when data violate the constant variance assumption. Here we synthesize data for which the variance is different for each observation. This is the only change in the simulation because the association of the outcome and the covariate is the same as before. We collect the test statistics of the coefficient of the covariate and create the graph as before.

```
. set seed 23984
. capture drop _all
. set obs 30
number of observations (_N) was 0, now 30
. foreach var in x1 x2 eta mu y {
  2.     quietly generate `var' = .
  3. }
. generate scale = 2*runiform()
. simulate z = r(z), reps(1000) nodots: Mysim, noconstant

      command:  Mysim, noconstant
            z:  r(z)

. label var z "Test statistic"
```

```
. histogram z, kdensity plot(function stdnorm = normalden(x,0,1),
> range(-3 3) lpattern(dash)) title(Distribution of test statistic)
> subtitle(Heteroskedastic Variance)
(bin=29, start=-2.9645872, width=.22543435)
```

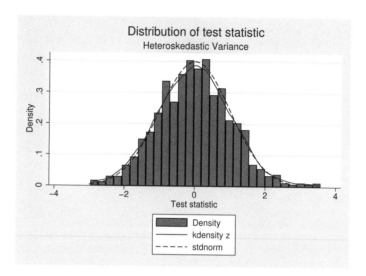

One can see that the histogram of the statistic as well as the kernel density estimate of the density of the statistic is close to the theoretical Gaussian density. This is the same result seen for data that did not violate the assumption.

Given these results for the statistic collected under conditions that violated the constant variance assumption, we conclude that the evaluation of coefficients using the test statistic for the Gaussian family with the identity link is robust to violation of the assumption of homoskedasticity.

22.3.2   Power analysis

Often, a researcher must present a power analysis to justify the costs of a research proposal. Here we outline how one might use simulation to determine power for hypotheses assessed by GLMs. Our simple example illustrates power to detect a difference from the null for a coefficient in linear regression. In our example, the standard deviation of the covariate and of the residuals is assumed to be one. In this way, the differences to be assessed are standardized (they are in terms of a standard deviation).

The usual justification accompanying grant proposals illustrates power to detect a small difference (0.2 standard deviations), a medium difference (0.5 standard deviations), and a large difference (0.8 standard deviations). Using software specifically designed for power analysis, we know that the power for these differences for one predictor in a sample of size 30 is given by 18%, 75%, and 99%. Often, such simple hypotheses are easy to specify and evaluate within software for power analysis. In other cases, the specifics required by the software can be confusing. One way to get around such specifics is to simulate the data and models that are planned.

```
. set seed 87232
. capture program drop Mysim
. program Mysim, rclass
  1.      args delta
  2.      quietly {
  3.          drop _all
  4.          set obs 30
  5.          generate x1  = rnormal(0)
  6.          generate eta = 0.00+(1.00+`delta´)*x1
  7.          generate mu  = eta
  8.          generate y = rnormal(mu)
  9.          glm y x1, family(gaussian) link(identity)
 10.          return scalar z = (_b[x1]-1)/_se[x1]
 11.      }
 12. end
```

The usual justification accompanying grant proposals illustrates power to detect a small difference (0.2 standard deviations), a medium difference (0.5 standard deviations), and a large difference (0.8 standard deviations). By using software specifically designed for power analysis, we know that the power for these differences for a single predictor in a sample of size 30 is given by 18%, 75%, and 99%, respectively.

In the following example, we present a simple program to run simulations for different parameter specifications. We use simulation to replicate the results from the power analysis software.

```
. capture program drop Mypow
. program Mypow
  1.      syntax, Vals(numlist min=1 >0 <1 sort)
  >         [Nobs(integer 30) Reps(integer 1000) ALPha(real .05)]
  2.      quietly {
  3.          noisily display
  4.          foreach delta in `vals´ {
  5.              capture drop _all
  6.              set obs 30
  7.              foreach var in x1 eta mu y {
  8.                  generate `var´ = .
  9.              }
 10.              simulate z = r(z), reps(`reps´) nodots: Mysim `delta´
 11.              count if abs(z) > invnormal(1-`alpha´/2)
 12.              local p = r(N)/`reps´
 13.              noisily display as txt "P(delete|delta=" as res %4.2f `delta´
  >                 as txt ") = " as res %6.4f `p´
 14.          }
 15.      }
 16. end
. Mypow, vals(0.2 0.5 0.8)
P(delete|delta=0.20) = 0.2020
P(delete|delta=0.50) = 0.7430
P(delete|delta=0.80) = 0.9750
```

Often, such simple hypotheses are easy to specify and evaluate within software for power analysis. In other cases, the specifics required for analysis by the software can be confusing. One way to get around providing such specifics is to simulate data and fit planned models. Power can then be estimated empirically.

22.3.3   Comparing fit of Poisson and negative binomial

In this example, we highlight a simple simulation. We investigate the comparison of Poisson and negative binomial models for data generated from a negative binomial population. In our simulation, the dispersion parameter of the negative binomial is small so that there is some chance for preferring the Poisson model. One of the ways in which researchers pursue model selection is with the AIC statistic. Our simulation generates 100 samples of size 30 and fits Poisson and negative binomial GLMs. The AIC statistic is saved for each of the models, and we count the number of times that the statistic correctly indicates preference for the negative binomial.

```
. capture program drop Mysim
. program Mysim, rclass
  1.      quietly {
  2.          drop _all
  3.          set obs 30
  4.          generate x1 = rnormal()
  5.          generate x2 = rnormal()
  6.          generate nby = rpoisson(rgamma(1/.5,.10)*exp(0.75*x1-1.25*x2+2))
  7.          capture glm nby x1 x2, family(nbinomial ml) iterate(10)
  8.          return scalar AICnb = e(aic)
  9.          capture glm nby x1 x2, family(poisson) iterate(10)
 10.          return scalar AICpo = e(aic)
 11.      }
 12. end
```

After defining the simulation program, we specify the particulars for our dataset, call the **simulate** program, and summarize the results.

```
. drop _all
. set obs  30
number of observations (_N) was 0, now 30
. set seed 88530
. quietly generate nby = .
. quietly generate x1  = .
. quietly generate x2  = .
```

```
. simulate negbin=r(AICnb) poisson=r(AICpo), reps(100): Mysim
        command:  Mysim
         negbin:  r(AICnb)
        poisson:  r(AICpo)
Simulations (100)
  ——+—— 1 ——+—— 2 ——+—— 3 ——+—— 4 ——+—— 5
..................................................           50
..................................................           100
. count if negbin < poisson
  94
```

The simulation could be more sophisticated, and the simulation program could produce and save additional statistics. The results of our simple investigation indicate good performance for the statistics. In a larger simulation, we would investigate other values for the dispersion parameter and other sample sizes. We leave those expansions of this investigation as exercises for the reader.

Further investigation of our simple simulation can shed light on the discrepancy of the two models. Knowing that the data are overdispersed, we can evaluate the test statistic of the coefficient in the Poisson model. Such an evaluation illustrates whether the statistic follows the assumed distribution.

Here we specify the simulation of a dataset for which the outcome variable should be modeled by the negative binomial distribution. However, the data are modeled assuming Poisson variance.

```
. capture program drop Mysim
. program Mysim, rclass
  1.      quietly {
  2.          drop _all
  3.          set obs 30
  4.          generate x1 = rnormal()
  5.          generate x2 = rnormal()
  6.          generate nby = rpoisson(rgamma(1/.5,.10)*exp(0.75*x1-1.25*x2+2))
  7.          capture glm nby x1 x2, family(poisson) iterate(10)
  8.          return scalar z = (_coef[x1]-0.75)/_se[x1]
  9.      }
 10. end
```

Next, we create the variables necessary for our simulation and run the analysis. Once the results are collected, we overlay the estimated and theoretical densities. For each simulated dataset, the test statistics of one of the regression coefficients is saved so that the distribution of the statistic can be superimposed on top of the theoretical (Gaussian) distribution.

```
. set seed 45323
. simulate z=r(z), reps(1000) nodots: Mysim
        command:  Mysim
              z:  r(z)

. label var z "Test statistic assuming Poisson variance"
```

```
. histogram z, kdensity plot(function stdnorm = normalden(x,0,1),
> range(-3 3) lpattern(dash)) title(Distribution of test statistic)
> subtitle(Negative Binomial Variance)
(bin=29, start=-9.5264149, width=.66415178)
```

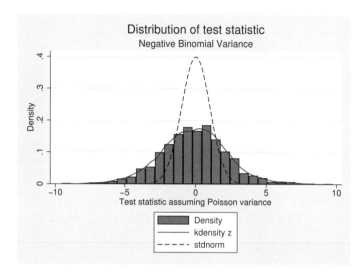

The figure clearly indicates that the distribution of the test statistic does not match the theoretical distribution. This will lead to an inflation of the type I error rate. One solution to overdispersion is to fit a quasi-Poisson model (discussed in chapter 11). In the quasi-Poisson model, we scale the variance using the Pearson dispersion statistic. Effectively, this should increase the estimated standard error and produce a statistic with the correct density.

First, we alter the simulation program to use the quasivariance when constructing the test statistic.

```
. capture program drop Mysim

. program Mysim, rclass
  1.      quietly {
  2.          drop _all
  3.          set obs 30
  4.          generate x1 = rnormal()
  5.          generate x2 = rnormal()
  6.          generate nby = rpoisson(rgamma(1/.5,.10)*exp(0.75*x1-1.25*x2+2))
  7.          capture glm nby x1 x2, family(poisson) iterate(10) scale(x2)
  8.          return scalar z = (_coef[x1]-0.75)/_se[x1]
  9.      }
 10. end
```

Next, we initiate the simulation and collect the statistics. Afterward, we display the appropriate graph.

```
. set seed 23553
. simulate z=r(z), reps(1000) nodots: Mysim
      command:  Mysim
            z:  r(z)
. label var z "Test statistic assuming quasi-Poisson variance"
. histogram z, kdensity plot(function stdnorm = normalden(x,0,1),
> range(-3 3) lpattern(dash)) title(Distribution of test statistic)
> subtitle(Negative Binomial Variance)
(bin=29, start=-5.9986701, width=.38076896)
```

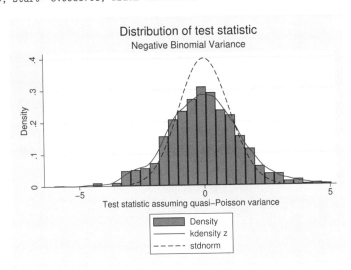

22.3.4 **Effect of missing covariate on** R^2_{Efron} **in Poisson regression**

In section 4.6.3.1, a pseudo-R^2 measure was defined based on the percentage variance of the dependent variable explained by a model; see Efron (1978). We are interested in the effect on the R^2_{Efron} measure as the association increases for a missing covariate.

In our simple evaluation, we define a Poisson outcome dependent on three covariates. The first two covariates, x_1 and x_2, are binary, and the last covariate, x_3, follows a standard normal distribution. When we fit a model, we include only the binary covariates.

We investigate the R^2_{Efron} measure as the coefficient of the third covariate increases. First, we create a program for use with the **simulate** program. This program creates the specified covariates with dependence a of the continuous covariate specified through an option.

$$\eta = 1.00x_1 + 2.00x_2 + ax_3 \qquad (22.5)$$

Finally, the R^2_{fron} value is calculated and made available in the $\mathbf{r()}$ vector.

```
. capture program drop Mysim
. program Mysim, rclass
  1.        syntax [, n(integer 30) a(real 0.0)]
  2.        quietly {
  3.            drop _all
  4.            set obs `n´
  5.            generate x1 = runiform() < .5
  6.            generate x2 = runiform() < .5
  7.            generate x3 = rnormal()
  8.            generate y  = rpoisson(exp(x1+2*x2+`a´*x3))
  9.            summarize y
 10.            local den = r(Var)*(r(N)-1)
 11.            glm y x1 x2, family(poisson) iterate(10)
 12.            predict yhat
 13.            generate num = (y-yhat)^2
 14.            summarize num
 15.            return scalar r2=1-r(sum)/`den´
 16.        }
 17. end
```

Subsequently, we create a loop using the **forvalues** command to calculate R^2_{Efron} under increasingly strong effects of the missing covariate. Each iteration simulates data 100 times and then presents the average value of the statistic across the 100 simulated datasets.

```
. set seed 4982
. forvalues a=0(10)100 {
  2.        quietly {
  3.            local alpha = `a´/100
  4.            simulate r2=r(r2), reps(100) nodots: Mysim, n(30) a(`alpha´)
  5.            summarize r2
  6.            noisily display "Efron Pseudo-R2[a=" %3.2f (`alpha´) "]="
  >                %6.4f r(mean)
  7.        }
  8. }
Efron Pseudo-R2[a=0.00]=0.8856
Efron Pseudo-R2[a=0.10]=0.8716
Efron Pseudo-R2[a=0.20]=0.8372
Efron Pseudo-R2[a=0.30]=0.7898
Efron Pseudo-R2[a=0.40]=0.7086
Efron Pseudo-R2[a=0.50]=0.6538
Efron Pseudo-R2[a=0.60]=0.5718
Efron Pseudo-R2[a=0.70]=0.5121
Efron Pseudo-R2[a=0.80]=0.4796
Efron Pseudo-R2[a=0.90]=0.3853
Efron Pseudo-R2[a=1.00]=0.3570
```

One could expand this simple simulation study to investigate changes in sample size or differences in covariates. Our aim was more modest in that we illustrated a known property of the goodness-of-fit measure and illustrated simple Stata programming.

A Tables

Table A.1. Variance functions

Family (distribution)	Variance $V(\mu)$	Range restrictions	$\partial V(\mu)/\partial \mu$
Gaussian	1	$\begin{cases} \mu \in \Re \\ y \in \Re \end{cases}$	0
Bernoulli	$\mu(1-\mu)$	$\begin{cases} 0 < \mu < 1 \\ 0 \le y \le 1 \end{cases}$	$1 - 2\mu$
Binomial(k)	$\mu(1-\mu/k)$	$\begin{cases} 0 < \mu < k \\ 0 \le y \le k \end{cases}$	$1 - 2\mu/k$
Poisson	μ	$\begin{cases} \mu > 0 \\ y \ge 0 \end{cases}$	1
Gamma	μ^2	$\begin{cases} \mu > 0 \\ y > 0 \end{cases}$	2μ
Inverse Gaussian	μ^3	$\begin{cases} \mu > 0 \\ y > 0 \end{cases}$	$3\mu^2$
Negative binomial(α)	$\mu + \alpha\mu^2$	$\begin{cases} \mu > 0 \\ y \ge 0 \end{cases}$	$1 + 2\alpha\mu$
Power(k)	μ^k	$\begin{cases} \mu > 0 \\ k \ne 0, 1, 2 \end{cases}$	$k\mu^{k-1}$
Quasi	$V(\mu)$		$\dfrac{\partial V(\mu)}{\partial \mu}$

Table A.2. Link and inverse link functions ($\eta = \mathbf{X}\boldsymbol{\beta} + \text{offset}$)

Link name	Link function $\eta = g(\mu)$	Inverse link $\mu = g^{-1}(\eta)$	Range of $\widehat{\mu}$
Identity	μ	η	$\widehat{\mu} \in \Re$
Logit	$\ln\left\{\mu/(1-\mu)\right\}$	$e^{\eta}/(1+e^{\eta})$	$\widehat{\mu} \in (0,1)$
Log	$\ln(\mu)$	$\exp(\eta)$	$\widehat{\mu} > 0$
Negative binomial(α)	$\ln\{\mu/(\mu+1/\alpha)\}$	$e^{\eta}/\{\alpha(1-e^{\eta})\}$	$\widehat{\mu} > 0$
Log-complement	$\ln(1-\mu)$	$1-\exp(\eta)$	$\widehat{\mu} < 1$
Log-log	$-\ln\{-\ln(\mu)\}$	$\exp\{-\exp(-\eta)\}$	$\widehat{\mu} \in (0,1)$
Complementary log-log	$\ln\{-\ln(1-\mu)\}$	$1-\exp\{-\exp(\eta)\}$	$\widehat{\mu} \in (0,1)$
Probit	$\Phi^{-1}(\mu)$	$\Phi(\eta)$	$\widehat{\mu} \in (0,1)$
Reciprocal	$1/\mu$	$1/\eta$	$\widehat{\mu} \in \Re$
Power($\alpha = -2$)	$1/\mu^2$	$1/\sqrt{\eta}$	$\widehat{\mu} > 0$
Power(α)	$\begin{cases} \alpha \neq 0 \\ \alpha = 0 \end{cases} \begin{cases} \mu^{\alpha} \\ \ln(\mu) \end{cases}$	$\begin{cases} \eta^{1/\alpha} \\ \exp(\eta) \end{cases}$	$\widehat{\mu} \in \Re$
Odds power(α)	$\begin{cases} \alpha \neq 0 \\ \alpha = 0 \end{cases} \begin{cases} \dfrac{\mu/(1-\mu)^{\alpha}-1}{\alpha} \\ \ln\left(\dfrac{\mu}{1-\mu}\right) \end{cases}$	$\begin{cases} \dfrac{(1+\alpha\eta)^{1/\alpha}}{1+(1+\alpha\eta)^{1/\alpha}} \\ \dfrac{e^{\eta}}{1+e^{\eta}} \end{cases}$	$\widehat{\mu} \in (0,1)$

Table A.3. First derivatives of link functions ($\eta = \mathbf{X}\boldsymbol{\beta} + \text{offset}$)

Link name	Link $\eta = g(\mu)$	First derivatives $\triangle = \partial\eta/\partial\mu$
Identity	μ	1
Logit	$\ln\{\mu/(1-\mu)\}$	$1/\{\mu(1-\mu)\}$
Log	$\ln(\mu)$	$1/\mu$
Negative binomial(α)	$\ln\{\alpha\mu/(1+\alpha\mu)\}$	$1/(\mu + \alpha\mu^2)$
Log-complement	$\ln(1-\mu)$	$-1/(1-\mu)$
Log-log	$-\ln\{-\ln(\mu)\}$	$-1/\{\mu\ln(\mu)\}$
Complementary log-log	$\ln\{-\ln(1-\mu)\}$	$\{(\mu-1)\ln(1-\mu)\}^{-1}$
Probit	$\Phi^{-1}(\mu)$	$1/\phi\{\Phi^{-1}(\mu)\}$
Reciprocal	$1/\mu$	$-1/\mu^2$
Power($\alpha = -2$)	$1/\mu^2$	$-2/\mu^3$
Power(α) $\begin{cases} \alpha \neq 0 \\ \alpha = 0 \end{cases}$	$\begin{cases} \mu^\alpha \\ \ln(\mu) \end{cases}$	$\begin{cases} \alpha\mu^{\alpha-1} \\ 1/\mu \end{cases}$
Odds power(α) $\begin{cases} \alpha \neq 0 \\ \alpha = 0 \end{cases}$	$\begin{cases} \dfrac{\{\mu/(1-\mu)\}^\alpha - 1}{\alpha} \\ \ln\left(\dfrac{\mu}{1-\mu}\right) \end{cases}$	$\begin{cases} \dfrac{\mu^{\alpha-1}}{(1-\mu)^{\alpha+1}} \\ \dfrac{1}{\mu(1-\mu)} \end{cases}$

Table A.4. First derivatives of inverse link functions ($\eta = \mathbf{X}\boldsymbol{\beta} + \text{offset}$)

Link name	Inverse link $\mu = g^{-1}(\eta)$	First derivatives $\nabla = \partial\mu/\partial\eta$
Identity	η	1
Logit	$e^{\eta}/(1+e^{\eta})$	$\mu(1-\mu)$
Log	$\exp(\eta)$	μ
Negative binomial(α)	$e^{\eta}/\{\alpha(1-e^{\eta})\}$	$\mu + \alpha\mu^2$
Log-complement	$1 - \exp(\eta)$	$\mu - 1$
Log-log	$\exp\{-\exp(-\eta)\}$	$-\mu\ln(\mu)$
Complementary log-log	$1 - \exp\{-\exp(\eta)\}$	$(\mu-1)\ln(1-\mu)$
Probit	$\Phi(\eta)$	$\phi(\eta)$
Reciprocal	$1/\eta$	$-\mu^2$
Power($\alpha = -2$)	$1/\sqrt{\eta}$	$-\mu^3/2$
Power(α) $\begin{cases} \alpha \neq 0 \\ \alpha = 0 \end{cases}$	$\begin{cases} \eta^{1/\alpha} \\ \exp(\eta) \end{cases}$	$\begin{cases} \frac{1}{\alpha}\mu^{1-\alpha} \\ \mu \end{cases}$
Odds power(α) $\begin{cases} \alpha \neq 0 \\ \alpha = 0 \end{cases}$	$\begin{cases} \dfrac{(1+\alpha\eta)^{1/\alpha}}{1+(1+\alpha\eta)^{1/\alpha}} \\ \dfrac{e^{\eta}}{1+e^{\eta}} \end{cases}$	$\begin{cases} \dfrac{\mu(1-\mu)}{1+\alpha\eta} \\ \mu(1-\mu) \end{cases}$

Table A.5. Second derivatives of link functions where $\eta = \mathbf{X}\boldsymbol{\beta} + \text{offset}$ and $\triangle = \partial\eta/\partial\mu$

Link name	Link $\eta = g(\mu)$	Second derivatives $\partial^2\eta/\partial\mu^2$
Identity	μ	0
Logit	$\ln\{\mu/(1-\mu)\}$	$(2\mu-1)\triangle^2$
Log	$\ln(\mu)$	$-\triangle^2$
Negative binomial(α)	$\ln\{\alpha\mu/(1+\alpha\mu)\}$	$-\triangle^2(1+2\alpha\mu)$
Log-complement	$\ln(1-\mu)$	$-\triangle^2$
Log-log	$-\ln\{-\ln(\mu)\}$	$\{1+\ln(\mu)\}\triangle^2$
Complementary log-log	$\ln\{-\ln(1-\mu)\}$	$-\{1+\ln(1-\mu)\}\triangle^2$
Probit	$\Phi^{-1}(\mu)$	$\eta\triangle^2$
Reciprocal	$1/\mu$	$-2\triangle/\mu$
Power($\alpha=-2$)	$1/\mu^2$	$-3\triangle/\mu$
Power(α)	$\begin{cases} \alpha \neq 0 \\ \alpha = 0 \end{cases}$ $\begin{cases} \mu^\alpha \\ \ln(\mu) \end{cases}$	$\begin{cases} (\alpha-1)\triangle/\alpha \\ -\triangle^2 \end{cases}$
Odds power(α)	$\begin{cases} \alpha \neq 0 \\ \alpha = 0 \end{cases}$ $\begin{cases} \dfrac{\{\mu/(1-\mu)\}^\alpha - 1}{\alpha} \\ \ln\left(\dfrac{\mu}{1-\mu}\right) \end{cases}$	$\begin{cases} \triangle\left(\dfrac{1-1/\alpha}{1-\mu}+\alpha+1\right) \\ \mu\triangle^2 \end{cases}$

Table A.6. Second derivatives of inverse link functions where $\eta = \mathbf{X}\boldsymbol{\beta} + \text{offset}$ and $\nabla = \partial\mu/\partial\eta$

Link name	Inverse link $\mu = g^{-1}(\eta)$	Second derivatives $\partial^2\mu/\partial\eta^2$
Identity	η	0
Logit	$e^\eta/(1 + e^\eta)$	$\nabla(1 - 2\mu)$
Log	$\exp(\eta)$	∇
Negative binomial(α)	$e^\eta/\{\alpha(1 - e^\eta)\}$	$\nabla(1 + 2\alpha\mu)$
Log-complement	$1 - \exp(\eta)$	∇
Log-log	$\exp\{-\exp(-\eta)\}$	$\nabla\{1 + \ln(\mu)\}$
Complementary log-log	$1 - \exp\{-\exp(\eta)\}$	$\nabla\{1 + \ln(1 - \mu)\}$
Probit	$\Phi(\eta)$	$-\nabla\eta$
Reciprocal	$1/\eta$	$-2\nabla\mu$
Power$(\alpha = -2)$	$1/\sqrt{\eta}$	$3\nabla^2/\mu$
Power(α) $\begin{cases} \alpha \neq 0 \\ \alpha = 0 \end{cases}$	$\begin{cases} \eta^{1/\alpha} \\ \exp(\eta) \end{cases}$	$\begin{cases} \nabla(\frac{1}{\alpha} - 1)/\mu^\alpha \\ \nabla \end{cases}$
Odds power(α) $\begin{cases} \alpha \neq 0 \\ \alpha = 0 \end{cases}$	$\begin{cases} \dfrac{(1 + \alpha\eta)^{1/\alpha}}{1 + (1 + \alpha\eta)^{1/\alpha}} \\ \dfrac{e^\eta}{1 + e^\eta} \end{cases}$	$\begin{cases} \nabla\left(1 - 2\mu - \dfrac{\alpha}{1 + \alpha\eta}\right) \\ \nabla(1 - 2\mu) \end{cases}$

Table A.7. Log likelihoods

Family	$\mathcal{L}$
Gaussian	$-\dfrac{1}{2}\sum\left\{\dfrac{(y_i-\mu_i)^2}{\phi}+\ln(2\pi\phi)\right\}$
Bernoulli	$\phi\sum_{i:y_i=0}\ln\left(1-\dfrac{\mu_i}{2}\right)+\phi\sum_{i:y_i\neq0}\ln\left(\dfrac{\mu_i}{2}\right)$
Binomial(k)	$\phi\sum\left\{\ln\Gamma(k_i+1)-\ln\Gamma(y_i+1)-\ln\Gamma(k_i-y_i+1)\right.$ $\left.+y_i\ln\left(\dfrac{\mu_i}{k_i}\right)+(k_i-y_i)\ln\left(1-\dfrac{\mu_i}{y_i}\right)\right\}$
Poisson	$\phi\sum\left\{-\mu_i+y_i\ln(\mu_i)-\ln\Gamma(y_i+1)\right\}$
Gamma	$-\dfrac{1}{\phi}\sum\left\{\dfrac{y_i}{\mu_i}-\ln\left(\dfrac{\phi}{\mu_i}\right)-\dfrac{\phi-1}{\phi}\ln(y_i)+\dfrac{1}{\phi}\ln\Gamma(\phi)\right\}$
Inverse Gaussian	$-\dfrac{1}{2}\sum\left\{\dfrac{(y_i-\mu_i)^2}{y_i\mu_i^2\phi}+\ln\left(\phi y_i^3\right)+\ln(2\pi)\right\}$
Negative binomial(α)	$\phi\sum\left\{\ln\Gamma\left(\dfrac{1}{\alpha}+y_i\right)-\ln\Gamma(y_i+1)-\ln\Gamma\left(\dfrac{1}{\alpha}\right)\right.$ $\left.-\dfrac{1}{\alpha}\ln(1+\alpha\mu_i)+y_i\ln\left(\dfrac{\alpha\mu_i}{\alpha\mu_i+1}\right)\right\}$

Table A.8. Weight functions (kernels) for weighted sandwich variance estimates

Estimator	Maximum lag G	Sandwich weights $\omega(z)$
Newey–West	q	$\begin{cases} 1 - \lvert z \rvert & \text{if } \lvert z \rvert \leq 1 \\ 0 & \text{otherwise} \end{cases}$
Gallant	q	$\begin{cases} 1 - 6z^2 + 6\lvert z \rvert^3 & \text{if } 0 \leq \lvert z \rvert \leq 1/2 \\ 2(1 - \lvert z \rvert)^3 & \text{if } 1/2 \leq \lvert z \rvert \leq 1 \end{cases}$
Anderson	q	$\dfrac{3}{(6\pi z/5)^2} \left\{ \dfrac{\sin(6\pi z/5)}{6\pi z/5} - \cos(6\pi z/5) \right\}$

Table A.9. Pearson residuals

Family (variance function)	Pearson residual (r_P)
Gaussian	$y_i - \widehat{\mu}_i$
Bernoulli	$\dfrac{y_i - \widehat{\mu}_i}{\sqrt{\widehat{\mu}_i(1 - \widehat{\mu}_i)}}$
Binomial(k)	$\dfrac{y_i - \widehat{\mu}_i}{\sqrt{\widehat{\mu}_i(1 - \widehat{\mu}_i/k_i)}}$
Poisson	$\dfrac{y_i - \widehat{\mu}_i}{\sqrt{\widehat{\mu}_i}}$
Gamma	$\dfrac{y_i - \widehat{\mu}_i}{\sqrt{\widehat{\mu}_i^2}}$
Inverse Gaussian	$\dfrac{y_i - \widehat{\mu}_i}{\sqrt{\widehat{\mu}_i^3}}$
Negative binomial(α)	$\dfrac{y_i - \mu_i}{\sqrt{\widehat{\mu}_i + \alpha\widehat{\mu}_i^2}}$
Power(k)	$\dfrac{y_i - \widehat{\mu}_i}{\sqrt{\widehat{\mu}_i^k}}$

Table A.10. Anscombe residuals

Family (variance function)	Anscombe residual (r_A)
Gaussian	$y_i - \widehat{\mu}_i$
Bernoulli	$\dfrac{1.5\left\{y_i^{2/3}\,_2F_1(y_i) - \mu_i^{2/3}\,_2F_1(\mu_i)\right\}}{(\widehat{\mu}_i - \widehat{\mu}_i^2)^{1/6}}$
Binomial(k)	$\dfrac{1.5\left\{y_i^{2/3}\,_2F_1(y_i/k_i) - \mu_i^{2/3}\,_2F_1(\mu_i/k_i)\right\}}{(\widehat{\mu}_i - \widehat{\mu}_i^2/k_i)^{1/6}}$
Poisson	$\dfrac{\frac{3}{2}(y_i^{2/3} - \widehat{\mu}_i^{2/3})}{\widehat{\mu}_i^{1/6}}$
Gamma	$\dfrac{3(y_i^{1/3} - \widehat{\mu}_i^{1/3})}{\widehat{\mu}_i^{1/3}}$
Inverse Gaussian	$\dfrac{\ln y_i - \ln \widehat{\mu}_i}{\sqrt{\widehat{\mu}_i}}$
Negative binomial(α)	$\dfrac{1.5\left\{y_i^{2/3}\,_2F_1(-\alpha y_i) - \mu_i^{2/3}\,_2F_1(-\alpha\mu_i)\right\}}{(\widehat{\mu}_i + \alpha\widehat{\mu}_i^2)^{1/6}}$
Power(k)	$\dfrac{3\widehat{\mu}_i^{-k/6}\left(\widehat{\mu}_i^{1-k/3} - y_i^{1-k/3}\right)}{k - 3}$

Table A.11. Squared deviance residuals and deviance adjustment factors $\rho_3(\theta)$

Family (variance function)	Squared deviance residual $\widehat{d}_i^2$	$\rho_3(\theta)$
Gaussian	$(y_i - \widehat{\mu}_i)^2$	0
Bernoulli	$\begin{cases} -2\ln(1 - \widehat{\mu}_i) & \text{if } y_i = 0 \\ -2\ln(\widehat{\mu}_i) & \text{if } y_i = 1 \end{cases}$	†
Binomial(k)	$\begin{cases} 2k_i \ln\left(\frac{k_i}{k_i - \widehat{\mu}_i}\right) & \text{if } y_i = 0 \\ 2y_i \ln\left(\frac{y_i}{\widehat{\mu}_i}\right) + 2(k_i - y_i)\ln\left(\frac{k_i - y_i}{k_i - \widehat{\mu}_i}\right) & \text{if } 0 < y_i < k_i \\ 2k_i \ln\left(\frac{k_i}{\widehat{\mu}_i}\right) & \text{if } y_i = k_i \end{cases}$	$\frac{1 - 2\widehat{\mu}_i/k_i}{\sqrt{k_i \widehat{\mu}_i (1 - \widehat{\mu}_i)}}$
Poisson	$\begin{cases} 2\widehat{\mu}_i & \text{if } y_i = 0 \\ 2\left\{ y_i \ln\left(\frac{y_i}{\widehat{\mu}_i}\right) - (y_i - \widehat{\mu}_i) \right\} & \text{otherwise} \end{cases}$	$\frac{1}{\sqrt{\widehat{\mu}_i}}$
Gamma	$-2\left\{ \ln\left(\frac{y_i}{\widehat{\mu}_i}\right) - \frac{y_i - \widehat{\mu}_i}{\widehat{\mu}_i} \right\}$	$\frac{2}{\sqrt{\widehat{\phi}}}$
Inverse Gaussian	$\frac{(y_i - \widehat{\mu}_i)^2}{\widehat{\mu}_i^2 y_i}$	$3\widehat{\mu}_i^{7/2}/\widehat{\phi}^2$
Negative binomial(α)	$\begin{cases} 2\ln(1 + \alpha\widehat{\mu}_i)/\alpha & \text{if } y_i = 0 \\ 2y_i \ln\left(\frac{y_i}{\widehat{\mu}_i}\right) - \frac{2}{\alpha}(1 + \alpha y_i)\ln\left(\frac{1 + \alpha y_i}{1 + \alpha\widehat{\mu}_i}\right) & \text{otherwise} \end{cases}$	‡
Power(k)	$\frac{2y_i}{(1 - k)(y_i^{1-k} - \widehat{\mu}_i^{1-k})} - \frac{2}{(2 - k)(y_i^{2-k} - \widehat{\mu}_i^{2-k})}$	‡

† Not recommended for Bernoulli. In general, not recommended for binomial families with $k < 10$.
‡ Not available.

Table A.12. Kullback–Leibler (K–L) divergence (Cameron and Windmeijer 1997)

Family	K–L divergence
Gaussian	$\displaystyle\sum \frac{(y_i - \mu_i)^2}{\sigma^2}$
Bernoulli	$\displaystyle -2\sum \{y_i \ln \mu_i + (1 - y_i)\ln(1 - \mu_i)\}$
Binomial(k)	$\displaystyle 2\sum \left\{ y_i \ln \frac{y_i}{\mu_i} + (k_i - y_i)\ln \frac{k_i - y_i}{k_i - \mu_i} \right\}$
Poisson	$\displaystyle 2\sum \left\{ y \ln\left(\frac{y_i}{\mu_i}\right) - (y_i - \mu_i) \right\}$
Gamma	$\displaystyle -2\phi \sum \left\{ \ln\left(\frac{y_i}{\mu_i}\right) + \frac{y_i - \mu_i}{\mu_i} \right\}$
Inverse Gaussian	$\displaystyle \sum \frac{(y_i - \mu_i)^2}{\mu_i^2 y_i}$

Table A.13. Cameron–Windmeijer (1997) R^2

Family	$R^2_{\text{Cameron\&Windmeijer}}$
Gaussian	$1 - \dfrac{\sum (y_i - \widehat{\mu}_i)^2}{\sum (y_i - \overline{y})^2}$
Bernoulli	$1 - \dfrac{\sum \widehat{\mu}_i \ln(\widehat{\mu}_i) + (1 - \widehat{\mu}_i) \ln(1 - \widehat{\mu}_i)}{n \left\{ \overline{y} \ln(\overline{y}) + (1 - \overline{y}) \ln(1 - \overline{y}) \right\}}$
Binomial(k)	$1 - \dfrac{\sum \widehat{\mu}_i \ln(\widehat{\mu}_i) + (k_i - \widehat{\mu}_i) \ln(k_i - \widehat{\mu}_i)}{n \left\{ \overline{y} \ln(\overline{y}) + (k_i - \overline{y}) \ln(k_i - \overline{y}) \right\}}$
Poisson	$1 - \dfrac{\sum y_i \ln(y_i / \widehat{\mu}_i) - (y_i - \widehat{\mu}_i)}{\sum y_i \ln(y_i / \overline{y})}$
Gamma	$1 - \dfrac{\sum y_i \ln(y_i / \widehat{\mu}_i) + (y_i - \widehat{\mu}_i) / \widehat{\mu}_i}{\sum \ln(y_i / \overline{y})}$
Inverse Gaussian	$1 - \dfrac{\sum (y_i - \widehat{\mu}_i)^2 / (\widehat{\mu}_i^2 y_i)}{\sum (y_i - \overline{y})^2 / (\overline{y}^2 y)}$

Table A.14. Interpretation of power links

Gaussian
link(power 1) canonical identity (standard OLS regression)
link(power 0) lognormal

Binomial
link(power 0) log-binomial; used with incidence rates

Poisson
link(power 0) canonical log link
link(power 1) identity link

Gamma
link(power -1) canonical inverse link
link(power 0) log link
link(power 1) identity link

Inverse Gaussian
link(power -2) canonical inverse quadratic
link(power 0) log link
link(power 1) identity link

Negative binomial
link(power 0) log link (common parameterization of negative binomial)

Continuous distributions
link(power 2) square link
link(power .5) square root link

References

Aitkin, M., B. Francis, J. Hinde, and R. Darnell. 2009. *Statistical Modelling in R.* Oxford: Oxford University Press.

Akaike, H. 1973. Information theory and an extension of the maximum likelihood principle. In *Second International Symposium on Information Theory*, ed. B. N. Petrov and F. Csaki, 267–281. Budapest: Akailseoniai–Kiudo.

Albright, R. L., S. R. Lerman, and C. F. Manski. 1977. Report on the development of an estimation program for the multinomial probit model. Preliminary report prepared by the Federal Highway Commission.

Anderson, J. A. 1984. Regression and ordered categorical variables. *Journal of the Royal Statistical Society, Series B* 46: 1–30.

Andrews, D. W. K. 1991. Heteroskedasticity and autocorrelation consistent covariance matrix estimation. *Econometrica* 59: 817–858.

Anscombe, F. J. 1953. Contribution of discussion paper by H. Hotelling "New light on the correlation coefficient and its transforms". *Journal of the Royal Statistical Society, Series B* 15: 229–230.

Atkinson, A. C. 1980. A note on the generalized information criterion for choice of a model. *Biometrika* 67: 413–418.

Bai, Z. D., P. R. Krishnaiah, N. Sambamoorthi, and L. C. Zhao. 1992. Model selection for log-linear models. *Sankhyā, Series B* 54: 200–219.

Barndorff-Nielsen, O. 1976. Factorization of likelihood functions for full exponential families. *Journal of the Royal Statistical Society, Series B* 38: 37–44.

Belsley, D. A., E. Kuh, and R. E. Welsch. 1980. *Regression Diagnostics: Identifying Influential Data and Sources of Collinearity.* New York: Wiley.

Ben-Akiva, M., and S. R. Lerman. 1985. *Discrete Choice Analysis: Theory and Application to Travel Demand.* Cambridge, MA: MIT Press.

Berkson, J. 1944. Application of the logistic function to bio-assay. *Journal of the American Statistical Association* 39: 357–365.

Berndt, E. K., B. H. Hall, R. E. Hall, and J. A. Hausman. 1974. Estimation and inference in nonlinear structural models. *Annals of Economic and Social Measurement* 3/4: 653–665.

Binder, D. A. 1983. On the variances of asymptotically normal estimators from complex surveys. *International Statistical Review* 51: 279–292.

Bliss, C. I. 1934a. The method of probits. *Science* 79: 38–39.

———. 1934b. The method of probits—a correction. *Science* 79: 409–410.

Boes, S., and R. Winkelmann. 2006. Ordered response models. *Allgemeines Statistisches Archiv* 90: 167–181.

Bollen, K. A., and R. W. Jackman. 1985. Regression diagnostics: An expository treatment of outliers and influential cases. *Sociological Methods & Research* 13: 510–542.

Brant, R. 1990. Assessing proportionality in the proportional odds model for ordinal logistic regression. *Biometrics* 46: 1171–1178.

Breslow, N. E. 1990. Tests of hypotheses in overdispersed Poisson regression and other quasi-likelihood models. *Journal of the American Statistical Association* 85: 565–571.

———. 1996. Generalized linear models: Checking assumptions and strengthening conclusions. *Statistica Applicata* 8: 23–41.

Brown, D. 1992. A graphical analysis of deviance. *Journal of the Royal Statistical Society, Series C* 41: 55–62.

Cameron, A. C., and P. K. Trivedi. 2010. *Microeconometrics Using Stata*. Rev. ed. College Station, TX: Stata Press.

———. 2013. *Regression Analysis of Count Data*. 2nd ed. Cambridge: Cambridge University Press.

Cameron, A. C., and F. A. G. Windmeijer. 1997. An R-squared measure of goodness of fit for some common nonlinear regression models. *Journal of Econometrics* 77: 329–342.

Carroll, R. J., S. Wang, D. G. Simpson, A. J. Stromberg, and D. Ruppert. 1998. The sandwich (robust covariance matrix) estimator. Technical report. http://www.stat.tamu.edu/ftp/pub/rjcarroll/sandwich.pdf.

Cleves, M. A., W. W. Gould, and Y. V. Marchenko. 2016. *An Introduction to Survival Analysis Using Stata*. Rev. 3 ed. College Station, TX: Stata Press.

Collett, D. 2003. *Modelling Binary Data*. 2nd ed. London: Chapman & Hall/CRC.

Consul, P. C., and F. Famoye. 1992. Generalized Poisson regression model. *Communications in Statistics—Theory and Methods* 21: 89–109.

Cordeiro, G. M., and P. McCullagh. 1991. Bias correction in generalized linear models. *Journal of the Royal Statistical Society, Series B* 53: 629–643.

Cox, D. R. 1983. Some remarks on overdispersion. *Biometrika* 70: 269–274.

Cox, D. R., and E. J. Snell. 1968. A general definition of residuals. *Journal of the Royal Statistical Society, Series B* 30: 248–275.

Cragg, J. G., and R. S. Uhler. 1970. The demand for automobiles. *Canadian Journal of Economics* 3: 386–406.

Cummings, T. H., J. W. Hardin, A. C. McLain, J. R. Hussey, K. J. Bennett, and G. M. Wingood. 2015. Modeling heaped count data. *Stata Journal* 15: 457–479.

Daganzo, C. 1979. *Multinomial Probit: The Theory and Its Application to Demand Forecasting*. New York: Academic Press.

Davidian, M., and R. J. Carroll. 1987. Variance function estimation. *Journal of the American Statistical Association* 82: 1079–1091.

Dean, C., and J. F. Lawless. 1989. Tests for detecting overdispersion in Poisson regression models. *Journal of the American Statistical Association* 84: 467–472.

Doll, R., and A. B. Hill. 1966. Mortality of British doctors in relation to smoking: Observations on coronary thrombosis. *Journal of the National Cancer Institute, Monographs* 19: 205–268.

Dyke, G. V., and H. D. Patterson. 1952. Analysis of factorial arrangements when the data are proportions. *Biometrics* 8: 1–12.

Efron, B. 1978. Regression and ANOVA with zero-one data: Measures of residual variation. *Journal of the American Statistical Association* 73: 113–121.

———. 1981. Nonparametric estimates of standard error: The jackknife, the bootstrap and other methods. *Biometrika* 68: 589–599.

———. 1992. Poisson overdispersion estimates based on the method of asymmetric maximum likelihood. *Journal of the American Statistical Association* 87: 98–107.

Famoye, F. 1995. Generalized binomial regression model. *Biometrical Journal* 37: 581–594.

———. 2010. On the bivariate negative binomial regression model. *Journal of Applied Statistics* 37: 969–981.

Famoye, F., and E. H. Kaufman, Jr. 2002. Generalized negative binomial regression model. *Journal of Applied Statistical Sciences* 11: 289–296.

Fisher, R. A. 1922. On the mathematical foundations of theoretical statistics. *Philosophical Transactions of the Royal Society of London, Series A* 222: 309–368.

———. 1934. Two new properties of mathematical likelihood. *Proceedings of the Royal Society of London, Series A* 144: 285–307.

Francis, B., M. Green, and C. Payne, ed. 1993. *The GLIM System*. New York: Oxford University Press.

Fu, V. K. 1998. sg88: Estimating generalized ordered logit models. *Stata Technical Bulletin* 44: 27–30. Reprinted in *Stata Technical Bulletin Reprints*, vol. 8, pp. 160–164. College Station, TX: Stata Press.

Gail, M. H., W. Y. Tan, and S. Piantadosi. 1988. Tests for no treatment effect in randomized clinical trials. *Biometrika* 75: 57–64.

Gallant, A. R. 1987. *Nonlinear Statistical Models*. New York: Wiley.

Gamerman, D., and H. F. Lopes. 2006. *Markov Chain Monte Carlo: Stochastic Simulation for Bayesian Inference*. 2nd ed. Boca Raton, FL: Chapman & Hall/CRC.

Ganio, L. M., and D. W. Schafer. 1992. Diagnostics for overdispersion. *Journal of the American Statistical Association* 87: 795–804.

Gould, W. W., J. S. Pitblado, and B. P. Poi. 2010. *Maximum Likelihood Estimation with Stata*. 4th ed. College Station, TX: Stata Press.

Green, P. J. 1984. Iteratively reweighted least squares for maximum likelihood estimation, and some robust and resistant alternatives. *Journal of the Royal Statistical Society, Series B* 46: 149–192.

Greene, W. H. 2002. *LIMDEP Version 8.0 Econometric Modeling Guide*. 2nd ed. Plainview, NY: Econometric Software.

———. 2008. Functional forms for the negative binomial model for count data. *Economics Letters* 99: 585–590.

———. 2012. *Econometric Analysis*. 7th ed. Upper Saddle River, NJ: Prentice Hall.

Hannan, E. J., and B. G. Quinn. 1979. The determination of the order of an autoregression. *Journal of the Royal Statistical Society, Series B* 41: 190–195.

Hardin, J. W. 2003. The sandwich estimate of variance. In *Advances in Econometrics, Vol. 17: Maximum Likelihood Estimation of Misspecified Models: Twenty Years Later*, ed. T. B. Fomby and R. C. Hill, 45–73. Bingley, UK: Emerald.

Hardin, J. W., and R. J. Carroll. 2003. Measurement error, GLMs, and notational conventions. *Stata Journal* 3: 329–341.

Hardin, J. W., and M. A. Cleves. 1999. sbe29: Generalized linear models: Extensions to the binomial family. *Stata Technical Bulletin* 50: 21–25. Reprinted in *Stata Technical Bulletin Reprints*, vol. 9, pp. 140–146. College Station, TX: Stata Press.

Hardin, J. W., and J. M. Hilbe. 2013. *Generalized Estimating Equations*. 2nd ed. Boca Raton, FL: Taylor & Francis.

———. 2014a. Estimation and testing of binomial and beta-binomial regression models with and without zero inflation. *Stata Journal* 14: 292–303.

———. 2014b. Regression models for count data based on the negative binomial(P) distribution. *Stata Journal* 14: 280–291.

———. 2015. Regression models for count data from truncated distributions. *Stata Journal* 15: 226–246.

Harris, T., J. M. Hilbe, and J. W. Hardin. 2014. Modeling count data with generalized distributions. *Stata Journal* 14: 562–579.

Harris, T., Z. Yang, and J. W. Hardin. 2012. Modeling underdispersed count data with generalized Poisson regression. *Stata Journal* 12: 736–747.

Hastie, T., and R. Tibshirani. 1986. Generalized additive models. *Statistical Science* 1: 297–318.

Hausman, J. A. 1978. Specification tests in econometrics. *Econometrica* 46: 1251–1271.

Hilbe, J. M. 1992. LNB, GLMLNB, and other negative binomial and geometric regression functions. GLM Library, XploRe Statistical Computing Environment, MD*Tech, Humboldt University, Berlin, Germany.

———. 1993a. Log-negative binomial regression as a generalized linear model. Technical Report 26, Graduate College, Arizona State University.

———. 1993b. sg16: Generalized linear models. *Stata Technical Bulletin* 11: 20–28. Reprinted in *Stata Technical Bulletin Reprints*, vol. 2, pp. 149–159. College Station, TX: Stata Press.

———. 1994. Generalized linear models. *American Statistician* 48: 255–265.

———. 2000. sg126: Two-parameter log-gamma and log-inverse Gaussian models. *Stata Technical Bulletin* 53: 31–32. Reprinted in *Stata Technical Bulletin Reprints*, vol. 9, pp. 273–275. College Station, TX: Stata Press.

———. 2005. hplogit: Stata module to estimate Poisson-logit hurdle regression. Statistical Software Components S456405, Department of Economics, Boston College. https://ideas.repec.org/c/boc/bocode/s456405.html.

———. 2009. *Logistic Regression Models*. Boca Raton, FL: Taylor & Francis.

———. 2011. *Negative Binomial Regression*. 2nd ed. Cambridge: Cambridge University Press.

———. 2014. *Modeling Count Data*. New York: Cambridge University Press.

———. 2016. *Practical Guide to Logistic Regression*. Boca Raton, FL: Taylor & Francis.

Hilbe, J. M., and W. H. Greene. 2008. Count response regression models. In *Handbook of Statistics, Volume 27: Epidemiology and Medical Statistics*, ed. C. R. Rao, J. P. Miller, and D. C. Rao, 210–251. Oxford: Elsevier.

Hilbe, J. M., and J. W. Hardin. 2008. Generalized estimating equations for longitudinal panel analysis. In *Handbook of Longitudinal Research: Design, Measurement, and Analysis*, ed. S. Menard. Burlington, MA: Elsevier.

Hilbe, J. M., and B. A. Turlach. 1995. Generalized linear models. In *XploRe: An Interactive Statistical Computing Environment*, ed. W. Härdle, S. Klinke, and B. A. Turlach, 195–222. New York: Springer.

Hinkley, D. V. 1977. Jackknifing in unbalanced situations. *Technometrics* 19: 285–292.

Hosmer, D. W., Jr., S. Lemeshow, and R. X. Sturdivant. 2013. *Applied Logistic Regression*. 3rd ed. Hoboken, NJ: Wiley.

Hotelling, H. 1953. New light on the correlation coefficient and its transforms". *Journal of the Royal Statistical Society, Series B* 15: 193–232.

Huber, P. J. 1967. The behavior of maximum likelihood estimates under nonstandard conditions. In Vol. 1 of *Proceedings of the Fifth Berkeley Symposium on Mathematical Statistics and Probability*, 221–233. Berkeley: University of California Press.

Hurvich, C. M., and C.-L. Tsai. 1989. Regression and time series model selection in small samples. *Biometrika* 76: 297–307.

Irwin, J. O. 1968. The generalized Waring distribution applied to accident theory. *Journal of the Royal Statistical Society, Series A* 131: 205–225.

Jain, G. C., and P. C. Consul. 1971. A generalized negative binomial distribution. *SIAM Journal on Applied Mathematics* 21: 501–513.

Joe, H., and R. Zhu. 2005. Generalized Poisson distribution: The property of mixture of Poisson and comparison with negative binomial distribution. *Biometrical Journal* 47: 219–229.

Kendall, M. G., and A. Stuart. 1979. *The Advanced Theory of Statistics, Vol. 2: Inference and Relationship*. 4th ed. London: Griffin.

Kish, L., and M. R. Frankel. 1974. Inference from complex samples. *Journal of the Royal Statistical Society, Series B* 36: 1–37.

Lambert, D. 1992. Zero-inflated Poisson regression, with an application to defects in manufacturing. *Technometrics* 34: 1–14.

Lambert, D., and K. Roeder. 1995. Overdispersion diagnostics for generalized linear models. *Journal of the American Statistical Association* 90: 1225–1236.

Lawless, J. F. 1987. Negative binomial and mixed Poisson regression. *Canadian Journal of Statistics* 15: 209–225.

Lerman, S. R., and C. F. Manski. 1981. On the use of simulated frequencies to approximate choice probabilities. In *Structural Analysis of Discrete Data with Econometric Applications*, 305–319. Cambridge, MA: MIT Press.

Liang, K.-Y., and S. L. Zeger. 1986. Longitudinal data analysis using generalized linear models. *Biometrika* 73: 13–22.

Lindsey, J. K. 1997. *Applying Generalized Linear Models*. New York: Springer.

Lindsey, J. K., and B. Jones. 1998. Choosing among generalized linear models applied to medical data. *Statistics in Medicine* 17: 59–68.

Long, D. L., J. S. Preisser, A. H. Herring, and C. E. Golin. 2014. A marginalized zero-inflated Poisson regression model with overall exposure effects. *Statistics in Medicine* 33: 5151–5165.

Long, J. S. 1997. *Regression Models for Categorical and Limited Dependent Variables*. Thousand Oaks, CA: Sage.

Long, J. S., and L. H. Ervin. 1998. Correcting for heteroskedasticity with heteroskedasticity consistent standard errors in the linear regression model: Small sample considerations. http://www.indiana.edu/~jslsoc/files_research/testing_tests/hccm/98TAS.pdf.

Long, J. S., and J. Freese. 2000. sg145: Scalar measures of fit for regression models. *Stata Technical Bulletin* 56: 34–40. Reprinted in *Stata Technical Bulletin Reprints*, vol. 10, pp. 197–205. College Station, TX.

———. 2014. *Regression Models for Categorical Dependent Variables Using Stata*. 3rd ed. College Station, TX: Stata Press.

Lumley, T., and P. Heagerty. 1999. Weighted empirical adaptive variance estimators for correlated data regression. *Journal of the Royal Statistical Society, Series B* 61: 459–477.

Machado, J. A. F., and J. M. C. Santos Silva. 2005. Quantiles for counts. *Journal of the American Statistical Association* 100: 1226–1237.

MacKinnon, J. G., and H. White. 1985. Some heteroskedasticity-consistent covariance matrix estimators with improved finite sample properties. *Journal of Econometrics* 29: 305–325.

Maddala, G. S. 1983. *Limited-Dependent and Qualitative Variables in Econometrics*. Cambridge: Cambridge University Press.

Maddala, G. S., and K. Lahiri. 2009. *Introduction to Econometrics*. 4th ed. Chichester, UK: Wiley.

Marquardt, D. W. 1963. An algorithm for least-squares estimation of nonlinear parameters. *SIAM Journal on Applied Mathematics* 11: 431–441.

Marshall, A. W., and I. Olkin. 1985. A family of bivariate distributions generated by the bivariate Bernoulli distribution. *Journal of the American Statistical Association* 80(390): 332–338.

McCullagh, P., and J. A. Nelder. 1989. *Generalized Linear Models*. 2nd ed. London: Chapman & Hall/CRC.

McDowell, A. 2003. From the help desk: Hurdle models. *Stata Journal* 3: 178–184.

McFadden, D. 1974. Conditional logit analysis of qualitative choice behavior. In *Frontiers of Econometrics*, ed. P. Zarembka, 105–142. New York: Academic Press.

McKelvey, R. D., and W. Zavoina. 1975. A statistical model for the analysis of ordinal level dependent variables. *Journal of Mathematical Sociology* 4: 103–120.

Milicer, H., and F. Szczotka. 1966. Age at menarche in Warsaw girls in 1965. *Human Biology* 38: 199–203.

Miller, R. G. 1974. The jackknife—a review. *Biometrika* 61: 1–15.

Miranda, A. 2006. qcount: Stata program to fit quantile regression models for count data. Statistical Software Components S456714, Department of Economics, Boston College. https://ideas.repec.org/c/boc/bocode/s456714.html.

Mullahy, J. 1986. Specification and testing of some modified count data models. *Journal of Econometrics* 33: 341–365.

Nelder, J. A., and D. Pregibon. 1987. An extended quasi-likelihood function. *Biometrika* 74: 221–232.

Nelder, J. A., and R. W. M. Wedderburn. 1972. Generalized linear models. *Journal of the Royal Statistical Society, Series A* 135: 370–384.

Neuhaus, J. M. 1992. Statistical methods for longitudinal and clustered designs with binary responses. *Statistical Methods in Medical Research* 1: 249–273.

Neuhaus, J. M., and N. P. Jewell. 1993. A geometric approach to assess bias due to omitted covariates in generalized linear models. *Biometrika* 80: 807–815.

Neuhaus, J. M., J. D. Kalbfleisch, and W. W. Hauck. 1991. A comparison of cluster-specific and population-averaged approaches for analyzing correlated binary data. *International Statistical Review* 59: 25–35.

Newey, W. K., and K. D. West. 1987. A simple, positive semi-definite, heteroskedasticity and autocorrelation consistent covariance matrix. *Econometrica* 55: 703–708.

———. 1994. Automatic lag selection in covariance matrix estimation. *Review of Economic Studies* 61: 631–653.

Newson, R. 1999. sg114: rglm—Robust variance estimates for generalized linear models. *Stata Technical Bulletin* 50: 27–33. Reprinted in *Stata Technical Bulletin Reprints*, vol. 9, pp. 181–190. College Station, TX: Stata Press.

Nyquist, H. 1991. Restricted estimation of generalized linear models. *Journal of the Royal Statistical Society, Series C* 40: 133–141.

Oakes, D. 1999. Direct calculation of the information matrix via the EM algorithm. *Journal of the Royal Statistical Society, Series B* 61: 479–482.

O'Hara Hines, R. J., and E. M. Carter. 1993. Improved added variable and partial residual plots for detection of influential observations in generalized linear models. *Journal of the Royal Statistical Society, Series C* 42: 3–20.

Pan, W. 2001. Akaike's information criterion in generalized estimating equations. *Biometrics* 57: 120–125.

Papke, L. E., and J. M. Wooldridge. 1996. Econometric methods for fractional response variables with an application to 401(k) plan participation rates. *Journal of Applied Econometrics* 11: 619–632.

Parzen, E. 1957. On consistent estimates of the spectrum of a stationary time series. *Annals of Mathematical Statistics* 28: 329–348.

Pierce, D. A., and D. W. Schafer. 1986. Residuals in generalized linear models. *Journal of the American Statistical Association* 81: 977–986.

Poisson, S. D. 1837. *Recherches sur la probabilité des jugements en matière criminelle et en matière civile: précédées des règles générales du calcul des probabilités.* Paris: Bachelier.

Pregibon, D. 1980. Goodness of link tests for generalized linear models. *Applied Statistics* 29: 15–24.

———. 1981. Logistic regression diagnostics. *Annals of Statistics* 9: 705–724.

Quenouille, M. H. 1949. Approximate tests of correlation in time-series. *Journal of the Royal Statistical Society, Series B* 11: 68–84.

Rabe-Hesketh, S., A. Pickles, and C. Taylor. 2000. sg129: Generalized linear latent and mixed models. *Stata Technical Bulletin* 53: 47–57. Reprinted in *Stata Technical Bulletin Reprints*, vol. 9, pp. 293–307. College Station, TX: Stata Press.

Rabe-Hesketh, S., and A. Skrondal. 2012. *Multilevel and Longitudinal Modeling Using Stata.* 3rd ed. College Station, TX: Stata Press.

Raftery, A. E. 1995. Bayesian model selection in social research. In Vol. 25 of *Sociological Methodology*, ed. P. V. Marsden, 111–163. Oxford: Blackwell.

Rasch, G. 1960. *Probabilistic Models for Some Intelligence and Attainment Tests.* Copenhangen: Danmarks Pædogogiske Institut.

Robert, C. P., and G. Casella. 2004. *Monte Carlo Statistical Methods.* 2nd ed. New York: Springer.

Rodríguez-Avi, J., A. Conde-Sánchez, A. J. Sáez-Castillo, M. J. Olmo-Jiménez, and A. M. Martínez-Rodríguez. 2009. A generalized Waring regression model for count data. *Computational Statistics & Data Analysis* 53: 3717–3725.

Rogers, W. 1992. sg8: Probability weighting. *Stata Technical Bulletin* 8: 15–17. Reprinted in *Stata Technical Bulletin Reprints*, vol. 2, pp. 126–129. College Station, TX: Stata Press.

Royall, R. M. 1986. Model robust confidence intervals using maximum likelihood estimators. *International Statistical Review* 54: 221–226.

Schwarz, G. 1978. Estimating the dimension of a model. *Annals of Statistics* 6: 461–464.

Sklar, A. 1959. Fonctions de repartition a n dimensiones et leurs marges. *Publicacion de l'Institut de Statistique de l'Universite de Paris* 8: 229–231.

———. 1973. Random variables, joint distribution functions, and copulas. *Kybernetika* 9: 449–460.

Smith, P. J., and D. F. Heitjan. 1993. Testing and adjusting for departures from nominal dispersion in generalized linear models. *Journal of the Royal Statistical Society, Series C* 42: 31–41.

Sugiura, N. 1978. Further analysis of the data by Akaike's information criterion and the finite corrections. *Communications in Statistics—Theory and Methods* 7: 13–26.

Tanner, M. A., and W. H. Wong. 1987. The calculation of posterior distributions by data augmentation (with discussion). *Journal of the American Statistical Association* 82: 528–550.

Veall, M. R., and K. F. Zimmermann. 1992. Pseudo-R^2's in the ordinal probit model. *Journal of Mathematical Sociology* 16: 333–342.

von Bortkiewicz, L. 1898. *Das Gesetz der Kleinen Zahlen*. Leipzig: Teubner.

Wacholder, S. 1986. Binomial regression in GLIM: Estimating risk ratios and risk differences. *American Journal of Epidemiology* 123: 174–184.

Wahrendorf, J., H. Becher, and C. C. Brown. 1987. Bootstrap comparison of non-nested generalized linear models: Applications in survival analysis and epidemiology. *Journal of the Royal Statistical Society, Series C* 36: 72–81.

Wedderburn, R. W. M. 1974. Quasi-likelihood functions, generalized linear models, and the Gauss–Newton method. *Biometrika* 61: 439–447.

Wheatley, G. A., and G. H. Freeman. 1982. A method of using the proportion of undamaged carrots or parsnips to estimate the relative population densities of carrot fly (*Psila rosae*) larvae, and its practical applications. *Annals of Applied Biology* 100: 229–244.

White, H. 1980. A heteroskedasticity-consistent covariance matrix estimator and a direct test for heteroskedasticity. *Econometrica* 48: 817–838.

Williams, D. A. 1982. Extra-binomial variation in logistic linear models. *Journal of the Royal Statistical Society, Series C* 31: 144–148.

————. 1987. Generalized linear model diagnostics using the deviance and single case deletions. *Journal of the Royal Statistical Society, Series C* 36: 181–191.

Williams, R. 2006a. oglm: Stata module to estimate ordinal generalized linear models. Statistical Software Components S453402, Department of Economics, Boston College. https://ideas.repec.org/c/boc/bocode/s453402.html.

————. 2006b. Generalized ordered logit/partial proportional odds models for ordinal dependent variables. *Stata Journal* 6: 58–82.

Winkelmann, R., and K. F. Zimmermann. 1995. Recent developments in count data modeling: Theory and applications. *Journal of Economic Surveys* 9: 1–24.

Wolfe, R. 1998. sg86: Continuation-ratio models for ordinal response data. *Stata Technical Bulletin* 44: 18–21. Reprinted in *Stata Technical Bulletin Reprints*, vol. 8, pp. 149–153. College Station, TX: Stata Press.

Wu, C. F. J. 1986. Jackknife, bootstrap and other resampling methods in regression analysis. *Annals of Statistics* 14: 1261–1350 (including discussions and rejoinder).

Xekalaki, E. 1983. The univariate generalized Waring distribution in relation to accident theory: Proneness, spells or contagion? *Biometrics* 39: 887–895.

Xu, X., and J. W. Hardin. 2016. Regression models for bivariate count outcomes. *Stata Journal* 16: 301–315.

Yang, Z., J. W. Hardin, C. L. Addy, and Q. H. Vuong. 2007. Testing approaches for overdispersion in Poisson regression versus the generalized Poisson Model. *Biometrical Journal* 49: 565–584.

Zeger, S. L., and M. R. Karim. 1991. Generalized linear models with random effects; a Gibbs sampling approach. *Journal of the American Statistical Association* 86: 79–86.

Zeger, S. L., K.-Y. Liang, and P. S. Albert. 1988. Models for longitudinal data: A generalized estimating equation approach. *Biometrics* 44: 1049–1060.

Author index

Subject index